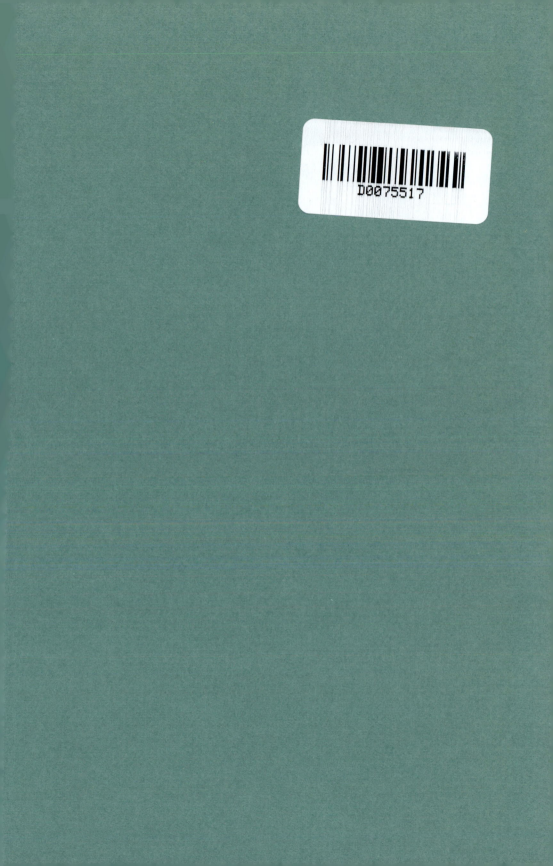

Environmental Chemistry of Lakes and Reservoirs

ADVANCES IN CHEMISTRY SERIES **237**

Environmental Chemistry of Lakes and Reservoirs

Lawrence A. Baker, EDITOR
Arizona State University

Developed from a symposium sponsored
by the Division of Environmental Chemistry, Inc.,
at the 201st National Meeting
of the American Chemical Society,
Atlanta, Georgia,
April 14–19, 1991

American Chemical Society, Washington, DC 1994

Library of Congress Cataloging-in-Publication Data

Environmental chemistry of lakes and reservoirs / Lawrence A. Baker, editor.

    p.   cm.—(Advances in chemistry series, 0065–2393; 237)

    "Developed from a symposium sponsored by the Division of Environmental Chemistry, Inc., at the 201st National Meeting of the American Chemical Society, Atlanta, Georgia, April 14–19, 1991."

    Includes bibliographical references and index.

    ISBN 0–8412–2526–5

    1. Limnology—Congresses.   2. Biogeochemical cycles—Congresses. 3. Environmental chemistry—Congresses.

    I. Baker, Lawrence A. II. American Chemical Society. Division of Environmental Chemistry. III. American Chemical Society. Meeting (201st: 1991: Atlanta, Ga.) IV. Series.

QD1.A335 no. 237
[GB1601.2]
551.48'2—dc20

                                        93–31891
                                           CIP

FOREWORD

The ADVANCES IN CHEMISTRY SERIES was founded in 1949 by the American Chemical Society as an outlet for symposia and collections of data in special areas of topical interest that could not be accommodated in the Society's journals. It provides a medium for symposia that would otherwise be fragmented because their papers would be distributed among several journals or not published at all.

Papers are reviewed critically according to ACS editorial standards and receive the careful attention and processing characteristic of ACS publications. Volumes in the ADVANCES IN CHEMISTRY SERIES maintain the integrity of the symposia on which they are based; however, verbatim reproductions of previously published papers are not accepted. Papers may include reports of research as well as reviews, because symposia may embrace both types of presentation.

ABOUT THE EDITOR

LAWRENCE BAKER is an environmental engineer who studies water quality in natural systems. He received his M.S. in civil and environmental engineering at Utah State University in 1979 and his Ph.D. in environmental engineering sciences at the University of Florida in 1984. After a stint as a postdoctoral and research associate at the University of Minnesota, he moved to the Environmental Protection Agency's (EPA) Environmental Research Laboratory in Corvallis, Oregon, as an on-site university cooperator. There he led a small group of scientists in the interpretation of the EPA's extensive lake and stream acidification surveys and served as a technical contributor to *Integrated Assessment*, the National Acid Precipitation Assessment Program's report to the U.S. Congress on 10 years of acid-deposition research.

His 40+ publications have dealt with biogeochemical processes that control the alkalinity of surface waters, the geochemisty of dilute "seepage" lakes, sediment chemistry, the interpretation of water-quality trends, regional analysis of water quality, modeling lake eutrophication, lake management, reservoir water quality, and nonpoint source pollution. He recently joined the faculty of the Department of Civil Engineering at Arizona State University.

CONTENTS

PREFACE

THIS BOOK EXPLORES A BROAD RANGE OF RESEARCH dealing with the environmental chemistry of lakes and reservoirs. Both it and the symposium on which it is based were developed with four goals in mind. The first was to include a wide spectrum of topics, ranging from trace metal cycling and nutrient biogeochemistry to organic geochemistry. Second, there was an emphasis on timeliness, with an effort to include only work that was coming into full fruition, neither preliminary nor overwrought. Third, the book was to have a distinctly practical orientation reflecting the background of the editor, an environmental engineer. Finally, this volume was intended to reach a broad audience, including not only chemists, but also environmental engineers and biologists; scientists involved in practical aspects of water pollution, as well as those with a theoretical bent. To prepare for wide readership, the authors were encouraged to write their chapters in a more didactic manner than is customary for journal articles. The introductions include extensive background on the topic for the reader who may be reaching into new territory, and discussion sections often set the stage for future research or delve into policy implications.

The chapters are divided into four groups. The first section emphasizes methodological advances in studies of lake geochemistry. It includes two chapters that use paleolimnological approaches to examine historical changes in the chemistry of lakes, two chapters that describe whole-lake manipulations (an acidification experiment and a nutrient-enrichment experiment), and one chapter on the use of stable oxygen isotopes to develop hydrologic budgets for lake–groundwater systems.

The second section examines the cycling and distribution of major elements (C, N, S, O, and P) in aquatic systems. It begins with a state-of-the-art description of the chemical composition of dissolved organic matter in aquatic systems. Two chapters examine the nature of phosphorus; one examines the nature of organic phosphorus compounds in lakes and the other describes a comprehensive study of the vertical transport of phosphorus in Lake Michigan. Two other chapters examine sulfur cycling; one examines the nitrogen saturation issue on a regional scale and the other summarizes an extensive series of studies on peroxide chemistry in freshwater systems.

The third section focuses on the behavior of trace metals, with an emphasis on processes that control their solubility and transport. It includes a chapter on mercury cycling in softwater lakes and two chapters that examine the solubility and transport of trace metals in two very different environments: a eutrophic and a mine-contaminated river–reservoir system. The final chapter in this section examines the controls on manganese solubility in a southeastern reservoir.

The last section deals with several organic contaminants. The first chapter is an analysis of the behavior of surfactants in the context of a comprehensive risk-assessment analysis. The next two chapters examine the fate of PCBs, as influenced by algal uptake in the water column, volatilization, and weathering in sediments.

Acknowledgments

I thank Dean Adams, the acting program chairman for the ACS Environmental Chemistry Division in 1991, for inviting me to develop the symposium from which this book was developed.

I also thank the authors for their contributions and for accepting editorial criticism intended to improve consistency among chapters. Steve Eisenreich, the co-editor of an earlier Advances volume (*Fate and Effects of Aquatic Pollutants*) was very helpful in providing advice to a novice editor on how to develop a book of this type. Finally, I would like to thank Cheryl Shanks and Colleen Stamm of the ACS Books Department for their guidance and support.

This book was developed during my tenure as an on-site university cooperator at the EPA's Environmental Research Laboratory in Corvallis, Oregon.

LAWRENCE A. BAKER
Department of Civil Engineering
Arizona State University

August 3, 1993

GEOCHEMICAL ANALYSIS

Coring for a paleolimnological analysis of lake acidification in Big Moose Lake, NY.
Photo courtesy of Donald R. Whitehead.

Often theoretical scientific advances are limited by our inability to see the world clearly. Thus, major theoretical breakthroughs are often preceded by improvements in analytical capability. Although many chapters in this book include methodological advances, the chapters in this section focus on new methods for studying aquatic systems. Several chapters represent advances in cross-disciplinary analysis, illustrating the fact that advances in environmental science are often made by bridging the treacherous gap between disciplines.

Long-Term Chemical Changes in Lakes

Quantitative Inferences from Biotic Remains in the Sediment Record

Donald F. Charles[1] and John P. Smol[2]

[1]Patrick Center for Environmental Research, Academy of Natural Sciences of Philadelphia, Philadelphia, PA 19103
[2]Paleoecological Environmental Assessment and Research Laboratory, Department of Biology, Queen's University, Kingston, Ontario K7L 3N6 Canada

One of the best ways, and often the only way, to obtain long-term data on lake-water chemistry is by inference from stratigraphic remains of aquatic biota preserved in sediment cores. Techniques are available for making accurate inferences of a variety of historical water chemistry characteristics (e.g., pH, aluminum, total phosphorus, and salinity). Many groups of biota can be used, including diatoms, chrysophytes, chironomids, and Cladocera. Inference techniques are based on the strong relationships that exist between the contemporary distributions of taxa and water chemistry characteristics. Recent advances in paleolimnological protocols, taxonomy, interpretations of ecological data, computer technology, and development of new statistical and multivariate techniques allow inferences of ever-increasing accuracy and precision. In our view, canonical correspondence analysis and weighted averaging regression and calibration are currently the best techniques available for exploring relationships between biota and chemistry and for making quantitative inferences of water chemistry, respectively. Computer-intensive techniques, such as bootstrapping, are available to estimate errors of prediction associated with inferred values.

MANY ENVIRONMENTAL PROBLEMS that involve chemical characteristics of lakes could be understood and managed better if we knew background (i.e., preimpact or "natural") environmental conditions and how environmental variables have changed over time. Policy analysts, decision makers, managers, and scientists all need information about past conditions (1). What were lakes like in their natural (i.e., preanthropogenic) states? What is the range of natural variability? Has lake-water chemistry changed? If so, when and by how much? Can our current understanding of the plausible mechanisms account for the observed changes? Answers to these questions are crucial for quantifying the amount of change that has occurred, determining the cause(s) of the changes, and assessing the potential for recovery of the system(s) under study.

Past lake-water chemistry changes can be determined by comparison of historical with current measurements; hindcasts using empirical and dynamic computer models; space-for-time substitutions (i.e., comparisons between the affected lakes and similar but unaffected lakes); and paleolimnological studies. Of these, paleolimnology is the only approach that does not require the existence of historical data or a thorough understanding of all important watershed and in-lake biogeochemical processes.

Vast information on past lake conditions is contained in sediment records, considering the large number of sediment characteristics that can now be analyzed (2–5) and the hundreds of thousands of lakes, reservoirs, ponds, wetlands, and estuaries that have sediments suitable for study. The sediment archive should not be considered only as a repository for environmental information describing conditions at single points in time. It records dynamic changes in lakes and should be viewed as the "notes" recording the results of full-scale natural and anthropogenic experiments, providing considerable insight into how lake and watershed systems evolve and respond to environmental change.

One of the most accurate, precise, robust, versatile, and widely applicable approaches for obtaining long-term water chemistry data from the sediment record is to infer them from the remains of biota. Although this approach has evolved over several decades, development of techniques has accelerated especially rapidly in the past 5–10 years. It is now possible to infer past chemical conditions for many variables, averaged over one to a few years, with an accuracy comparable to that of a modern chemical monitoring program. Most recently, paleolimnological approaches have been used widely in North America and Europe to reconstruct trends in lake acidification, particularly as caused by acidic deposition. They have also been used to provide information on changes in trophic state, organic loading, and salinity and other climatically related factors.

Geochemical analyses of dated sediment cores can also provide important information on past environmental changes (e.g., ref. 6). However, the geochemical record of many water chemistry variables of interest (such as lake-water pH, total phosphorus, monomeric aluminum, and salinity) is often difficult to decipher (for example, because of postdepositional processes such as the mobility of metal ions within sediments) (7).

Paleolimnological inferences from biological remains are based on the strong relationships between biota and many limnological characteristics and the fact that many groups of organisms leave identifiable remains in sediment strata that can be dated by using radiometric and other techniques. The groups used most extensively to infer past chemical conditions include algae with siliceous cell walls (diatoms and chrysophytes), nonsiliceous algae (e.g., vegetative structure of green algae and pigment degradation products), chironomids (midge larvae), and other insects and invertebrates (e.g., *Cladocera*, ostracods, and sponge spicules).

Quantitative paleolimnological reconstructions involve two basic steps: establishing calibration or transfer functions (also called predictive or inference models), and then using these functions to infer environmental variables from fossil assemblages. To accomplish the first step, a calibration (also called a training or reference) data set is created for a group of lakes (usually about 50, but the more the better) in a geographic region, the biological indicators of interest are identified and quantified in the recently deposited surface sediments (e.g., 0–1 cm), and the chemistry of the overlying lake water is measured. The relationships between the surface-sediment assemblage data and the overlying water chemistry are quantified by using a variety of statistical and multivariate procedures, and equations based on these relationships are developed to infer water chemistry values from the biotic remains. In the second step, sediment cores are collected from the study lakes, depth–time profiles are established (by using, for example, ^{210}Pb chronology); the biotic remains in the dated sediment slices are identified and counted; and the inference equations, developed with the calibration set of lakes in step one, are used to reconstruct past chemistry.

These techniques have made it possible to establish the onset of acidification, and later recovery, resulting from sulfur emissions from smelters in Sudbury, Ontario (8) (Figure 1); the progressive acidification of a lake in Sweden and the temporary rise in pH caused by liming (9) (Figure 2); the increase in monomeric aluminum in Adirondack lakes and its relationship to declines in fish populations (10) (Figure 3); and changes in salinity of a North Dakota lake in response to climate change and upstream regulation of reservoirs (11) (Figure 4). In all four of these examples, inferred values generally agree well with measurements, providing further evidence of the overall accuracy of the paleolimnological techniques and results.

This chapter provides an overview of the present status of paleolimnological techniques for quantitatively inferring water chemistry from biological remains preserved in lake sediments. After presenting the background and rationale of the approach and a discussion of biotic remains in sediments, it moves to a discussion of the most recent advances for some of these groups,

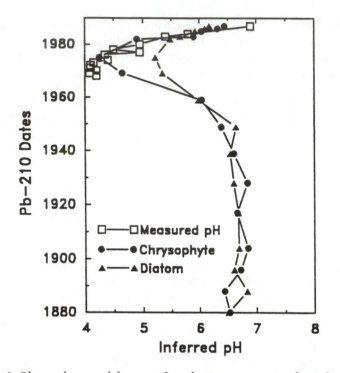

Figure 1. Chrysophyte- and diatom-inferred pH reconstruction for Baby Lake, Sudbury, Ontario, Canada. Recent lake-water pH measurements are shown on the graph as open squares. The figure shows the decrease of lake pH attributed to the emission of sulfates from smelters in the Sudbury area, and the recovery after 1970 following major reductions of emissions. (Reproduced with permission from reference 8. Copyright 1992 Kluwer Academic Publishers.)

emphasizing diatoms and chrysophytes, the most widely used water chemistry indicators. Finally, we provide some example applications that focus on the issues of lake acidification and eutrophication. These approaches can be used to infer a variety of limnological characteristics, but we limit our discussion to water chemistry.

Several recent papers and books provide further background for the general paleolimnological approaches described in this chapter (12–18). Those dealing most directly with the methods described here include publications by ter Braak (19), ter Braak and Barendregt (20), ter Braak and van Dam (21), Birks et al. (17, 22), Stevenson et al. (23), Battarbee (15), and Dixit et al. (24). These references contain a more thorough explanation of concepts and terminology than is provided in this chapter. Reading the most current literature is important because paleolimnological approaches continue to be modified and are improving at a rapid rate.

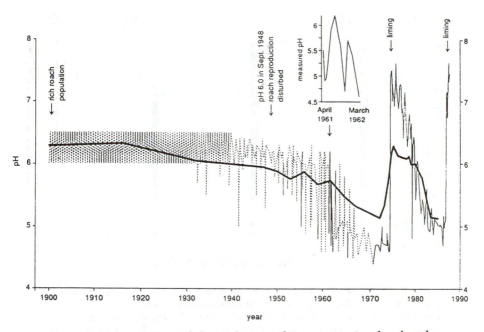

Figure 2. Reconstruction of the pH history of Lysevatten, Sweden, based on historical data and inference from the fossil diatoms in the sediment. Historical data are pH measurements (thin solid line), indirect data from fish reports, and data from other similar lakes (thin broken line). The insert, showing pH variations from April 1961 to March 1962, is based on actual water chemistry measurements. Diatom-inferred values (thick solid line) were obtained by weighted averaging. (Reproduced with permission from reference 9. Copyright 1992 Ministry of Supply and Services Canada.)

Biological Remains Used To Infer Water Chemistry

Ever since the first remains of biota were discovered in lake sediment (*see* review by Frey, ref. 25), paleolimnologists have been investigating long-term changes in lakes. Aquatic organisms are good indicators of water chemistry because they are in direct contact with water, and their occurrence is strongly affected by the chemical composition of their surroundings. Paleolimnological approaches, unlike "snapshot" water chemistry analyses, are integrative in that the biota preserved in lake sediments provide information on average or typical conditions. Most indicator groups grow quickly and are short-lived, so assemblage composition responds rapidly to changing chemical conditions. Biota are useful for paleolimnological investigations because many groups leave identifiable remains in sediments, often in quantities sufficient for rigorous statistical analysis. The main groups of sedimentary biota used for inferring water chemistry conditions are diatoms, chrysophytes, chironomids, *Cladocera*, ostracods, sponge spicules, and

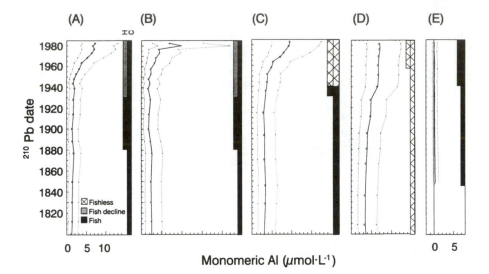

Figure 3. Reconstructions of (A) diatom-based and (B) chrysophyte-based monomeric Al for Big Moose Lake, and diatom-based monomeric Al for (C) Deep Lake, (D) Upper Wallface Pond, and (E) Windfall Pond in the Adirondack Mountains, New York. Reconstructions are bounded by bootstrapping estimates of the root mean-squared error of prediction for each sample. Bars to the right of each reconstruction indicate historical (H) and Chaoborus-based (C) reconstructions of fishery resources. The historical fish records are not continuous, unlike the paleolimnological records. Intervals older than ~1884 are dated by extrapolation. (Reproduced with permission from reference 10. Copyright 1992 Ministry of Supply and Services Canada.)

bryophytes. Unfortunately, the records of some biota are not preserved, and the parts that are preserved may allow identification only to the genus or family level. Some remains are not abundant or are present only in certain types of lakes. One of the best approaches for dealing with these limitations is to examine as many indicators as possible; the most reliable assessments of environmental change are based on inferences from several taxonomic groups.

The groups most often used for quantitative inferences are siliceous algae, primarily diatoms and chrysophytes. These groups are useful largely because:

1. Remains of many taxa from these groups are recoverable from the sediment, thereby providing comprehensive and abundant ecological information.

2. Diatom valves and chrysophyte scales are usually well preserved.

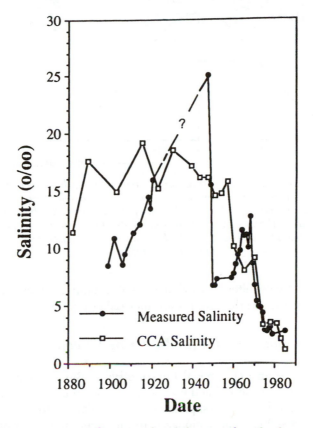

Figure 4. A comparison of measured and diatom-inferred salinity in Devil's Lake, North Dakota, for the period of historic record. No salinity measurements were made between 1923 and 1948, hence the dotted line and question mark between these dates. (Reproduced with permission from reference 11. Copyright 1990 American Society of Limnology and Oceanography.)

3. Their remains can be identified to a low taxonomic level (e.g., species level or lower).

4. Concentrations of remains are high.

5. They occur in a wide range of lakes and habitats.

6. There is a large volume of data on their ecological characteristics.

Most importantly, research efforts have shown that the water chemistry optima (e.g., for pH, monomeric aluminum, and total phosphorus) and tolerances or amplitudes (Figure 5) of diatoms and chrysophytes can be esti-

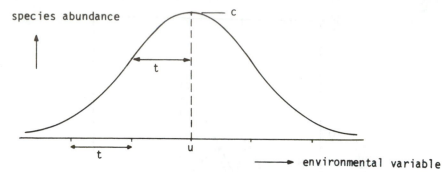

Figure 5. Unimodal (Gaussian) response curve with its three ecologically important parameters: maximum (c), optimum (u), and tolerance (t). Vertical axis: species abundance. Horizontal axis: environmental variable. The range of occurrence of the species is seen to be about 4t. (Reproduced with permission from reference 45. Copyright 1987 PUDOC-Centre for Agricultural Publishing and Documentation.)

mated quantitatively, often to a high degree of certainty. Nonetheless, efforts to develop the use of other groups are increasing rapidly.

Approaches and Methods for Inferring Chemistry

Background and Historical Perspective. Biological data from sediment cores can be analyzed and interpreted in several ways to infer chemistry. All methods begin with separation of the indicators from the sediment matrix; reference 2 contains summaries of the techniques most often used. The indicators can then be identified (usually by using high-resolution microscopy) and enumerated. Early inference techniques were straightforward but qualitative; more recent techniques are quantitative, mathematically sophisticated, sometimes computer-intensive, and capable of providing inferred chemistry values (and error estimates) with surprising accuracy and precision.

The simplest approach is to interpret the relative abundances of the common taxa and make qualitative determinations of conditions based on the known ecological characteristics of the taxa encountered, often as reported in published sources (e.g., "The presence of these taxa suggests that the lake was oligotrophic."). Sometimes the information provided by the most common taxa is of limited value, because the taxa occur over broad environmental gradients and so convey little indication of specific chemical conditions. A closely related approach is to focus on indicator taxa, which usually occur over a narrow range of ecological conditions (e.g., "This taxon occurs only in highly saline environments."). These two approaches are qualitative, and therefore difficult to use either to test hypotheses or to make management decisions requiring other than very general information. Furthermore, these approaches use only some of the taxa in an assemblage for interpretations, and so much information remains unused.

Several decades ago ecologists began creating categories for classifying biota (e.g., polysaprobic, eutrophic, and acidophilic) (26, 27). Each set of categories was created by arbitrarily dividing an ecological continuum into segments. Taxa were then assigned to one of the categories on the basis of available ecological knowledge, although some taxa could not be assigned because of insufficient data. This approach has been used widely in studies of trophic state (e.g., 28), organic pollution (e.g., 29), and acidification (e.g., 30–32). With this approach, the percentage of all taxa in the ecological categories, plus unknowns, totals 100%, and changes in the relative percentages of each category can be used to indicate the direction of water chemistry changes. For example, an increase in the percentage of acidobiontic forms (i.e., those taxa with optimum distributions at pH ≤5.5) and acidophilic forms (i.e., those taxa with widest distribution at pH <7.0) indicates that a lake has become more acidic.

However, because it is difficult to assess the meaning of changes in three or more categories over time, ratios of the categories, or indices based on the categories, were developed so that the conditions represented by a single assemblage could be expressed as a single univariate index, and changes in that single number among samples from various depths in a single core could then be interpreted as indicating chemical trends over time (Battarbee et al. (33) present a historical review). These techniques evolved to a point at which index values for the assemblages could be related directly to measured chemical parameters (34).

This relationship was accomplished by first creating a calibration data set, also called a reference set or training set. The calibration set consists of two basic types of data for a suite of lakes: assemblage count data for surface sediment samples (usually the top 1.0 cm of sediment, which includes the past few years of sediment accumulation) and measured water chemistry values. The relationship between the assemblage data (expressed as percents of ecological categories, ratios, indices, or in some other manner) is correlated to chemistry measurements. A predictive equation (31) or transfer function (35) is developed, which then can be used to infer water chemistry on the basis of assemblage data (e.g., fossil assemblages).

The potential usefulness of the equation is indicated by the strength of the correlation between observed and inferred values characterized by the coefficient of determination (r^2), as well as the standard error and the 95% confidence intervals associated with the regression. The overriding value of the relationship is that it can be used to infer past lake-water chemistry characteristics, with quantitative error estimates (e.g., "The lake-water pH value, inferred from the sediment deposited at the 5.0-cm interval, is 6.3 with an estimated standard error of 0.3 pH units."). To base inferred values only on the percent abundance of a limited number of categories is wasteful of ecological information because many taxa are clumped into large categories. In addition, this approach assumes a priori a strong quantitative relationship between the reconstructed variable and biotic composition.

Recently the methods of canonical correspondence analysis (CCA) and weighted averaging (WA) regression and calibration have been applied to paleoecological data to overcome the shortcomings of the techniques described. CCA, developed by ter Braak (19, 36), is an extension of correspondence analysis (also known as reciprocal averaging), a technique widely used by many ecologists. It is built upon the DECORANA program (detrended correspondence analysis) developed by Hill (37). Papers by ter Braak (e.g., refs. 19 and 36) are good references to consult for background on CCA; papers describing the specific application of CCA for reconstruction of water chemistry include Birks et al. (17, 22), Stevenson et al. (23), Dixit et al. (38–40), and Fritz (11). Weighted averaging regression and calibration can be implemented by using the CANOCO program (36), or more easily with WACALIB (41).

The techniques of CCA and WA are currently considered to be state of the art for inferring water chemistry characteristics from biota. CCA is used to explore, simplify, and express underlying patterns and relationships between assemblage composition and measured water chemistry data. It can also be used to assess which environmental variables explain the largest amounts of variability in assemblage composition, and therefore gives an indication of which variables can, in theory, be reconstructed. WA regression and calibration can then be used to develop equations to infer the water chemistry characteristics from the biological data in sediment core profiles. Rigorous error-estimation techniques have been developed to provide an accurate assessment of the error associated with the reconstruction (17, 22).

Both CCA and WA require a calibration data set, as already described. For best results, these data sets should be developed carefully and with considerable forethought. The lakes chosen should include the range of limnological conditions that investigators anticipate inferring from the stratigraphic data from cored lakes. Several chemical and other environmental characteristics likely to influence the distribution of taxa should also be measured accurately on as many samples as possible to characterize temporal variability. Insufficient or inadequate chemistry data is an important source of error associated with inferred values.

Quality assurance and database management deserve special attention. Study designs should include replicate cores, subsampling, and counts (e.g., refs. 42 and 43). Taxonomy should be given a high priority, and representative specimens should be documented with photographs and reference slides. All data necessary for interpretation of results should be documented (44). Because combinations of calibration and stratigraphic data sets are complex, relatively sophisticated database management programs are necessary for efficient data handling and analysis.

Canonical Correspondence Analysis.   Calibration data sets developed to infer water chemistry can be used to evaluate a number of important questions such as: What measured environmental variables appear most influential in determining the distribution of taxa? Are relationships between biota and certain chemical characteristics strong enough to develop useful predictive relationships? Are there potential problems with inference procedures because unmeasured factors also influence assemblage composition? In our view, CCA is presently the best approach for addressing these questions; it provides convincing evidence that the biota–chemistry relationships of interest are strong and that there is a sound basis for using these relationships to infer past water chemistry.

CCA, like all ordination techniques, orders samples and species along axes. As with other ordination techniques, CCA axes are constructed to maximize the dispersion of the samples or species. Thus, when points representing samples or species are plotted along an axis, those most similar to each other are grouped together, whereas those most different from each other are more distant. For example, if phosphorus concentration strongly determines species distributions, a sample assemblage from a lake with high total phosphorus will plot at the opposite end of an axis from an assemblage from a lake with low phosphorus concentrations.

CCA, a direct gradient analysis technique, differs from other ordination and multivariate techniques such as polar ordination, principal components, and canonical correlation analysis. An important underlying assumption of CCA is that the abundance of individual taxa follows a unimodal curve over long environmental gradients (45). An example of a unimodal curve is a Gaussian curve (Figure 5). A number of other ordination techniques assume a linear response of taxa abundance to environmental gradients. The biggest advantage of CCA over these techniques is that the position of a species or sample on an axis is not only determined by the taxa present in the samples, but is also a function (linear combination) of a defined set of environmental variables. Thus CCA provides visual and mathematical expression of the patterns of variation among sample assemblages and also enables evaluation of the role and importance of environmental variables in explaining that pattern of variation.

Ordination axes are determined sequentially. The first axis is calculated to account for the most prominent variance in the samples or species; subsequent axes are calculated to account for variability not explained by the earlier axes. It is useful to visualize the distribution of samples or species as a function of their scores on the two or three most significant axes (as represented by their eigenvalues and testable by permutation tests). Plots of one ordination axis versus another (called bivariate plots or biplots) can be used to examine patterns of similarity and difference among species and samples.

An example of a CCA biplot is shown in Figure 6. It was developed to explore the relationships between trophic state, several water chemistry

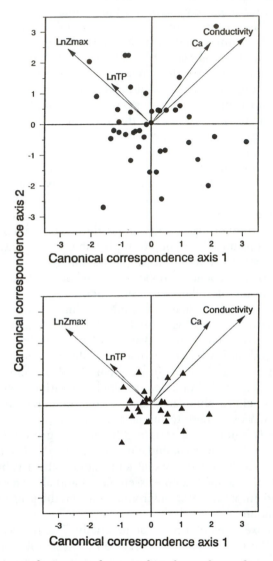

Figure 6. Canonical correspondence analysis for surface sediments of 41 lakes in British Columbia, Canada, that encompass a broad range of trophic states. Circles represent lakes and triangles represent the 25 most abundant diatom taxa. Arrows indicate environmental variables that correlate most strongly with the distribution of diatom taxa and lake-water chemistry, as detected by forward selection. Maximum depth (Z_{max}) and total phosphorus (TP) were transformed by using the $\ln (x + 1)$ function. This analysis is discussed in detail in reference 46.

characteristics, and surface-sediment diatom assemblages from 41 lakes in British Columbia (46). Relationships among species (and samples) are shown on a biplot of the scores generated for axes 1 and 2. The arrows represent environmental axes, in this case the gradients of chemical characteristics. They extend on both sides of the origin, but are shown on only one (thus the origin is the mean of that variable). The length of each arrow indicates how important that variable is in determining the distribution of the lakes and the taxa in the ordination. The arrows point in the direction of the highest positive magnitude. The angle between the arrows and/or the or-dination axes shows how closely the two variables are correlated with each other. The distance between any two species points or between any two sample points indicates how similar the pairs of points are to each other. The CCA analysis shows that conductivity, maximum depth, calcium, and total phosphorus can explain most of the variation in diatom assemblage composition and lake-water chemistry in this training set. It also shows that total phosphorus and lake depth are closely related and that there is a wide spread of axis scores along these gradients. It makes clear, however, that factors related to the amount of dissolved minerals also have an influence.

CCA can be used to examine long-term trends in water chemistry by combining stratigraphic (i.e., fossil) sample data into a modern calibration diagram. This analysis is done so that all the stratigraphic points are posi-tioned along the axes on the basis of their overall similarity with the cali-bration samples. Of course, only the calibration samples are used to calculate direction and magnitude of the environmental arrows, as the fossil assem-blages do not have any associated water chemistry variables. A biplot can be made that shows the changing location of the stratigraphic data points, over time, with respect to several environmental variables on the same graph. For example, this type of analysis for Round Loch of Glenhead (Figure 7) shows some very distinct stages in the long-term acidification of the lake and how it relates to several characteristics. The fossil assemblage scores indicate that Round Loch had its highest pH, alkalinity, and calcium con-centrations following deglaciation (pre-10,000–9200 B.P.). The lake gradually became more acidic, and dissolved organic carbon (DOC) concentrations fluctuated, apparently in response to changes in percent of conifer vegetation and mire development in the watershed. From about 1864 to 1985, pH, alkalinity, and calcium decreased rapidly. Aluminum concentration in-creased during this period as a result of greater atmospheric deposition of strong acids (22). The overall merit of this use of CCA is its efficiency in examining relative changes in several water chemistry characteristics si-multaneously.

Weighted Averaging.   WA regression and calibration is a robust, computationally simple, and straightforward method for reconstructing en-vironmental variables. It provides a more accurate and precise inference

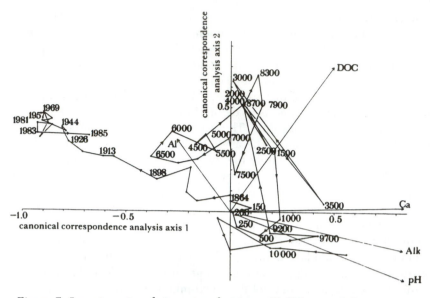

Figure 7. Long-term trends in water chemistry, 10,000 years B.P. to present for Round Loch of Glenhead (RLGH), United Kingdom, as determined qualitatively from CCA. Samples are connected in chronological order. A CCA analysis was first run using a calibration data set of 131 lakes in England, Norway, Scotland, Sweden, and Wales, to determine the positions of environmental arrows (DOC, Ca, Alk, pH, Al) on the biplot. The CCA was run a second time with the RLGH stratigraphic data as passive samples that did not influence the environmental arrows. These data are plotted on the graph, with slight rescaling. (Reproduced with permission from reference 22. Copyright 1990 Cambridge University Press.)

than methods based on ecological categories (*17*). It is effective mainly because it uses information provided by each individual taxon, and not just percent abundance of categories, as with some other methods. The computer program WACALIB (*41*) provides a convenient procedure for doing WA calculations.

The fundamental assumption of the WA technique is that the weighted average of a taxon represents the conditions for which this taxon is most abundant (Figure 8 shows typical distribution data). This optimum condition (*see* Figure 5) for each taxon can be calculated as the average of mean values for the environmental characteristics (e.g., water chemistry) at the sites in which it is found, weighted by the abundance of the taxon at the sites (Figure 9), namely

$$\hat{u}_k = \frac{\sum\limits_{i=1}^{n} y_{ik} x_i}{\sum\limits_{i=1}^{n} y_{ik}} \qquad (1)$$

where $\hat{u}_k$ is the abundance-weighted mean or optimum of taxon k; x_i is the water chemistry characteristic x (e.g., salinity) for lake i; y_{ik} is the percent abundance of taxon k in the surface-sediment assemblage of lake i; and n is the number of lakes in the calibration set. Thus, the WA indicates the weighted centroid of a taxon's distribution along an environmental gradient. The tolerance of a taxon $(\hat{t}_k)$ can be represented by a simple weighted standard deviation, which gives an indication of the range of environmental gradient over which the taxon is likely to occur (Figure 5) and is estimated by

$$\hat{t}_k = \left[\frac{\sum_{i=1}^{n} y_{ik}(x_i - \hat{u}_k)^2}{\sum_{i=1}^{n} y_{ik}} \right]^{1/2} \tag{2}$$

Other assumptions made in developing and using equations to infer water chemistry are (based on refs. 49 and 17)

1. The taxa to be used are systematically related to the chemical characteristics to be reconstructed, and these characteristics are ecologically important or linearly related to some component that is significant.

2. The taxa in the calibration set include those found in fossil data sets to which inference equations will be applied.

3. The ecological characteristics of taxa have not changed over the time period represented by the fossil data.

4. Ecological factors other than those to be reconstructed do not have a strong influence on the taxa–chemistry relationships or, if they do, the relationship is the same for the calibration and fossil data sets.

   Once WA values for an environmental characteristic (e.g., water chemistry) have been calculated for taxa in a calibration data set, the information can be used to infer that characteristic from sediment core samples and consequently to reconstruct past conditions. The first step is to determine the percent abundance of each taxon in the sediment core assemblages. The taxon abundance is then multiplied by the WA value for that taxon (determined from the calibration data set). These products are summed for all taxa and are standardized by the sum of the relative abundances of the taxa in that sample to obtain an inferred value, namely

$$\hat{x}_i = \frac{\sum_{k=1}^{m} y_{ik}\hat{u}_k}{\sum_{k=1}^{m} y_{ik}} \tag{3}$$

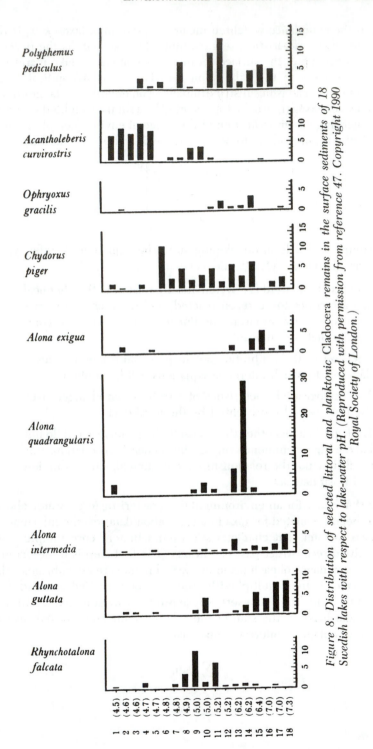

Figure 8. Distribution of selected littoral and planktonic Cladocera remains in the surface sediments of 18 Swedish lakes with respect to lake-water pH. (Reproduced with permission from reference 47. Copyright 1990 Royal Society of London.)

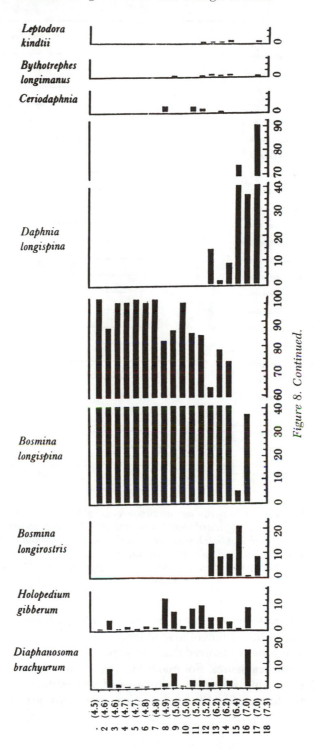

Figure 8. Continued.

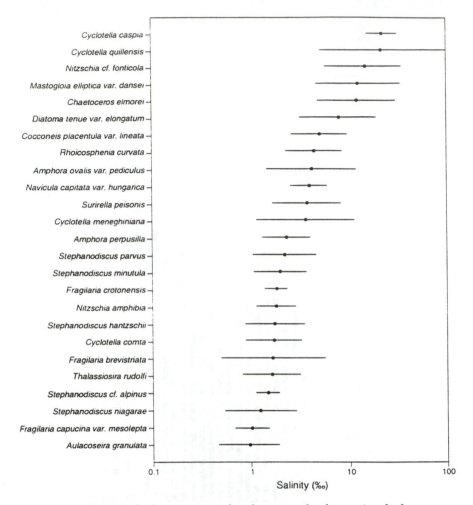

Figure 9. Estimated salinity optima (abundance-weighted means) and tolerance (abundance-weighted standard deviations) for selected diatom taxa from 55 lakes in North and South Dakota and Saskatchewan, Canada. Formulae for calculating optima and tolerance are given in the text (eqs 1 and 2). (Reproduced with permission from reference 48. Copyright 1991 McMillan Publishing Co.)

where $\hat{x}_i$ is the water chemistry characteristic being inferred and m is the number of taxa in the sediment assemblage.

The process is more involved than this in practice, and steps are required to deal with statistical issues. For example, "rogue" or "outlier" lakes should be identified to avoid unrealistic estimates of taxa optima (*see* ref. 17 for objective procedures to identify rogues). Also, steps must be taken to account for the shrinkage of the environmental variable being reconstructed. In WA regression and calibration, averages are calculated twice, first to calculate an optimum value for individual taxa and second to calculate the inferred characteristics. This procedure results in the reconstructed values being biased toward the mean value for the calibration data set (i.e., the overall gradient of potential inferred values has shrunk). To minimize this effect, a simple classical or inverse regression (sometimes called a deshrinking step) can be used (refs. 17 and 21 give more detailed discussion). For example, the classical deshrinking process performs a simple linear regression to determine the relationship between the inferred values (x_{inf}) and actual measured values (x_{meas}):

$$\text{initial } x_{inf} = a + bx_{meas} + \epsilon \tag{4}$$

where a is the intercept, b is the slope, and ϵ is an error term. The terms from this regression equation are then used to calculate a final (corrected or deshrunk) value.

$$\text{final } x_{inf} = \frac{\text{initial } x_{inf} - a}{b} \tag{5}$$

Inverse deshrinking involves the same process, except that the measured values are regressed on the initial inferred values instead of vice versa. Classical deshrinking moves inferred values farther from the mean than inverse deshrinking, and the former is best if the values to be inferred lie near the ends of the environmental gradient. Inverse deshrinking minimizes the root mean-squared error of the predicted versus measured regression relationships, and therefore may lead to more accurate inferred values over the entire range of the environmental gradient.

Another statistical issue is the relationship between the composition in the calibration set used to derive the transfer functions and the lakes for which the transfer functions will be applied. Calibration data sets should be modified if used to reconstruct chemistry of different types of lakes. A subset of calibration lakes can be selected that does not contain lakes so different that they might unduly influence optimum environmental values for a taxon (for example, saline lakes can be removed from a calibration data set to be used for generating data to infer trophic-state change in low-conductivity lakes).

An option to consider in using the WA technique is tolerance weighting. The rationale for this approach is that taxa occurring over a narrow range of an environmental gradient should be better indicators than taxa with broader tolerances. Consequently, taxa with narrower tolerances should be weighted more heavily in the WA calculations. A tolerance-weighted estimate would be calculated as:

$$\hat{x}_i = \frac{\sum_{k=1}^{m} \dfrac{y_{ik}\,\hat{u}_k}{\hat{t}_k^{\,2}}}{\sum_{k=1}^{m} \dfrac{y_{ik}}{\hat{t}_k^{\,2}}} \tag{6}$$

Tolerance weighting does not always work as well as expected in diatom reconstructions (17), perhaps because calibration data sets have not been large enough to provide accurate estimates of tolerance. Comparisons of tolerance-weighted versus unweighted approaches have shown that the simplest approach works best with diatoms (17, 22), but chrysophyte inference models perform best when tolerances are included (50, 51).

Error Analysis and Quantification of Uncertainty.   The error associated with paleolimnological inferences must be understood. Two sources of error worthy of special attention are the predictive models (transfer functions) developed to infer chemistry and inferences generated by using those equations with fossil samples in sediment strata. Much of the following discussion is based on the pioneering work reviewed by Sachs et al. (35) and by Birks et al. (17, 22), among others. We emphasize error analysis here because it is not covered in detail in most of the general review articles cited earlier.

Relevant questions about the development and use of predictive models include the following: Which transfer functions will provide the most accurate reconstructions, and what is the associated error? How much of the error is attributable to the size and composition of the calibration data set? Typically, predictive models are assessed and compared on the basis of objective statistical criteria derived from the relationship between measured and inferred values for the training set (e.g., Figure 10). These criteria include the coefficient of determination (r^2), the standard error (SE), the root mean-squared error (RMSE) of prediction, and/or the RMSE of bootstrapped samples. The r^2 value provides an indication of the strength of the predictive relationship; typically, r^2 values for most predictive models are >0.8. The RMSEs provide the most useful criteria for assessing the predictive ability of transfer functions because they indicate a specific amount of variability to expect when applying the predictive equations to new assemblage data. Oehlert (52) used a Bayesian approach for examining sources of error inherent in the development and use of inference equations.

In some studies the r^2 and SE are calculated by using with data only from the original calibration data set. Thus, the inferred values used in the correlation analysis are for the same lakes used to develop the equations.

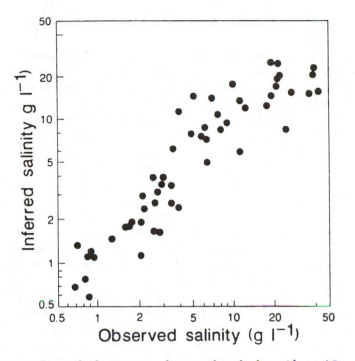

Figure 10. Observed salinity versus diatom-inferred salinity (derived from WA regression) for 55 calibration set lakes in North and South Dakota and Saskatchewan ($r^2 = 0.83$; standard error = 0.481 ln salinity, ref. 11). Equations used for the WA calculations are given in the text (eq 3). (Reproduced with permission from reference 48. Copyright 1991 McMillan Publishing Co.)

Although this approach is still used, it is undesirable for statistical reasons; error calculations underestimate the true uncertainty associated with the equations (17, 21). A better approach is to use the equations developed for one set of lakes to infer chemistry values from counts of taxa from a second set of lakes (i.e., cross-validation). The extra time and effort required to develop the additional data for the test set is a major limitation to this approach. Computer-intensive techniques, such as jackknifing or bootstrapping, can produce error estimates from the original training set (53), without having to collect data for additional lakes.

Jackknifing involves removing one sample from the calibration set, deriving the inference equations based on the remaining set of lakes (i.e., $n - 1$), and then using the new inference equation to derive an inferred value for the one sample that was removed (i.e., providing an independent error estimate). These steps are repeated until all samples have been left out once from the calibration process and used to calculate a new inferred value. The set of new inferred values is then used in conjunction with the measured values to calculate statistical measures of uncertainty (e.g., r^2, SE, and RMSE).

Bootstrapping, as demonstrated by Birks et al. (17) for paleolimnological models, is a computer-intensive resampling procedure. Within each of many bootstrap cycles, a new training set the same size as the original set is selected randomly with replacement. Because samples are selected with replacement, some of the original lakes will not be represented in the bootstrap set. A transfer function is then developed from this bootstrap set and is used to infer chemistry values for the samples from the original set that were not included in the bootstrap set (called the test set). This process is repeated in an iterative fashion (typically 1000 times) and the error estimates are developed from the distribution of samples in the test set. Besides providing a more accurate estimate of the true error associated with the equation's ability to infer chemistry values, bootstrapping can partition error resulting from variability in the size and makeup of the calibration set due to estimation of the weighted average value of a chemical characteristic for each taxon, and the inherent variability in the data set.

Two relevant uncertainty questions about chemistry values can be inferred from sediment assemblages: How accurate are the inferred values? What are realistic errors associated with the inferred values? Errors associated with inferred chemistry values for stratigraphic samples are best estimated by using the bootstrap method. The procedure is the same as described, except that stratigraphic data are used in addition to the training set. During each bootstrap cycle the transfer function is used to infer chemistry values for all of the stratigraphic assemblages in addition to the test samples (17). A RMSE is then calculated for each individual stratigraphic sample. It will vary, depending in part on how accurately the calibration data set estimates the abundance-weighted mean of the taxa in the stratigraphic sample. If the common taxa in a stratigraphic assemblage are not adequately represented (but are not absent) in the calibration set, error estimates will tend to be larger than if they are well represented.

Other questions regarding uncertainty of inferred values are important: What is the nature and magnitude of the error that might be attributable to differences between assemblage composition of the calibration data set and the core strata assemblage? More specifically, are taxa in the stratigraphic samples well represented in the calibration set? Are there any stratigraphic rogues with respect to the calibration data set, having no appropriate modern analog? The simplest way to compare composition of calibration and stratigraphic assemblages is a visual search for abundant taxa that are not represented in the calibration set. A more sophisticated approach is analog matching (17, 22, 54), whereby each stratigraphic sample is compared with each calibration sample by using a dissimilarity measure (e.g., chi-square distance and chord distance). If the dissimilarity value is less than a minimum value established for the calibration, the stratigraphic sample is considered to have a close analog in the calibration set. If the dissimilarity value is greater than that value, there is no close analog and any inferred chemistry value should be treated cautiously.

Another procedure is to measure the lack of fit of an assemblage with respect to inference of a particular chemical variable. This calculation is done by analyzing the square residual distance between a fossil sample and the CCA environmental axis (arrow) of the chemical variable being reconstructed (e.g., Figure 6). First the residual distance from each calibration sample point to the environmental axis is calculated for the most extreme 5% of the points. This step determines the lack-of-fit cutoff. Then the residual distance is calculated from all stratigraphic sample points to the environmental axis; any distances that are further than the cutoff distance are determined to have poor or very poor fits (*17, 22*). Computer programs are currently available to perform these analyses (*see* refs. 17 and 22). A more direct technique to assess the accuracy of inferred values is comparison of paleolimnological inferences to long-term water chemistry data. This procedure has been followed with, for example, pH (*8, 9, 24*) and salinity reconstructions (*11*) (Figures 1, 2, and 4).

Applications of Techniques

Paleoecological approaches are used increasingly to provide valuable data for understanding and managing lakes and their watersheds, and the volume of literature on this topic has accumulated rapidly over the past 5–10 years. Several papers summarized some of these applications (*5, 15, 18, 24, 38, 55–57*), so only a few representative examples are mentioned briefly here.

Many of the recent advances in quantitative paleolimnology were driven by funding of large studies of lake acidification. These studies include the Paleoecological Investigation of Recent Lake Acidification projects [PIRLA I and II (*58, 59*)] and the Paleolimnological Programme of the Surface Water Acidification Project (SWAP) in the United Kingdom and Scandinavia (*16*). These and other studies showed that many low-alkalinity lakes have become more acidic primarily because of acidic atmospheric deposition resulting from combustion of fossil fuels (*55*). Analysis of sediment diatom assemblages has proven particularly useful (*13, 55*). New applications include improved methods for reconstructing pH (*17, 22, 23*), organic content (*60, 61*), and metals (*39, 50, 51*); use of open-water samples of extant biota to create calibration data sets (*62*); use of museum archives of extant biota to infer past chemistry (*63*); statistical assessment of acidification of lakes in an entire region [Adirondack Park, NY (*64, 65*)]; and documented evidence of recovery (*8, 24, 40, 66*). Paleolimnological approaches have also been used to reconstruct detailed long-term pH records that are very effective in placing recent trends in perspective, especially with respect to natural processes and variability (*67, 68*) (Figure 11). Assessment of the sensitivity of lake pH to acidic

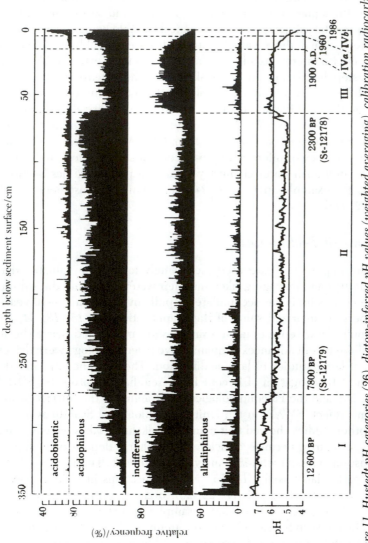
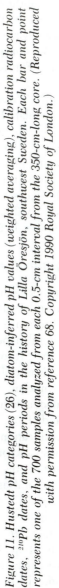

Figure 11. Hustedt pH categories (26), diatom-inferred pH values (weighted averaging), calibration radiocarbon dates, ^{210}Pb dates, and pH periods in the history of Lilla Öresjön, southwest Sweden. Each bar and point represents one of the 700 samples analyzed from each 0.5-cm interval from the 350-cm-long core. (Reproduced with permission from reference 68. Copyright 1990 Royal Society of London.)

deposition, based on diatom-inferred pH changes, can be used to help select lakes for long-term monitoring of acidification trends (69). Yet another application is to help evaluate computer models of lake and watershed processes by comparing paleolimnological reconstructions with computer model hindcasts of the same or closely related environmental variables. This procedure has been followed with fruitful results in lake acidification studies (70–72).

Changes in trophic state, including total phosphorus, chlorophyll *a*, and transparency, have been inferred by using several sediment characteristics (28, 46, 73, 74). Newly developed models for predicting total phosphorus from diatom assemblages by using WA look particularly promising (e.g., Figure 12). Paleolimnological approaches have been used to examine eutrophication trends in several lakes (75–78). In addition to analysis of changes in species composition, change in morphometry of single species is being further explored as an indicator of trophic status (79). Diatoms have been used to infer changes in lake-water salinity (11, 48, 80, 81) and temperature (82), and many paleolimnological techniques are available for inferring climate and related factors (83–86).

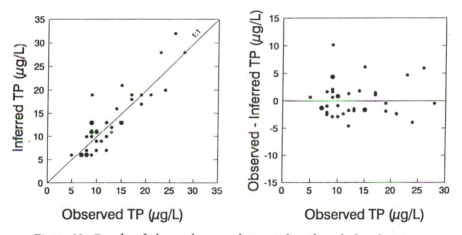

Figure 12. Graphs of observed versus diatom-inferred total phosphorus concentrations (TP) and observed minus diatom-inferred TP (i.e., a residual analysis) are based on weighted averaging regression and calibration models and classical deshrinking. The large circles indicate two coincident values. This analysis is discussed in detail in reference 46.

Summary and Conclusions

Long-term data on changes in water chemistry are valuable and often necessary for understanding lakes and for making sound management decisions. Paleolimnological approaches are one of the few ways to obtain this information. Historical trends of many chemical characteristics can now be interpreted, with a relatively high degree of precision and accuracy, by examining the remains of biota such as siliceous algae (diatoms and chrysophytes), zooplankton (*Cladocera*), midge larvae (chironomids), and other groups archived in lake sediments. By analyzing the past assemblages from dated cores, paleolimnologists can reconstruct changes in lake acidity, trophic state, salinity, climate-related factors, toxic chemicals, and other characteristics. Knowledge and techniques necessary for reconstructing these characteristics have advanced rapidly in the past 5–10 years, particularly in the areas of taxonomy, acquisition of ecological data, multivariate and statistical data analysis, data management, and quality assurance. Canonical correspondence analysis and weighted averaging regression and calibration have proven particularly useful for developing statistically robust and ecologically relevant equations for inferring chemical conditions from biological data. These advances offer many new opportunities for gaining important insights to better understand and manage lake resources.

Acknowledgments

Preparation of this paper by the first author was funded in large part by the U.S. Environmental Protection Agency (Cooperative Agreement CR–813933 with Indiana University). This manuscript has not been subjected to the Agency's peer and administrative review and approved for publication. We thank Larry Baker, John Birks, Brian Cumming, Roland Hall, Allen Uutala, and two anonymous reviewers for their comments on the manuscript, and Roland Hall for providing copies of unpublished figures.

References

1. Smol, J. P. *J. Aq. Ecosystem Health* **1992**, *1*, 49–58.
2. *Handbook of Holocene Palaeoecology and Palaeohydrology*; Berglund, B. E., Ed.; John Wiley and Sons: Chichester, England, 1986; pp 1–869.
3. Gray, J. *Paleolimnology: Aspects of Freshwater Paleoecology and Biogeography*; Elsevier: Amsterdam, Netherlands, 1988.
4. *Methods in Quaternary Ecology*; Warner, B., Ed.; Geological Association of Canada: St. John's, Canada, 1990.
5. Smol, J. P.; Glew, J. R. In *Encyclopedia of Earth System Science*; Academic: San Diego, CA, 1992; Vol. 3, pp 551–564.
6. Engstrom, D. R. In *Environmental Chemistry of Lakes and Reservoirs*; Baker, L. A., Ed.; Advances in Chemistry 237; American Chemical Society: Washington, DC, 1994; Chapter 2.
7. Engstrom, D. R.; Wright, H. E., Jr. In *Lake Sediments and Environmental History*; Haworth, E. Y.; Lund, J. W. G., Eds.; Leicester University Press: Leicester, England, 1984; pp 11–67.
8. Dixit, A. S.; Dixit, S. S.; Smol, J. P. *Water Air Soil Pollut.* **1992**, *62*, 75–87.
9. Renberg, I.; Hultberg, H. *Can. J. Fish. Aquat. Sci.* **1992**, *49*, 65–72.

10. Kingston, J. C.; Birks, H. J. B.; Uutala, A. J.; Cumming, B. F.; Smol, J. P. *Can. J. Fish. Aquat. Sci.* **1992**, *49*, 116–127.
11. Fritz, S. *Limnol. Oceanogr.* **1990**, *35*, 1771–1781.
12. Huttunen, P.; Meriläinen, J. *Hydrobiologia* **1983**, *103*, 91–97.
13. Smol, J. P.; Battarbee, R. W.; Davis, R. B.; Meriläinen, J., Eds.; *Diatoms and Lake Acidity;* Dr. W. Junk: Dordrecht, Netherlands; 1986.
14. Davis, R. B. *Quat. Sci. Rev.* **1987**, *6*, 147–163.
15. Battarbee, R. W. In *Quaternary Landscapes;* Shane, L. C. K.; Cushing, E. J., Eds; University of Minnesota Press: Minneapolis, MN, 1991; pp 129–174.
16. Battarbee, R. W.; Renberg, I. *Philos. Trans. R. Soc. London B* **1990**, *327*, 227–232.
17. Birks, H. J. B.; Line, J. M.; Juggins, S.; Stevenson, A. C.; Ter Braak, C. J. F. *Philos. Trans. R. Soc. London B* **1990**, *327*, 263–278.
18. Charles, D. F.; Smol, J. P.; Engstrom, D. R. In *Biological Monitoring of Freshwater Ecosystems;* Loeb, S.; Spacie, A., Eds.; Lewis: Boca Raton, FL, in press.
19. Ter Braak, C. J. F. *Ecology* **1987**, *67*, 1167–1179.
20. Ter Braak, C. J. F.; Barendregt, L. G. *Math. Biosci.* **1986**, *78*, 57–72.
21. Ter Braak, C. J. F.; van Dam, H. *Hydrobiologia* **1989**, *178*, 209–223.
22. Birks, H. J. B.; Juggins, S.; Line, J. M. In *The Surface Waters Acidification Programme;* Mason, B. J., Ed.; Cambridge University Press: Cambridge, United Kingdom, 1990; pp 301–313.
23. Stevenson, A. C.; Birks, H. J. B.; Flower, R. J.; Battarbee, R. W. *Ambio* **1989**, *18*, 229–233.
24. Dixit, S. S.; Smol, J. P.; Kingston, J. C.; Charles, D. F. *Environ. Sci. Technol.* **1992**, *26*, 23–32.
25. Frey, D. G. *Mitt. Int. Ver. Theor. Angew. Limnol.* **1969**, *17*, 7–18.
26. Hustedt, F. *Arch. Hydrobiol. Suppl.* **1939**, *16*, 274–394.
27. Lowe, R. L. *Environmental Requirements and Pollution Tolerance of Freshwater Diatoms;* EPA/670/4–74/005; U.S. Environmental Protection Agency: Washington, DC, 1974.
28. Whitmore, T. J. *Limnol. Oceanogr.* **1989**, *34*, 882–895.
29. Watanabe, T.; Asai, K.; Houki, A. In *Encyclopedia of Environmental Control Technology: Hazardous Waste Containment and Treatment;* Cheremisinoff, P. N., Ed.; Gulf Publishing: Tokyo, Japan, 1990; Vol. 4, pp 252–281.
30. Huttunen, P.; Meriläinen, J. In *Diatoms and Lake Acidity;* Smol, J. P.; Battarbee, R. W.; Davis, R. B.; Meriläinen, J., Eds.; Dr. W. Junk: Dordrecht, Netherlands, 1986; pp 201–211.
31. Battarbee, R. W. *Philos. Trans. R. Soc. London B* **1984**, *305*, 451–477.
32. Charles, D. F.; Smol, J. P. *Limnol. Oceanogr.* **1988**, *33*, 1451–1462.
33. Battarbee, R. W.; Smol, J. P.; Meriläinen, J. In *Diatoms and Lake Acidity;* Smol, J. P.; Battarbee, R. W.; Davis, R. B.; Meriläinen, J., Eds.; Dr. W. Junk: Dordrecht, Netherlands, 1986; pp 5–14.
34. Meriläinen, J. *Ann. Bot. Fenn.* **1967**, *4*, 51–58.
35. Sachs, H. M.; Webb, T., III; Clark, D. R. *Annu. Rev. Earth Planet. Sci.* **1977**, *5*, 159–178.
36. Ter Braak, C. J. F. *CANOCO: A FORTRAN Program for Canonical Community Ordination;* TNO Inst. Appl. Computer Sci.: Wageningen, Netherlands, 1987.
37. Hill, M. O. *DECORANA: A FORTRAN Program for Detrended Correspondence Analysis and Reciprocal Averaging;* Section of Ecology and Systematics, Cornell University: Ithaca, NY, 1979.
38. Dixit, S. S.; Cumming, B. F.; Smol, J. P.; Kingston, J. C. In *Ecological Indicators;* McKenzie, D. H.; Hyatt, D. E.; MacDonald, V. J., Eds.; Elsevier: Amsterdam, Netherlands, 1992; Vol. 2, pp 1135–1155.

39. Dixit, S. S.; Dixit, A. S.; Smol, J. P. *Can. J. Fish. Aquat. Sci.* **1989**, *46*, 1–10.
40. Dixit, S. S.; Dixit, A. S.; Evans, R. D. *Environ. Sci. Technol.* **1989**, *23*, 110–115.
41. Line, J. M.; Birks, H. J. B. *J. Paleolimnol.* **1990**, *3*, 170–173.
42. Kreis, R. G., Jr. In *Paleoecological Investigation of Recent Lake Acidification: Methods and Project Description;* Charles, D. F.; Whitehead, D. R., Eds.; Report EA–4906; Electric Power Research Institute: Palo Alto, CA, 1986; pp 17–19.
43. Charles, D. F.; Dixit, S. S.; Cumming, B. F.; Smol, J. P. *J. Paleolimnol.* **1991**, *5*, 267–284.
44. Charles, D. F. *J. Paleolimnol.* **1990**, *3*, 175–178.
45. Jongman, R. G. H.; ter Braak, C. J. F.; van Tongeren, O. F. R. *Data Analysis in Community and Landscape Ecology;* Centre for Agricultural Publishing and Documentation; Wageningen: Netherlands, 1987.
46. Hall, R. I.; Smol, J. P. *Freshwater Biol.* **1992**, *27*, 417–434.
47. Nilssen, J. P.; Sandoy, S. *Philos. Trans. R. Soc. London B* **1990**, *327*, 299–309.
48. Fritz, S. C.; Juggins, S.; Battarbee, R. W.; Engstrom, D. R. *Nature (London)* **1991**, *352*, 706–708.
49. Imbrie, J.; Webb, T., III. In *Climatic Variations and Variability: Facts and Theories;* Berger, A., Ed.; D. Reidel: Dordrecht, Netherlands, 1981; pp 125–134.
50. Cumming, B. F.; Smol, J. P.; Birks, H. J. B. *Nord. J. Bot.* **1991**, *11*, 231–242.
51. Cumming, B. F.; Smol, J. P.; Birks, H. J. B. *J. Phycol.* **1992**, *28*, 162–178.
52. Oehlert, G. W. *Can. J. Stat.* **1988**, *16*, 51–60.
53. Efron, B.; Gong, G. *Am. Stat.* **1983**, *37*, 36–48.
54. Overpeck, J. T.; Webb, T., III; Prentice, I. C. *Quat. Res. N.Y.* **1985**, *23*, 87–108.
55. Charles, D. F.; Battarbee, R. W.; Renberg, I.; van Dam, H.; Smol, J. P. In *Acidic Precipitation: Soils, Aquatic Processes, and Lake Acidification;* Norton, S. A.; Lindberg, S. E.; Page, A. L., Eds.; Springer–Verlag: New York, 1989; Vol. 4., pp 207–276.
56. Smol, J. P. *Mem. Ist. Ital. Idrobiol. Dott. Marco de Marchi* **1990**, *47*, 253–276.
57. Walker, I. R. In *Biomonitoring and Freshwater Invertebrates;* Rosenburg, D. M.; Resh, V. H., Eds.; Routledge, Chapman, and Hall: New York, 1993; pp 304–343.
58. Charles, D. F.; Whitehead, D. R. *Hydrobiologia* **1986**, *143*, 13–20.
59. Charles, D. F.; Smol, J. P. *Verh. Int. Ver. Theor. Angew. Limnol.* **1990**, *24*, 474–480.
60. Davis, R. B.; Anderson, D. S.; Berge, F. *Nature (London)* **1985**, *316*, 436–438.
61. Kingston, J. C.; Birks, H. J. B. *Philos. Trans. R. Soc. London B* **1990**, *327*, 279–288.
62. Siver, P. A.; Hamer, J. S. *Can. J. Fish. Aquat. Sci.* **1990**, *47*, 1339–1347.
63. Arzet, K.; van Dam, H. In *Proceedings of the Eighth International Diatom Symposium;* Ricard, H., Ed.; Koeltz: Koenigstein, 1986; pp 748–749.
64. Sullivan, T. J.; Charles, D. F.; Smol, J. P.; Cumming, B. F.; Selle, A. R.; Thomas, D. Bernert, J. A.; Dixit, S. S. *Nature (London)* **1990**, *345*, 54–58.
65. Cumming, B. F.; Smol, J. P.; Kingston, J. C.; Charles, D. F.; Birks, H. J. B.; Camburn, K. E.; Dixit, S. S.; Uutala, A. J.; Selle, A. R. *Can. J. Fish. Aquat. Sci.* **1992**, *49*, 128–141.
66. Battarbee, R. W.; Flower, R. J.; Stevenson, A. C.; Jones, V. J.; Harriman, R.; Appleby, P. G. *Nature* **1988**, *332*, 530–532.
67. Flower, R. J.; Cameron, N. G.; Rose, N.; Fritz, S. C.; Harriman, R.; Stevenson, A. C. *Philos. Trans. R. Soc. London B* **1990**, *327*, 427–433.
68. Renberg, I. *Philos. Trans. R. Soc. London B* **1990**, *327*, 357–361.
69. Young, T. C. *Water Resour. Res.* **1991**, *27*, 317–326.

70. Wright, R. F.; Cosby, B. J.; Hornberger, G. M.; Galloway, J. N. *Water Air Soil Pollut.* **1986**, *30*, 367–380.
71. Jenkins, A.; Whitehead, P. G.; Cosby, B. J.; Birks, H. J. B. *Philos. Trans. R. Soc. London B* **1990**, *327*, 435–440.
72. Sullivan, T. J.; Turner, R. S.; Charles, D. F.; Cumming, B. F.; Smol, J. P.; Schofield, C. L.; Driscoll, C. T.; Uutala, A. J.; Kingston, J. C.; Dixit, S. S.; Bernert, J. A.; Ryan, P. F.; Marmorek, D. R. *Environ. Pollut.* **1992**, *77*, 253–262.
73. Warwick, W. F. *Can. Bull. Fish. Aquat. Sci.* **1980**, *206*, 1–117.
74. Agbeti, M.; Dickman, M. *Can. J. Fish. Aquat. Sci.* **1989**, *46*, 1013–1021.
75. Bradbury, J. P. *Diatom Stratigraphy and Human Settlement in Minnesota;* Geological Society of America: Boulder, CO, 1975.
76. Brugam, R. B.; Vallarino, J. *Arch. Hydrobiol.* **1989**, *116*, 129–159.
77. Engstrom, D. R.; Swain, E. B.; Kingston. J. C. *Freshwater Biol.* **1985**, *15*, 261–288.
78. Stoermer, E. F.; Wolin, J. A.; Schelske, C. L.; Donley, D. J. *J. Phycol.* **1985**, *21*, 257–276.
79. Theriot, E.; Håkansson, H.; Stoermer, E. F. *Phycologia* **1988**, *27*, 485–493.
80. Juggins, S. Ph.D. Thesis ("A Diatom/Salinity Transfer Function for the Thames Estuary and Its Application to Waterfront Archaeology"), University College, London, 1988.
81. Radle, N.; Keister, C. M.; Battarbee, R. W. *J. Paleolimnol.* **1989**, *2*, 159–172.
82. Servant–Vildary, S. *Acta Geol. Acad. Sci. Hung.* **1982**, *25*, 179–210.
83. Gasse, F.; Fontes, J. C.; Plaziat, J. C.; Carbonel, P.; Kaczmarska, I.; De Deckker, P.; Soulié-Marsche, I.; Callot, Y.; Dupeuble, P. A. *Palaeogeogr. Palaeoclimatol. Palaeoecol.* **1987**, *60*, 1–46.
84. Smol, J. P.; Walker, I. R.; Leavitt, P. R. *Verh. Int. Ver. Theor. Angew. Limnol.* **1991**, *24*, 1240–1246.
85. Walker, I. R.; Mathewes. R. W. *J. Paleolimnol.* **1989**, *2*, 61–80.
86. Walker, I. R.; Smol, J. P.; Engstrom, D. R.; Birks, H. J. B. *Can. J. Fish. Aquat. Sci.* **1990**, *48*, 975–987.

RECEIVED for review April 6, 1992. ACCEPTED revised manuscript August 6, 1992.

Atmospheric Mercury Deposition to Lakes and Watersheds

A Quantitative Reconstruction from Multiple Sediment Cores

Daniel R. Engstrom[1], Edward B. Swain[2], Thomas A. Henning[3,4], Mark E. Brigham[3,5], and Patrick L. Brezonik[3]

[1]Limnological Research Center, University of Minnesota, Minneapolis, MN 55455
[2]Minnesota Pollution Control Agency, St. Paul, MN 55155
[3]Department of Civil and Mineral Engineering, University of Minnesota, Minneapolis, MN 55455

Historic increases in atmospheric mercury loadings caused by anthropogenic emissions are documented from sediment cores from seven remote headwater lakes in Minnesota and northern Wisconsin. Whole-basin changes in Hg accumulation, determined from lakewide arrays of ^{210}Pb-dated cores, show that regional atmospheric Hg deposition has increased by a factor of 3.7 since preindustrial times. The relative increase is consistent among lakes, although preindustrial Hg accumulation rates range from 4.5 to 9 $\mu g/m^2$ per year, and modern rates range from 16 to 32 $\mu g/m^2$ per year. The distribution of these rates is highly correlated with the ratio of catchment area to the lake surface area. Modern and preindustrial atmospheric deposition rates of 12.5 and 3.7 $\mu g/m^2$ per year are calculated from this relationship, along with the relative contribution of Hg from the terrestrial catchment. Release of atmospheric Hg from catchment soils accounts for 20–40% of the Hg loading to the lakes, depending on

[4]Current address: James M. Montgomery Consulting Engineers, 545 Indian Mound, Wayzata, MN 55391
[5]Current address: U.S. Geological Survey, District Office, 2280 Woodale Drive, Mounds View, MN 55112

0065–2393/94/0237–0033$09.50/0

catchment size. This export represents about 25% of the atmospheric Hg falling on the catchments under both modern and preindustrial conditions. The absence of geographic trends in either Hg deposition rates or their increase implies regional if not global sources for the Hg entering these lakes.

L AKE SEDIMENTS ARE EXCELLENT ARCHIVES for historic changes in the industrial discharge of potentially toxic metals such as Hg, Zn, Pb, Cu, Ni, Cd, and Co (*1–3*). Most heavy metals have short residence times in the water column and are quantitatively retained in the sediments, so stratigraphic interpretations are relatively straightforward (but *see* refs. 4 and 5). Sediment records have provided compelling evidence for the role of atmospheric deposition in contaminating aquatic environments far removed from direct (point-source) industrial influence (*6, 7*). These studies have been used to document the history of emissions, the effect of abatement, and the geochemical cycling of trace metals in the aquatic environment (*8–10*).

In most cases stratigraphic interpretations are limited to description of relative changes in metal loading; more (or less) of something is deposited each year at a given site on the lake bottom. Higher concentrations of a metal in surface sediments relative to deeper (uncontaminated) strata—often corrected for matrix effects by ratio of the metal to a silicate proxy such as TiO_2—represent the relative enrichment attributable to industrial discharge. Where fine-scale dating (e.g., ^{210}Pb) is available, actual deposition rates may be calculated. But because sediment deposition and composition are spatially variable, accumulation rates at a single core site cannot be automatically extrapolated to the entire lake bottom (*11–14*) and actual fluxes to the lake in a mass-balance sense cannot be calculated.

Few studies have exploited the full potential of lake sediments to obtain quantitative estimates of whole-lake metal fluxes, despite the fact that sediments often represent the only available record of past deposition. Moreover, whole-basin sediment retention may be critical to mass-balance calculations where contemporary fluxes are difficult to measure because of their spatial and temporal variability (*15, 16*). The primary limitation on obtaining whole-basin fluxes is the large effort required to analyze and date multiple cores representing the various depositional environments within a single basin. Most multiple-core studies of metal pollution have been limited to one lake (*17*) and do not provide a regional picture of deposition rates, transport pathways, or geographic trends. In the few studies where multiple cores were taken from several lakes, total anthropogenic sediment burdens were calculated without recourse to dating or sedimentation rates (*6, 18*), and actual metal fluxes could not be calculated.

In this study, however, we demonstrate that it is feasible to obtain whole-basin metal fluxes—in this case Hg—from a suite of lakes with an economy of dating and stratigraphic analysis. The principal outcome of this investigation is the reconstruction of preindustrial and anthropogenic Hg inputs to a suite of relatively undisturbed lake catchments in Minnesota and north-central Wisconsin. Mercury contamination has been a serious concern for lakes in this region and elsewhere for more than 20 years. Elevated Hg levels in fish from sites remote from point-source discharge (*19–21*), together with stratigraphic evidence for recent increases in Hg sedimentation (*7, 22–25*), indicate that anthropogenic emissions to the atmosphere are to blame. Two mechanisms for the observed increase in Hg inputs have been advanced:

1. increased Hg emissions, principally from coal combustion and waste incineration (*26*), or

2. changes in atmospheric chemistry (higher levels of SO_4, ozone, and other oxidants) that could enhance atmospheric removal or promote leaching from watershed soils (*27*).

Results from this study shed light on the relative importance of these processes. By comparing whole-lake Hg fluxes among a group of sites of differing hydrology, we are able to assess the relative contribution of watershed inputs versus direct atmospheric deposition to the lake surface. Finally, the calculated Hg accumulation rates provide a regional picture of Hg loading from which we are able to derive atmospheric deposition rates and infer possible geographic sources and mechanisms for enhanced deposition.

Study Sites

During the course of the study seven lakes were analyzed for whole-basin sedimentary Hg: an initial set of four lakes from the Superior National Forest in northeastern Minnesota (Thrush, Dunnigan, Meander, and Kjostad), two additional sites that expanded the geographic coverage to central and western Minnesota (Cedar and Mountain), and Little Rock Lake in northern Wisconsin, currently the site of a split-lake acidification experiment (Figure 1). The four sites in northeastern Minnesota are underlain by Precambrian crystalline bedrock and a thin veneer of noncalcareous glacial drift. In northwestern Wisconsin (Little Rock Lake) the glacial deposits are substantially thicker, and in western Minnesota (Cedar and Mountain lakes) the drift is highly calcareous. All of the sites except Mountain Lake lie in mixed deciduous–conifer forest. This vegetation becomes progressively more boreal toward the Minnesota–Canadian border. The watershed of Mountain Lake is covered with a mosaic of native tall-grass prairie and oak woodland.

The study lakes span a climatic gradient that becomes appreciably drier toward the southwest. Mean annual precipitation ranges from 75–80 cm in northeastern Minnesota and northwestern Wisconsin to 60 cm near Cedar

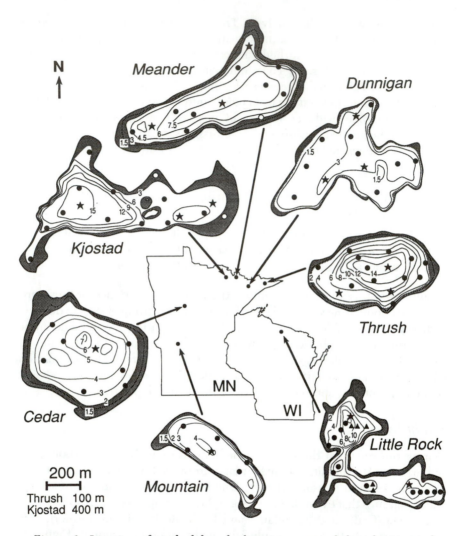

Figure 1. Location of study lakes, bathymetric maps of their basins, and approximate position of core sites; shaded contours represent nondepositional areas for fine-grained sediment. Key: star, stratigraphically detailed cores; ●, coarse-interval cores; ○, cores from nondepositional sites; and ▲, supplemental ^{210}Pb-dated cores from Little Rock. Depth contours are given in meters.

Lake and Mountain Lake; evaporative losses exceed precipitation in the west (Cedar and Mountain), but represent only about 60% of the precipitation falling in northern Wisconsin (Little Rock). Precipitation chemistry varies along the same gradient, with pH increasing from 4.6 in northern Wisconsin to 4.8 in northeastern Minnesota and 5.2 in western Minnesota. The corresponding values for wet sulfate deposition decrease from 15 kg/ha per year in northern Wisconsin to 5–8 kg/ha per year in northeastern and west-central Minnesota (28, 29).

All seven study lakes are located in primary watersheds and are fed principally by groundwater seepage and direct precipitation to the lake surface; two of the lakes (Meander and Kjostad) receive drainage from intermittent streams and possess permanent surface outlets. The lakes are relatively small (7–40 ha), except for Kjostad (168 ha), and shallow (Z_{max} = 4–16 m, Z_{mean} = 3–7 m) (Table I). Water residence times are 5–10 years, except for the two western sites (~30 years) where much of the inflow is lost through evaporation. The western Minnesota lakes are distinctly higher in dissolved solids (e.g., alkalinity is 2600 and 4200 μequiv/L) because of evaporative concentration and drainage from calcareous soils. The other five sites are dilute (alkalinity is 25–180 μequiv/L), and only Kjostad is appreciably colored by organic acids (30 Pt–Co units).

A primary selection criterion for all of these sites was the absence of significant land-use disturbance in the catchment, which might otherwise accelerate soil erosion and sediment flux during this century. Five of the sites (Thrush, Dunnigan, Kjostad, Cedar, and Little Rock) were logged to some extent in the early 1900s; the forests were allowed to regrow and have not been appreciably disturbed since then. Except for a few shoreline cabins on Kjostad and one on Cedar, none of the watersheds are presently inhabited. The Meander watershed was partially burned in the Little Sioux fire of 1971, and a small portion of the Mountain Lake catchment was farmed for a brief period following European settlement. Today the watershed of Mountain Lake is contained entirely within Glacial Lakes State Park, and it is one of the most pristine lakes remaining in the agricultural regions of western Minnesota.

Experimental Methods

Coring Strategy. Sediment cores were collected with a thin-walled polycarbonate tube fitted with a piston and operated from the lake surface by rigid drive rods (30). This device recovers the very loose uncompacted sediment surface as well as deeper strata without disturbance or displacement (core-shortening, cf. refs. 31 and 32). Core sections were extruded vertically from the top of the tube into polypropylene collection jars, transported on ice to the laboratory, and stored at 4 °C until analysis.

Because a large number of cores was required for this study, we chose to economize on the number of samples for ^{210}Pb and Hg analysis by sectioning most of the cores at coarse intervals. Historic trends in Hg deposition were provided by a few cores analyzed in stratigraphic detail, whereas the coarsely sectioned cores provided the spatial pattern in Hg accumulation across each basin at a few discrete time intervals. The samples from the coarse-interval cores were homogenized and analyzed for Hg and ^{210}Pb content in the same manner as the detailed cores. Detailed cores were collected with a 10-cm-diameter corer, and the coarse-interval cores were obtained with a 5-cm-diameter corer. The number of cores collected from each lake is shown in Table II.

Table I. Lake Location and Certain Morphometric, Hydrologic, and Chemical Characteristics

Lake	County	State	Alkalinity (μequiv/L)	Surface Area (ha)	Deposition Area (ha)	Maximum Depth (m)	Mean Depth (m)	Volume (10^5 m^3)	Catchment Area (ha)[a]	Residence Time (years)[b]
Thrush	Cook	MN	60	6.6	5.3	14.5	6.9	4.52	23.8	5.7
Dunnigan	Lake	MN	66	32.9	28.6	4.3	2.3	7.42	46.0	4.6
Meander	St. Louis	MN	70	39.6	28.7	7.6	4.8	19.01	126.6	6.1
Kjostad	St. Louis	MN	180	167.7	117.6	15.6	6.5	108.79	985.3	5.0
Cedar	Hubbard	MN	2600	39.1	27.9	8.4	3.8	14.76	87.9	34.4
Mountain	Pope	MN	4200	15.7	12.9	4.2	2.7	4.27	81.5	28.5
Little Rock	Vilas	WI	25	18.2	11.5	10.3	3.5	6.24	34.8	10.0

[a]Catchment area excludes lake surface.
[b]Calculated from inflow estimates.

Table II. Average Values for Sedimentary Parameters for All Cores from Each Lake Basin

Lake	% Organic Matter (core top)	Cumulative Unsup. ^{210}Pb (pCi/cm²)	Modern Hg Conc. (ng/g)	Preindustrial Hg Conc. (ng/g)	SEF $[(Hg_{(m)} - Hg_{(pi)})/Hg_{(pi)}]$	Sediment Accumulation[a] (g/m² per year)	Hg Flux Ratio (Modern/ Preindust.)	Number of Cores
Thrush	34.5 ± 6.4	18.6 ± 7.3	370 ± 188	106 ± 35	2.7 ± 1.7	97 ± 46	3.30	15
Dunnigan	61.5 ± 3.3	16.3 ± 3.4	198 ± 52	56 ± 11	2.5 ± 0.6	97 ± 22	3.58	12
Meander	32.0 ± 9.6	22.3 ± 11.5	213 ± 87	60 ± 30	3.2 ± 1.4	136 ± 40	3.34	11
Kjostad	24.1 ± 10.4	16.1 ± 8.3	255 ± 82	78 ± 22	2.3 ± 0.9	146 ± 60	3.22	14
Cedar	56.4 ± 3.4	14.3 ± 6.8	101 ± 20	55 ± 36	1.3 ± 1.2	227 ± 102 / 144 ± 66	3.37	8
Mountain	23.9 ± 1.8	15.7 ± 3.8	63 ± 19	21 ± 7	2.5 ± 2.1	550 ± 164 / 366 ± 76	4.88	7
Little Rock	53.1 ± 5.6	16.5 ± 4.4	255 ± 44	64 ± 17	3.2 ± 1.3	139 ± 52	4.03	14 (7)[b]

[a] Mean sediment-accumulation rates except Cedar and Mountain, for which preindustrial rates are shown.

[b] Dating parameters for Little Rock were provided by seven cores.

The exact sampling strategy varied somewhat among the study lakes. For the four lakes in northeastern Minnesota, sediment cores were collected from three locations within each basin (representing deep, shallow, and intermediate water depths) and analyzed at fine intervals (1–2 cm). These detailed sediment profiles guided the collection and sectioning of subsequent cores, which were extruded in three coarse intervals (~0–2, 2–45, and 45–60 cm, although the exact intervals varied from lake to lake). The topmost samples gave modern Hg concentrations, and the bottom intervals provided preindustrial values (>150 years ago for the Midwest). The long middle section was used to calculate whole-core ^{210}Pb burdens required for dating. This approach provided sediment accumulation rates for only the topmost interval. However, these values were used to calculate Hg flux for the lower sections, assuming a constant sediment accumulation rate for the entire core. Dating results from the detailed cores showed this to be a valid assumption for these relatively pristine lakes.

For Cedar and Mountain lakes, detailed sediment cores collected from the deeper regions of each basin showed increasing sedimentation rates up-core. Thus a slightly different approach was used to provide additional temporal detail from the coarsely sectioned cores. These subsequent cores were extruded into five intervals 5–20 cm long so that dates and sediment accumulation rates could be explicitly calculated for the deeper strata.

For Little Rock Lake, a single core from each of its two basins was analyzed and dated in stratigraphic detail. The remaining cores were analyzed for Hg content in three coarse intervals as described, but none of these profiles was actually dated. Instead the sedimentation rates were inferred from a series of five nearby cores that had been dated by ^{210}Pb for other purposes (16). The mean sedimentation rates from dated cores collected at similar depth in the same basin were used to calculate Hg accumulation for each undated profile.

Analytical Methods. *Loss on Ignition.* Wet density, dry mass density, organic content, and carbonate content were determined by the method of Dean (33). Volumetric samples (1.0 cm³) were dried overnight at 100 °C and ignited at 550 and 1000 °C for 1 h. Mass measurements were made on the wet sample and after each heating on an electronic analytical balance.

Lead-210 Dating. Sediment cores were analyzed at appropriate depth intervals for ^{210}Pb to determine age and sediment-accumulation rates for the past 100–150 years. Lead-210 was measured through its granddaughter product ^{210}Po, with ^{208}Po added as an internal yield tracer. The polonium isotopes were distilled from 1–10 g of dry sediment at 550 °C following pretreatment with concentrated HCl. They were plated directly (without HNO_3 oxidation) onto silver planchets from a 0.5 N HCl solution (modified from ref. 34). Activity was measured for $1-5 \times 10^5$ s with Si-depleted surface barrier detectors and an α-spectroscopy system (Ortec Adcam). Unsupported ^{210}Pb was calculated by subtracting supported ^{210}Pb (estimated from constant activity at depth) from total activity at each level. Dates and sedimentation rates were determined according to the constant rate of supply (c.r.s.) model (35), with confidence intervals calculated by first-order error analysis of counting uncertainty (36).

Mercury. Sediment samples were digested with a strong acid–permanganate–persulfate digestion technique. Total Hg analyses were conducted by cold-vapor atomic absorption spectrophotometry (37). Wet sediment samples (2–8 g) were treated with 10 mL of H_2SO_4 (conc.), 5 mL of HNO_3

(70%), and 2 mL of HCl (36%), and then heated for 2–3 h at 80–110 °C on a sand bath. After cooling the samples were treated with 15 mL of saturated $KMnO_4$ (~6% w/v), followed by 5 mL of $K_2S_2O_8$ (5% w/v). The next day excess oxidant was reduced with 10 mL of hydroxylamine hydrochloride (6% w/v) in 6% NaCl. Two hours later the digests were poured into 250-mL polypropylene jars and connected to a Hg purging system. Five milliliters of $SnCl_2$ (10% w/v in 2M H_2SO_4) was transferred by syringe to each jar, the contents of which was then vigorously stirred for 2 min. The headspace gas was swept through a drying tube ($CaCl_2$ desiccant) with prepurified carrier-grade N_2 and through a 10-cm absorption cell. Absorbance readings were based on peak height.

Three distilled, deionized reagent blanks ($d-H_2O$) and National Institute of Standards and Technology (NIST) standards (NBS 1646; estuarine sediment), and two sets of Hg standards (four or five different dilutions of 1000-ppm Alpha Hg standard) were digested and analyzed each day that samples were run. Reagent blanks were quite low in Hg, generally yielding absorbance values ≤0.002 above the N_2 baseline. If at least two of the three NIST standards were not within the certified range of acceptable concentration (63 ± 12 ng/g), the entire sample run was redigested and reanalyzed. No absorbance due to matrix effects—determined periodically by standard additions—was observed. Our detection limit (IUPAC method, ref. 38) for total Hg was about 6 ng, and the coefficient of variation on replicate samples was about 8%.

Calculations.   To calculate whole-basin Hg accumulation, a portion of the lake bottom was assigned to each core according to two methods based on either spatial proximity (polygon method) or lake depth (contour method). In both approaches core sites were located spatially on bathymetric lake-basin maps by visual approximation to shoreline features and other landmarks. For the polygon method the lake map was divided into tiles or Theissen polygons, which were generated by the perpendicular bisectors between each core site and its nearest neighbors. In the second method 1-m depth contours were used to subdivide the lake bottom, and core sites were assigned to these bathymetric regions according to their depths.

Nondepositional regions of the lake bottom—shallow areas where fine-grained sediments do not accumulate conformably—were excluded from these calculations. The nondepositional region of each lake was delimited by the depth contour above which silty or sandy sediments were found at or near the mud surface. Because the exact location of this boundary was uncertain in some lakes, depth contours 1 m above and below were used to estimate a range of values for the depositional region. The area of each polygon or contour was measured from bathymetric maps on a microcomputer–digitizer. The location of coring sites, depositional zones, and the bathymetry of each study lake are shown in Figure 1.

The terrestrial drainage basin for each lake was digitized from 1:24,000 U.S. Geological Survey topographical maps. The reported values (which exclude lake surface area) represent the average of two estimates, one that includes and one that excludes small wetlands with internal drainage and areas where watershed boundaries were uncertain.

Results

Sediment Lithology.   The organic content of sediments at most coring sites, between 20 and 60% of the dry mass, is typical for small north-temperate lakes in forested catchments. Dunnigan, Cedar, and Little Rock sediments are consistently above 50% organic matter, whereas Meander, Thrush, Kjostad, and Mountain are consistently less organic (20–40%). Carbonate, which accounts for 30–50% of Mountain Lake sediments and 5–10% of most Cedar Lake cores, is virtually absent in the other five lakes (Figure 2).

In most of the study lakes organic content is poorly correlated with lake depth, a feature that may be partially attributed to the fact that fine-grained sediments are only weakly focused in small shallow lakes (6, 18). Indeed, sediments do become somewhat more organic with depth in the two deepest lakes, Thrush and Kjostad. The perception of uniform sediment composition also results from sampling bias, because few littoral cores were actually retained for analysis. Visibly inorganic sediments were assumed to represent nondepositional sites for purposes of calculating a Hg flux and were simply mapped and discarded.

For the most part, sediments are also stratigraphically uniform, showing only a few percentage variation in lithologic composition. Cores from Mountain Lake, which consistently show up-core decreases in carbonate content (to about 60% that at depth), are the only exception. A number of shallow-water cores that contain a thin veneer of organic-rich sediments overlying silt and sand were also excluded from analysis. In most locations the spatial boundary between organic-rich profundal-type sediments and littoral deposits of coarse detritus or massive silt was clearly defined.

Dating and Sediment Accumulation.   *Stratigraphic Patterns.* Lead-210 profiles from profundal cores from each study lake show conformable declines in unsupported activity to an asymptote of supported ^{210}Pb typically below 40–60 cm deep (Figure 3). Supported activity, attained at shallower depths in Thrush (28 cm) and Kjostad (36 cm), indicates slower linear rates of sedimentation at these sites. The activity profiles for several lakes, most notably Thrush and Kjostad, are almost perfectly exponential and thus indicate nearly constant sediment accumulation rates. Others, such as Cedar and Mountain, show flat spots and kinks that probably represent shifts in sediment flux.

Sediment accumulation curves derived by c.r.s. calculations from these activity profiles show variable sedimentation rates in Dunnigan, Meander, and Little Rock (Figure 4). Although the sediment flux varies by more than a factor of 2 within each of these profiles, the changes are asynchronous among core sites in a given lake (other detailed cores not shown) and may therefore represent shifts in sediment deposition patterns within the basin, as opposed to changes in whole-lake sediment loading (16, 39). In Cedar and Mountain lakes, the systematic increase in sediment accumulation since 1930 is synchronous among cores within each lake, and thus constitutes a

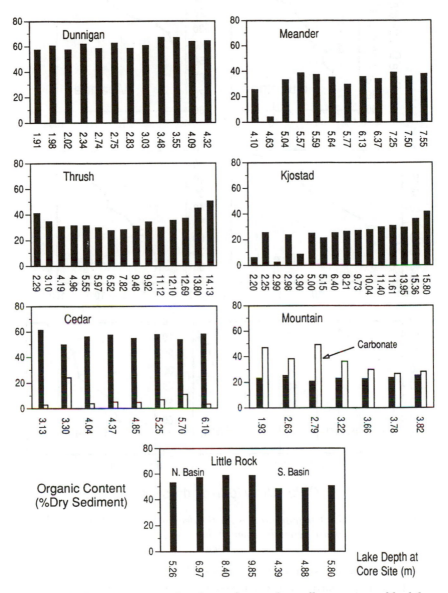

Figure 2. *Organic content of surface sediments from all cores arrayed by lake depth. Significant carbonate is present only in Cedar Lake and Mountain Lake sediments.*

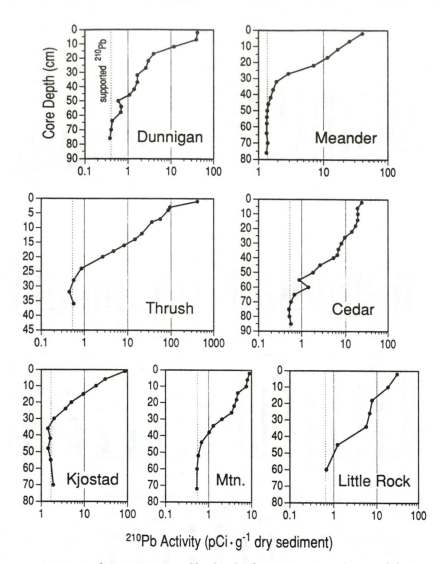

Figure 3. Lead-210 activity profiles for the deep-water cores from each basin. Counting errors are generally smaller than the plotted symbols and are not shown.

lakewide increase in sediment deposition. The higher accumulation in Mountain Lake is primarily an erosion signal (greater inorganic sedimentation) that is probably related to watershed disturbance following the development of Glacial Lakes State Park. Organic accumulation increases in Cedar Lake imply higher biological productivity.

These results indicate that our use of a single average sediment-accumulation rate for the cores from Dunnigan, Meander, Thrush, Kjostad,

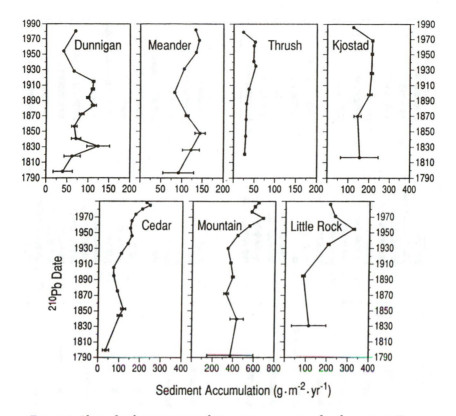

Figure 4. Plots of sediment-accumulation rate versus age for the cores in Figure 3. Error bars represent one standard deviation propagated from counting uncertainty. The apparent increase in recent sediment accumulation in Little Rock Lake is atypical of other cores from this site.

and Little Rock will not result in a systematic bias in calculation of Hg fluxes. However, the estimates will be less reliable than if we had done detailed dating on all cores. On the other hand, recent increases in sediment flux to Cedar and Mountain lakes are factored into our estimates of Hg deposition to compensate for greater dilution of Hg by the sediment matrix in modern times.

Spatial Variability. A summary of unsupported ^{210}Pb burdens for both coarse- and fine-interval cores (Figure 5) shows fairly similar sedimentary conditions within some basins (e.g., Dunnigan and Mountain) and substantial variability in others (e.g., Kjostad, Cedar, and Meander). The least variable lakes are the smallest and shallowest; this result is to be expected because sediment deposition patterns are least accentuated by wave and current action in small basins of uniform depth. Uniform sediment deposition in

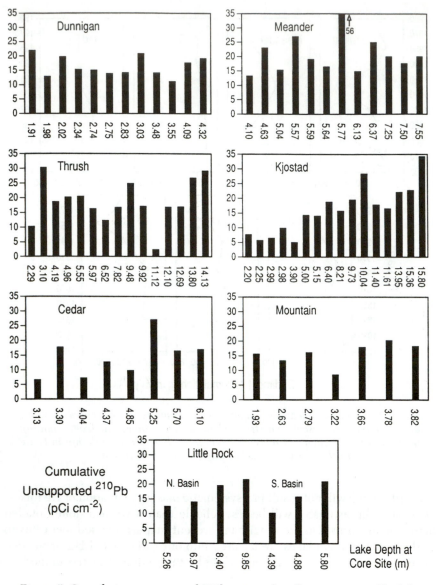

Figure 5. Cumulative unsupported ^{210}Pb activity for all cores, arrayed by lake depth.

Dunnigan and Mountain lakes is also indicated by the fact that most whole-core ^{210}Pb inventories are 15–20 pCi/cm^2. These values correspond to a calculated ^{210}Pb flux of 0.46–0.62 pCi/cm^2 per year, which is a reasonable estimate for mean atmospheric ^{210}Pb deposition for this region (*40*). Core sites with inventories substantially greater (less) than these values are likely to overestimate (underestimate) the average lakewide sediment accumulation because of resuspension and focusing of sediments from shallow to deep water. This pattern is clearly shown by Cedar Lake, where sediment accumulation rates are highest in cores with the largest ^{210}Pb inventories.

Kjostad Lake, the largest basin, shows increasing ^{210}Pb burdens with greater lake depth. The distribution of lead and organic content indicate that fine-grained ^{210}Pb-bearing sediments are highly focused in this lake. However, the poor correlation of ^{210}Pb burdens with lake depth in the other study lakes indicates that simple bathymetric models cannot reliably predict whole-lake ^{210}Pb deposition in most small basins. Two cores, notable by their very low ^{210}Pb inventories and low organic content (Kjostad cores at 2.20 and 2.99 m) were subsequently excluded from calculations of whole-basin sediment accumulation because they were considered to represent nondepositional sites.

Dry-mass sediment accumulation rates calculated by the c.r.s. dating model (Figure 6) are most uniform across the small flat-bottomed basins of Dunnigan and Mountain lakes, and most variable in Kjostad, Meander, and Cedar. Mean accumulation rates (Table II) are lowest in Thrush and Dunnigan ($\sim$100 g/m^2 per year) and only slightly higher in Kjostad, Little Rock, Meander (excluding cores at 4.63 and 5.77 m), and presettlement Cedar ($\sim$140 g/m^2 per year). Mountain Lake is the one site that stands out with substantially higher sedimentation rates than the other basins. Both modern and presettlement rates (550 and 360 g/m^2 per year, respectively) are strongly enhanced by carbonate precipitation, which does not contribute to sediment loading in the other six lakes. Mountain Lake is, nonetheless, a relatively pristine site. Its sedimentation rates are an order of magnitude lower than that measured in cores from agriculturally affected lakes in southern Minnesota (unpublished data).

Contrary to expectations of sediment focusing, only Thrush Lake showed any relationship between lake depth and sediment accumulation, and in this case deposition rates were lowest in profundal regions of the basin and highest toward the margins. Anderson (*41*) made similar observations from multiple cores from an equally small lake in Northern Ireland and concluded that sediment trapping by macrophytes and higher organic loads in shallow water could account for high littoral deposition rates, particularly if wind-induced currents were insufficient to move sediment offshore. Although macrophytes are not abundant in Thrush Lake, extensive beds of aquatic mosses (*Drepanocladus* and *Sphagnum*) extend to considerable depth in the lake's clear water and could inhibit sediment resuspension and act as a local source of organic detritus.

Problematic Cores. Two cores from Meander Lake exhibit sediment accumulation rates considerably higher than the basin average (cores at 4.63 and 5.77 m). The core at 4.63 m is highly inorganic (5% organic matter) and

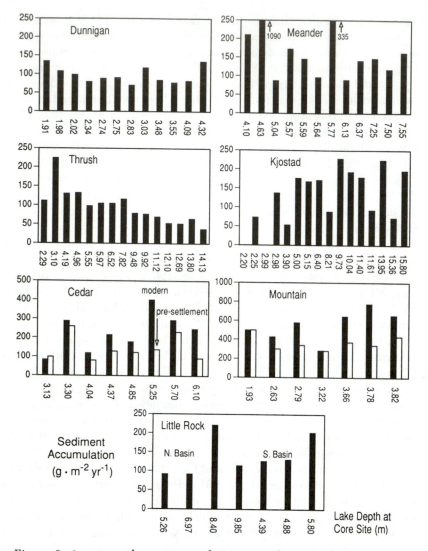

Figure 6. Average sediment accumulation rates for cores from Dunnigan, Meander, Thrush, Kjostad, and Little Rock. Modern (~post-1980) and pre-industrial (~pre-1850) rates are shown for Cedar and Mountain.

is located very near the depositional limit for fine-grained sediments. Its accumulation rate of more than 1 kg/m² per year could be an artifact of downward mixing of ²¹⁰Pb-bearing sediment into an otherwise erosional sand and silt deposit (42). The low surface activity in this core relative to other sites (6 pCi/g versus 35–50 pCi/g) tends to support this contention. On the other hand, the core site is located near a small inlet stream, and the high accumulation rate could represent offshore deposition of stream-borne clastic materials. However, for purposes of estimating lakewide Hg deposition, it makes little difference whether this core is included or excluded from calculations. The Hg concentrations are quite low, and hence the Hg flux is similar to the basinwide average.

The other atypical core at 5.77 m has a calculated sediment accumulation rate of 335 g/m^2 per year, but in this case a high ^{210}Pb inventory (56 pCi/cm^2) and surface activity (47 pCi/g) indicates that fine-grained sediments are currently accumulating at this site. However, the presence of highly inorganic silts (2–6% organic matter) beneath the ^{210}Pb-rich surface veneer indicates that until recently the site was nondepositional. Because such conditions clearly violate the assumption of a constant ^{210}Pb flux required by the c.r.s. model, the dating is considered to be unreliable and the core is excluded from further analysis.

Mercury Concentration.   *Stratigraphic Trends.*   Detailed profiles of Hg concentration from the deep-water cores from each of the study lakes are shown in Figure 7. Definite enrichment is seen in the upper sediments over the deeper strata in all cases. In Dunnigan, Meander, Thrush, Kjostad, and Little Rock, surficial Hg concentrations range from 200 to 400 ng/g of dry sediment, and background concentrations range from about 50 to 100 ng/g. Surface concentrations in Cedar Lake do not exceed 150 ng/g and in Mountain Lake 100 ng/g; background concentrations in Mountain Lake are about 20 ng/g. The distinctly lower concentrations in Mountain Lake can be attributed to dilution by high sediment inputs, particularly carbonates. In Cedar Lake, the relatively modest up-core increase in Hg concentration also results from dilution of Hg inputs by recent increases in lakewide sediment flux. The data suggest two distinct periods of increase in Hg deposition to lakes in this region. The first came between 1860 and 1890 and the second between 1920 and 1950. The other detailed Hg profiles (not shown here) are of generally similar shape and magnitude.

Similar trends and concentrations have been reported from sediment cores collected from other lakes in the region. Meger (24) found recent (core top) and background concentrations of about 110 and 40 ng/g for cores from two large lakes (Crane and Kabetogama, respectively) in nearby Voyageurs National Park. Rada et al. (7) observed surficial sediment concentrations between 90 and 190 ng/g and values between 40 and 70 ng/g in deeper strata in 11 seepage lakes in north-central Wisconsin. Somewhat higher concentrations (100–200 ng/g background and 200–500 ng/g surface) were found in cores from southern Ontario by Evans (6) and Johnson et al. (23). Most of these workers consider atmospheric deposition of Hg from industrial sources to be the likely cause of increasing Hg concentrations in lake sedi-

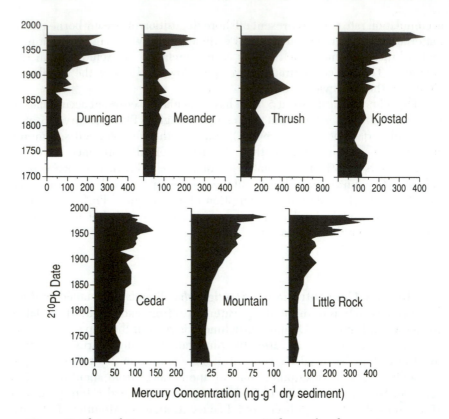

Figure 7. Plots of Hg concentration versus ^{210}Pb age for deep-water cores. Dates older than about 1800 are extrapolations based on mean dry-mass sediment accumulation rates.

ments that are remote from point-source discharge. Hence for discussion purposes, Hg levels from strata deposited before the mid-1800s are considered to represent preindustrial conditions, although anthropogenic emissions of Hg began earlier than this in some regions.

Enrichment Factors. Modern and preindustrial (background) Hg concentrations for both coarse- and fine-interval cores are illustrated in Figure 8. As shown by the detailed core stratigraphy, Hg concentrations in surficial sediments are significantly elevated above those in deeper strata in nearly every case. Surface Hg concentrations in Thrush Lake average 370 ng/g, whereas background values are ~100 ng/g (Table II). In Kjostad and Little Rock recent Hg concentrations are 255 ng/g, and background levels average around 80 and 60 ng/g, respectively. Hg levels are slightly lower in Meander and Dunnigan, where surface concentrations are about 200 ng/g and background values are ~60 ng/g. Mountain and Cedar lake sediments are con-

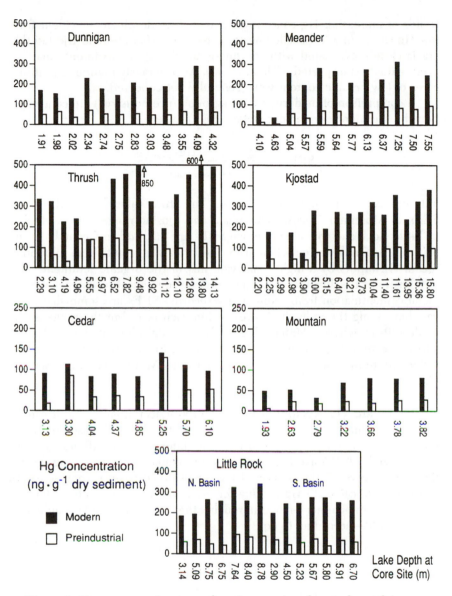

Figure 8. *Hg concentration in modern (core-top) and preindustrial (~pre-1850) sediments from all cores, arrayed by lake depth.*

sistently lowest in Hg; recent values are 100 and 60 ng/g and background concentrations are 55 and 20 ng/g, respectively.

In most of the study lakes Hg concentrations are spatially less variable than sediment accumulation rates, although the large range of modern Hg values for Thrush Lake is a notable exception. Although there are no obvious trends of increasing Hg concentration with lake depth, it is evident from low Hg values in a few shallow-water cores from Meander and Kjostad that Hg is closely associated with the fine-grained organic sediments that are preferentially transported offshore (6, 43). As previously mentioned, littoral cores of low organic content were generally not collected.

A comparison of modern and background (preindustrial) Hg concentrations by sediment enrichment factors (SEF), where

$$\text{SEF} = \frac{\text{Hg}_{\text{(modern)}} - \text{Hg}_{\text{(background)}}}{\text{Hg}_{\text{(background)}}} \tag{1}$$

indicates fairly similar increases among the study lakes (Table II). Mean SEFs for Mountain, Thrush, Dunnigan, and Kjostad range from 2.3 to 2.7; in Little Rock and Meander average SEFs are 3.2; and in Cedar Lake the SEF is 1.3. The lower enrichment for the latter site results from dilution of recent Hg inputs by a higher lakewide flux of the organic sediment. This fact illustrates the difficulty of relying solely on concentration data to evaluate metal contamination from sedimentary records. SEFs are somewhat more variable among the individual cores within each lake, but the range is still modest (0–7), whereas 80% of all cores have an enrichment of 1–4. A fairly narrow range of Hg enrichment (SEF = 0.8–2.8) was also noted in surface sediments from other remote lakes in northern Minnesota and Wisconsin (7, 24).

Lakewide Hg Accumulation.   Mercury accumulation rates can be calculated for individual core sites as the product of the ^{210}Pb-based sediment accumulation rate and Hg concentration in different strata. If enough cores are analyzed, whole-lake Hg inputs can be calculated by weighting the Hg flux of each core by the portion of the depositional basin it represents. In this study we calculate Hg loading for each lake on an areal basis for two time-stratigraphic units—modern (roughly the last decade) and preindustrial (before 1850)—according to eq 2:

$$Q_i = \frac{\sum_{j=1}^{n} R_{ij} \text{Hg}_{ij} A_j}{A_0} \tag{2}$$

where Q_i is the Hg flux in micrograms per square meter per year for time-stratigraphic interval i, R_{ij} is sediment accumulation in grams per square meter per year for interval i in core j, Hg_{ij} is Hg concentration in micrograms per gram for sediment interval i in core j, A_j is depositional zone in square meters represented by core j, A_0 is total lake surface area in square meters, and n is number of cores.

Depositional areas for each core were approximated by both Theissen polygons and depth contours, although the reported fluxes are those only from the polygon method. In practice the two approaches gave virtually the same results. However, we favor the polygon method because proximity to the core site should be a better predictor of Hg accumulation in basins such as these, where sediment deposition is not correlated with lake depth.

The basinwide flux calculations from the seven lakes show that pre-industrial Hg accumulation rates in the sediments ranged between 4.5 and 9.0 $\mu g/m^2$ per year, and the modern rates range between 16 and 32 $\mu g/m^2$ per year (Figure 9). More striking is the observation that the range in these rates is a function of the relative size of the terrestrial catchment surrounding each lake basin. Over 90% of the variation in modern Hg accumulation can be accounted for by the ratio of a lake's catchment area to its surface area ($A_d:A_0$). The correlation between preindustrial Hg accu-

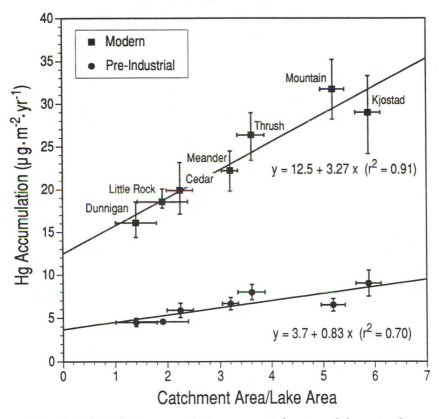

Figure 9. Lake-wide Hg accumulation rates as a function of the ratio of terrestrial catchment area to lake area. The error bars propagate maximum and minimum estimates of the depositional region of each lake and that of its functional catchment.

mulation and $A_d{:}A_0$ is also strong ($r^2 = 0.7$), but the range of Hg fluxes is substantially smaller, accounting for the weaker regression. None of the other hydrologic or limnologic variables can account for this distribution of Hg accumulation (Table I). Water residence time and ionic strength, for example, are highest in Mountain and Cedar lakes, yet Cedar has Hg accumulation rates similar to that of Little Rock, a dilute seepage lake. Mountain and Kjostad lakes have similar Hg fluxes, though one is surrounded by upland prairie and the other by conifer forest and muskeg.

These results strongly imply that a portion of the Hg inputs to the sediments in these lakes comes from their watersheds, and that the magnitude of this terrestrial component increases with catchment size. However, the relative increase in Hg accumulation from preindustrial rates is nearly constant among sites (Table II). The ratio of modern to background accumulation ranges from 3.2 to 4.9. This uniformity indicates that the change in Hg inputs since preindustrial times has been regionally similar for rural or remote areas of the upper Midwest.

Discussion

Mercury Fluxes and Sediment Records.   Our calculations of lake-wide Hg fluxes from more than 80 dated sediment cores from seven small headwater lakes reveal a regionally consistent increase in Hg inputs from preindustrial times to the present; the modern Hg flux to each of these lakes is about 3.7 times that of the early 1800s. Such increases are typical of that reported in other investigations of lake sediments from remote or rural sites in eastern North America (3, 6, 7, 10, 23). Most researchers have concluded that the increase is anthropogenic and that the Hg must be transported through the atmosphere and deposited on the lake and its terrestrial catchment.

Precise estimates of the anthropogenic component of the global Hg budget have been elusive (27, 44). Values for anthropogenic contributions range from 20% to 75% of total (natural and industrial) Hg emissions, but recent estimates seem to converge on the upper end of this scale (26, 45–49). The dominant industrial sources today are coal combustion ($\sim$65%) and waste incineration (25%), whereas volcanism, marine degassing, and terrestrial volatilization are the most important natural sources (44, 50). However, there is exceedingly little information on actual rates of Hg deposition at potentially sensitive sites remote from point-source discharge. Reliable measurements of Hg in wet deposition are available from only a few locations, and these are temporally limited to the past few years (51–54). Thus although the relative change in Hg inputs to lakes and landscapes is reasonably well documented, the actual magnitude of the increase is poorly known.

Results from this study can provide quantitative estimates of Hg deposition rates now and in the past. Because of the strong relationship between

Hg flux and watershed area, the relative importance of direct deposition and catchment contributions can be assessed. Such interpretations assume that lake sediments are stratigraphically and quantitatively reliable archives of Hg inputs to aquatic systems.

Experimental observations on the geochemical behavior of Hg in lacustrine environments indicate that sediments should retain most of the Hg entering a lake. Mercury in the water column is rapidly sorbed by particulates and efficiently removed to the sediments where the bulk of Hg in the aquatic environment is immobilized (55, 56). Mass-balance studies on one of our sites, Little Rock Lake, indicate that about 90% of the incoming Hg is deposited in the sediments; the remaining inputs are lost to gaseous evasion of Hg^0 (52, 57). In humic-stained lakes with very short residence times (one to a few months) outflow losses of Hg may also be significant (58). Within the sediment column, Hg sorption kinetics strongly favor the particulate phase so that postdepositional movement of Hg appears to be very limited (59). Well-preserved Hg stratigraphy with subsurface peaks has been noted in sediment cores from riverine systems after point-source Hg inputs were curtailed (60–62).

Direct Atmospheric Deposition.   Our calculation of whole-basin sediment deposition shows that Hg accumulation in a lake is directly proportional to the ratio of the catchment area to the lake surface area (Figure 9). The intercept of the regression line in this relationship predicts the Hg accumulation rate for a lake with no terrestrial catchment. In other words, it shows the net atmospheric deposition rate accounting for wet and dry deposition and losses due to gaseous evasion and outflow (groundwater or surface water) after deposition. The preindustrial atmospheric flux of Hg estimated from the intercept is 3.7 $\mu g/m^2$ per year, and the modern rate is 12.5 $\mu g/m^2$ per year. These values have a relatively high uncertainty because they are extrapolations beyond the data set. However, the modern rate is in good agreement with current measurements of Hg deposition in north-central North America. Glass et al. (54) found a 3-year average of 15 $\mu g/m^2$ per year for three Minnesota deposition sites, Mierle (63) measured 10.2 $\mu g/m^2$ per year for a 1-year study at a catchment in central Ontario, and Fitzgerald et al. (52) reported a mean of about 10 $\mu g/m^2$ per year for Little Rock Lake. The first two studies are for wet deposition only; dry deposition is thought to be as large as 50% of wet deposition (49). Fitzgerald et al. estimate 6.8 and 3.5 $\mu g/m^2$ per year for wet and dry deposition, respectively.

Our estimates of atmospheric deposition in preindustrial and modern times indicate that Hg inputs have increased by a factor of 3.4 in 130 years (3.7 to 12.5 $\mu g/m^2$ per year). Alternatively, a factor of 3.7 is obtained by averaging the increase factor from each lake (Table II). The 3.7-fold increase translates to an average increase of about 2.2% per year, compared to an annual increase of 1.5% measured in air over the north Atlantic Ocean for the period 1977–1990 (26).

Our results show that for the geographic region represented by the seven lakes, atmospheric Hg loading has increased by a relatively constant factor and that modern deposition rates are similar among sites (accounting for catchment size). The Hg burden of undisturbed forest soils is also relatively uniform across this region but becomes significantly higher to the east (64), presumably because of greater proximity to industrialized regions. Scandinavian researchers have documented a similar gradient of increasing Hg deposition toward industrialized areas (22, 53, 65, 66).

Catchment Contributions.   The slope of the regression lines in Figure 9 is the rate at which Hg is transported from the terrestrial catchment to the lake sediments (in units of micrograms of Hg per square meter of catchment per year). If one assumes that all of the Hg in the catchment is derived from the atmosphere, then the slope divided by the atmospheric deposition rate (the intercept) is the proportion of terrestrial Hg deposition that is transported to the lake. The slope will equal the rate of atmospheric deposition if the entire flux to the catchment is transported to the lake. On the other hand, the slope will equal zero—as observed for Pb by Dillon and Evans (18)—if direct deposition to the lake surface is the only significant source.

This simple model of Hg accumulation assumes that there are minimal losses of Hg from the lake by evasion or outflow, that dry deposition rates are similar for lake surfaces and terrestrial catchments, and that there are no significant mineral sources of Hg in the catchment. The good fit of the data to straight lines (r^2 = 0.91 for modern and 0.70 for preindustrial) indicates that none of these processes exerts a large effect. Furthermore, the gabbro and granite bedrock of northern Minnesota is poor in Hg, averaging 10 ng/g (67). The deeper mineral–soil horizons (75–100 cm) in this region contain an average of 14 ng/g (64) compared to an average preindustrial sediment concentration of 80 ng/g. The lake sediments contain 30–70% mineral matter, which includes diatom silica and authigenic iron as well as detrital silts and clays. Therefore, erosion of mineral soil contributed at most 5–12% of the preindustrial sedimentary Hg accumulation and 2–4% of the modern accumulation.

By using this model, we find that roughly the same proportion of atmospheric Hg has been transported from the catchments to the various lakes in modern and preindustrial times (26% and 22%, respectively). The balance of the Hg is either volatilized back to the atmosphere or retained by soils in the catchment. Because Hg has a high affinity for soil organic matter, it is not appreciably leached from soils even under acidic conditions, in contrast to other metals (68, 69). However, volatilization to the atmosphere from soils can be significant. In one experiment with undisturbed soil profiles, none of the Hg applied at the surface moved deeper than 20 cm after 19 weeks of irrigation and incubation, although 7–31% of the applied Hg was

lost through volatilization during this time (70). Similarly, Nater and Grigal
(64) found that forest soils in Minnesota, Wisconsin, and Michigan have
retained only a small proportion of the total Hg deposited from the atmosphere since deglaciation 10,000 years ago (165 years worth of deposition,
or about 2%). If about 25% of atmospheric deposition is transported to lakes
and ponds in the catchment, about 70–75% of deposited Hg must be re-
volatilized to the atmosphere. There appears to be very little net retention
of Hg in these soils.

Aastrup et al. (71) found in a Swedish study that about 80% of the Hg
deposited in a catchment was retained in the mor (organic surface soil), but
noted that an unknown proportion of the retained Hg might be lost to the
atmosphere. Similarly, in a mass-balance study of three catchments around
Harp Lake in central Ontario, Mierle (63) estimated that 84–92% of Hg
deposition was retained.

Although the retention of Hg in soils may be low, it is still important
to use relatively undisturbed lake catchments to assess atmospheric Hg
loading. Sites strongly affected by land-use changes (such as farming or
urbanization) will exhibit erosion rates many times that of undisturbed systems, and greater loading of soil-bound Hg will multiply the Hg accumulation
in the sediments. Thus the modest increase in soil erosion to Mountain Lake
in the mid-1900s may be responsible for an increase in Hg deposition—a
4.9-fold increase over preindustrial rates (Table II)—that is notably larger
than that at the other sites. More striking are preliminary results from an
agriculturally affected lake in southern Minnesota, which show modern Hg
accumulation rates in a single sediment core that are more than an order of
magnitude greater than the highest values reported here (72). Simola and
Lodenius (73) reached similar conclusions regarding the large increase in
sedimentary Hg accumulation that accompanied peatland ditching and af-
forestation of a lake catchment in northern Finland, although mobilization
of Hg by soil humic substances rather than particulate erosion was probably
responsible.

Mierle (63) and others (58, 74, 75) suggested that, because dissolved
organic matter (DOM) strongly complexes Hg, the export of Hg to lake
basins from their terrestrial watersheds may be controlled by the nature of
catchment soils and the movement of humic and fulvic acids. These observations offer a possible mechanism for the observed relationship between
Hg accumulation and catchment area in our study lakes. If most catchment-
derived Hg entered lakes by groundwater, these inputs should be only
weakly related to the size of the topographic watershed, especially for seepage lakes. On the other hand, if Hg export is linked to DOM, catchment
area is a logical correlate of Hg loading. The export of DOM from catchment
soils, which occurs largely in surficial drainage from upper soil horizons (76),
can be expected to increase with size of catchment area. Thus for a given
biogeographic region, the humic content of a lake is strongly related to the

relative size of its catchment (77, 78). The close correlation between Hg concentration and humic matter in surface waters (74, 75), the observation that peak concentrations of both Hg and DOM tend to occur during periods of high runoff (71, 74), and the experimental determination that Hg transport occurs primarily in the upper soil horizons (71) all support the conclusion that Hg export may be explained by factors regulating the export of fulvic and humic matter and by catchment area in particular.

Mercury in Fish.   Our results show that watershed contributions are neither dominant nor trivial in the total Hg budgets of small midwestern lakes, although the terrestrial inputs are ultimately atmospheric. Evans (6) reached similar conclusions following much the same approach used in this study. He observed a strong positive relationship between whole-lake Hg burdens calculated from multiple cores and $A_d:A_0$ from one of three lake districts in south-central Ontario. The study lacked sediment dating or flux calculations, which may account for weak correlations from the other two lake groups. Likewise, Suns and Hitchin (79) noted a strong correlation between Hg residues in yellow perch and the ratio of catchment area to lake volume. Mercury levels in fish are several steps removed from Hg deposition rates, and yet both the intercept and slope of this regression are significant positive terms. Evidently direct Hg deposition to the lake surface and wash-out from the catchment are important inputs to the lake.

Clearly Hg loading rates are not the only factor limiting Hg residues in fish. The production of methylmercury is probably the critical process controlling Hg bioaccumulation, but factors affecting methylation are not well understood. Empirical observations have identified pH, alkalinity, and dissolved organic carbon (DOC) as significant correlates of Hg in fish (19, 80–82). Possible pH effects include controls on Hg solubility, methylation rates, and the production of volatile species such as Hg^0 (52, 83, 84). Humic and fulvic components of DOC may enhance Hg loading and solubility, catalyze Hg methylation, or directly methylate Hg (85, 86). Alkalinity may be a proxy for calcium, which itself may inhibit Hg uptake by fish (87). Furthermore, recent evidence for the role of sulfate reduction in Hg methylation (88) implies that increased SO_4 loading may in some systems contribute to higher Hg levels in the biota.

Nonetheless, methylmercury production can be proportional to Hg concentration (83, 89) and may be limited in lakes by the flux of reactive Hg(II) species across the sediment–water interface (52). Thus an increase in total Hg deposition could produce an equivalent response in Hg bioaccumulation, all other factors being equal. It may be significant in this regard that the average rate of increase in Hg residues in fish in Minnesota is of the same magnitude as that calculated here for atmospheric loading, roughly 3% annually since 1930 (19).

Beyond Mercury

Other Applications of the Multiple-Core Approach.   The bulk of this chapter has dealt with the specific application of multiple-core methodology to questions of atmospheric Hg deposition. Whole-basin Hg accumulation rates for seven lakes, calculated from multiple sediment cores, were used in a simple mass-balance model to estimate atmospheric fluxes and Hg transport from catchment soils. This approach can be used to answer other limnological questions, and the model is not restricted to Hg or atmospheric deposition.

Multiple-core methods applied to individual lakes yield information on whole-basin fluxes, internal biogeochemical cycling, and variations in depositional processes in space and time. Specific applications include the calculation of catchment erosion rates (90, 91), studies of nutrient loading and retention (15), interpretations of mineral cycling (92) and the diagenesis of magnetic minerals (93), studies of sulfur accumulation and cycling (5, 16), reconstructions of productivity from fossil diatoms (14, 94, 95), and investigations of pollen recruitment and deposition (96, 97).

Multiple cores taken from a suite of lakes can be used to generate a regional picture of material fluxes, transport mechanisms, and geographic trends. The application of multiple cores on multiple lakes was pioneered by Evans and his co-workers to assess the atmospheric deposition of heavy metals including Hg, Pb, Zn, and Cd (6, 18, 98). These publications recognized the value of plotting whole-basin metal accumulation against the ratio of catchment to lake areas $(A_d:A_0)$, but did not describe the general meaning of the slope and intercept of this relationship.

As we have shown previously (99), the intercept is the accumulation rate of a given material for a lake with no catchment; for Hg it is the atmospheric deposition rate. For parameters with no appreciable atmospheric source, such as soil erosion, one would expect an intercept not significantly different from zero. The slope of the regression line is the rate at which the material is transported from the catchment to the lake sediments (grams per square meter of catchment per year). If the material has no significant source within the catchment except atmospheric deposition, then the slope divided by the intercept is the proportion of the atmospheric flux to the catchment that is transported to the lake. If the slope is zero, then there is no significant transport to the lake from the catchment. This simple model assumes that the ultimate fate of the substance is the sediments of the lake. Specifically, there must be no significant losses through degradation, diagenesis, evasion, or outflow in surface water or groundwater. Moreover, if catchment or atmospheric fluxes are regionally variable, this feature should be evident as a weak relationship between whole-basin accumulation and $A_d:A_0$.

How Many Cores?   Although the heterogeneity of lake-sediment accumulation is well known, most workers tend to rely on a single core to represent lakewide conditions. The single-core approach can reveal a good deal about lake history if uncertainties arising from spatial heterogeneity are taken into account. Changes in nutrient loading, erosion, or atmospheric deposition that alter the chemical composition of the sediments should register the same stratigraphic trends at different core sites (as shown in this study), although the actual flux and composition of the sediments may differ greatly among sites.

Even though single cores can document stratigraphic trends and trajectories, multiple cores are required to obtain quantitative data on whole-basin fluxes. Multiple-core studies might be more common were it not for the large effort required to collect, date, and stratigraphically analyze sediments from the various deposition regions of a lake basin. This study illustrates how, by reducing the number of stratigraphic units in each core, it is possible to economize on historical detail and increase the number of cores that can be analyzed.

The actual number of cores needed depends on the size and morphometry of the basin, the nature of the environmental signal under investigation, and the level of accuracy required by the study. Sediment deposition is spatially more variable in large deep lakes with irregular morphometry than in small basins of uniform depth. More cores are required from the former to attain the accuracy that a few cores would provide in the latter. Impoundments and lakes with major river inputs typically exhibit strong depositional gradients (*100*) and require a higher density of cores to accurately characterize sediment accumulation. Finally, only a few cores might be necessary to confirm that accumulation changes in one core are qualitatively representative of the entire lake, whereas a quantitative estimate of lakewide flux could require a dozen cores or more.

Results from our study can be used to address in a statistical sense the question of accuracy and core numbers. If we assume that our existing core data provide the true mean and variance for lakewide Hg accumulation, we can ask how likely we are to obtain, with a given number of cores, an estimate of Hg accumulation that is close to the true mean. We restrict this analysis to those sites with at least 12 dated cores (Dunnigan, Meander, Thrush, and Kjostad). Means and standard deviations are calculated with equal weighting of cores (as opposed to unequal weighting by depositional area). The probability of obtaining results that are within $\pm 10\%$ and $\pm 25\%$ of the true mean for lakewide Hg accumulation are calculated from a normal distribution for 2, 4, 8, and 16 cores (Figure 10).

These calculations show that the likelihood of estimating the true lakewide Hg accumulation rate from two cores is not very good if our criterion is $\pm 10\%$ of the true value ($p = 0.2$–0.35), but is substantially better if we lower our standards to $\pm 25\%$ ($p = 0.45$–0.75). If we increase the number

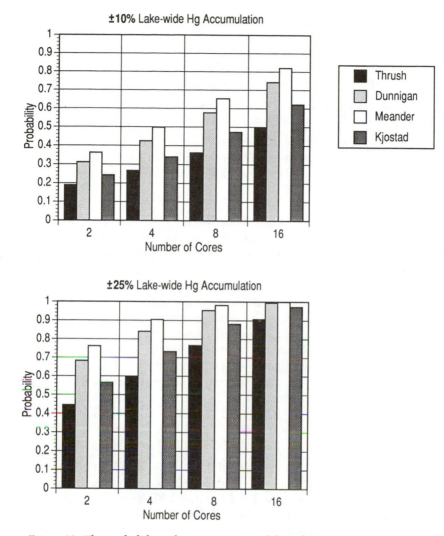

Figure 10. The probability of estimating mean lakewide Hg accumulation rates within ±10% and ±25% of the true mean as a function of the number of analyzed cores. The population mean and variance for each lake are approximated from the existing core data by equal weighting of cores, and probabilities are drawn from a normal distribution.

of cores to 16, our chances of getting within ±10% of the true value are as high as 0.75–0.8 for some sites (Dunnigan and Meander), but no better than 0.5–0.6 for others (Thrush and Kjostad). On the other hand, 16 cores virtually guarantees a correct answer ($p = 0.9$–0.99) if an estimate of ±25% of the true lakewide mean is acceptable. For some sites (Dunnigan and Meander) half that number of cores does nearly as well.

The conclusions that may be drawn from this exercise are

1. Many cores (>16) are required to obtain highly accurate estimates of lakewide accumulation rates, even in lakes with
   relatively uniform sediment deposition.

2. In certain lakes as few as four cores will give the desired result
   if somewhat lower accuracy is acceptable.

3. To attain the same level of accuracy, large or steep-sided basins
   with heterogeneous sediments may require 4 times the number of cores needed for small shallow lakes with more uniform
   sediments.

In practice, most multiple-core studies should yield greater accuracy than
is suggested by this analysis. Because accumulation rates vary greatly among
different depositional regions, a large number of cores would be required
to estimate the lakewide mean if core collection were actually random (as
assumed in our statistical exercise). However, coring strategies are usually
based on knowledge of basin morphometry and sedimentary processes, and
sites are selected to encompass a range of depositional environments. As
sediment deposition is nonrandom and neighboring cores from the same
environment are generally similar, a small selection of representative cores
from specific depositional regions, each weighted by the area of that region,
should provide a more accurate estimate of lakewide accumulation than a
strategy of random core collection.

Summary

This study demonstrates the use of multiple-core methods to obtain whole-
basin sediment fluxes from a suite of lakes and the application of these data
to questions of atmospheric metal deposition. Multiple-core data can be
economically produced by integrating longer core sections and reducing the
number stratigraphic units for analysis. As few as three ^{210}Pb analyses per
core can yield a modern accumulation rate; additional samples provide more
historical detail.

The number of cores needed to characterize accumulation in a lake basin
depends on hydrology, bathymetry, and degree of accuracy desired. The
fewest cores will be needed in small lakes of uniform depth that have no
significant inflowing streams.

If whole-basin accumulation rates for a substance are produced for multiple lakes in a geographic region, it is possible to use a simple mass-balance
model to estimate both the atmospheric deposition rate and transport from
the terrestrial catchment. The model was applied to both modern and preindustrial Hg accumulation in seven undisturbed lakes in the upper midwest

of the United States. Most of the variation in Hg fluxes among the lakes can be explained by the model, which incorporates the ratio of catchment to lake area. Local geological sources of Hg represent only a minor component of the total budget.

The atmospheric deposition rate in this midcontinental area, which has increased by a factor of about 3.7, suggests that natural Hg concentrations were only about 25% of modern levels. Current estimates of recent increases in global atmospheric Hg support this conclusion, and indicate that increased anthropogenic Hg emissions, rather than enhanced removal by atmospheric oxidants, are responsible for elevated Hg deposition. Moreover, the increase appears to be relatively uniform across our study area, implying regional if not global sources for the Hg falling on these remote sites.

About 25% of the Hg deposited to the terrestrial catchment is transported to the various lakes, probably in association with organic acids. Lakes with larger catchments (relative to their surface area) receive proportionally higher Hg loading. Among the seven lakes studied, between 40 and 80% of Hg inputs were made directly to the lake surface.

Acknowledgments

We thank M. Hora of the Minnesota Pollution Control Agency for initiating this project, D. Helwig for help in designing parts of the study, and H. Wiegner, W. Popp, and D. Verschuren for assistance in sediment coring. Support was provided by the Legislative Commission on Minnesota Resources, the Acid Deposition Program of the Minnesota Pollution Control Agency, and the University of Minnesota Water Resources Research Center. This chapter is Contribution No. 438 of the Limnological Research Center.

References

1. Goldberg, E. D.; Hodge, V. F.; Griffin, J. J.; Koide, M.; Edgington, D. N. *Environ. Sci. Technol.* **1981,** *15,* 466–471.
2. Rippey, B.; Murphy, R. J.; Kyle, S. W. *Environ. Sci. Technol.* **1982,** *16,* 23–30.
3. Norton, S. A.; Dillon, P. J.; Evans, R. D.; Mierle, G.; Kahl, J. S. In *Acidic Precipitation: Sources, Deposition, and Canopy Interactions;* Lindberg, S. E.; Page, A. L.; Norton, S. A., Eds.; Springer-Verlag: New York, 1990; Vol. 3, pp 73–102.
4. Carignan, R.; Tessier, A. *Science (Washington, D.C.)* **1985,** *228,* 1524–1526.
5. White, J. R.; Gubala, C. P.; Fry, B.; Owen, J.; Mitchell, M. J. *Geochim. Cosmochim. Acta* **1989,** *53,* 2547–2559.
6. Evans, R. D. *Arch. Environ. Contam. Toxicol.* **1986,** *15,* 505–512.
7. Rada, R. G.; Wiener, J. G.; Winfrey, M. R.; Powell, D. E. *Arch. Environ. Contam. Toxicol.* **1989,** *18,* 175–181.
8. Carignan, R.; Nriagu, J. O. *Geochim. Cosmochim. Acta* **1985,** *49,* 1753–1764.
9. Renberg, I. *Hydrobiologia* **1986,** *143,* 379–385.
10. Johnson, M. G. *Can. J. Fish. Aquat. Sci.* **1987,** *44,* 3–13.
11. Dearing, J. A. *Hydrobiologia* **1983,** *103,* 59–64.

12. Hilton, J.; Gibbs, M. M. *Chem. Geol.* **1984,** *47,* 57–83.
13. Downing, J. A.; Rath, L. C. *Limnol. Oceanogr.* **1988,** *33,* 447–458.
14. Anderson, N. J. *Limnol. Oceanogr.* **1990,** *35,* 497–508.
15. Evans, R. D.; Rigler, F. H. *Can. J. Fish. Aquat. Sci.* **1980,** *37,* 817–822.
16. Baker, L. A.; Engstrom, D. R.; Brezonik, P. L. *Limnol. Oceanogr.* **1992,** *37,* 689–702.
17. Evans, R. D.; Rigler, F. H. *Environ. Sci. Technol.* **1980,** *14,* 216–218.
18. Dillon, P. J.; Evans, R. D. *Hydrobiologia* **1982,** *91,* 121–130.
19. Swain, E. B.; Helwig, D. D. *J. Minn. Acad. Sci.* **1989,** *55,* 103–109.
20. Lathrop, R. C.; Rasmussen, P. W.; Knauer, D. R. *Water Air Soil Pollut.* **1991,** *56,* 295–307.
21. Wren, C. D.; Scheider, W. A.; Wales, D. L.; Muncaster, B. W.; Gray, I. M. *Can. J. Fish. Aquat. Sci.* **1991,** *48,* 132–139.
22. Johansson, K. *Verh. Int. Ver. Theor. Angew. Limnol.* **1985,** *22,* 2359–2363.
23. Johnson, M. G.; Culp, L. R.; George, S. E. *Can. J. Fish. Aquat. Sci.* **1986,** *43,* 754–762.
24. Meger, S. A. *Water Air Soil Pollut.* **1986,** *30,* 411–419.
25. Verta, M.; Tolonen, K.; Simola, H. *Sci Total Environ.* **1989,** *87/88,* 1–18.
26. Slemr, F.; Langer, E. *Nature (London)* **1992,** *355,* 434–437.
27. Lindberg, S.; Stokes, P. M.; Goldberg, E.; Wren, C. In *Lead, Mercury, Cadmium, and Arsenic in the Environment;* Hutchinson, T. C.; Meema, K. M., Eds.; John Wiley and Sons: New York, 1987; pp 17–33.
28. Orr, E. J.; Swain, E. B.; Strassman, R. L.; Bock, D. C. *Sulfur Emissions and Deposition in Minnesota: 1990 Biennial Report to the Legislature;* Minnesota Pollution Control Agency: St. Paul, MN, 1991; p 23.
29. National Atmospheric Deposition Program (NRSP–3)/National Trends Network, NADP/NTN Coordination Office, Colorado State University: Fort Collins, CO, 1992.
30. Wright, H. E., Jr. *J. Paleolimnol.* **1991,** *6,* 37–49.
31. Blomqvist, S. *Sedimentology* **1985,** *32,* 605–612.
32. Blomqvist, S. *Mar. Ecol. Prog. Ser.* **1991,** *72,* 295–304.
33. Dean, W. E., Jr. *J. Sediment. Petrol.* **1974,** *44,* 242–248.
34. Eakins, J. D.; Morrison, R. T. *Int. J. Appl. Radiat. Isot.* **1978,** *29,* 531–536.
35. Appleby, P. G.; Oldfield, F. *Catena* **1978,** *5,* 1–8.
36. Binford, M. W. *J. Paleolimnol.* **1990,** *3,* 253–267.
37. *Mercury: Methods of Sampling, Preservation, and Analysis;* Economic and Technical Review Report EPS 3–EC–81–4, Environment Canada, 1981.
38. Winefordner, J. D.; Long, G. L. *Anal. Chem.* **1983,** *55,* 712A–724A.
39. Engstrom, D. R.; Whitlock, C.; Fritz, S. C.; Wright, H. E., Jr. *J. Paleolimnol.* **1991,** *5,* 139–174.
40. Urban, N. R.; Eisenreich, S. J.; Grigal, D. F.; Schurr, K. T. *Geochim. Cosmochim. Acta* **1990,** *54,* 3329–3346.
41. Anderson, N. J. *J. Paleolimnol.* **1990,** *3,* 143–160.
42. Evans, R. D.; Rigler, F. H. *Can. J. Fish. Aquat. Sci.* **1983,** *40,* 506–515.
43. Wren, C. D.; MacCrimmon, H. R.; Loescher, B. R. *Water Air Soil Pollut.* **1983,** *19,* 277–291.
44. Nriagu, J. O.; Pacyna, J. M. *Nature (London)* **1988,** *333,* 134–139.
45. Andren, A. W.; Nriagu, J. O. In *The Biogeochemistry of Mercury in the Environment;* Nriagu, J. O., Ed.; Elsevier/North–Holland Biomedical: Amsterdam, Netherlands, 1979; pp 1–21.
46. Fitzgerald, W. F.; Gill, G. A.; Kim, J. P. *Science (Washington, D.C.)* **1984,** *224,* 597–599.

47. Lindberg, S. E. In *Lead, Mercury, Cadmium, and Arsenic in the Environment*; Hutchinson, T. C.; Meema, K. M., Eds.; John Wiley and Sons: New York, 1987; pp 89–106.
48. Nriagu, J. O. *Nature* **1989**, *338*, 47–49.
49. Lindqvist, O.; Johansson, K.; Aastrup, M.; Andersson, A.; Bringmark, L.; Hovsenius, G.; Håkanson, L.; Iverfeldt, Å.; Meili, M.; Timm, B. *Water Air Soil Pollut.* **1991**, *55*, 1–261.
50. Nriagu, J. O.; Soon, Y. K. *Geochim. Cosmochim. Acta* **1985**, *49*, 823–834.
51. Fitzgerald, W. F. In *The Role of Air-Sea Exchange in Geochemical Cycling*; Baut-Menard, P., Ed.; D. Reidel: Boston, MA, 1986; pp 363–408.
52. Fitzgerald, W. F.; Mason, R. P.; Vandal, G. M. *Water Air Soil Pollut.* **1991**, *56*, 745–767.
53. Iverfeldt, Å. *Water Air Soil Pollut.* **1991**, *56*, 251–265.
54. Glass, G. E.; Sorensen, J. A.; Schmidt, K. W.; Rapp, G. R., Jr.; Yap, D.; Fraser, D. *Water Air Soil Pollut.* **1991**, *56*, 235–249.
55. Rodgers, J. S.; Huang, P. M.; Hammer, U. T.; Liaw, W. K. *Verh. Int. Ver. Theor. Angew. Limnol.* **1984**, *22*, 283–288.
56. Schindler, D. W.; Hesslein, R. H.; Wagemann, R.; Broecker, W. S. *Can. J. Fish. Aquat. Sci.* **1980**, *37*, 373–377.
57. Vandal, G. M.; Mason, R. P.; Fitzgerald, W. M. *Water Air Soil Pollut.* **1991**, *56*, 791–803.
58. Meili, M. *Water Air Soil Pollut.* **1991**, *56*, 719–727.
59. Henning, T. A. M.S. Thesis, University of Minnesota, Minneapolis, MN, 1989.
60. Smith, J. N.; Loring, D. H. *Environ. Sci. Technol.* **1981**, *15*, 944–951.
61. Breteler, R. J.; Bowen, V. T.; Schneider, D. L.; Henderson, R. *Environ. Sci. Technol.* **1984**, *18*, 404–409.
62. Lodenius, M. *Water Air Soil Pollut.* **1991**, *56*, 323–332.
63. Mierle, G. *Environ. Toxicol. Chem.* **1990**, *9*, 843–851.
64. Nater, E. A.; Grigal, D. F. *Nature (London)* **1992**, *358*, 139–141.
65. Steinnes, E.; Andersson, E. M. *Water Air Soil Pollut.* **1991**, *56*, 391–404.
66. Jensen, A.; Jensen, A. *Water Air Soil Pollut.* **1991**, *56*, 769–777.
67. Smith, E.A. M.S. Thesis, University of Minnesota, Duluth, MN, 1990.
68. Schuster, E. *Water Air Soil Pollut.* **1991**, *56*, 667–680.
69. Lodenius, M.; Autio, S. *Arch. Environ. Contam. Toxicol.* **1989**, *18*, 261–267.
70. Hogg, T. J.; Stewart, J. W. B.; Bettany, J. R. *J. Environ. Qual.* **1978**, *7*, 440–445.
71. Aastrup, M.; Johnson, J.; Bringmark, E.; Bringmark, I.; Iverfeldt, Å. *Water Air Soil Pollut.* **1991**, *56*, 155–167.
72. Brigham, M. E. M.S. Thesis, University of Minnesota, Minneapolis, MN, 1992.
73. Simola, H.; Lodenius, M. *Bull. Environ. Contam. Toxicol.* **1982**, *29*, 298–305.
74. Mierle, G.; Ingram, R. *Water Air Soil Pollut.* **1991**, *56*, 349–357.
75. Johansson, K.; Iverfeldt, Å. *Verh. Int. Ver. Theor. Angew. Limnol.* **1991**, *24*, 2200–2204.
76. Cronan, C. S.; Aiken, G. R. *Geochim. Cosmochim. Acta* **1985**, *49*, 1697–1705.
77. Engstrom, D. R. *Can. J. Fish. Aquat. Sci.* **1987**, *44*, 1306–1314.
78. Rasmussen, J. B.; Godbout, L.; Schallenberg, M. *Limnol. Oceanogr.* **1989**, *34*, 1336–1343.
79. Suns, K.; Hitchin, G. *Water Air Soil Pollut.* **1990**, *50*, 255–265.
80. Håkanson, L.; Nilsson, A.; Andersson, T. *Environ. Pollut.* **1988**, *49*, 145–162.
81. McMurtry, M. J.; Wales, D. L.; Scheider, W. A.; Beggs, G. L.; Diamond, P. E. *Can. J. Fish. Aquat. Sci.* **1989**, *46*, 426–434.
82. Grieb, T. M.; Driscoll, C. T.; Gloss, S. P.; Schofield, C. L.; Bowie, G. L.; Porcella, D. B. *Environ. Toxicol. Chem.* **1990**, *9*, 919–930.

83. Xun, L.; Campbell, N. E. R.; Rudd, J. W. M. *Can. J. Fish. Aquat. Sci.* **1987**, *44*, 750–757.
84. Wiener, J. G.; Fitzgerald, W. F.; Watras, C. J.; Rada, R. G. *Environ. Toxicol. Chem.* **1990**, *9*, 909–918.
85. Nagase, H.; Ose, Y.; Sato, T.; Ishikawa, T. *Sci. Total Environ.* **1982**, *24*, 133–142.
86. Meili, M.; Iverfeldt, Å.; Håkanson, L. *Water Air Soil Pollut.* **1991**, *56*, 439–453.
87. Rodgers, D. W.; Beamish, F. W. H. *Can. J. Fish. Aquat. Sci.* **1983**, *40*, 824–828.
88. Kerry, A.; Welbourn, P. M.; Prucha, B.; Mierle, G. *Water Air Soil Pollut.* **1991**, *56*, 565–575.
89. Jensen, S.; Jernelov, A. *Nature (London)* **1969**, *223*, 753–754.
90. Dearing, J. A.; Elner, J. K.; Happey-Wood, C. M. *Quat. Res.* **1981**, *16*, 356–372.
91. Foster, I. D. L.; Dearing, J. A.; Simpson, A.; Carter, A. D.; Appleby, P. G. *Earth Surf. Processes Landforms* **1985**, *10*, 45–68.
92. Engstrom, D. R.; Swain, E. B. *Hydrobiologia* **1986**, *143*, 37–44.
93. Anderson, N. J.; Rippey, B. *Limnol. Oceanogr.* **1988**, *33*, 1476–1492.
94. Battarbee, R. W. *Philos. Trans. R. Soc. London B* **1978**, *281*, 303–345.
95. Anderson, N. J. *J. Ecol.* **1989**, *77*, 926–946.
96. Davis, M. B.; Brubaker, L. B. *Limnol. Oceanogr.* **1973**, *3*(27), 635–646.
97. Davis, M. B.; Ford, M. S. *Limnol. Oceanogr.* **1982**, *27*, 137–150.
98. Evans, H. E.; Smith, P. J.; Dillon, P. J. *Can. J. Fish. Aquat. Sci.* **1983**, *40*, 570–579.
99. Swain, E. B.; Engstrom, D. R.; Brigham, M. E.; Henning, T. A.; Brezonik, P. L. *Science (Washington, D.C.)* **1992**, *257*, 784–787.
100. Håkanson, L.; Jansson, M. *Principles of Lake Sedimentology;* Springer-Verlag: New York, 1983; p 316.

RECEIVED for review April 1, 1992. ACCEPTED revised manuscript September 11, 1992.

Use of Oxygen-18 and Deuterium To Assess the Hydrology of Groundwater–Lake Systems

David P. Krabbenhoft[1], Carl J. Bowser[2], Carol Kendall[3], and Joel R. Gat[4]

[1]U.S. Geological Survey, Madison, WI 53719
[2]Department of Geology and Geophysics, University of Wisconsin–Madison, Madison, WI 53706
[3]U.S. Geological Survey, Menlo Park, CA 94025
[4]Weizmann Institute of Science, Rehovot, Israel

A thorough understanding of a lake's hydrology is essential for many lake studies. In some situations the interactions between groundwater systems and lakes are complex; in other cases the hydrology of a multilake system needs to be quantified. In such places, stable isotopes offer an alternative to the more traditional piezometer networks, which are costly to install and time-consuming to maintain. The stable-isotope mass-balance relations presented here can be used to estimate groundwater exchange rates for individual lakes and geographically clustered lakes. These relations also can be used to estimate other hydrological factors, such as average relative humidity. In places where the groundwater system is unstable (e.g., where flow reversals occur), natural solute tracers may provide a better alternative than stable isotopes for estimating rates of groundwater flow to and from lakes.

M OST LAKES ARE IN DIRECT HYDRAULIC communication with the contiguous groundwater system. Lakes may serve as recharge or discharge areas for the local groundwater system, or as both if the lake is a surface expression of groundwater flowing down the hydraulic gradient. The hydrological setting is complex because of seasonal variations in rainfall and evaporation, as well as in topographic variability and aquifer heterogeneity. Natural variations in these elements of a groundwater–lake system can complicate groundwater–lake interactions and can make it difficult to quantify the groundwater contribution in a lake's hydrological budget.

Hydrogeologists take a variety of approaches in quantifying the rates of groundwater exchange with lakes. These approaches can be physically or chemically based; isotopic approaches are one example of the chemical methods. The application of stable isotopes to the study of groundwater–lake systems is the primary focus of this chapter.

Quantifying Rates of Exchange

Physical Approaches. Groundwater-exchange rates with lakes are traditionally estimated by careful measurements of hydraulic potentials within the groundwater system, followed by application of Darcy's law in the form of flow-net analysis or numerical modeling. However, these measurements can be time-consuming and costly, and can require monthly to weekly measurements at many piezometers to examine the three-dimensional nature of the hydraulic-potential field. In addition, characterization of the hydraulic conductivity of the aquifer is critical to physical approaches and typically leads to results with large uncertainties (1, 2).

Numerical groundwater-flow models are commonly used to synthesize physical measurements and to estimate groundwater–lake exchange rates (3–6). These models typically require a substantial amount of time to construct, and they usually require subjective adjustments of model parameters such as hydraulic conductivity and recharge. In spite of these shortcomings, these models are now standard tools in modern hydrogeological investigations; uncertainties associated with such approaches are the focus of much research in modern hydrogeology.

Chemical Approaches. Because evaporation, chemical precipitation and dissolution, and in-lake biological processes affect the solute composition of waters, researchers have attempted to use chemical tracers to quantify hydrological budgets (7). However, complications caused by solute contamination from human activities such as road salting (8), failure to quantify the role of mineral dissolution (weathering) reactions and precipitation of phases within either the aquifer or the lake, and lack of information about evaporative concentration of waters have made the general application of solute-based models difficult. Such concerns limit the broad application of solute-based groundwater–lake exchange estimates, in spite of the recognized significance of groundwater in controlling the solute composition of lakes (9). Clearly, a spatially and temporally integrative approach that would allow reasonable estimates of groundwater flow and be less costly than physically based approaches is needed.

Isotopes of oxygen and hydrogen are useful tracers of water sources because they are constituents of the water molecule itself and because they are conservative in aquifers at near-surface temperatures. Isotopic techniques take advantage of the fact that lakes and their surrounding groundwater systems are usually isotopically distinct. Applications of stable isotopes for the study of lakes were first described by Dincer (*10*) and were discussed in several subsequent review articles (*11–14*). Most applications of isotopic techniques to lake systems are designed for the determination of water balances, nutrient-uptake studies, and paleotemperature reconstructions.

This chapter demonstrates the usefulness of stable isotopes in investigating groundwater–lake systems. The discussion emphasizes isotopic applications to groundwater–lake systems characteristic of the temperate glaciated regions of the north-central and northeastern United States. Thus, it is also applicable to similar systems in other glaciated parts of the world, such as the Scandinavian peninsula and northern Asia. The applications stem from our experience with lake systems in the lake district of north-central Wisconsin. As such, we restrict our discussion to shallow groundwater systems that are hydraulically connected to freshwater lakes.

Background

Oxygen has three naturally occurring stable isotopes whose atomic masses are 16, 17, and 18 (designated ^{16}O, ^{17}O, and ^{18}O). Hydrogen occurs as three isotopes whose masses are 1, 2, and 3 (^{1}H, deuterium, and tritium, respectively) (*15, 16*). Tritium is a radioisotope (unstable isotope) whose half-life is 12.26 years (*15*). In contrast to *radioisotopes*, which spontaneously emit alpha or beta particles and sometimes gamma rays during the disintegration of their nuclei, *stable isotopes* do not undergo nuclear transformations over time. Although radioisotopes such as tritium can be valuable research tools for hydrological studies (*15*), we discuss only the use of stable isotopes in this chapter.

Different isotopes of the same element differ slightly in chemical and physical properties because of their mass differences. For elements with low atomic masses, these mass differences are large enough for many physical, chemical, and biological reactions to fractionate or change the relative proportions of different isotopes of the same element in various compounds. Thus, a particular water or mineral may have a unique isotopic composition (ratio of the isotopes of an element) that indicates its source or the process that formed it. Two different processes—equilibrium and kinetic isotope effects—cause *isotope fractionation*.

Equilibrium isotope-exchange processes involve the redistribution of isotopes of an element among various species or compounds. At equilibrium, the forward and backward reaction rates of any particular isotope are identical. Isotopic equilibrium between two compounds does not mean that their isotopic compositions are identical, but only that the ratios of the different isotopes in each compound are constant. At equilibrium, the difference between the isotopic compositions of any two compounds is largely a function of their chemical composition and temperature. In general, the difference between the equilibrium isotopic compositions of any two compounds decreases as temperature increases. This phenomenon results from the fact that as temperature increases the relative difference in vibrational frequencies between isotopic species decreases (16).

Forward and backward reaction rates are not identical in systems that are not at isotopic equilibrium. Reactions may, in fact, be unidirectional if reaction products become physically isolated from the reactants. The reaction rates depend on the ratios of the masses of the isotopes and their vibrational energies; hence, such reactions result in *kinetic isotope fractionations*. The magnitude of a kinetic isotope fractionation depends on the reaction pathway and the relative energies of the bonds being severed or formed by the reaction. The kinetic fractionation factor is typically larger than the equilibrium fractionation factor for the same reaction. As a rule, bonds of light isotopes are broken more easily than equivalent bonds of heavier isotopes. Hence, light isotopes react faster than heavy isotopes.

The bond energies of the isotopically light species of water, $H_2^{16}O$, are weaker than those of the isotopically heavier species $H_2^{18}O$ and HDO (where D is deuterium, ^2H). Thus, the light species has a higher vapor pressure and diffusivity than the heavy species. The difference in vapor pressures is sufficient to cause lake waters to become isotopically heavier (i.e., enriched in $H_2^{18}O$ and HDO) during evaporation because of selective removal of the light isotopic species. The degree of enrichment is a function of climatic conditions, including temperature, relative humidity, rain and lake evaporation rates, and degree of mixing of the water body.

The fractionation of stable isotopes between two substances A and B can be expressed by use of the isotope fractionation factor alpha (α),

$$\alpha_{A-B} = \frac{R_A}{R_B} \qquad (1)$$

where R_x is the ratio of the heavier isotope to the lighter isotope (i.e., D/H, $^{18}O/^{16}O$, etc.); α values generally are close to 1. The D/H ratio is often written $^2H/^1H$. Other common formulations include the fractionation factor, α^*,

$$\alpha^* = \frac{1}{\alpha} = \alpha_{B-A} = \frac{R_B}{R_A} \qquad (2)$$

and ϵ, which is a convenient way to express the equilibrium fraction in parts per thousand:

$$\epsilon_{A-B} = (\alpha_{A-B} - 1)1000 \tag{3}$$

Stable-isotope ratios are commonly reported in delta (δ) values in parts per thousand (denoted as ‰, or per mil) that quantify enrichments or depletions relative to a standard of known composition. For example,

$$\delta^{18}O(‰) = \left(\frac{R_{sample}}{R_{standard}} - 1 \right) 1000 \tag{4}$$

with R defined as for eq 1. Positive values indicate that the sample contains a higher ratio of the heavy to the light isotope than does the standard; negative values indicate that the sample is depleted in the heavy isotope relative to the standard.

Various isotope standards are used for reporting isotopic compositions; the composition of each of the standards has been defined as 0 per mil. Oxygen and hydrogen isotopic compositions are commonly reported relative to standard mean ocean water (SMOW or V-SMOW) (*17*).

Oxygen-18 and Deuterium in the Hydrological Cycle

Lakes represent only a small part of the hydrological cycle; however, a thorough understanding of the entire hydrological cycle, in terms of both water transfer and isotopic fractionation, is necessary to comprehend the intricacies of isotope hydrology as applied to lakes. Components of the hydrological cycle that are critical to isotope hydrology of lakes are shown in Figure 1. The hydrological and isotopic significance of each of these components to the hydrology of lakes is described in the following sections.

Precipitation.   Meteoric waters acquire different isotopic compositions by a variety of processes, the most important of which are temperature during condensation and degree of rainout in the air mass (*17, 18*). On the basis of hundreds of analyses, Craig (*17*) noted a linear relation between the δD and $\delta^{18}O$ values of most meteoric waters; specifically,

$$\delta D = 8\delta^{18}O + 10 \tag{5}$$

This equation describes the linear distribution of data points on a plot of δD versus $\delta^{18}O$ that is commonly referred to as the global meteoric-water line (GMWL). The zero intercept for this line, defined as the *deuterium excess* (*18*), differs for waters from different source areas. A worldwide network of stations has been established for the collection of monthly bulk rain samples (*19*) from which regionally applicable equations have been determined. Continuous sampling at a specific site for several years will establish a local meteoric-water line (LMWL), which, as will be shown, is essential for any application of stable-isotope hydrology.

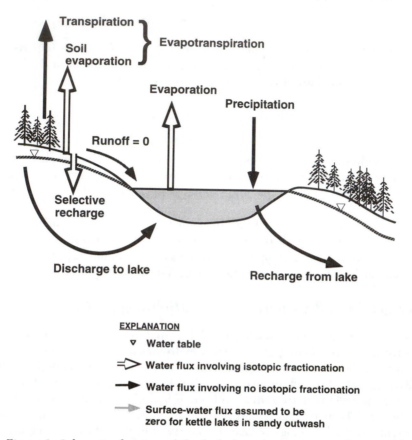

Figure 1. Schematic drawing of the hydrological components of a ground-water–lake system and their respective isotopic fractionation characteristics.

The isotopic composition of precipitation varies significantly over the year, principally because of seasonal temperature differences but also partly because of differences in moisture sources for a particular area. Short-term variations in the isotopic composition of precipitation within storms are well recognized, but such variations tend to average out if monthly means are determined (20). Because isotopic fractionation during evaporation and condensation of water is inversely related to temperature, precipitation is isotopically enriched in the heavy isotopic species of water during warm summer months and correspondingly depleted during cold winter months. Seasonal warming and cooling result in a sinusoidal annual variation in the isotopic composition of precipitation. Although the isotopic composition of precipitation for any particular location can vary widely on an annual basis (12–18), spread about the LMWL generally is minimal. Coefficients of variation (R^2) commonly exceed 0.95 for a regression analysis of δD versus $\delta^{18}O$ in precipitation samples.

Given that evaporation and condensation are complex processes, it is difficult to explain why the observed relationship between $\delta^{18}O$ and δD is so strong. One explanation is that the formation of water droplets in clouds is an equilibrium-controlled process (because of 100% or greater water-vapor saturation in the air mass) and the fractionation is only a function of the ambient temperature. However, other processes can cause precipitation samples from the same locality to plot along slightly different meteoric-water lines, thereby resulting in decreased R^2 values for the entire data set (*20*). Changes in the sources of local air masses can cause rain samples to plot along slightly different deuterium excess lines because of differences in the humidity of the air masses at their oceanic origins (*18*). Additionally, evaporation of falling raindrops, as well as large contributions of re-evaporated water to the local atmosphere, may cause departures from the LMWL. The isotopic composition of atmospheric moisture also plots along the meteoric-water line. However, because equilibrium fractionations between vapor and liquid water result in depletions of 11.7‰ ($\delta^{18}O$) and 112‰ (δD) at 0 °C (*21*), atmospheric vapor is depleted relative to the corresponding rain samples, but plots along the same meteoric-water line.

Distribution of isotopic compositions of groundwater, precipitation, and lake water on a plot of δD versus $\delta^{18}O$ for the Sparkling Lake area, north-central Wisconsin, is shown in Figure 2. Although precipitation samples have a wide range in isotopic composition, groundwater is much more homogeneous. This phenomenon results from the conservative nature of the isotopic species of water in groundwater systems and various mixing processes in the saturated and unsaturated zones.

Evaporation and Evaporative Fractionation of Water.   Evaporation from standing water bodies is the principal fractionation mechanism in most hydrological systems. Evaporative isotopic enrichment is a function of numerous factors (e.g., temperature, salinity, and relative humidity) that cause considerable variation in the $^{18}O/^{16}O$ and D/H ratios of natural surface waters. Craig and Gordon (*22*) evaluated isotopic effects on precipitation and evaporation in the ocean–atmosphere system. Much of what was developed in that work is directly applicable to the freshwater systems discussed here.

Isotopic fractionation resulting from evaporation from standing water bodies can be described in terms of equilibrium and nonequilibrium fractionation effects. Equilibrium fractionation occurs when the isotopic composition of the evaporated water or lake evaporate is in thermodynamic equilibrium with the lake water (*23*). Equilibrium fractionation, however, can occur only when the water vapor in the air mass above the lake is 100% saturated. The process of equilibrium isotopic fractionation is described by Raleigh fractionation. The isotopic composition of water vapor in equilibrium with liquid water at any time is given by

$$R_v = \alpha * R_1 \qquad (6)$$

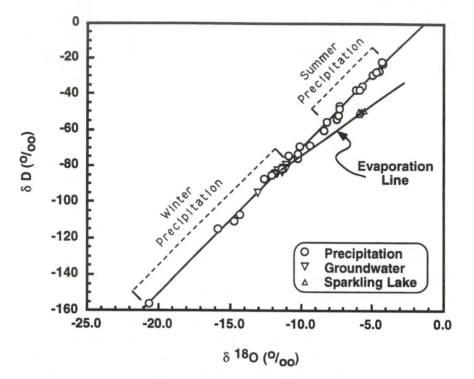

Figure 2. Isotopic compositions (δD versus δ¹⁸O) of precipitation, ground-water, and Sparkling Lake water. Precipitation and groundwater samples were collected near Sparkling Lake, WI.

where R_v and R_l are the isotopic ratios ($\delta^{18}O$ or δD) of the vapor and liquid, respectively, and α^* is the equilibrium fractionation factor defined in eq 2.

Kinetically controlled isotopic fractionation is a significant factor in most evaporation processes that lead to an isotopic fractionation exceeding that predicted by equilibrium isotope fractionation in eq 6. Kinetically controlled fractionation results from the fact that light isotopic species have greater free-air diffusion rates than do heavier isotopes; thus, the light species are preferentially transported away during the diffusion-controlled stage of evaporation (20). The evaporative fractionation process was formalized by Craig and Gordon (22) and later summarized (20) as

$$\frac{d \ln \dfrac{N_i}{N}}{d \ln N} = \frac{d \ln R_l}{d \ln N} = \frac{h(R_l - R_v)\dfrac{1}{R_l} = \epsilon^* - \Delta\epsilon}{(1 - h) + \Delta\epsilon} \tag{7}$$

where N and N_i are the number of moles of the abundant (light) and the heavy isotopic species, respectively; $\Delta\epsilon$ is the kinetic fractionation factor;

$\epsilon^* = 1 - \alpha^*$ is the equilibrium fractionation factor (or 1000 times the small deviation from unity that α values generally have); and h is the relative humidity normalized to the temperature of the liquid phase. The kinetic fractionation factor, $\Delta\epsilon$, is a function of the relative humidity and is given by the relation

$$\Delta\epsilon = K(1 - h) \tag{8}$$

where the factor K has been empirically determined to be 14.3‰ and 12.5‰ for oxygen-18 and deuterium, respectively (*13*). Examination of eq 7 reveals the critical dependence of evaporative isotope fractionation on the relative humidity. In temperate to humid environments, where humidities approach 100%, the $(1 - h)$ term in the denominator becomes small and the evaporative fractionation process becomes increasingly sensitive to variations in humidity.

The net result of evaporative isotopic fractionation of lake water is a water mass that is not in isotopic equilibrium with its water sources. Therefore, the isotopic composition of a lake water should plot below the LMWL, as shown in Figure 2. The line that passes through the isotopic composition of a lake water and its volume-weighted input from precipitation and groundwater is referred to as an *evaporation line*. This line represents the isotopic evolution of lake water from its source waters. The slope of an evaporation line is a function of humidity, temperature, and the isotopic composition of atmospheric moisture above the lake; the slope generally ranges from about 3.5 to 6.0 (*23*).

Evapotranspiration.   *Evapotranspiration* refers to the combined moisture evaporated from vegetation and underlying soil. It generally includes leaf-moisture loss through stomatal conductance, stem- and leaf-moisture loss from precipitation that falls directly on the plant, and soil-moisture loss. As in evaporative fractionation of surface waters, evaporation of soil water can result in isotopically enriched residual waters and depleted vapor. If all the infiltrating water evaporates from the soil, the water vapor produced is isotopically identical to the original soil water. Uptake of soil water by tree roots is not an isotope-fractionating process (*24–27*). Therefore, transpired water vapor removed from the ground by plants is virtually identical to the local soil water or groundwater. The net effect of evapotranspiration in most cases is a small-to-negligible change in the isotopic composition of ambient vapor. Therefore there is generally a small difference between the average isotopic compositions of groundwater and precipitation.

The seasonal effects of evapotranspiration on the transfer of water from precipitation to the saturated zone can be pronounced. Temperate climates with two distinct periods of vegetative cover (leafout in the spring and leaf-down in the fall) are prevalent in a broad zone extending from subtropical to subpolar environments. In forested regions with a significant proportion of deciduous vegetation, evapotranspiration is much greater in summer than in winter; in regions with substantial winter snow cover, evapotranspiration may be virtually zero. Even in coniferous forests where evapotranspiration generally occurs year-round, the difference between winter and summer evapotranspiration is marked, and the dominant moisture loss occurs during the summer.

Recharge to Groundwater. Groundwater derived from infiltration of precipitation may become isotopically distinct from precipitation because of the combined processes of selective recharge and fractionation. The process whereby all precipitation events contribute water to a lake, and only large events result in groundwater recharge, is referred to as *selective recharge*. Regions subject to considerable snow accumulation effectively store moisture above ground, and spring snowmelt (prior to vegetative leafout) can be a dominant source of recharge to the local groundwater system. Such is the case in the glaciated terrain of northern Wisconsin, where permeable sandy soils are conducive to infiltration of spring snowmelt. Rapid infiltration of snowmelt is responsible for the 0.6‰ difference in $\delta^{18}O$ between local groundwater and average annual precipitation in northern Wisconsin (28).

Fractionation arises if infiltrating soil water is partially evaporated (29), if exchange with atmospheric vapor in the soil is significant (C. Kendall, U.S. Geological Survey, oral communication, 1991), or if precipitation intercepted by the tree canopy is fractionated by evaporation. Evaporation of water from the soil generally affects the isotopic composition of soil water only in arid climates (27, 30). Observations in northern Wisconsin (*see* the isotopic composition of groundwater plots on the LMWL, Figure 2) and other areas (24) indicate that fractionation caused by soil-water evaporation is minimal in temperate climates and can be neglected for the glacial-lake systems. The literature shows that atmospheric vapor exchange in the soil zone and evaporative fractionation caused by canopy wetting have not been examined fully in a wide range of environments and require additional research.

Groundwater–Lake Systems

Although lakes are more complex, many can be considered as surface expressions, or outcrops, of the water table. Depending on the distribution of hydraulic heads surrounding a lake, groundwater–lake systems can be described as one of the following types:

- *recharge systems*, where the lake surface is higher than the surrounding water table and water flows to the groundwater system;

- *discharge systems*, where the lake level is lower than the surrounding water table and water flows to the lake; and

- *flow-through systems*, where the lake gains water from the groundwater system in some parts and loses water to the groundwater system in other parts.

Both recharge and discharge lakes can change in character over short periods of time (9, *31*). Thus, it is critical to characterize the temporal variations in groundwater flow around lakes when estimating solute loads. At Crystal Lake, WI, which would be classified as a recharge lake on the basis of annual water budgets, short-term inflow of groundwater after spring snowmelt nearly doubled the concentration of dissolved reactive silica and led to a bloom of diatoms within the lake (9). Although the flux of water was small, the inflowing groundwater had a significantly higher concentration of silica than the lake water, and the result was a substantial mass flux of dissolved silica to the lake.

Lakes have also been classified as *seepage* or *drainage* lakes. Such classifications were determined by whether the lakes have surface-water inflows or outflows. From the perspective of understanding interactions between groundwater systems and lakes, these classifications are of little use because they do not define the lakes with regard to water source. Seepage lakes could be recharge lakes, discharge lakes, or flow-through lakes; the criteria that help in understanding the solute budgets and overall chemical character of lakes are undefined. Therefore, use of such terms as seepage and drainage is discouraged.

Further classification of lakes relates to their position within the regional groundwater-flow system. Terminal-lake systems are defined as lakes that function as the discharge point of the regional groundwater-flow system. For *terminal lakes*, water is removed by evaporation and sometimes through surface outflow. These lakes typically evolve into saline lake systems characteristic of the semiarid or arid regions of the world (*32*).

Use of Oxygen-18 and Deuterium in Groundwater–Lake Assessments

Groundwater components of water budgets for lakes are commonly calculated as the residual of average precipitation and evaporation fluxes and changes in lake storage, and this practice leads to considerable uncertainty in calculated values (*33*). This type of budget calculation provides only an estimate for net groundwater flow, because the calculation never separates the inflow and outflow components. Such budgets combine the net groundwater fraction (inflow–outflow) of the lake budget with the errors associated with other components of the lake budget. In many cases, these errors are comparable in magnitude to the individual groundwater-flow components; relative errors greater than 100% are possible. For solute-loading estimates, errors of this magnitude are unacceptable and prevent further understanding of groundwater–lake systems.

Stable-Isotope Mass-Balance Method. The equations presented in this section apply to groundwater–lake systems that are at hydrological and isotopic steady states. Equations that describe isotopic mass balances for non-steady-state systems and forms that pertain to the estimation of evaporation from lakes have been presented by other authors (13, 14).

The water budget of a lake is given by

$$\frac{dV_1}{dt} = P + G_i + S_i - E - G_o - S_o \qquad (9)$$

where V_1 is the lake volume, t is time, P is the precipitation rate, G_i is the groundwater-inflow rate, S_i is the surface-water-inflow rate (including run-off), E is the evaporation rate, G_o is the groundwater-outflow rate, and S_o is the surface-water-outflow rate. An equivalent expression for the isotopic mass budget of a lake is given by

$$\frac{d(\delta_1 V_1)}{dt} = \delta_P P + \delta_{G_i} G_i + \delta_{S_i} S_i - \delta_E E - \delta_{G_o} G_o - \delta_{S_o} S_o \qquad (10)$$

where all the terms have been multiplied by their respective isotopic compositions, given in delta notation.

This chapter applies the isotope mass-balance method to several lakes in northern Wisconsin where streamflows and overland flows are insignificant. Such lakes are typical of the poorly integrated drainage of glaciated regions underlain by moderate to thick glacial deposits. Under these conditions, eq 10 can be simplified to include only terms for precipitation, evaporation, and groundwater. By restricting our analysis to lakes that are at isotopic steady state [i.e., $d(\delta_1 V_1) \sim \delta_1 d(V_1)$] and by assuming that groundwater outflow is isotopically the same as lake water ($\delta_{G_o} = \delta_l$), we can equate eqs 9 and 10 to derive the following expression for the groundwater-inflow rate:

$$G_i = \frac{P(\delta_1 - \delta_P) + E(\delta_E - \delta_l)}{\delta_{G_i} - \delta_1} \qquad (11)$$

In this expression, all of the terms except the isotopic composition of the lake evaporate (δ_E) are directly measurable. The average isotopic composition of the lake evaporate can be calculated from the relation formulated by Craig and Gordon (22),

$$\delta_E = \frac{\alpha^* \delta_l - h\delta_a - \epsilon_T}{1 - h + 10^{-3}\Delta\epsilon} \tag{12}$$

where δ_a is the isotopic composition of local atmospheric moisture, $\epsilon_T = (1 - \alpha^*) + \Delta\epsilon$ is the total fractionation factor, h is the relative humidity normalized to the surface temperature of the lake, and all δ and ϵ values are expressed in per mil.

Sparkling Lake, a groundwater flow-through lake in north-central Wisconsin (Figure 3), is in a nearly ideal situation to test the isotope mass-balance method for estimating the groundwater component of a lake's hydrological budget. Both the isotope mass-balance method and a numerical groundwater-flow model were used (*31, 34*); rates calculated by the two methods were comparable. The major results of the isotope mass-balance study are summarized here.

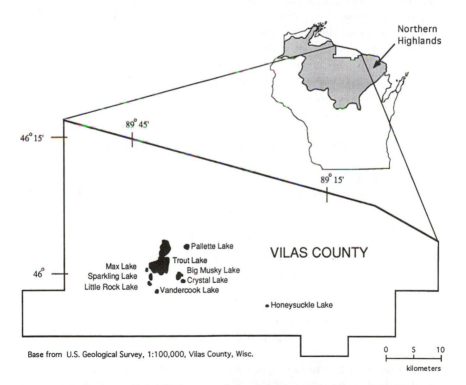

Base from U.S. Geological Survey, 1:100,000, Vilas County, Wisc.

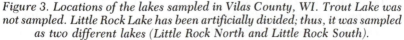

Figure 3. Locations of the lakes sampled in Vilas County, WI. Trout Lake was not sampled. Little Rock Lake has been artificially divided; thus, it was sampled as two different lakes (Little Rock North and Little Rock South).

In the sandy outwash area of northern Wisconsin, permeable soils promote the exchange of water between groundwater systems and lakes. In addition, Sparkling Lake occupies a topographically low position in the local groundwater system; thus, it receives a consistent flow of groundwater, which constitutes a substantial part of the lake's hydrological budget. The average lake depth is 10 m (lake volume/surface area ratio), and the hydraulic residence time (lake volume/total outflow rate ratio) is relatively long (about 10 years). All of these characteristics of Sparkling Lake help to satisfy the assumptions for use in eq 11 and result in accurate groundwater-flow estimates.

In the mass-balance study, the isotopic composition of precipitation at Sparkling Lake was shown to vary sinusoidally, with nearly a 17‰ difference in $\delta^{18}O$ between summer rains and the winter snowpack (-5 to $-22‰$, respectively; *see* Figure 2). The average volume-weighted $\delta^{18}O$ of precipitation was estimated to be $-10.9‰$. Several atmospheric moisture samples were collected during the ice-free periods. Analysis of these samples showed that the local atmospheric moisture is in isotopic equilibrium with precipitation except during July and August, the warmest months. The disequilibrium during these 2 months is believed to be the result of water-vapor contributions from the many nearby lakes, including Lake Superior.

The $\delta^{18}O$ composition of the lake water was virtually invariant during the 2-year sampling period: the average $\delta^{18}O$ value was $-5.75‰$ and the standard deviation was 0.1‰. These values were calculated from samples collected during semiannual turnover periods, when the lake is thermally and chemically homogeneous. During maximum summer and winter thermal stratification, however, epilimnetic waters were observed to be slightly fractionated: the summer $\delta^{18}O$ value was $-5.6‰$ and the winter $\delta^{18}O$ value was $-5.9‰$.

The monthly isotopic composition ($\delta^{18}O$) of lake evaporate from Sparkling Lake was estimated by use of eq 12. Monthly evaporation rate estimates were then used to calculate the weighted average annual $\delta^{18}O$ of lake evaporate, $-16.9‰$.

Ambient groundwater in the Sparkling Lake area is isotopically homogeneous; average $\delta^{18}O$ is $-11.5 \pm 0.3‰$. Downgradient from the lake, however, an easily identifiable plume of isotopically enriched lake water provided substantiating evidence for assumed flow paths based on hydraulic-head measurements. The 0.6‰ difference between average precipitation and groundwater was attributed to the selective recharge of isotopically depleted spring snowmelt compared to isotopically enriched summer precipitation.

By use of eq 11, this isotopic information, and the average annual precipitation and evaporation rates (0.79 and 0.52 m/year, respectively), the average annual groundwater inflow rate to Sparkling Lake was estimated to be 0.27 m/year (expressed as the volumetric flow rate divided by the surface area of the lake). Because Sparkling Lake has no surface-water inflows, the groundwater outflow rate could be estimated as the residual in the hydrological budget, 0.50 m/year. These results were consistent with the results from a three-dimensional groundwater-flow and solute-transport model of the Sparkling Lake system, from which the groundwater-inflow and -outflow rates were estimated to be 0.20 and 0.52 m/year, respectively (28).

Index-Lake Method.   The most difficult aspect of using stable isotopes for estimating hydrological-budget components of lakes is determining the evaporation rate (E) and the weighted-average isotopic composition of the lake evaporate (δ_E) (31). The isotopic composition of water vapor that evaporates from the surface of a lake can be estimated by use of eq 12. This expression shows that the isotopic composition of lake evaporate is controlled by the interactions of the lake with the overlying atmosphere. Measurements of air and water temperatures, relative humidity, and the isotopic composition of ambient atmospheric moisture are needed. Sampling of ambient atmospheric moisture is a tedious and time-consuming process and is rarely done. When atmospheric-moisture measurements are made, as they were at Sparkling Lake (31), it is theoretically possible to extrapolate the results for use on nearby lake systems and to assume that these lakes are affected by the same atmosphere. This kind of extrapolation, whereby the results from a lake whose isotopic balance has been carefully determined are used to estimate hydrological-budget components of nearby lakes, is referred to as the *index-lake method* (10, 35).

The lake district of north-central Wisconsin is a particularly appropriate area for applying the index-lake method. This area contains more than 3000 lakes situated in sandy, glacial outwash soils (36, 37). The region is topographically homogeneous, consisting of a mosaic of similar low-relief watersheds that yield little or no overland runoff to lakes and streams. Therefore, recharge on a regional scale should be relatively uniform and result in a groundwater system with a uniform isotopic composition (28); this uniformity is an underlying assumption of the index-lake method.

The application of the index-lake method presented here is only for lakes that are at hydraulic and isotopic steady states. A lake's steady-state isotopic composition is determined by the long-term averages of δ_a, δ_P, h, P, E, and water and air temperatures, which can vary greatly by the day and season and can vary to some degree annually. Therefore, it is only proper to apply the index-lake method to lakes of similar hydraulic residence time, during which time the averages of these controlling factors are determined.

As discussed previously, the isotopic composition of evaporating water bodies on plots of $\delta^{18}O$ versus δD lie on evaporation lines. The intersection of any evaporation line with the LMWL corresponds to the average composition of water entering the lake. Geographically clustered lakes that meet the requirement of being at isotopic steady state should plot along the same evaporation line, provided they are indeed affected by the same atmosphere and have about the same hydraulic-residence times. Groundwater-rich lakes should fall along the line closer to the meteoric-water line, whereas ground-water-poor lakes containing highly evaporated water should plot farther along the line.

The isotopic compositions of four lakes (*see* Figure 3 for lake locations) in northern Wisconsin that are within 10 km of each other and that have hydraulic-residence times of about 10 years are plotted in Figure 4. The four lakes (Crystal, Pallette, Big Musky, and Sparkling) are groundwater flow-through lakes and have no surface inflows or outflows. Thus they have the same hydrological-budget components (they receive water from precip-

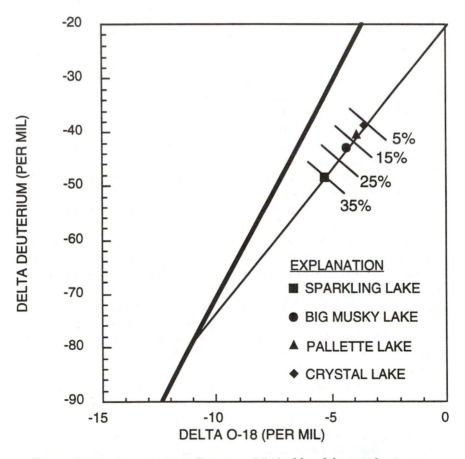

Figure 4. Isotopic compositions (δD versus $\delta^{18}O$) of four lakes in Vilas County, WI. The thick line is the local meteoric-water line (LMWL) determined from the data shown in Figure 2; the thin line is the evaporation line determined by the best-fit line for the four lakes shown here. Line segments plotted on the evaporation line represent specified ratios [G_i/P] calculated using eq 13.

itation and groundwater inflow, and lose water to evaporation and ground-water outflow). A best-fit evaporation line for these four lakes has an $R^2 = 0.997$ and intersects the LMWL at $\delta^{18}O = -11.0‰$, a value close to the measured input-water composition for the index lake (Sparkling Lake), $-11.1‰$. This close agreement indicates that the hydrological budgets of the lakes chosen for this application can be determined by the index-lake method.

Under the assumption of isotopic steady state, eqs 9 and 10 can be combined and solved for the total surface- and groundwater-outflow rate (O) relative to the evaporation rate (E), as follows:

$$\frac{O}{E} = \frac{\dfrac{S_i}{E}(\delta_{G_i} - \delta_{S_i}) + \dfrac{P}{E}(\delta_{G_i} - \delta_P) + \delta_E - \delta_{G_i}}{\delta_{G_i} - \delta_l} \tag{13}$$

where O is $G_o + S_o$. For lakes with no surface flows (S_i and S_o are equal to 0), this expression is simplified. If long-term averages of P and E for the index lake are assumed to be the same for those of the other lakes, the only remaining unknown in each lake's hydrological budget, G_i, can easily be determined by setting eq 9 equal to 0. In the same manner, one can specify a groundwater-inflow rate and solve eq 13 for δ_l to estimate the steady-state isotopic composition of a lake. Specified ratios of annual groundwater inflow to annual precipitation rates (G_i/P) that approximately bracket the compositions of these three lakes and Sparkling Lake are plotted on the evaporation line in Figure 4. Groundwater-inflow rates for these lakes were estimated by use of eq 11 and are listed in Table I.

The accuracy of the estimates for Big Musky, Pallette, and Crystal Lakes depends on the accuracy of the groundwater-inflow estimate for Sparkling Lake (the index lake) and the validity of the assumptions for the index-lake method. The estimated error for groundwater inflow to Sparkling Lake, ±7 cm/year (*31*), represents a minimum value for the errors associated with the estimates for the other lakes. Therefore, for lakes that receive relatively small amounts of estimated groundwater inflow, the relative error associated with the estimate increases and the utility of the method is reduced.

Table I. Comparison of Groundwater Inflow Rates

Lake	Isotope Method	Solute Method
Sparkling Lake	0.29	0.24
Big Musky Lake	0.15	0.14
Pallette Lake	0.10	0.13
Crystal Lake	0.07	0.03
Honeysuckle Lake	NA[a]	0.01

NOTE: All values are given in meters per year.
[a]Not available.

Reliable estimates of relative humidity are critical for use of the isotope mass-balance method; however, humidity data are difficult to interpret and commonly not available for a specific study area. The index-lake method provides a means for checking the accuracy of these data and the validity of their use in isotope hydrology. The equation that describes the steady-state isotopic composition of a lake (35) is

$$\delta_1{}^s - \delta_l = \frac{E[h(\delta_a - \delta_l) + \epsilon]}{I(1 - h) + hE} = \Delta\delta_1{}^s \tag{14}$$

where $\delta_1{}^s$ is the steady-state isotopic composition of the lake; I is the total inflow rate from precipitation, groundwater, and streams; and δ_l is the properly weighted isotopic composition of all the inflows. If equations for both δD and $\delta^{18}O$ are developed from eq 14, an expression for the slope of the evaporation line under steady-state conditions for all of the parameters in eq 14 can be derived, as follows:

$$E_{slope}\frac{\Delta\delta^s_{l,D}}{\Delta\delta^s_{l,18}} = \frac{h(\delta_a - \delta_l)_D + \epsilon_D}{h(\delta_a - \delta_l)_{18} + \epsilon_{18}} \tag{15}$$

where E_{slope} is the slope of the evaporation line, and the subscripts D and 18 refer to deuterium and oxygen-18 values, respectively.

Equation 15 can be used to estimate the average relative humidity over a given lake by solving this equation for various values of relative humidity and atmospheric moisture composition (Figure 5). The three lines in Figure 5 were calculated for δ_a (oxygen-18 and deuterium) equal to $(-19.9, -150)$, $(-20.4, -154)$, and $(-20.6, -156)$, which represent moisture compositions in equilibrium with average annual precipitation, the measured moisture composition at Sparkling Lake (31), and a moisture composition in equilibrium with local groundwater, respectively. It is assumed that this range of δ_a values brackets the actual value. The slope of the evaporation line from Figure 4 is 5.3, which, if plotted on Figure 5, gives a range for the average annual humidity of 0.80–0.83; this computed range agrees well with the field-measured value, 0.82 (31). This close agreement also corroborates the isotope mass-balance calculations, which depend on accurate humidity data.

Application of Isotopic Methods to Non-Steady-State Groundwater–Lake Systems. Isotopic compositions of lakes whose hydraulic-residence times are relatively short (about 2 years or less) vary seasonally (32). Seasonal response occurs whenever a significant mass of water of a different isotopic composition is either added to or removed from the lake. Seasonal variations in P, E, G_i, δ_P, δ_a, and δ_E are the principal driving forces behind observed variations in δ_l for isotopically non-steady-state systems. In northern Wisconsin, many lakes are isotopically non-steady-state.

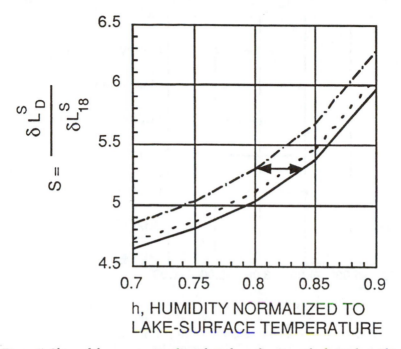

Figure 5. *Slope of the evaporation line plotted as a function of relative humidity by use of eq 15. The three curves shown are for different assumed average isotopic compositions of local atmospheric moisture. The assumed conditions are equilibrium with average precipitation (-19.9 [$\delta^{18}O$], -150 [δD]), average of the measured values at Sparkling Lake (-20.4 [$\delta^{18}O$], -154 [δD]), and equilibrium with local groundwater (-20.6 [$\delta^{18}O$], -156 [δD]). The arrow shows the range for average annual relative humidity (during the ice-free season) for the three atmospheric humidity scenarios if the slope of the evaporation line is 5.3 (from Figure 4).*

Observed $\delta^{18}O$ variation for several shallow lakes in northern Wisconsin and Sparkling Lake (isotopically steady-state) are shown in Figure 6. The most negative value for each of these lakes represents an early spring water sample, whereas the least negative value is from the fall. These data demonstrate the significant seasonal variations in isotopic composition that may arise in relatively shallow lakes when compared to isotopically invariant lakes such as Sparkling Lake. Attempts to apply the isotope mass-balance method for estimating annual groundwater-exchange rates for these lakes would be challenging, because determination of the average annual δ_1 would be difficult. On the other hand, seasonal variations in the isotopic compositions of lakes can provide valuable insight into processes (such as exchanges of water with the atmosphere) that would otherwise be imperceptible in isotopically steady-state systems.

ENVIRONMENTAL CHEMISTRY OF LAKES AND RESERVOIRS

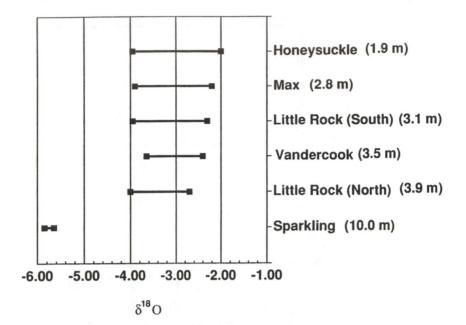

Figure 6. Observed range of $\delta^{18}O$ values for five non-steady-state lake systems and the steady-state Sparkling Lake system. Mean depth of each lake is shown in parentheses after its name.

Solute-Tracer Mass-Balance Method

In northern Wisconsin, and presumably elsewhere, groundwater systems near lakes that receive relatively little groundwater inflow commonly undergo flow reversals. Flow reversals of a few days or weeks occur seasonally, whereas reversals lasting months or years can occur during drought (*31, 38*). During a flow reversal, lake water seeps into the groundwater system; then, following a significant recharge event, the hydraulic gradient reverses and groundwater discharges to the lake. Because the water that seeps into the aquifer has the same isotopic composition as lake water, the mass-balance method will not account for this water flux. Thus, another method must be applied to account for this exchange of water between the groundwater system and the lake.

In some situations natural-solute tracers can be used in the same manner as stable-isotope tracers to estimate hydrological-budget components for lakes (*7*). As dilute recharge waters from precipitation enter the aquifer, dissolution reactions result in net additions of dissolved solids. In northern Wisconsin, where glacial outwash sediments are lacking in carbonate minerals, silicate hydrolysis is the dominant dissolution reaction (*39*). These reactions result in net additions of major cations (Ca^{2+}, Mg^{2+}, Na^+, and K^+) and bicarbonate (HCO_3^-) to water. Through this process, groundwater discharge becomes the dominant source of cations and alkalinity for northern Wisconsin lakes (38). Lake water that seeps into the aquifer would immediately begin increasing in dissolved solute concentrations. This increase could be accounted for by the solute mass-balance method during the next flow reversal. Therefore, by measuring solute concentrations in precipitation, groundwater, and lake water, the solute mass-balance equation that is directly analogous to eq 11 can be derived:

$$G_i = \frac{P(C_1 - C_P) - E(C_l)}{C_{G_i} - C_1} \qquad (16)$$

where C_P, C_{G_i}, and C_1 are the solute concentrations in precipitation, groundwater inflow, and the lake, respectively. This relation is only applicable to systems that are compositionally at steady state and in which the solute acts nearly conservatively within the lake.

Use of solute tracers has one particular advantage over the use of stable isotopes, in that the solute concentration in lake evaporate is assumed to be equal to zero and thus is not a variable in eq 15. This assumption means that the errors associated with estimating the isotopic composition of lake evaporate do not apply to the solute tracer method. On the other hand, the chemistry of groundwater is much more heterogeneous than its isotopic composition (28). In addition, although contamination problems are of concern for solutes at low concentrations, isotope samples are virtually unaffected by contamination. Thus, estimates of average isotopic composition of a groundwater system are much more accurate reflections of the system than are average chemical compositions. The strengths and weaknesses of these two methods are complimentary, and determination of the best method for a particular site should be left to the investigator.

Groundwater-inflow rates as calculated by the solute and isotope mass-balance methods for several northern Wisconsin lakes are listed in Table I. Dissolved calcium was used as the solute tracer because it is the constituent whose concentration differs the most between groundwater and precipitation, the two input components to be separated by the method. In addition, calcium is nearly conservative* in the soft-water, moderately acidic to circum-neutral lakes in northern Wisconsin. Results from the two methods agree relatively well, except for Crystal Lake, where groundwater-flow reversals are frequent.

---

*Most of the dissolved calcium in groundwater in northern Wisconsin is the result of silicate hydrolysis of the aquifer materials. The assumption of conservancy is accurate only because of the relatively slow rates of silicate dissolution. The presence of more soluble calcium-containing minerals, such as calcite or gypsum, would invalidate assumptions of conservancy and would lead to significant errors in solute budgets.

Equation 16 can be set equal to zero (i.e., no groundwater inflow) and solved for C_1, as follows:

$$C_1^0 = \frac{PC_P}{P - E} \tag{17}$$

where C_1^0 is the solute concentration for a lake with no groundwater inflow. Substitution of values for northern Wisconsin [$P = 0.79$ m, $E = 0.52$ m, and $C_P = 0.2$ mg/L of Ca (28)] in eq 17 yields a dissolved calcium concentration of 0.7 mg/L. This value is close to that of Honeysuckle Lake (0.6 mg/L of Ca), a nearby lake that has no groundwater inflow (William Rose, U.S. Geological Survey, oral communication; 1991). The close agreement between the calculated and the measured values also increases confidence in the values of P, E, and C_P used in the solute mass-balance method.

Summary and Conclusions

Experience gained from studies in northern Wisconsin indicates that stable isotopes can be valuable tools for assessing the hydrology of groundwater–lake systems. Several conclusions are evident:

- Mass-flux calculations require knowledge of exchange rates between groundwater systems and lakes. Even though the groundwater-inflow rate may represent a small fraction of the hydrological budget, it can dominate the solute budget.

- In the simplest case, groundwater-flow rates for lakes at isotopic steady state (or those with relatively long hydraulic-residence times) can be estimated from data on average annual precipitation rates; average annual evaporation rates; the isotopic compositions of precipitation, lake water, and inflowing groundwater; and relative humidity and lake temperature.

- Where lakes are geographically clustered, the index-lake method can be useful for estimating the groundwater components of additional lake budgets after calculating the budget for the index lake.

- Annual isotopic variation is substantial in shallow lakes with relatively short residence times. These variations reflect the dominance of summer evaporation and spring snowmelt. In addition, these variations may prove useful for examining atmospheric exchange rates with lakes and for quantifying periods of episodic water input.

- The isotope mass-balance method is not as useful for estimating groundwater-flow rates for groundwater-poor lakes as it is for lakes that receive substantial quantities of groundwater. Solute tracers, such as dissolved calcium, may be useful in assessing groundwater-poor lakes and may yield reliable results if the assumption of conservancy is maintained.

- In the lakes investigated in northern Wisconsin, groundwater-inflow rates ranged from negligible to nearly 25% of the total inflow (G_i / Σ inflow terms).

The purpose of this chapter is not to promote the replacement of traditional physically based methods of assessing groundwater–lake systems with isotopic methods, but rather to demonstrate the utility of isotopic techniques. Physically based methods can provide more detailed information on the spatial and temporal variability of a groundwater–lake system than isotopic approaches can provide. Regardless of the method chosen, however, an adequate number of piezometers is necessary to ensure that groundwater samples are collected from upgradient areas.

Ideally, groundwater–lake investigations would include both isotopic and physically based approaches. In areas where results of the two methods agree, considerable confidence can be placed on the interpretations; in areas where they disagree, future research could be focused on explaining the discrepancies. The use of isotopic and other geochemically based methods in combination with physical methods could substantially improve our understanding of the interaction of lakes with groundwater systems and of some of the chemical mechanisms that operate within these systems.

Acknowledgments

Support for the research reported here was provided by the U.S. Geological Survey Water Energy and Biogeochemical Budget Program (WEBB) and the National Science Foundation, Long-Term Ecological Research Program (LTER), North Temperate Lakes. We gratefully acknowledge Blair Jones of the U.S. Geological Survey for contributing some of the isotopic data.

References

1. Bear, J. *Hydraulics of Groundwater;* McGraw-Hill: New York, 1979; p 569.
2. Freeze, R. A.; Cherry, J. A. *Groundwater;* Prentice-Hall: Englewood Cliffs, NJ, 1979; p 604.
3. McBride, M. S.; Pfannkuch, H. O. *J. Res. U.S. Geol. Surv.* 1975, *3*(5), 505–512.
4. Munter, J. A.; Anderson, M. P. *Ground Water* 1981, *19*(6), 608–616.
5. Winter, T. C.; Pfannkuch, H. O. *J. Hydrol. (Amsterdam)* 1984, *75*, 239–253.
6. Krabbenhoft, D. P.; Anderson, M. P. *Ground Water* 1986, *24*(1), 49–55.
7. Stauffer, R. E. *Environ. Sci. Technol.* 1985, *19*, 405–411.
8. Bowser, C. J.; Krabbenhoft, D. P. *EOS; Trans. Am. Geophy. Union* 1987, *68*(44).
9. Hurley, J. P.; Armstrong, D. E.; Kenoyer, G. J.; Bowser, C. J. *Science (Washington, D.C.)* 1985, *227*, 1576–1578.
10. Dincer, T. *Water Resour. Res.* 1968, *4*(6), 1289–1306.
11. *IAEA Isotopes in Lake Studies;* International Atomic Energy Agency: Vienna, Austria, 1979; p 290.

12. Gat, J. R. In *Stable Isotope Hydrology;* Gat, J. R.; Gonfiantini, R., Eds.; International Atomic Energy Agency: Vienna, Austria, 1981; pp 203–222.
13. Gilath, C.; Gonfiantini, R. In *Guidebook on Nuclear Techniques in Hydrology;* International Atomic Energy Agency: Vienna, Austria, 1983; pp 129–161.
14. Gonfiantini, R. In *Handbook of Environmental Isotope Geochemistry;* Fritz, P.; Fontes, J. Ch., Eds.; Elsevier Scientific: Amsterdam, Netherlands, 1986; Vol. 2, pp 113–167.
15. Fritz, P.; Fontes, J. Ch. In *Handbook of Environmental Isotope Geochemistry;* Fritz, P.; Fontes, J. Ch., Eds.; Elsevier Scientific: Limerick, Ireland, 1980; Vol. 1, pp 1–19.
16. Hoefs, J. *Stable Isotope Geochemistry;* Springer-Verlag: Berlin, Germany, 1980.
17. Craig, H. *Science (Washington, D.C.)* **1961,** *133,* 1833–1834.
18. Dansgaard, W. *Tellus* **1964,** *16,* 436–468.
19. Yurtsever, Y.; Gat, J. R. In *Stable Isotope Hydrology;* Gat, J. R.; Gonfiantini, R., Eds.; International Atomic Energy Agency: Vienna, Austria, 1981; pp 103–142.
20. Gat, J. R. In *Handbook of Environmental Isotope Geochemistry;* Fritz, P.; Fontes, J. Ch., Eds.; Elsevier Scientific: Limerick, Ireland, 1980; Vol. 1, pp 21–47.
21. Majobe, M. *J. Chem. Phys.* **1971,** *197,* 1423–1436.
22. Craig, H.; Gordon, L. I. In *Stable Isotopes in Oceanographic Studies and Paleotemperatures, Spoleto;* Tongiorgi, E., Ed.; Consiglio Nazionale delle Ricerche: Pisa, Italy, 1965; pp 9–130.
23. Gat, J. R.; Gonfiantini, R. *Stable Isotope Hydrology;* International Atomic Energy Agency: Vienna, Austria, 1981.
24. Zimmermann, U. In *Isotopes in Hydrology;* International Atomic Energy Agency: Vienna, Austria, 1967; pp 567–585.
25. Allison, G. B.; Barnes, C. J.; Hughes, M. W.; Leaney, F. W. J. In *Isotope Hydrology;* International Atomic Energy Agency: Vienna, Austria, 1983; pp 105–123.
26. White, J. W. C.; Cook, E. R.; Lawrence, J. R.; Broecker, W. S. *Geochim. Cosmochim. Acta* **1985,** *49,* 237–249.
27. Turner, J. V.; Arad, A.; Johnston, C. D. *J. Hydrol. (Amsterdam)* **1987,** *94,* 89–107.
28. Krabbenhoft, D. P.; Bowser, C. J.; Anderson, M. P.; Valley, J. W. *Water Resour. Res.* **1990,** *26*(10), 2445–2453.
29. Kennedy, V. C.; Kendall, C.; Zellweger, G. W.; Wyerman, T. A.; Avanzino, R. J. *J. Hydrol. (Amsterdam)* **1986,** *84,* 107–140.
30. Barnes, C. J.; Allison, G. B. *J. Hydrol. (Amsterdam)* **1984,** *74,* 119–135.
31. Webster, K. E.; Newell, A. D.; Baker, L. A.; Brezonik, P. L. *Nature (London)* **1990,** *347*(6291), 374–376.
32. Gat, J. R. In *Lakes II;* Lerman, A., Ed.; Springer-Verlag: Berlin, Germany, in press.
33. Winter, T. C. *Water Resour. Bull.* **1981,** *17*(1), 82–115.
34. Krabbenhoft, D. P.; Anderson, M. P.; Bowser, C. J. *Water Resour. Res.* **1990,** *26*(10), 2455–2462.
35. Gat, J. R. *Water Resour. Res.* **1971,** *7*(4), 980–993.
36. Frey, D. G. *Limnology in North America;* University of Wisconsin Press: Madison, WI, 1966; p 255.
37. Attig, J., Jr. *Pleistocene Geology of Vilas County, Wisconsin;* Information Circular 50; Wisc. Geol. Nat. Hist. Surv.: Madison, WI, 1985; p 32.
38. Kenoyer, G. J.; Anderson, M. P. *J. Hydrol. (Amsterdam)* **1989,** *109,* 287–306.
39. Kenoyer, G. J.; Bowser, C. J. *Water Resour. Res.* **1992,** *28,* 579–589.

RECEIVED for review September 26, 1991. ACCEPTED revised manuscript April 20, 1992.

Ecosystem-Scale Experiments

The Use of Stable Isotopes in Fresh Waters

George W. Kling

Department of Biology, University of Michigan, Ann Arbor, MI 48109–1048

Experimental studies using additions of stable isotopes of nitrogen and carbon to an arctic lake indicated that new primary production rather than terrestrial detritus supports most animals in the planktonic food web and to a lesser degree in the benthic food web. The lake was divided by a curtain, and one half was fertilized with N and P through a 6-week experiment. $^{15}NH_4Cl$ was added to both sides to label algae; terrestrial detritus remained unlabeled. Although nutrients cycled more quickly in the fertilized treatment, the trophic pathways of nitrogen flow were unaltered by fertilization. The retention time of nitrogen in the ecosystem was about 3 years in both control and fertilized treatments. ^{13}C-leucine additions to mesocosms indicated that phytoplankton make direct use of amino acids and that some macrozooplankton derive nutrition from the microbial food web.

THE SCIENCE OF ECOLOGY emerged at the turn of the last century and brought with it the experimental approaches that were already central to the study of physiology (*1–3*). Manipulations of whole aquatic ecosystems—excluding aquaculture, which dates back 2500 years (*4*)—developed more slowly, mainly because of difficulties associated with increased biotic complexity and physical scale in larger systems. One technique initially used to overcome the problems of complexity, scale, and replicability was creation of controlled microcosms that embodied a more or less natural representation of the whole system (*5, 6*).

The earliest large-scale experiments began in the 1940s. These experiments included perturbations, in which fertilizers or poisons were added in an attempt to alter entire food webs (7–12), and tracers, in which minute additions of radioisotope were used to study the cycling of phosphorus in lakes (13, 14). In the following years, as ecologists recognized the power of these experiments to integrate processes across trophic levels and as our need to understand the effects of major environmental perturbations such as eutrophication or acid rain became acute, more ecosystem-scale experiments were performed (15–21).

A clear understanding of processes in natural systems, which is critical to the interpretation of many such disturbance studies, can be difficult to achieve. Although radioisotopes facilitate the nonintrusive study of biogeochemical processes in undisturbed ecosystems (22–25), radioisotope applications may be impossible for a variety of operational and political reasons.

This chapter discusses an alternative approach that uses additions of stable isotopes as chemical tracers of biological and geochemical processes in ecosystems. The approach is illustrated by two experiments; the first uses ^{15}N additions to compare the importance of terrestrial detritus as nutrition for pelagic and benthic organisms in both a fertilized and a control setting, and the second uses ^{13}C additions to test the role of the microbial food web in passing carbon and nitrogen to higher trophic levels. The knowledge gained from these initial large-scale experiments with stable isotopes has been expanded to include at least four other similar studies:

1. a 2-month continuous addition of ^{15}N to a fourth-order river in arctic Alaska (Kling, G. W.; Peterson, B. J.; unpublished data);

2. a series of pulsed additions of ^{15}N to a 10-ha hardwood forest catchment in Maine (26);

3. a pulsed addition of ^{34}S to intact sediment microcosms (27); and

4. a continuous addition of ^{13}C to mesocosms in the Baltic Sea (Hobbie, J. E.; Fry, B.; unpublished data) and in an arctic lake (Hobbie, J. E.; Kling, G. W.; unpublished data).

Stable Isotopes in Ecological and Ecosystem Studies

Stable isotopes of H, He, and O, as well as experimental additions of radioisotopes and neutron-activatable halogens, are widely applied as tracers in hydrology or hydrodynamics. Their use is described more fully elsewhere (28–33).

Understanding of biogeochemical cycling and trophic interactions is often hindered by the nature of budgetary approaches to ecosystem study (16). For example, many estimates of element flux are made by difference (e.g., between inflow and outflow) or are inferred from indirect evidence such as carbon-to-nitrogen ratios. Direct measurement of element contents in all important pools in an ecosystem is at best time-consuming and methodologically difficult.

Naturally occurring stable isotopes of C, N, and S have been used extensively for over a decade as direct tracers of element cycling in marine and terrestrial food webs (34–39). Carbon and sulfur isotopes fractionate very little between food and consumer; thus their measurement indicates which primary producers or detrital pools are sources of C and S for consumers. For example, a study of plants and animals in Texas sand dunes showed that insect species had $\delta^{13}C$ values either like those of C_3 plants or like those of C_4 plants (–27 and –13‰, respectively). Rodent species had intermediate values near –20‰ that indicated mixed diets of both C_3 and C_4 plants (40). The ^{13}C measurements, used in simple linear mixing models, proved to be quick and reliable indicators of which plant sources provided the carbon assimilated by higher trophic levels.

Subsequent marine studies have shown that there are often several important plant sources of carbon, and a multiple stable-isotope approach has been widely adopted to resolve this more complex situation. Stable isotopes of sulfur, successfully combined with carbon measurements, have been particularly valuable for tracing the importance of detrital foods that are difficult to identify or quantify visually (41). The information gained in these isotopic studies has allowed ecologists to powerfully verify or refute hypotheses about the importance to consumers of certain food sources (42).

Much of our present knowledge concerning nitrogen use and recycling in aquatic systems is based on ^{15}N isotope-tracer techniques performed in bottles or small enclosures (43, 44). Natural-abundance measurements of ^{15}N have similarly expanded our understanding of trophic interactions and food webs as well as nitrogen flow (45–48). Studies with stable nitrogen isotopes show a ^{15}N content in consumers that consistently increases with increasing trophic level. Preferential loss of ^{15}N-depleted nitrogen in urine and feces typically results in the animal becoming enriched in ^{15}N relative to the diet; this enrichment averages 3.3‰ in many systems (47).

The degree of ^{15}N enrichment between particulate organic matter (POM) and consumers can be used to accurately estimate trophic level (47, 49). Thus isotopic studies permit identification of consumer groups that form nutritional guilds via ^{13}C and ^{34}S analyses, while establishing trophic-level interactions within these groups via ^{15}N analyses. To date, very few combined ^{13}C and ^{15}N analyses of trophic structure in freshwater systems have been published (47, 50–52).

The foregoing review is centered on natural-abundance measurements of stable isotopes. In many cases, however, several potential foods or prey items are indistinguishable in their natural-abundance isotopic ratios, and thus isotopic analyses provide poor resolution. One strategy for overcoming this problem is to purposefully manipulate the isotopic ratios in one or more of the element or biomass pools of interest. This manipulation is most easily accomplished by labeling a nutrient or substrate pool that is required for a biological process. For example, ^{15}N-enriched NH_4^+ or NO_3^- can be used to label primary producers, ^{13}C-enriched amino acids can be used to label bacteria, or ^{34}S-enriched SO_4^{2-} can be used to label the pathways and end products of bacterial sulfate reduction.

The natural abundance of the stable isotopes ^{13}C, ^{15}N, and ^{34}S in the environment relative to ^{12}C, ^{14}N, and ^{32}S is very low (1.12, 0.36, and 4.20%, respectively). In addition, the commercial production of nearly pure ^{13}C, ^{15}N, and ^{34}S (99+%) is relatively easy. Therefore increasing the signal of the heavy isotope in an entire ecosystem can be accomplished economically by adding only tracer amounts. For example, the ^{15}N isotope required for the Lake N2 experiment described here cost $300 and resulted in an increase of 30‰ in the $\delta^{15}N$ value of phytoplankton. Similarly, the ^{15}N used for the Alaskan whole-river experiment cost less than $4000 and resulted in an increase of 850‰ in the $\delta^{15}N$ value of filamentous algae. Such large enrichments provide greatly increased resolution in tracing the pathways of nitrogen flow over time and distance.

Fueling the Food Web

Algal Production. The abundance of inorganic nutrients sets the general level of productivity in most aquatic systems, and fertilization has profound effects on the fate of nutrients and the functioning of ecosystems. Responses of algae to increased nutrients are usually clear and dramatic in lakes and streams (17, 53, 54). The link between algal production and the response of zooplankton and benthos has received somewhat less attention, and the results are more variable. Some systems show increased secondary production after fertilization (55–57); in other systems this response is absent or much delayed (11, 58–61). Still fewer studies have examined the extent to which new algal production is passed further up the food chain (62, 63), or the "top-down" effects of higher level consumers on nutrient cycles (64–66). In general, it remains difficult to predict whether nutrients will move along one pathway rather than another or how the partitioning of nutrients between components of the ecosystem will be regulated.

Terrestrial Detritus. Variability in ecosystem response to fertilization may be attributed in part to the interaction of aquatic and terrestrial ecosystems. In contrast to the many aquatic ecosystems in which higher trophic levels are fueled almost entirely by organic matter originating in the water column, other systems are driven by inputs of particulate and dissolved organic matter from land. The importance of this land–water interaction in regulating system metabolism has been obvious to stream ecologists for some time (*67, 68*). It has stimulated much recent work, especially in boreal lakes (*69–73*) and estuaries (*41, 74*) where loading rates of terrestrial organic matter are high.

In some boreal lakes up to 80% of body carbon in zooplankton may originate from allochthonus detritus (*72*). Evidence suggests that inputs of terrestrial detritus to lowland, coastal arctic lakes support much of the secondary productivity and biomass carbon in aquatic food webs (*62, 75*). In upland arctic lakes far from the coast the importance of terrestrial inputs to organisms is less well known (*49, 76*), although inputs of dissolved inorganic carbon from tundra do influence carbon cycling and the flux of CO_2 and CH_4 from surface waters to the atmosphere (*77*). The following stable-isotope experiments were designed to test the importance of terrestrial organic matter to the food web in an upland arctic lake and to examine the transfer of organic matter from bacteria up the food chain.

Site Description

Lake N2 lies in the northern foothills of the Brooks Range on the North Slope of Alaska (68°38'N, 149°36'W) about 0.5 km northwest of the Toolik Lake Research Station (elevation 724 m above sea level; *see* ref. *78*). The lake is oligotrophic and has a surface area of 1.8 ha, a watershed area of 9 ha, and a maximum depth of 10.7 m (Figure 1). Two channelized inlet streams to the lake flow continuously for several weeks after snowmelt in the spring; flow later in the summer is very low and dependent on episodic rain events. The lake was divided by a curtain in 1985. The smaller of the two inlets enters the western or fertilized side, and the larger inlet enters the eastern or control side of the lake. The outlet is on the fertilized side of the lake. Ice, which covers the lake for 8 months of the year, usually melts in early June during or just following peak runoff. The lake is thermally stratified from late June through August with a thermocline 3–5 m deep and maximum surface-water temperatures of 18 °C (Figure 2).

Surface-water chlorophyll concentrations average only about 1 μg/L during the summer, and the phytoplankton are mostly flagellates belonging to the Chrysophyceae, Cryptophyceae, and Dinophyceae. Bacteria numbers, around 2×10^6/mL, are similar to those found in nearby waters (*79*). Four species of macrozooplankton dominate the open water: *Daphnia longiremis, Bosmina longirostris, Diaptomus pribilofensis,* and *Cyclops scutifer; Polyphemus pediculus* is found in littoral areas. The microplankton consist mainly of several genera of oligotrichs and the rotifers *Keratella cochlearis, Kellicottia longispina,* and *Polyarthra vulgaris* (*80*). The benthos in Lake N2, which is dominated by the snail *Lymnaea elodes,* includes a sparse community of chironomids, tricopterans, and sphaerids (*81*). The benthic-feeding fish, slimy sculpin (*Cottus cognatus*), and the planktivorous arctic grayling (*Thymallus arcticus*) are the only fish in the lake (*82*).

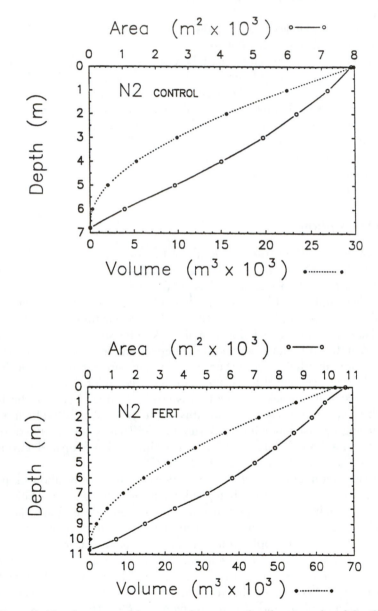

Figure 1. Depth–area and hypsographic curves for the control and fertilized sides of Lake N2.

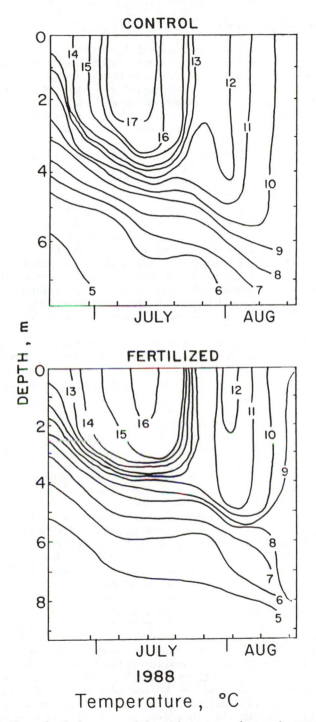

Figure 2. Time–depth diagrams of thermal structure during the 1988 experiment for the control and fertilized sides of Lake N2.

Experimental Methods

In 1985 Lake N2 was divided by a polyethylene curtain as part of a long-term fertilization experiment (83). Each year until 1991 nutrients were added continuously to one side of the lake by using a solar-powered peristaltic pump, beginning on July 1 and ending in mid-August. NH_4NO_3 was added at a rate of 2.91 mmol of N/m^2 per day, and phosphoric acid was added at a rate of 0.23 mmol of P/m^2 per day. The N:P ratio of added nutrients was similar to the Redfield ratio, and areal loading rates were about 5 times the natural loading rate of nearby Toolik Lake (84).

In 1988, 0.335 g of ^{15}N (as $^{15}NH_4Cl$) was added per week to the fertilized side of the lake and 0.113 g of ^{15}N was added per week to the control side. A larger mass of ^{15}N was added to the fertilized side to compensate for the larger water volume and for the N fertilizer additions; the NH_4NO_3 fertilizer was assumed to have a $\delta^{15}N$ of around 0‰. The $^{15}NH_4Cl$ was added continuously to the epilimnion by using a peristaltic pump or a Mariotte bottle during the entire 6-week experiment. The total amount of N in the $^{15}NH_4Cl$ added to the control side over the entire experiment, divided by the average volume of the epilimnion (0–4-m depth; Figure 1), increased the concentration of nitrogen in the epilimnion by <0.002 μmol/L. Because the background concentration of dissolved inorganic nitrogen (DIN; $NH_4^+ + NO_3^-$) in the epilimnion was always much greater than 0.1 μmol/L (the detection limit of the analysis), no nitrogen fertilization effect was attributable to the isotope addition in the control side of the lake.

Physical and chemical measurements were made weekly at a central station in each side of the lake. Water samples were filtered through Whatman GF/C or Gelman A/E glass-fiber filters (1.0-μm pore size). NO_3^- was measured by reduction to NO_2^- in a cadmium column and formation of a pink azo dye, NH_4^+ was measured by using a phenol–hypochlorite method, and soluble reactive phosphate was measured by a molybdenum blue method. After 1990 nutrients were measured by using similar methods on a Technicon Auto Analyzer (83).

Water for dissolved inorganic carbon (DIC) analysis was collected without exposure to the atmosphere and preserved with $HgCl_2$. Samples were acidified with phosphoric acid and purged with nitrogen gas into a vacuum line. CO_2 was collected by cryogenic distillation, and gas pressures were measured with an electronic manometric gauge (total precision ±5 μmol of CO_2/kg). The distilled CO_2 was then analyzed isotopically. Concentrations of dissolved CO_2 (CO_{2diss}) were calculated from DIC, pH, and temperature data by using dissociation constants (85) and carbonate species relations (86). Water for dissolved organic carbon (DOC) was filtered through 0.22-μm filters that were preleached in 50% purified HCl. Samples were acidified with purified HCl to pH 2 and stored refrigerated in glass-stoppered bottles. Aliquots were analyzed for DOC by high-temperature combustion using a platinum catalyst (Ionics).

Particulate organic matter (POM) was collected by filtration of open lake water at 4-m depth onto Whatman quartz or glass-fiber filters (QM-A or GF/C, effective wet pore size ~1.0 μm). The POM is considered to be mainly phytoplankton, although some microheterotrophs and terrestrial detritus will be retained by the filters. In the epilimnion (0–4-m depth) the average molar C:N ratios were 12.4 ± 1.12 SE (standard error of the mean, $N = 16$) in the control side and 8.2 ± 0.65 SE ($N = 28$) in the fertilized side. C:N ratios tended to be higher in the control side, perhaps because of a greater influence of detrital inputs from the main lake inlet or greater nitrogen limitation compared to the fertilized side of the lake. Macrozooplankton were collected from vertical tows with a 100- or 335-μm mesh net, and the adults of major species were separated by hand under a dissecting scope.

Zooplankton population samples for isotope analysis were composites of 50–200 individuals. Population samples are less variable in isotope composition than are samples of individuals. Replicate isotope analyses of composite samples of zooplankton or POM collected at different locations within the lake varied by no more than 0.5‰. Larger organisms such as molluscs, insects, and fish were analyzed individually. Molluscs were soaked in dilute HCl to remove carbonates and then rinsed copiously with distilled water. Fish muscle was analyzed. Sediment trap material was collected in replicate cylinders (11.4-cm diameter, 76.2-cm length) suspended at 4.5-m depth. All isotope samples were dried at 60 °C before analysis.

In 1989 a second stable-isotope-addition experiment was performed in 10-m^3 polyethylene limnocorrals placed in both sides of the lake and not exposed to the sediments. The limnocorrals were large enough so that weekly sampling of macrozooplankton for isotope analyses depleted the animal numbers by less than 10% by the end of the experiment. In this experiment uniformly labeled $^{13}C_6$-leucine and $^{15}NH_4Cl$ were added to the water every 3 days for the first 3 weeks and then once weekly for the remaining 3 weeks of the experiment. The leucine addition was designed to label the bacteria with ^{13}C; the phytoplankton would be labeled by the $^{15}NH_4$. During the 6-week experiment a total of 26.7 and 13.3 mg of $^{13}C_6$-leucine at 85% isotopic enrichment was added to the fertilized and control corrals, respectively. Twice as much leucine was added to the fertilized corral under the assumption that bacterial growth was double that in the control corral. The amount of leucine added increased the background concentration by <0.5 nmol/L per day, and so stimulation of bacterial growth by the added leucine was probably negligible. Inorganic N and P were added to the fertilized limnocorral at the same volumetric loading rates as to the whole lake.

Measurements of carbon and nitrogen stable-isotope ratios were made by using a Finnigan MAT 251 or a Delta S isotope-ratio mass spectrometer. Results are reported versus atmospheric nitrogen (N) or PeeDee Belemnite (C) as standards and calculated as:

$$\delta^{13}C \text{ or } \delta^{15}N \ (‰) = \left(\frac{R_{sample}}{R_{standard}} - 1 \right) \times 10^3$$

where R is $(^{15}N/^{14}N)$ or $(^{13}C/^{12}C)$. Duplicate determinations on the same sample usually differed by <0.2‰.

Results and Discussion

Phytoplankton. The experimental design of the ^{15}N addition to Lake N2 is shown in Figure 3. The background $\delta^{15}N$ value of particulate terrestrial detritus averages 1.1‰ (*49*); the background value of dissolved terrestrial detritus is unknown but assumed to be similar. Because new detrital inputs from the catchment remained unlabeled with ^{15}N, and assuming that terrestrial detritus suspended in the lake did not drastically increase in $\delta^{15}N$ value because of adsorption of $^{15}NH_4Cl$, the relative importance of detrital

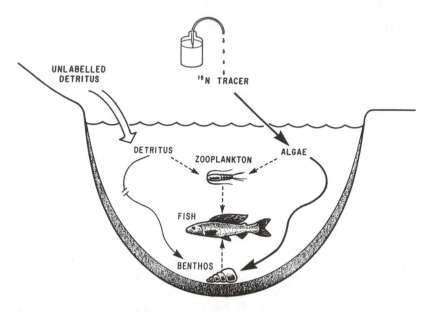

Figure 3. Experimental design. Diagram showing additions of nutrients and tracer amounts of ^{15}N into the divided arctic lake, N2.

inputs versus new algal production could be measured by tracing the isotopic signal through the food web. Phytoplankton uptake of the added $^{15}NH_4$ occurred at a roughly linear rate in both the control and fertilized sides of the lake. By the end of the summer the $\delta^{15}N$ values of POM had increased from a background level of 3–4‰ to around 25–30‰ (Figure 4).

Labeling of the phytoplankton nitrogen pool in this linear fashion was contrary to the expected result. The solid line in Figure 4 shows the expected asymptotic increase in ^{15}N that was calculated by using the DIN concentrations in the epilimnion, the addition rate of $^{15}NH_4$, and the assumption of a 1-week nitrogen-turnover time (87) in the phytoplankton. A similar asymptotic increase in ^{15}N content of phytoplankton, with the phytoplankton reaching isotopic equilibrium in 1 week, was measured in an isotope-addition experiment done in limnocorrals in nearby Toolik Lake (Kipphut, G. W.; Whalen, S. C.; unpublished data).

The observed slow increase in ^{15}N content of POM at the start of the experiment was probably caused by isotopic dilution from unlabeled dissolved inorganic nitrogen (DIN) brought into the lake from the catchment or input from the hypolimnion via cross-thermocline mixing. Horizontal mixing in the epilimnion is rapid in this small lake; samples of moss and periphyton from next to the ^{15}N drippers and on the opposite shores of the lake were equally labeled.

Inputs from the hypolimnion can be estimated by using measured temperatures (Figure 2) to calculate Brünt–Väisälä frequencies (N^2); N^2 values

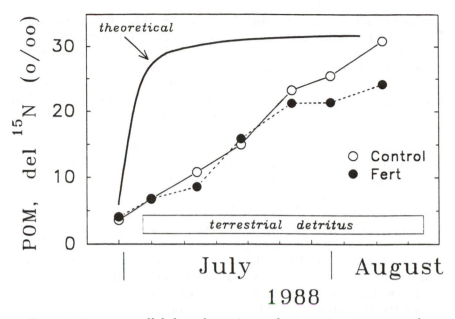

Figure 4. Time course of labeling of POM (particulate organic matter, primarily phytoplankton) with tracer additions of ^{15}N in the fertilized and control sides of Lake N2. The solid line represents the predicted labeling, given the turnover time of phytoplankton. Terrestrial detritus remained unlabeled throughout the experiment.

were $1.4–7.8 \times 10^{-3}$ s^{-2}. These frequencies were then related to vertical diffusion coefficients (K_z) by using an empirical relationship developed for lakes similar in size to Lake N2 (88, 89). Calculated values of K_z for mass were $1.7–6.2 \times 10^{-4}$ cm^2/s across the thermocline in both sides of the lake. With the maximum gradient of DIN concentrations across the thermocline during the experiment (1.23 μmol/L per meter; 83) and the maximum K_z, the total flux of DIN into the epilimnion over the course of the experiment was 3.56 mol of N. Assuming that this unlabeled N had a $\delta^{15}N$ value of 0‰, the dilution effect would decrease the observed 30‰ value in the POM by less than 2‰.

This scenario should be considered as a maximum effect of the hypolimnetic contribution because it does not include regeneration of ^{15}N in the hypolimnion from sinking phytoplankton and the subsequent transport of this ^{15}N back into the upper water column. Clearly, this kind of dilution is insufficient to account for the slow labeling of POM at the start of the experiment.

The second and more probable cause of the unexpected linear increase in $\delta^{15}N$ of POM is the input of unlabeled N from the catchment. Because a water budget for Lake N2 is not available, only a rough estimate of the catchment contribution is possible. With an average precipitation of 2 mm/day, a 50% loss of water to evapotranspiration and soil storage, and a DIN concentration of 10 μM in the input water (78), a total of about 45 mol of N could have been added to Lake N2 during the experiment. Such an input would decrease a 30‰ POM value by nearly 12‰. Thus inputs of N from the catchment probably are the right order of magnitude to explain the observed linear increase in ^{15}N labeling of the phytoplankton seen in Figure 4.

Although the observed rate of labeling of POM with ^{15}N (the slope of the line indicates the increases) was similar between the control and fertilized sides, the rate may be affected by different mechanisms within the lake. For example, large differences in the input of unlabeled nitrogen from the catchment between the two sides could differentially affect the rate of ^{15}N uptake. Similarly, differences in the degree of N versus P limitation could affect the uptake rate by algae. If the control side was always strongly P-limited then the rate of nitrogen uptake may be lower than that expected in the fertilized side, where additions of N plus P at the Redfield ratio would relieve P limitation. Because data on hydrological budgets or nutrient-limitation experiments are unavailable for Lake N2, caution should be used in interpreting the similar rates of ^{15}N labeling of the phytoplankton.

Zooplankton.   Two distinct patterns of nitrogen isotope content are observed among the zooplankton during the summer. At the start of the summer, prior to the ^{15}N additions, the relative ordering of the zooplankton with respect to ^{15}N content resulted from the inherent trophic fractionation of ^{15}N; organisms higher in the food chain are enriched in ^{15}N (Figure 5). *Diaptomus* is exclusively herbivorous (49, 82), but *Daphnia* is also known to consume bacteria in nearby Toolik Lake. The slight variations in δ^{15}N values among the herbivorous zooplankton may be due to selective feeding on phytoplankton species with different ^{15}N contents (90). The magnitude of this potential effect in Lake N2 is unknown because individual species of phytoplankton were not analyzed.

Although *Cyclops* is the most widespread copepod in lakes of this region (78), its diet is incompletely understood. *Cyclops* had the highest ^{15}N content of all zooplankton, and in the fertilized side of the lake *Cyclops* was nearly one full trophic level enriched (~3‰) above *Daphnia*. Thus *Cyclops* may function as an omnivore or a carnivore in this lake, and a second experiment was performed to address this question. Overall, these natural-abundance isotopic values are similar to the δ^{15}N values reported for zooplankton in other arctic lakes (49, 52).

By the end of the summer all zooplankton were enriched in ^{15}N, and *Daphnia* had δ^{15}N values similar to those of POM (Figures 4 and 5). Because zooplankton are labeled on a time scale that depends on their tissue-turnover time, the final ^{15}N values represent equilibration with the ^{15}N content of

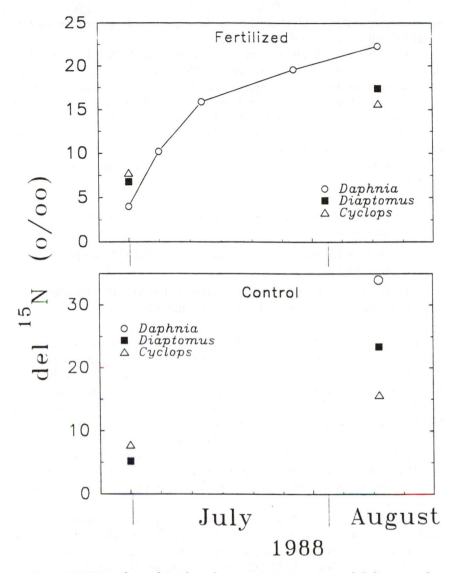

Figure 5. δ¹⁵N *values of* Daphnia longiremis, Diaptomus pribilofensis, *and* Cyclops scutifer *during the course of the 1988 experiment. Initial sampling on July 1 occurred just before the* ¹⁵N *additions began.*

POM at some earlier time in the summer. Exact measures of the fraction of POM in the zooplankton diet or, conversely, measures of tissue-turnover time in the zooplankton are possible only if the ¹⁵N pool in POM has come to equilibrium. Even though this equilibrium in the POM was not reached, it is clear that the growth and maintenance of these zooplankton depended mainly on new algal production during the experiment rather than on terrestrial detritus washed into the lake.

Two of the genera, *Diaptomus* and *Cyclops*, had $\delta^{15}N$ values less than those of the phytoplankton at the end of the experiment. In this case, an assessment of the importance of phytoplankton versus detritus as food must consider several competing explanations for the patterns of isotope content in the zooplankton. In both control and fertilized sides, the relative order of increasing ^{15}N content in the zooplankton was reversed at the end of the summer, compared to the order at the start of the experiment (Figure 5). *Daphnia* now had the highest $\delta^{15}N$ values, followed by *Diaptomus* and then *Cyclops*. This reversal may be due to

1. species-specific differences in growth and nitrogen-turnover time in the tissues,

2. the time lag involved in moving nitrogen through the planktonic food web, or

3. the differential reliance on phytoplankton versus unlabeled foods such as terrestrial detritus for nutrition.

It is unlikely that the first possibility can account for much of the variance because the specific growth rates of arctic zooplankton are fairly similar and, if anything, *Cyclops* is the fastest growing (91). The second possibility is based on the fact that organisms higher in the food chain take longer to receive the ^{15}N from lower trophic levels. This explanation is consistent with the preexperiment assignment of trophic level based on $\delta^{15}N$ values although, as before, feeding on phytoplankton species with differing ^{15}N contents could produce a similar result. The third possibility is that *Cyclops* and *Diaptomus* feed less on phytoplankton and more on terrestrial detritus or on another unlabeled food source like microheterotrophs that filter mostly bacteria (79).

Analyses of carbon isotopic composition were performed as an independent measure of potential food sources for zooplankton. The $\delta^{13}C$ values of terrestrial detritus and of littoral zone emergents such as *Carex* ranged from −26.0 to −29.2‰ (Table I; 49, 92). By comparison, the zooplankton were quite depleted in ^{13}C; their $\delta^{13}C$ values were more similar to phytoplankton and ranged from −32.1‰ in *Daphnia* to −41.0‰ in *Cyclops* (Table I).

A trend of ^{13}C enrichment developed in POM as the summer progressed, especially in the fertilized side, where $\delta^{13}C$ values increased to around −25‰ (Table I). The ^{13}C content of zooplankton in the fertilized side mirrored this trend, but the absolute ^{13}C enrichment was less than that found in POM. The trend in POM is caused in small part by increasing $\delta^{13}C$-DIC values over the same time period (Table II). More importantly, increasing algal biomass and uptake of CO_2 lowered dissolved CO_2 concentrations to

Table I. $\delta^{15}N$ and $\delta^{13}C$ Values for Plants and Animals in Lake N2

Sample		Control			Fertilized		
		Date (1988)	$\delta^{15}N$ (‰)	$\delta^{13}C$ (‰)	Date (1988)	$\delta^{15}N$ (‰)	$\delta^{13}C$ (‰)
POM (4-m depth)		Jun 30	3.6	−35.9	Jun 30	4.1	−35.6
		Jul 5	6.8	−33.1	Jul 5	6.8	−33.6
		Jul 12	10.8	−[a]	Jul 12	8.6	−31.2
		Jul 19	15.0	−31.7	Jul 19	15.9	−27.9
		Jul 27	23.3	−29.0	Jul 27	21.3	−25.2
		Aug 2	25.5	−30.0	Aug 2	21.4	−26.4
		Aug 10	30.9	−28.5	Aug 10	24.2	−24.5
Daphnia		Jun 30	−[a]	−35.2	Jun 30	4.0	−36.9
					Jul 5	10.2	−38.9
					Jul 12	15.9	−37.4
					Jul 27	19.6	−35.3
		Aug 10	34.0	−33.3	Aug 10	22.3	−32.1
Diaptomus		Jun 30	5.2	−34.9	Jun 30	6.8	−38.4
		Aug 10	23.4	−35.2	Aug 10	17.4	−33.2
Cyclops		Jun 30	5.9	−38.0	Jun 30	7.8	−41.0
		Aug 10	14.6	−37.5	Aug 10	15.7	−38.2
Hydridae		Jun 30	6.4	−36.6			
		Aug 16	24.7	−37.5			
Periphyton (curtain)					Jun 30	1.9	−29.9
		Aug 16	13.2	−31.6	Aug 16	3.4	−23.5
Calliergon		Aug 10	9.2	−35.7	Aug 10	10.7	−31.9
Carex		Aug 10	1.0	−26.4	Aug 10	0.9	−29.2
Lymnaea	TL^b = 12 mm				Aug 10	6.5	−29.0
	TL^b = 16 mm	Aug 10	1.5	−30.7			
Valvata	TL^b = 5 mm	Jun 25	6.4	−28.9			
Chironomid					Jun 30	3.8	−36.2
Sculpin	TL^b = 55 mm	Jun 28	7.5	−31.1	Aug 9	9.3	−29.4
	TL^b = 58 mm	Aug 1	8.8	−31.8			
Grayling	TL^b = 249 mm				Jun 30	7.2	−32.7
	TL^b = 225 mm	Jun 30	7.1	−32.3			
	TL^b = 211 mm	Aug 6	6.6	−31.5			
	TL^b = 197 mm				Aug 8	14.7	−30.8

NOTE: Blank spaces indicate that no sample was collected.
[a]Not analyzed.
[b]Total length.

less than 6 μM after July 13 in the fertilized side (77; Table II). Although the chemical and physiological isotope fractionation effects associated with this depleting pool of CO_2, including the effect of active transport of HCO_3^-, are somewhat complex, they could account for much of the observed ^{13}C enrichment in the POM (93–95).

The difference between zooplankton and POM $\delta^{13}C$ values at the end of the experiment could be explained by lags in tissue-turnover time of

Table II. Concentrations of Dissolved Inorganic Carbon, Dissolved CO_2, and Dissolved Organic Carbon in Surface Waters of Lake N2

	Control				Fertilized			
Date (1988)	DIC (μM)	CO_{2diss} (μM)	$\delta^{13}C$-DIC (‰)	DOC (μM)	DIC (μM)	CO_{2diss} (μM)	$\delta^{13}C$-DIC (‰)	DOC (μM)
Jun 30	1159	29.6	−5.0		1550	25.5	−4.9	
Jul 5	1214	23.7	−5.3		1555	19.0	−4.9	
Jul 12	1161	22.6	—[a]		1433	8.4	−4.4	
Jul 19	1158	23.3	−5.1		1375	5.6	−4.4	
Jul 27	1298	27.8	—[a]		1437	5.7	−3.6	
Aug 2	1347	29.0	−4.1		1443	4.9	−3.2	
Aug 10	1364	30.0	−3.8		1508	5.9	—[a]	
Jun 29				560				460
Jul 10				470				550

NOTE: Blank spaces indicate that no sample was collected.
[a]Not analyzed.

zooplankton as discussed, high lipid content in zooplankton, or the fact that POM represents a mixture of potential foods. High lipid content can deplete an organism in ^{13}C relative to its food, although an analysis of lipid content in zooplankton from nearby lakes indicated that bulk isotopic values are lowered by only 1–2‰ from this effect (49). In addition, 1because the POM is a mixture of detritus and plankton, and because $\delta^{13}C$ values of terrestrial material were less negative than POM, the autochthonous portion of POM must be more negative (96) and thus more similar to the zooplankton values. Because the difference in carbon isotopic values between trophic levels is typically small, the disparity in $\delta^{13}C$ values between zooplankton and detritus indicate that detritus is an unimportant food source for *Daphnia*, *Diaptomus*, and *Cyclops*.

The final consideration in zooplankton nutrition is whether the microbial food web plays a role in supplying carbon and nitrogen to the macrozooplankton, and especially to *Cyclops* because its $\delta^{13}C$ values are the most different from POM. Obtaining measurements of isotopic composition of bacteria or microheterotrophs is difficult, and so a second experiment was designed to label natural bacteria with ^{13}C-leucine and phytoplankton with $^{15}NH_4Cl$ in mesocosms, and to follow the transfer of ^{13}C and ^{15}N through the food web. Microheterotrophs (20–100-μm length) would consume the labeled bacteria, and *Cyclops* would eat the microheterotrophs.

Soon after the leucine experiment began on July 2, 1989, the POM in both the control and fertilized corrals was enriched in ^{15}N, as were the grazers *Daphnia* and *Diaptomus* and, to a lesser extent, *Cyclops* (Table III). The high initial $\delta^{15}N$-POM values on June 29 were caused by uptake of $DI^{15}N$ regenerated from the previous year's ^{15}N whole-lake addition experiment (*see* Figure 7). But the most startling result was that the POM was

Table III. δ¹³C and δ¹⁵N Values of POM and Zooplankton from ¹³C-Leucine and ¹⁵NH₄ Addition Experiment in Lake N2 Limnocorrals, Summer 1989

		Control		Fertilized	
Sample	Date (1989)	$\delta^{13}C$ (‰)	$\delta^{15}N$ (‰)	$\delta^{13}C$ (‰)	$\delta^{15}N$ (‰)
POM	Jun 29	−30.6	12.0	−31.3	22.2
	Jul 2	−29.5	6.8	−30.4	10.3
	Jul 6	−11.9	30.4	—[a]	—[a]
	Jul 10	3.9	78.2	−12.7	77.2
	Jul 19	—[a]	—[a]	−17.6	78.7
	Jul 26	14.1	63.6	—[a]	—[a]
	Aug 2	22.5	86.1	−18.2	116.6
Daphnia	Jul 13	9.4	34.8	−15.4	60.4
	Jul 27	19.0	25.6	—[a]	—[a]
Diaptomus	Jul 13	3.3	38.9	−23.4	68.5
Cyclops	Jul 13	−30.4	18.6	−37.8	23.3
	Jul 19	—[a]	—[a]	−33.2	23.6
	Jul 27	−10.8	25.4	—[a]	—[a]

[a]Not analyzed

also enriched in ¹³C by over 50‰ in the control corral. The simplest interpretation of this enrichment is that phytoplankton take up ¹³C-leucine, although the enrichment could be a result of contamination. The potential importance of this contamination can be determined by using the following mass-balance arguments.

The enrichment of algae by ¹³C-leucine might stem from the conversion of added ¹³C-leucine to inorganic ¹³CO₂ rather than the leucine being sequestered as organic carbon in plankton, and then the algae consume the labeled CO_2 during photosynthesis. But even in the unlikely event that all of the ¹³C-leucine was converted to ¹³CO_2, the maximum enrichment of the DIC pool available to phytoplankton would be about 10‰. In this case the final POM $\delta^{13}C$ value would be around −20‰ rather than the 50‰ value observed. The other potential source of contamination is bacteria (nearly all are free-living in both sides of the lake) or microplankton retained on the POM filters.

A worst-case scenario might assume that 50% of the ¹³C-leucine is incorporated by bacteria, 50% of all incorporated ¹³C is returned to the dissolved pool as ¹³C-leucine so that it may be taken up again, and 50% of the bacteria are retained on the POM filters. With the maximum concentration of bacterial carbon observed in the corrals (6.7 μmol of C/L; 3.4 × 10⁶ cells/mL) and a minimum POM concentration (control corral, 50 μmol of C/L), the final $\delta^{13}C$ value of POM including the bacterial contamination would be about −13‰. Additional ¹³C enrichment from microheterotrophs, whose mean carbon concentration for control and fertilized sides in 1989 was 1.25

μmol of C/L (ciliates + rotifers + nauplii; 80), was calculated with the assumption that their maximum $\delta^{13}C$ value was 100‰. There are no direct measurements of bacterial or microheterotroph $\delta^{13}C$ values. This value of 100‰ is the enrichment expected to result from the amount of ^{13}C-leucine added to the bags, the range of bacterial production reported in nearby Toolik Lake (79), and the fact that the microheterotrophs would have $\delta^{13}C$ values similar to those of their bacterial food. The level of contamination from microheterotrophs is similar to the level of contamination from bacteria, and it would result in a final POM value of about −16‰.

Even considering these maximum contaminations, it is apparent that the uptake of added leucine, and perhaps of dissolved organic matter (DOM) in general, must be considered as a source of carbon for phytoplankton and subsequently for herbivorous zooplankton, at least in the control mesocosm. Use of DOM by autotrophs has been previously reported in marine algae, although the widespread occurrence, rates of uptake, and ecological signif- icance of this use is still in question (97, 98). It is tempting to ascribe the larger ^{13}C enrichment of POM in the control corral to lower nutrient or labile DOC concentrations relative to the fertilized corral. This assessment could be misleading, however, because actual rates of bacterial uptake and therefore the amount of ^{13}C-leucine ultimately available to phytoplankton are unknown. Further long-term experiments of this kind, coupled with new methods designed to measure the isotopic composition of nucleic and fatty acids in bacteria and phytoplankton, will help to clarify the importance of DOM uptake by autotrophs in ecosystem metabolism.

The relatively smaller ^{15}N enrichments of *Cyclops* compared to phyto- plankton and the other zooplankton in the leucine experiment are consistent with the idea that *Cyclops* feeds on microheterotrophs that are less labeled with ^{15}N than the phytoplankton. As with the ^{15}N, *Cyclops* was less enriched in ^{13}C than phytoplankton. But the ^{13}C proved less useful as a tracer because phytoplankton took up the label and because the actual isotopic compositions of bacteria and microheterotrophs were unknown.

A series of feeding experiments using fluorescently labeled microspheres (2-μm diameter) confirmed that *Cyclops* can feed on microheterotrophs. In several experiments *Daphnia* consumed the microspheres although *Cyclops* and *Diaptomus* did not (Rublee, P. A.; Kling, G. W.; unpublished data). Microheterotrophs (mainly ciliates) alone were observed to consume the microspheres. When *Cyclops* was added to a sample, the *Cyclops* became fluorescently labeled, clearly through the consumption of labeled ciliates. Thus all feeding experiments and isotope analyses are consistent with the idea that *Cyclops* derives at least part of its nutrition directly through the microbial food web.

Use of DOM and Terrestrial Detritus.   The importance of ter- restrially derived organic matter (dissolved and particulate) to zooplankton nutrition is now well-known in many subarctic and boreal forest lakes (*70*, *73*). Animals in lowland arctic lakes on the coastal plain may also derive nutrition from inputs of terrestrial peat (*62*). In comparison, the results from arctic Lake N2 are different in that a much smaller proportion of planktonic secondary productivity depended on terrestrial organic matter. It is unclear, however, exactly how the microbial food web is linked to dissolved and particulate organic matter from land. Certainly the relative importance of allochthonus organic matter to pelagic organisms varies with the magnitude of external inputs. This variation is seen from considerations of the ratio of open water to land–water interface in lakes of increasing size (*99*, *100*) and of the nature of the coniferous forest soils and peatland catchments surrounding boreal or alpine lakes (*101*, *102*).

Even within the subset of boreal lakes there is probably a direct relationship between external inputs of organic matter and their importance to zooplankton (Meili, M.; Fry, B.; Kling, G. W.; unpublished data). In the case of Lake N2 and other upland arctic lakes, thermokarst processes and active erosion of shoreline peat banks are much less important than they are in coastal plain lakes (*62*, *75*, *103*). In addition, DOC made up less of the total organic carbon in Lake N2 than it did in the humic lake studied by Hessen (*72*); the ratio of DIC:DOC:POC in Lake N2 was 25:8:1 (Table II), whereas in the humic lake the ratio was 1.6:21:1. The lower loading rates of particulate carbon and the smaller relative amounts of DOM in Lake N2 may explain the observation that pelagic productivity depended mainly on new algal production.

Scaling the use of terrestrial organic matter in fresh waters to the magnitude of inputs is somewhat of an oversimplification, however. For example, in an arctic stream with active bank erosion the carbon and nitrogen budgets are dominated by terrestrial material (*92*). But the reliance of stream insects on this material varies widely with species and feeding mode, and the biomass in fish is derived mostly from autochthonus production (*17*, *92*). A second consideration is the fact that there are seasonal shifts of up to 40% in the use and significance of littoral and terrestrial inputs (*76*, *104*). Finally, the chemical character or quality of organic matter varies greatly with its origin and pathways of degradation (*73*, *100*). Tracing the flow and biogeochemical transformation of organic matter with stable isotopes provides a means of easily integrating the details of specific processes into patterns of overall use or importance in an ecosystem.

Benthic Food Web.   In contrast to the planktonic community, there was a much smaller enrichment of ^{15}N in plants and animals of the benthic food web. The moss *Calliergon* showed an enrichment in ^{15}N up to about 10‰, which indicates that these plants are drawing fixed nitrogen directly from the water column (Table I; natural-abundance $\delta^{15}N$ values of *Calliergon* are around 1–2‰). *Periphyton* scraped from the control side of the curtain dividing the lake had $\delta^{15}N$ values similar to those of the moss, but on the fertilized side the *Periphyton* was less enriched (Table I). This difference was probably caused by the thick mat of *Periphyton* from previous years' growth that had accumulated on the fertilized side. This older unlabeled material was included in the sampling and diluted the ^{15}N signal.

The ^{15}N content of snails, caddis flies, and sculpin increased only slightly over the summer in 1988 (Figure 6). This slower response indicates that new algal production is initially unimportant in the diets of benthos, and that there is a lag between production and consumption. In spring 1989, 10 months after stopping the ^{15}N additions, most animals had higher ^{15}N contents than at the end of the experiment in 1988. Thus for most benthic organisms there was a time lag of at least one growing season in their incorporation of nitrogen from phytoplankton.

Part of this time lag is caused by the delay in new planktonic production settling to the bottom of the lake. This delay was empirically determined by analyzing the ^{15}N content of material collected in sediment traps suspended near the mean lake depth on both sides. On August 3, 1988 the $\delta^{15}N$ value of this material was 12.6 and 11.4‰ in the control and fertilized sides, respectively. A second collection was made 2 weeks later on August 16, and the respective $\delta^{15}N$ values had increased to 19.8 and 15.5‰. Because the ^{15}N labeling of POM was roughly linear over time (Figure 4), the average settling time was calculated by extrapolating these isotopic values back to when POM in the water column had similar values. For the first collection date the observed settling time was 18 days in both sides. For the second collection date the time was 23 and 27 days in the control and fertilized sides, respectively. The longer settling times near the end of the experiment are expected because of the partial breakdown of stratification and the associated increase in turbulent mixing (Figure 2).

Another part of this time lag in ^{15}N incorporation by the benthos is explained by the dilution effect of unlabeled detritus deposited in previous years. In a similar tracer study using ^{14}C in marine sediments, benthic grazers highly selective for phytodetritus became labeled similarly to phytoplankton within 2 months; other, less selective, grazers were labeled at a level only 10 to 30% of that of phytoplankton after 5 months (*105*). In Lake N2 the rates of incorporation of new production were similar to those of the less selective grazers in the ^{14}C study. Apparently the benthos in Lake N2 did not readily distinguish new algal production from old phytodetritus or terrestrial detritus in the sediments.

An exception to the slow and steady labeling of benthic animals during the first 2 years was the isotopic shifts in the pill clam *Pisidium*. *Pisidium* use their siphons to filter phytoplankton from the water column, and so became labeled rapidly with ^{15}N during the first summer. By the following spring, however, *Pisidium* had lost nearly all of its ^{15}N label (Figure 6). This pattern of changing ^{15}N content in *Pisidium* was similar in the fertilized and

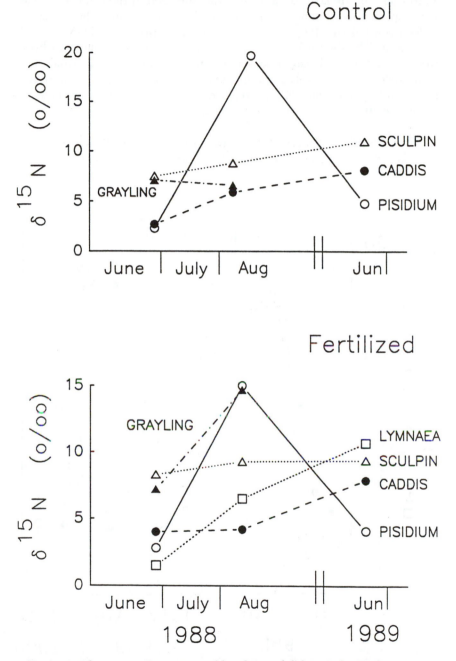

Figure 6. Changes in ^{15}N content of benthos and fish in Lake N2 from June 1988 to June 1989.

control sides of the lake. Because all tissue nitrogen in the clams must have been replaced to account for the loss of ^{15}N, *Pisidium* must have changed its food source over the winter. The following spring the POM was still enriched in ^{15}N (Figure 7), so even if the clams remained inactive or frozen over the winter and began to feed again in early summer they would have been labeled. A more likely scenario is that during winter these clams burrow into the interstitial water in unfrozen sediment and feed on bacteria or fine particulates that were unlabeled with ^{15}N.

Fluxes and Pathways of N Flow. In 1988 the fertilized side of the lake showed distinct increases in the rates of primary production in the water

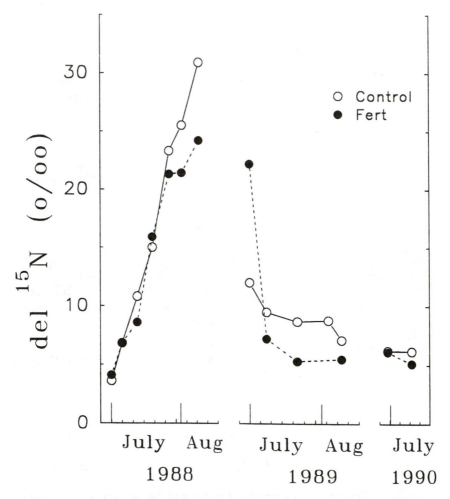

Figure 7. ^{15}N values of POM taken from 4-m depth in control and fertilized sides of Lake N2 from 1988 to 1990.

column and in the surface sediments, compared to production in the control side of the lake (83). In addition, there were increases in densities of the zooplankton except for *Cyclops* and of the benthos except for chironomids and sculpin (83). Thus the overall response to fertilization was an increase in the flux of nitrogen through the pelagic and benthic food webs, as has been found in other whole-lake fertilization studies (18).

The mass of a nutrient moved through different pools is only part of the ecosystem response, and a second question is whether the pathways of nutrient flow or the trophic interactions were altered by the fertilization. In both sides of Lake N2 there were similar patterns and rates of ^{15}N enrichment in pelagic and benthic organisms (Figures 4–8). An exception to this was the fact that a grayling from the fertilized treatment was more enriched in ^{15}N by the end of the summer than a similarly sized grayling taken from the control side. Grayling of this size are zooplanktivorous in these lakes (82), and therefore it is unlikely that the control fish was less enriched in ^{15}N because it fed on benthos (lower in ^{15}N) rather than zooplankton (higher in ^{15}N). Although there is only one sample, it may be that fish growth was faster in the fertilized treatment and thus more of the fish tissue was labeled with ^{15}N. Overall, however, the similar rates and patterns of isotopic labeling suggest that the pathways of nitrogen flow in the ecosystem were unaltered by fertilization.

Nitrogen Retention Time. Discussion of the ^{15}N isotope results so far has been in the context of a continuous-addition experiment. But the ^{15}N addition in 1988 can also be viewed as a pulse experiment when considering longer time scales. In both sides of Lake N2 the ^{15}N content of POM was high in the following early summer of 1989, 10 months after the ^{15}N addition stopped, and was even slightly above background the following summer in 1990 (Figure 7). Several mechanisms of removal contribute to the decreases in $\delta^{15}N$ values of POM over time. Early in the summer the most important of these mechanisms is flushing from snowmelt and runoff, and dilution of ^{15}N by the continued addition of unlabeled fertilizer (Figure 7). Throughout the remainder of the growing season the mechanisms of biological uptake, sedimentation and burial, and chemical removal through denitrification may also be important.

Determining nitrogen-retention times for entire ecosystems by measuring standing stocks and fluxes of nitrogen in all important pools is difficult, and the propagation of errors in such an exercise would reduce confidence in the final estimate. If, however, phytoplankton integrate the amount of ^{15}N remaining and available in the water column, then the calculation of a nitrogen-retention time is much simpler and potentially more accurate.

This method of establishing retention time is valid only if isotopic fractionation effects during nitrogen transformations are unimportant. For example, nitrification results in a ^{15}N enrichment of the residual NH_4^+ pool

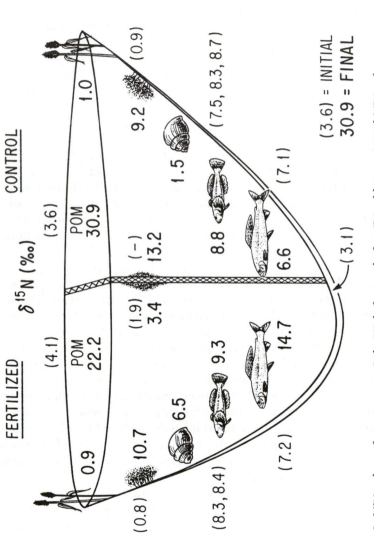

Figure 8. δ^{15}N values of organisms in Lake N2 before and after ^{15}N additions. Initial δ^{15}N values are given in parentheses, and boldface numbers are final values after the 6-week experiment. The largest increases in ^{15}N content occurred in phytoplankton (POM) and submerged mosses.

and relocates ^{15}N into a pool of NO_3^- that may be less rapidly used by phytoplankton. In Lake N2, both NH_4^+ and NO_3^- were depleted to very low levels by phytoplankton during 1988–1990 (*83*). Even if there were isotopic fractionations, the effects on the final $\delta^{15}N$ values of POM would be negligible. A second potentially important process is denitrification, which leaves a residual pool of NO_3^- enriched in ^{15}N. Although nitrogen is lost from the system during this process, the retention time would be over-estimated because of the preferential accumulation of ^{15}N in the residual NO_3^- pool. In Lake N2, as in nearby Toolik Lake, rates of denitrification appear to be low (*83, 106*). These fractionation effects, although potentially important in this study where the $\delta^{15}N$ value of phytoplankton was increased by only 30‰, would have a negligible effect if the phytoplankton enrichment was, for example, 300‰ or an order of magnitude greater than the maximum expected fractionation.

In early 1989, the year following the ^{15}N addition, the initial $\delta^{15}N$ value of POM was greater in the fertilized side than in the control side of the lake (Figure 7). This difference most likely resulted from greater inputs of nitrogen from the main lake inlet to the control side during spring runoff. Additions of nitrogen fertilizer began on July 1, and the dramatic decrease in $\delta^{15}N$ of POM in the fertilized side during the first week of July in 1989 was caused by this dilution with unlabeled nitrogen. By the third year after the ^{15}N addition the $\delta^{15}N$ values of POM had returned to near-background levels in both sides of the lake (Figure 7). Thus it appears that in Lake N2 the nitrogen retention time is about 3 years in both the control and fertilized sides of the lake. In Toolik Lake the retention time of nitrogen is much shorter, on the order of 1 year (*84*), probably because Toolik has a much greater ratio of watershed area to lake area than does Lake N2 (66:1 compared to 5:1, respectively).

The calculated retention time was very similar between the control and the fertilized treatment. Thus nutrient enrichment had little effect on the retention time of nitrogen in this ecosystem. The greater delivery of nitrogen to the fertilized side must have been fully balanced by compensating loss mechanisms. Loss from flushing would increase simply because of the greater stock of nitrogen in the water column, and loss from sedimentation would increase because of the increases in algal production. It is unknown, however, what the response time of these compensating mechanisms is and whether this balance was achieved rapidly or only after 3 years of fertilization. Although the amount of nitrogen in the water column and eventually in the entire system is ultimately set by the physical delivery rate, the compensating mechanisms that control retention time appear to be biological. Additionally, the controlling effect of these mechanisms seems independent of the amount of nutrient added, at least within the range of conditions encountered in Lake N2. In general, biological mechanisms will become more important as the physical flushing decreases.

Summary

Value of Isotopic Labeling.   The experiments described demonstrate the usefulness of stable-isotope additions in understanding trophic interactions and biogeochemical fluxes in whole ecosystems. Using stable isotopes as chemical tracers in natural, undisturbed systems is especially helpful in interpreting the results of perturbation experiments.

The ^{15}N content of phytoplankton increased rapidly following the $^{15}NH_4$ additions to the water column of Lake N2, and by the end of the experiment $\delta^{15}N$ values of POM were enriched to 25–30‰. Similar labeling of pelagic zooplankton with ^{15}N indicated that these organisms relied mainly on new algal production as a source of nitrogen and carbon rather than on unlabeled terrestrial detritus. New algal production is initially unimportant in the diets of benthos; the time lag involved results from a delay in the transport of POM to the sediment and the dilution of ^{15}N label by older phytodetritus and terrestrial detritus.

Nutrient enrichment stimulated the nitrogen-cycling rate and thus the mass of nitrogen moving through the ecosystem. However, the trophic pathways of nitrogen flow were unaltered by fertilization. A nitrogen-retention time in the ecosystem of about 3 years was calculated by using phytoplankton as integrators of the ^{15}N remaining in the water column each spring. The retention time was similar between control and fertilized treatments, a result indicating that nutrient enrichment had little effect on nitrogen-retention time.

Feeding experiments coupled with ^{13}C-leucine additions indicated that the zooplankton *Cyclops* derived some nutrition directly from the microbial food web. In addition, there was indirect evidence of DOM uptake by phytoplankton, and the subsequent transfer of that carbon to macrozooplankton.

Considerations in Stable-Isotope Experiments.   Isotopic tracer additions should label the specific pool of interest quickly, before chemical or biological transformations of no interest distribute the tracer throughout the ecosystem. For calculation of transfers of material from one pool to another the isotope must be well mixed within the ecosystem in the compound it is supposed to trace, and the isotopic content between both pools should be given time to reach equilibrium.

Small enrichments of stable isotopes can be used for detailed studies of specific processes such as denitrification. If these processes are important in the flux of material in the ecosystem from an isotopic mass-balance standpoint, then process rates and associated isotopic fractionations must be measured independently. Alternatively, sufficient isotope must be added to enrich the pool of interest to a level at which the magnitude of fractionation effects is negligible. For example, an enrichment of ammonium by 1 order of magnitude over the maximum fractionation effect expected during nitrification (about 30‰) would result in at most a 10% error in the isotopic mass balance if the entire ammonium pool were nitrified during the experiment.

Acknowledgments

The field work and most of the isotope analyses were done while I was a research fellow at The Ecosystems Center, Marine Biological Laboratory, Woods Hole, MA. Brian Fry and Bruce Peterson were especially helpful in providing comments and sharing their knowledge and insights about isotopes during this time. I thank John O'Brien, Mike Miller, George Kipphut, Anne Hershey, Mike McDonald, Parke Rublee, Ed Rastetter, and John Hobbie for critical discussions and for generously providing samples or unpublished data. I also appreciate the comments of three anonymous reviewers. Bernie Moller, Carolyn Bauman, Kristi Hanson, Val Barber, and Bill Rowe helped with collecting and processing field samples. Meredith Hullar, Bob Michener, Hap Garritt, and Kris Tholke helped with methods development and the laboratory analyses. This work was supported by NSF grants BSR–8702328, DPP–8722015, and DPP–8320544, and the A. W. Mellon Foundation.

References

1. Tansley, A. G. *J. Ecol.* **1917**, *5*, 173–179.
2. Pearl, R. *The Biology of Population Growth;* Knopf: New York, 1925; p 260.
3. Elton, C. *Animal Ecology;* Sidgwick and Jackson: London, 1927.
4. Matena, J.; Berka, R. In *Managed Aquatic Ecosystems;* Michael, R. G., Ed.; Elsevier: Amsterdam, Netherlands, 1987; pp 3–27.
5. Mortimer, C. H. *J. Ecol.* **1941–1942**, *29*, 280–329; *30*, 147–201.
6. Beyers, R. J. *Ecol. Monogr.* **1963**, *33*, 281–306.
7. Smith, M. W. *Trans. Am. Fish. Soc.* **1945**, *75*, 165–174.
8. Orr, A. P. *Proc. R. Soc. Edinburgh. B* **1947**, *68*, 3–20.
9. Langford, R. R. *Trans. Am. Fish. Soc.* **1948**, *78*, 133–144.
10. Pratt, D. M. *Sears Found. J. Mar. Res.* **1949**, *8*, 36–59 (reprinted as Contribution no. 466, Woods Hole Oceanographic Institute, 1949).
11. Nelson, P. R.; Edmondson, W. T. *U. S. Fish. Wildl. Serv. Fisheries Bulletin* **1955**, *56*, 414–436.
12. Hayes, F. R.; Livingstone, D. A. *J. Fish. Res. Board. Can.* **1955**, *12*, 618–35.
13. Hutchinson, G. E.; Bowen, V. T. *Proc. Natl. Acad. Sci. U.S.A.* **1947**, *33*, 148–153.
14. Coffin, C. C.; Hayes, F. R.; Jodrey, L. H.; Whiteway, S. G. *Can. J. Res. D* **1949**, *27*, 207–222.
15. Bormann, F. H; Likens, G. E. *Pattern and Process in a Forested Ecosystem;* Springer-Verlag: New York, 1979; p 253.
16. *An Ecosystem Approach to Aquatic Ecology;* Likens, G. E., Ed.; Springer-Verlag: New York, 1986; p 516.

17. Peterson, B. J.; Hobbie, J. E.; Hershey, A. E.; Lock, M. A.; Ford, T. E.; Vestal, J. R.; McKinley, V. L.; Hullar, M. A. J.; Ventullo, R. M.; Volk, G. S. *Science (Washington, D.C.)* **1985**, *229*, 1383–1386.
18. Schindler, D. W. In *Ecosystem Experiments*; Mooney, H. A.; Medina, E; Schindler, D. W.; Schulze, E.-D; Walker, B. H., Eds.; Wiley: Chichester, United Kingdom, 1991; pp 121–140.
19. Stockner, J. G.; Shortreed, K. S. *Can. J. Fish. Aquat. Sci.* **1985**, *42*, 649–658.
20. Carpenter, S. R. *Ecology* **1989**, *70*, 453–463.
21. Mooney, H. A; Medina, E; Schindler, D. W.; Schulze, E.-D; Walker, B. H., Eds.; *Ecosystem Experiments*; Wiley: Chichester, United Kingdom, 1991; p 268
22. Hayes, F. R.; McCarter, J. A.; Cameron, M. L.; Livingstone, D. A. *J. Ecol.* **1952**, *40*, 202–216.
23. Ball, R. C.; Hooper, F. F. In *Radioecology*; Schultz, V.; Klement, A. W., Jr., Eds; Reinhold: New York, 1963.
24. Elwood, J. W.; Nelson, D. J. *Oikos* **1972**, *23*, 295–303.
25. Hesslein, R. H.; Broecker, W. S.; Quay, P. D.; Schindler, D. W. *Can. J. Fish. Aquat. Sci.* **1980**, *37*, 454–463.
26. Nadelhoffer, K.; Fry, B. In *Stable Isotopes in Ecology*; Lajtha, K.; Michener, B., Eds.; Blackwell: London, in press.
27. Giblin, A. E.; Peterson, B. J.; Fry, B.; Dornblaser, M.; Tucker, J.; Regan, K. Abstract. American Society of Limnology and Oceanography; Feb. 1992, Sante Fe.
28. Likens, G. E.; Hasler, A. D. *Science (Washington, D.C.)* **1960**, *131*, 1676.
29. Torgersen, T.; Top, Z.; Clarke, W. B.; Jenkins, W. J.; Broecker, W. S. *Limnol. Oceanogr.* **1977**, *22*, 181–193.
30. Emerson, S.; Broecker, W. S.; Schindler, D. W. *J. Fish. Res. Board. Can.* **1973**, *30*, 1475–1484.
31. *Isotopes in Lake Studies*; International Atomic Energy Agency: Vienna, Austria, 1979; p 290.
32. *Isotope Techniques in Water Resources Development*; International Atomic Energy Agency: Vienna, Austria, 1987; p 813.
33. Krabbenhoft, D. P.; Bowser, C. J.; Kendall, C.; Gat, J. R. In *Environmental Chemistry of Lakes and Reservoirs*; Baker, L. A., Ed.; Advances in Chemistry 237; American Chemical Society: Washington, DC, 1994; Chapter 3.
34. Nriagu, J. O. In *Sulphur in the Environment*, Part II; Nriagu, J. O., Ed.; Wiley: New York, 1978; pp 1–59.
35. DeNiro, M. J.; Epstein, S. *Geochim. Cosmochim. Acta* **1978**, *42*, 495–506.
36. Fry, B.; Sherr, E. B. *Contrib. Mar. Sci.* **1984**, *27*, 13–47.
37. *Stable Isotopes: Natural and Anthropogenic Sulphur in the Environment*; Krouse, H. R.; Grinenko, V. A., Eds.; Wiley: Chichester, United Kingdom, 1991; p 440.
38. Rundel, P. W.; Ehleringer, J. R.; Nagy, K. A., Eds.; *Stable Isotopes in Ecological Research*; Springer-Verlag: New York, 1989; p 525.
39. *Stable Isotopes in Ecology*; Lajtha, K.; Michener, B., Eds.; Blackwell: London, in press.
40. Fry, B.; Jeng, W. L.; Scalan, R. S.; Parker, P. L. *Geochim. Cosmochim. Acta* **1978**, *42*, 1299–1302.
41. Peterson, B. J.; Howarth, R. W. *Limnol. Oceanogr.* **1987**, *32*, 1195–1213.
42. Kitting, C. L.; Fry, B.; Morgan, M. D. *Oecologia (Berlin)* **1984**, *62*, 145–149.
43. Dugdale, R. C. In *The Ecology of the Seas*; Dugdale, R. C.; Cushing, D. H.; Walsh, J. J., Eds.; Saunders: Philadelphia, PA, 1976; pp 141–172.
44. Harrison, W. G. In *Nitrogen in the Marine Environment*; Carpenter, E. J.; Capone, D. G., Eds.; Academic: New York, 1983; pp 763–808.

45. Miyake, Y.; Wada, E. *Rec. Oceanogr. Works Jpn.* **1967,** *9,* 32–53.
46. DeNiro, M. J.; Epstein, S. *Geochim. Cosmochim. Acta* **1981,** *45,* 341–351.
47. Minagawa, M.; Wada, E. *Geochim. Cosmochim. Acta* **1984,** *48,* 1135–1140.
48. Peterson, B. J.; Fry, B. *Annu. Rev. Ecol. Syst.* **1987,** *18,* 293–320.
49. Kling, G. W.; Fry, B.; O'Brien, W. J. *Ecology* **1992,** *73,* 561–66.
50. Estep, M. L. F.; Vigg, S. *Can. J. Fish. Aquat. Sci.* **1985,** *42,* 1712–1719.
51. Yoshioka, T.; Wada, E.; Saijo, Y. *Verh. Int. Ver. Limnol.* **1988,** *23,* 573–578.
52. Fry, B. *Ecology* **1992,** *72,* 2293–2297.
53. *Eutrophication: Causes, Consequences, Correctives;* National Academy of Sciences: Washington, DC, 1969; p 661.
54. Schindler, D. W. *Science (Washington, D.C.)* **1974,** *184,* 897–899.
55. Hall, D. J.; Cooper, W. E.; Werner, E. E. *Limnol. Oceanogr.* **1970,** *15,* 839–928.
56. O'Brien, W. J.; deNoyelles, F., Jr. *Hydrobiologia* **1974,** *44,* 105–25.
57. Nixon, S. W.; Oviatt, C. A.; Frithsen, J.; Sullivan, B. *J. Limnol. Soc. South Afr.* **1986,** *12,* 43–71.
58. Malley, D. F.; Chang, P. S. S.; Schindler, D. W. *Decline of Zooplankton Populations following Eutrophication of Lake 227, Experimental Lakes Area, Ontario: 1969–1974;* Dept. Fisheries & Oceans, Freshwater Institute, Winnipeg, Manitoba, 1977.
59. Smith, S. V.; Kimmerer, W. J.; Laws, E. A.; Brock, R. E.; Walsh, T. W. *Pac. Sci.* **1981,** *35,* 279–395.
60. Holmgren, S. K. *Int. Rev. Gesamten. Hydrobiol.* **1984,** *69,* 781–817.
61. O'Brien, W. J.; Hershey, A. E.; Hobbie, J. E.; Kipphut, G. W.; Miller, M. C.; Moller, B.; Vestal, J. R. *Hydrobiologia* **1992,** *240,* 143–188.
62. Schell, D. M. *Science (Washington, D.C.)* **1983,** *219,* 1068–1071.
63. Hyatt, K. D.; Stockner, J. G. *Can. J. Fish. Aquat. Sci.* **1985,** *42,* 320–331.
64. Taylor, W. D. *Can. J. Fish. Aquat. Sci.* **1984,** *38,* 1316–1321.
65. Mazumder, A.; McQueen, D. J.; Taylor, W. D.; Lean, D. R. S. *Limnol. Oceanogr.* **1988,** *33,* 421–430.
66. D'Elia, C. F. In *Concepts of Ecosystem Ecology;* Pomeroy, L. R.; Alberts, J. J., Eds.; Springer-Verlag: New York, 1988; pp 195–230.
67. Fisher, S. G.; Likens, G. E. *Ecol. Monogr.* **1973,** *43,* 421–439.
68. Vannote, R. L.; Minshall, G. W.; Cummins, K. W.; Sedell, J. R.; Cushing, C. E. *Can. J. Fish. Aquat. Sci.* **1980,** *37,* 130–137.
69. Wetzel, R. G. *Mitteilungen int. Ver. Limnol.* **1968,** *14,* 261–273.
70. Salonen, K.; Hammar, T. *Oecologia (Berlin)* **1986,** *68,* 246–253.
71. Tranvik, L. J. *Microb. Ecol.* **1988,** *16,* 311–322.
72. Hessen, D. O.; Andersen, T.; Lyche, A. *Limnol. Oceanogr.* **1990,** *35,* 84–99.
73. Salonen, K.; Kairesalo, T.; Jones, R. I., Eds. *Hydrobiologia (Dissolved Organic Matter in Lacustrine Ecosystems: Energy Source and System Regulator)* **1992,** *229,* 1–291.
74. Findlay, S.; Pace, M. L.; Lints, D.; Cole, J. J.; Caraco, N. F.; Peierls, B. *Limnol. Oceanogr.* **1991,** *36,* 268–278.
75. Schell, D. M.; Ziemann, P. J. In *Permafrost* (4th Int. Conf.); National Academy Press: Washington, DC, 1983; pp 1105–1110.
76. Schell, D. M.; Ziemann, P. J. In *Stable Isotopes in Ecological Research;* Rundel, P. W.; Ehleringer, J. R.; Nagy, K. A., Eds.; Springer-Verlag: New York, 1989; pp 230–251.
77. Kling, G. W.; Kipphut, G. W.; Miller, M. C. *Science (Washington, D.C.)* **1991,** *251,* 298–301.
78. Kling, G. W.; O'Brien, W. J.; Miller, M. C.; Hershey, A. E. *Hydrobiologia* **1992,** *240,* 114.

120 ENVIRONMENTAL CHEMISTRY OF LAKES AND RESERVOIRS

79. Hobbie, J. E.; Helfrich, J. V. K., III. *Arch. Hydrobiol. Beih.* **1988,** *31,* 281–288.
80. Rublee, P. A. *Hydrobiologia* **1992,** *240,* 133–141.
81. Hershey, A. E. *Holarctic Ecol.* **1985,** *8,* 39–48.
82. O'Brien, W. J.; Buchanan, C.; Haney, J. *Arctic* **1979,** *32,* 237–247.
83. O'Brien, W. J.; Hershey, A. E.; Kipphut, G. W.; Kling, G. W.; Miller, M. C. *Ecology,* in review.
84. Whalen, S. C.; Cornwell, J. C. *Can. J. Fish. Aquat. Sci.* **1985,** *42,* 797–808.
85. Plummer, L. N.; Busenberg, E. *Geochim. Cosmochim. Acta* **1982,** *46,* 1011–1040.
86. Stumm, W.; Morgan, J. J. *Aquatic Chemistry,* 2nd ed.; John Wiley and Sons: New York, 1981; p 780.
87. Miller, M. C.; Hater, G. R.; Spatt, P.; Westlake, P.; Yeakel, D. *Arch. Hydrobiol. Suppl.* **1986,** *74,* 97–131.
88. Quay, P. D.; Broecker, W. S.; Hesslein, R. H.; Schindler, D. W. *Limnol. Oceanogr.* **1980,** *25,* 201–218.
89. Kling, G. W.; Giblin, A. E.; Peterson, B. P.; Fry, B. *Limnol. Oceanogr.* **1991,** *36,* 106–122.
90. Montoya, J. P. "Natural Abundance of ^{15}N in Marine and Estuarine Plankton: Studies of Biological Isotopic Fractionation and Plankton Processes"; Ph.D. Thesis, Harvard University, Cambridge, MA, 1990.
91. Stross, R. G.; Miller, M. C.; Daley, R. J. In *Limnology of Tundra Ponds;* Hobbie, J. E., Ed.; Dowden, Hutchinson and Ross: Stroudsburg: PA, 1980; pp 251–296.
92. Peterson, B. J.; Hobbie, J. E.; Corliss, T. L. *Can. J. Fish. Aquat. Sci.* **1986,** *43,* 1259–1270.
93. McKenzie, J. A. In *Chemical Processes in Lakes;* Stumm, W., Ed.; John Wiley: New York, 1985; pp 99–118.
94. Fogel, M. L.; Cifuentes, L. A.; Velinsky, D. J.; Sharp, J. H. *Mar. Ecol. Prog. Ser.,* in press.
95. Goericke, R.; Montoya, J. P.; Fry, B. In *Stable Isotopes in Ecology;* Lajtha, K.; Michener, B., Eds.; Blackwell: London, in press.
96. LaZerte, B. D. *Can. J. Fish. Aquat. Sci.* **1983,** *40,* 1658–1666.
97. Wheeler, P. A.; North, B. B.; Stephens, G. C. *Limnol. Oceanogr.* **1974,** *19,* 249–258.
98. Palenik, B.; Morel, F. M. M. *Limnol. Oceanogr.* **1990,** *35,* 260–269.
99. Wetzel, R. G. *Verh. Int. Ver. Theor. Angew. Limnol.* **1990,** *24,* 6–24.
100. Wetzel, R. G. *Hydrobiologia* **1992,** *229,* 181–198.
101. Baron, J.; McKnight, D.; Denning, S. *Biogeochemistry,* in press.
102. Meili, M. *Hydrobiologia* **1992,** *229,* 23–42.
103. *Guidebook to Permafrost and Related Features along the Elliott and Dalton Highways, Fox to Prudhoe Bay, Alaska* (4th Int. Conf. on Permafrost, July 18–22, 1983, University of Alaska, Fairbanks); Brown, J.; Kreig, R. A., Eds.; State of Alaska: Juneau, AK, 1983; p 230.
104. Kairesalo, T.; Lehtovaara, A.; Saukkonen, P. *Hydrobiologia* **1992,** *229,* 199–224.
105. Rudnick, D. T. *Mar. Ecol. Prog. Ser.* **1989,** *50,* 231–240.
106. Klingensmith, K. M.; Alexander, V. *Appl. Environ. Microbiol.* **1983,** *46,* 1084–1092.

RECEIVED for review May 22, 1992. ACCEPTED revised manuscript August 11, 1992.

Effects of Acidification on Chemical Composition and Chemical Cycles in a Seepage Lake

Inferences from a Whole-Lake Experiment

Carolyn J. Sampson[1], Patrick L. Brezonik[1,*], and Edward P. Weir[2]

[1]Department of Civil and Mineral Engineering, University of Minnesota, Minneapolis, MN 55455
[2]Minnesota Pollution Control Agency, Rochester, MN 55904

Responses of the major ion and nutrient chemistry of Little Rock Lake, Wisconsin, to experimental acidification are described, and the underlying processes affecting the responses are inferred. Total base cations increased from 90 to 140 μequiv/L over the pH range of the experiment (6.1–4.7). The order of increase was $Ca^{2+} \gg Mg^{2+} > K^+ > Na^+$. Loss of exchangeable cations from the upper few centimeters of sediment by H^+-cation exchange can account for the increase in water-column cations, and evidence indicates that the surficial sediments became acidified. The necessary reversal of this effect will probably slow the recovery of water-column alkalinity. Several reactor-based models that describe internal alkalinity-generation processes are presented to predict rates of alkalinity recovery in the water column. Major impacts on lake nutrient chemistry were not observed to pH 4.7, but small differences were found between treatment and reference basins of the lake for summer epilimnetic averages of silica and inorganic nitrogen at pH 4.7. Total N was lower in the treatment basin at pH 5.1 and 4.7. N_2-fixation was inhibited in the treatment basin at pH 4.7 but, in contrast to observations on other acidified lakes, under-ice nitrification was not inhibited.

*Corresponding author

0065–2393/94/0237–0121$10.75/0

ACIDIFICATION CAUSES A WIDE VARIETY of chemical and biological changes in lake ecosystems. An understanding of response mechanisms is needed to correctly model the effects of acidification on water chemistry and the time to recovery under reduced acid loading. Accurate models provide policy-makers with interpretations of the current situation and help in evaluation of the need for legislated reductions of acid-producing emissions.

We report results of a 6-year acidification experiment designed to verify and quantify the changes resulting from acidification in a small seepage lake. Both laboratory and whole-lake experiments were used. Experimental manipulation of whole-lake ecosystems offers several advantages over other approaches in evaluating the effects of acidification (1). Whereas data from lake surveys are useful in generating hypotheses, well-designed whole-lake manipulations can be used to test hypotheses and verify cause–effect relationships. In addition, lake manipulations provide opportunities to interpret the mechanisms of acidification and develop deterministic models.

This chapter summarizes water chemistry changes and effects of acidification on biogeochemical processes. We focus on major ions and nutrients, discuss internal alkalinity generation and sediment ion-exchange processes, and present preliminary recovery models. Results for trace and minor metals and other chemical constituents are presented elsewhere (2–4).

Background Information

Site Description.   Little Rock Lake (LRL) is a soft-water oligotrophic seepage lake located in the Northern Highland Lake District in north-central Wisconsin (Vilas County; 45°59′55″N, 89°42′15″W). Public access to the site, which is in the Northern Highlands State Forest, has been restricted since the study began.

The lake lies in an uninhabited forested watershed (mixed conifers and hardwood) with highly permeable sandy soil (Sayner Rubicon Sand) that overlies ~40 m of noncalcareous glacial material that was deposited during the last glaciation when most of the lakes in the region, including LRL, were formed. This deposit overlies Precambrian igneous–metamorphic bedrock. The lake has two main basins that are separated by a narrow constriction (Figure 1). The two basins are similar in surface area, but the south basin has smaller average and maximum depths than the north. Stratification in the south basin is weak and ephemeral, but the deeper north basin forms a small hypolimnion that becomes anoxic by midsummer.

LRL is a groundwater recharge system (no surface inlets or outlets) and receives 98–100% of its water from precipitation directly onto the lake surface. Groundwater seepage accounts for about 35% of the water output from the lake, and evaporation accounts for the remaining 65%. Water residence time, τ_w, is about 9–11 years. Most of LRL is situated above the regional

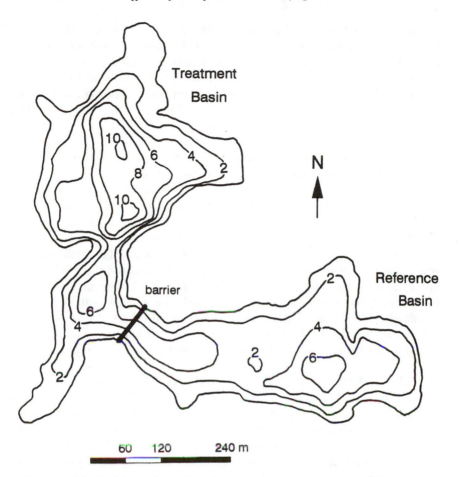

Figure 1. *Bathymetric map of Little Rock Lake, WI. Contours are given in meters. Treatment-basin surface area was 9.8 ha; mean depth was 3.8 m; and maximum depth was 10.3 m. Reference-basin surface area was 8.1 ha; mean depth was 3.1 m; and maximum depth was 6.5 m.*

surficial (unconfined) groundwater table, which slopes downward from southeast to northwest. As a result, groundwater inflow occurs only in the extreme southeast corner of the lake. During years with relatively high precipitation (such as 1984–1986) about 2% of the inflow was contributed by shallow groundwater. During the drought in 1987–1989 the water table declined below the lake level even at the southeast corner; no groundwater inflow occurred, and the lake level declined by nearly 1 m. Near-normal precipitation was not sufficient to raise the water table in 1990, and the lake continued to receive all of its water directly from the atmosphere until 1991.

The chemistry of LRL reflects the provenance of its water; specific conductance is very low ($\sim$12 μS/cm), near that of local rainfall. Total base-cation concentration is ~91 µequiv/L, alkalinity is ~25 µequiv/L, and nutrient levels also are low. Major sources of the elements and ions found in the lake are wet and dry atmospheric deposition, but litterfall from surrounding hardwoods and conifers also provides nutrients and some ions. Sinks for chemical constituents include groundwater recharge, sedimentation of particulate matter, and diffusion of ions into the sediment, with subsequent diagenetic conversion to solid forms.

Experimental Design.   The two basins were monitored during 1983–1985 to establish preacidification limnological conditions. In August 1984 a Dacron-reinforced polyvinyl barrier was installed at the narrows dividing the two basins (Figure 1). Results through May 1985 showed that the basins were nearly identical with respect to chemical conditions (5, 6). The south basin was left as a reference throughout the experiment.

Stepwise acidification of the north basin began in May 1985: three 2-year treatments with technical-grade sulfuric acid lowered the pH from the initial value of 6.1 to target values of 5.6, 5.1, and 4.7. The acid was added from a fiberglass boat and mixed by the turbulence of an outboard motor. Monitoring of epilimnetic pH at numerous locations showed that rapid mixing was achieved. The basin pH was monitored at least weekly. Acid was added as necessary during the ice-free season to maintain the pH near the target level.

Table I shows the pH values (summer average and the summer and winter maximum and minimum) for the acidified basin during each treatment period. Background values (after barrier installation but before acid addition) are shown for comparison. On the whole, target values were achieved. The average summer pH values are within 0.18 of the target values, and summer minima are not more than 0.21 pH unit below the target. Different biogeochemical processes may respond primarily to extreme values or to long-term values. For example, biological processes may be most sensitive to the minimum pH value, whereas chemical processes such as ion exchange may be more sensitive to the average pH value. To simplify the discussion, we will refer to each treatment period by its target pH.

Experimental Methods

Water Chemistry.   Sampling and analytical procedures, as well as quality control and assurance information for chemical and biological parameters, are described in detail elsewhere (e.g., 7–9). A brief summary of sampling and analytical methods is provided here. Water samples were collected biweekly during the ice-free season and every 5 weeks under the ice at depths of 0, 4, 6, 8, and 9 m in the north basin and 0, 4, and 6 m in the south basin. Water was pumped from depth by using a peristaltic pump and Tygon tubing and collected in prewashed polyethylene bottles. Alkalinity and pH were measured on unfiltered, unpreserved samples.

Table I. pH Averages, Maxima, and Minima

Measure	Background 1984–1985	pH 5.6		pH 5.1		pH 4.7	
		1985–1986	1986–1987	1987–1988	1988–1989	1989–1990	1990–1991
Summer ave	6.15	5.61	5.61	5.24	5.15	4.79	4.88
Summer max	6.33	5.80	5.74	5.39	5.33	5.02	5.04
Summer min	6.00	5.39	5.41	5.05	5.03	4.68	4.73
Winter max	6.13	5.50	5.63	5.22	5.27	5.12	4.90
Winter min	5.39	5.36	5.34	4.73	5.05	4.81	4.74

The pH was measured with a radiometer (PHM 84) or Beckman (ϕ41) closed-cell pH meter within 4 h of sampling. Acid-neutralizing capacity (ANC), often referred to as alkalinity, was calculated from titration data by the Gran procedure. Major cations, anions, and inorganic nutrients were measured on samples filtered through in-line 0.4-μm membrane filters (Nuclepore) in the field. Major cations (Ca^{2+}, Mg^{2+}, K^+, and Na^+) were measured by flame atomic absorption spectrophotometry (AAS); anions (Cl^-, SO_4^{2-}, and F^-) were determined by ion chromatography. Ammonium was analyzed manually by the indophenol method (10), and nitrate was measured by the automated cadmium reduction method or ion chromatography. Total nitrogen was measured by converting all nitrogen forms to nitrate by alkaline persulfate oxidation (11) and subsequent analysis of nitrate by the automated cadmium reduction method (12). Soluble reactive phosphate was analyzed manually by the ascorbic acid–molybdenum blue method, and total phosphorus was determined similarly after persulfate digestion (12). Soluble reactive silica was determined manually by the molybdosilicate heteropoly blue method (12). Nitrogenase activity (N_2-fixation) in periphyton and benthic algae was estimated by the acetylene reduction assay, as described by Dierberg and Brezonik (13) and Flett et al. (14).

H^+-Cation Exchange. Sediment samples used to measure exchangeable cations were collected from 5-m sites in both basins of LRL in October of 1987 (the first year at pH 5.1). Samples were obtained by a box core, which was subcored by using Plexiglas tubes that were subsequently extruded into 2- or 4-cm sections. A fractionation scheme, based on techniques developed by Tessier et al. (15) and Suhr and Ingamells (16), was developed to determine the contribution of organic matter, ion exchange, and mineral phase to the total base-cation content of the sediments (17). The exchangeable fraction of base cations (Ca^{2+}, Mg^{2+}, Na^+, and K^+) was determined by the ammonium saturation method (18). The hydrogen peroxide digestion method of Tessier et al. (15) was used to determine the amounts of base cations associated with the organic fraction of the sediments. The total base-cation content of the mineral fraction was determined by lithium metaborate (LMB) fusion (16), followed by analysis by direct current plasma emission spectroscopy. The mineralogy of LRL sediments was determined by X-ray diffraction.

Ca^{2+}–NH_4^+ Exchange. Sediment samples used to determine the Ca^{2+}–NH_4^+ exchange coefficient were collected by a Ponar dredge at a depth of 5 m. Sediment from three collections was mixed manually in a large plastic tub to obtain a homogeneous sample. The sediment was transferred to 250-mL round bottles and centrifuged at 1500 rpm for 30 min, after which the supernatant was removed. Prior to the analysis, samples were kept sealed air-tight at 4 °C. Portions of unaltered sediment were converted to Ca^{2+}-saturated or NH_4^+-saturated forms by repeated washings with calcium chloride or ammonium acetate after the method of Chapman (19). A third portion was washed with distilled water for determination of total exchangeable bases by using LiCl (to include ammonium as an exchangeable base).

Exchangeable acidity was determined by the $BaCl_2$–TEA (triethanolamine) method (20). The equilibrium calcium–ammonium exchange coefficient (Gapon coefficient, K_G) was determined on the Ca^{2+}-saturated, NH_4^+-saturated portions and unaltered sediment by methods described by Baes and Bloom (21). This determination was accomplished by allowing exchange solutions with compositions similar to LRL pore waters (Ca^{2+} = 6.6–52.9 μequiv/L; NH_4^+ = 8.7–44

μequiv/L; total cation concentrations = 45.1–62.5 μequiv/L) to equilibrate with the treated sediments. An aliquot of supernatant obtained by centrifugation was analyzed for Ca^{2+}, Mg^{2+}, Na^+, and K^+ by flame atomic absorption spectrometry (AAS) and for NH_4^+ by the indophenol method. The pH was measured on the sediment samples before and after equilibration to determine the competition of H^+ for exchange sites.

Results and Discussion

Water-Column Responses. *Major Ions.* Table II shows average concentrations of the major ions for the reference basin and the treatment basin for pretreatment and acidification to pH 5.6, 5.1, and 4.7. As expected, calcium was the base cation most responsive to acidification (5, 6). It increased steadily from the first treatment period (pH 5.6) until it reached an average of 82 μequiv/L at pH 4.7, nearly double the reference-basin average. Calcium is important biologically as it may buffer the toxic effects of H^+, aluminum, and heavy metals (22, 23). Potassium increased at pH 5.6 and 5.1, when it reached an average of 18.3 μequiv/L (~30% higher than in the reference basin), but it did not increase further at pH 4.7. Magnesium did

Table II. Average Concentrations of Major Ions for Each Treatment Period

Ion	Basin	Reference pH 6.1 1983–1990	Preacid pH 6.1 1983–1985	Treatment pH 5.6 4/85–4/86	Treatment pH 5.1 4/87–4/88	Treatment pH 4.7 4/89–4/90
Ca^{2+}	ave	45.5	44.9	48.3	67.7	81.9
	s	7.3	4.6	4.8	14.1	17.0
	n	95	21	12	13	13
Mg^{2+}	ave	24.8	24.5	23.3	29.4	33.0
	s	5.2	4.1	2.6	6.7	3.5
	n	95	21	12	13	13
K^+	ave	14.6	14.3	16.0	18.3	18.2
	s	2.5	1.6	2.3	2.6	6.5
	n	92	21	11	12	13
Na^+	ave	7.2	6.6	5.4	8.4	6.8
	s	2.8	1.4	0.7	3.8	1.0
	n	92	21	11	12	13
SO_4^{2-}	ave	56.0	53.1	74.4	116.2	147.4
	s	10.0	6.0	5.5	12.9	50.4
	n	88	20	12	10	13
Cl^-	ave	8.4	7.5	6.5	9.0	9.0
	s	3.3	1.1	2.2	2.8	8.4
	n	87	20	12	11	13

NOTE: All averages are in microequivalents per liter (s is the standard deviation; n is the number of data points). Reference (south basin) values are averaged over the entire experiment. Preacid (north basin) values are averaged over the prebarrier and pretreatment years; treatment (north basin) averages are obtained over the second year at the indicated pH.

not respond at pH 5.6, but increased at pH 5.1 and 4.7; it reached an average of 33 μequiv/L (~40% higher than in the reference basin). Sodium showed no trend during the experiment. The increase in water-column base cations may contribute to the generation of alkalinity. Sources of base cations and their role in internal alkalinity generation are discussed in more detail in separate sections later in this chapter.

Because sulfuric acid was used to acidify the treatment basin, the increase in SO_4^{2-} was expected. However, only about 50% of the sulfate added as sulfuric acid remained in the water column of the treatment basin; the remainder was lost to outseepage and in-lake processes. For example, had the added sulfate remained in the water column, $[SO_4^{2-}]$ at pH 4.7 would have been ~257 μequiv/L (versus the measured $[SO_4^{2-}] = 147$ μequiv/L). The loss of sulfate by reduction may contribute to the generation of alkalinity; this possibility is discussed in the Internal Alkalinity section. Chloride showed no significant trend with decreasing pH.

Nitrogen.   Seasonal patterns of near-surface concentrations of NO_3^- and NH_4^+ in both basins did not change over the entire study. Concentrations of both ions peaked under the ice during late winter each year and declined rapidly after ice-out. NH_4^+ accumulated in the hypolimnion of the acidified basin each year. Peaks in NH_4^+ at 9 m ranged from 670 to 960 μg/L prior to 1988, but during 1988–1990 peaks in NH_4^+ at 9 m reached as high as 2000 μg/L. In contrast, the vertical patterns in NO_3^- were not consistent from year to year. No acidification trend was apparent in the seasonal patterns of NO_3^- in the bottom water.

Plots of near-surface summer averages and winter maxima show some interesting patterns (Figure 2), but no statistically significant differences between basins appear. Reference-basin NO_3^- and NH_4^+ summer averages both were greater than treatment-basin values until pH 4.7, when the treatment-basin averages exceeded those of the reference basin. Winter maxima for NH_4^+ in the reference basin increased linearly from 1986 (second year at pH 5.6) through 1990 (second year at pH 4.7), but there was no corresponding trend in the treatment basin. Winter maxima for NO_3^- in the treatment basin varied little until the first year at pH 4.7, when the maximum was less than one-third that of the previous years. The reference-basin NO_3^- maximum also declined (but only by ~25%) the same winter. The "tentative" treatment effect was not repeated in year 2 at pH 4.7, when the treatment basin had the highest winter maximum in NO_3^- observed over the study. Total nitrogen (TN) showed smaller intra- and interannual variations than inorganic N forms, and reference summer averages were consistently higher than treatment values at pH 5.6 or lower (Figure 3). The difference between basins (almost 100 μg/L) was statistically significant ($\alpha < 0.05$, Wilcoxon signed rank test) beginning in the second year at pH 5.1.

A one-time areal survey of acetylene reductase activity was made during the second summer at pH 4.7. The rates of ethylene produced by acetylene

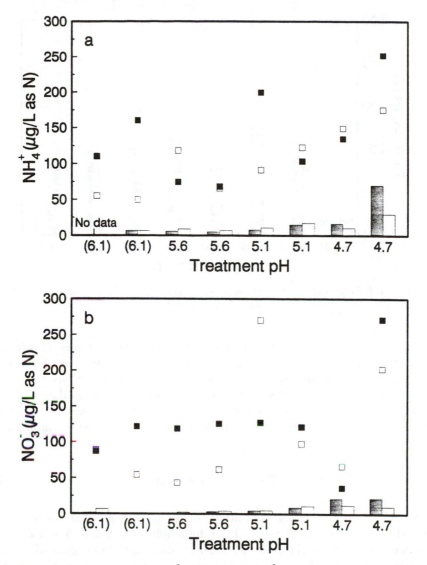

Figure 2. Summer averages and winter maxima of inorganic nitrogen species: NH_4^+ (a) and NO_3^- (b). Bars are summer averages: shaded, treatment; and unshaded, reference. Squares are winter maxima: black, treatment; and white, reference.

reduction are proportional to the rates of molecular nitrogen, N_2, fixed.

$$HC\equiv CH + E_{NR} \longrightarrow E_{NR}-HC=CH_2 \longrightarrow H_2C-CH_2 + E_{NR} \quad (1)$$

$$N\equiv N + E_{NR} \longrightarrow E_{NR}-N=NH \longrightarrow E_{NR}-HN-NH_2 \longrightarrow$$

$$E_{NR} + 2NH_3 \quad (2)$$

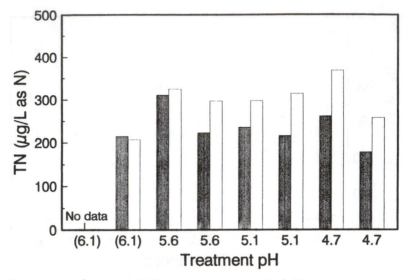

Figure 3. Total nitrogen (TN) summer averages: shaded bars, treatment; and unshaded, reference.

where E_{NR} is the enzyme dinitrogen reductase. In the reference basin 15 out of 17 samples showed a strong positive response. Only two trace responses were observed out of 28 samples from the treatment basin (4).

Rudd et al. (24) reported disruption of several processes in the nitrogen cycle during experimental acidification of Lakes 223 and 302S in Ontario's Experimental Lakes Area (ELA). In particular, they reported the inhibition of nitrification.

$$NH_4^+ + 2O_2 \longrightarrow NO_3^- + H_2O + 2H^+ \qquad (3)$$

This inhibition was apparently caused by the inability of nitrifying bacteria to adapt to pH below 5.4–5.7. Prior to acidification below 5.4–5.7, ammonium showed a slight under-ice accumulation and nitrate showed a substantial increase that peaked just before spring thaw. After acidification to <5.4–5.7, the lake had a large accumulation of ammonium under ice cover and little increase in nitrate.

Garrison et al. (25) reported similar findings for Max Lake (Vilas Co., WI), which is close to Little Rock Lake and has similar hydrology and geological setting (i.e., it is a seepage lake that receives ~100% of its water from the atmosphere). Max Lake was treated with groundwater in an experiment to evaluate the effectiveness of this approach in mitigating the effects of atmospheric acid deposition. Prior to groundwater addition (at pH ~5.1), under-ice accumulation of ammonium and little increase in nitrate were observed. Following additions of groundwater to pH above 5.3, under-ice concentrations of ammonium were reduced and nitrate concentrations increased. This change suggested the reestablishment of nitrifying bacteria.

LRL exhibited natural under-ice depressions to pH 5.4 in the reference basin (every year) and in the treatment basin (through treatment to pH 5.6) (*4, 17, 26*). Thus, disruption of nitrification was not expected to occur at least until the pH 5.1 treatment period. In contrast to the ELA and Max Lake experiments, however, nitrification appears to be unaffected in LRL down to pH 4.7. To explain why some acidic systems sustain a population of nitrifying bacteria, Rudd et al. (*24*) suggested that nitrifying bacteria find refuge in nonacidified microenvironments. During year 2 at pH 5.1 and year 1 at pH 4.7 in LRL, vertical profiles of nitrate suggest that nitrifiers may have existed primarily near the sediments (i.e., the highest NO_3^- concentrations were found at 9 m), but there is no evidence of this configuration for other years. In addition, the only known microenvironments with pH > 5.1 were within the anoxic hypolimnion and sediments, which cannot be considered a refuge for obligately aerobic nitrifiers. Acid tolerance apparently has developed to some degree in the LRL population of nitrifying bacteria.

In LRL, assimilation by phytoplankton is the most important uptake mechanism for ammonium and nitrate (*27, 28*). This importance is evidenced by their rapid depletion after the ice cover melts. The decrease in TN mentioned earlier, coupled with an increase in soluble inorganic forms, may indicate an overall decrease in biomass and perhaps primary productivity; that is, the difference may be attributable to fewer algal cells (particulate matter). Thus nitrogen cycle processes were disrupted in at least one way (possibly two) in LRL at pH 4.7: nitrogen fixation was inhibited, and planktonic uptake of inorganic N ions was slightly diminished, as suggested by a large decrease in TN and a slight increase in inorganic N (possibly indicating a general decline in productivity).

Phosphorus. No obvious differences in seasonal patterns of near-surface concentrations of soluble reactive phosphorus (SRP) or total phosphorus (TP) were found in the surface waters of the two basins during the 6 years of acidification (Figure 4). However, the winter maxima and summer averages for SRP show a possible treatment effect at pH 4.7; the treatment averages were slightly higher (~20%) and the winter maxima were much higher than those of the reference basin. No trend was observed for TP averages and maxima. Similarly, phosphorus concentrations in ELA lakes 223 and 302S (*29, 30*) were unaffected by acidification to pH 4.5.

Early studies of acidified lakes often reported lower chlorophyll *a* and greater transparency. The resulting "oligotrophication hypothesis" (*31*) stated that lower rates of organic-matter decomposition and coverage of bottom sediments by mats of acidophilic algae or *Sphagnum* would reduce the circulation of nutrients and thence decrease primary productivity. Og-

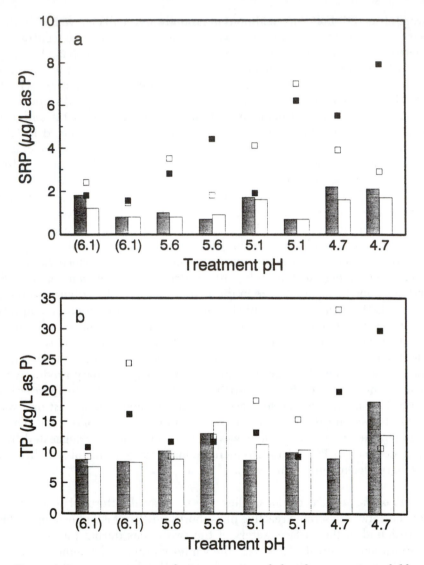

Figure 4. Summer averages and winter maxima of phosphorus species: soluble reactive P (SRP) (a) and total P (TP) (b). Bars are summer averages: shaded, treatment; and unshaded, reference. Squares are winter maxima: black, treatment; and white, reference.

burn (32) and Ogburn and Brezonik (33) examined this hypothesis with regard to phosphorus cycling for McCloud Lake, a small, acidic (pH 4.5–4.6), oligotrophic seepage lake in northern Florida. They concluded that the only factor that was significantly affected by acidification is the sorption (or desorption) of P to (from) the lake sediments.

Another hypothesis suggests that acidification by H_2SO_4 may increase the release of inorganic P from the sediments (*34*). This hypothesis states that increased SO_4^{2-} (which stimulates SO_4^{2-} reduction, resulting in increased H_2S) may increase the precipitation of Fe sulfides and thus reduce the amount of iron precipitating as Fe (hydr)oxide, which is known to co-precipitate phosphorus (*35, 36*). However, <3% of the inorganic P in LRL sediments is occluded in Fe (hydr)oxides and about 70% is readily exchangeable (*37*). Therefore, increased sulfate and resulting increased Fe sulfide precipitation is not likely to be important in the P cycle in LRL.

Detenbeck (*37*) and Detenbeck and Brezonik (*38, 39*) examined the effect of pH on phosphorus sorption for LRL sediments. Their results suggested that the flux of inorganic P from sediments could be diminished by as much as 90% if the pH of sediments decreased from 6.0 to 4.5. However, there was no observed treatment effect for TP and an apparent increase in SRP summer averages at pH 4.7 (Figure 4). Therefore, chemical sorption–desorption processes probably do not control phosphorus levels in LRL. The direction of response at lower pH implies that the balance between biotic uptake, deposition to sediments, and release from organic detritus by decomposition most likely controls SRP levels in the water column.

Silica. Seasonal patterns of dissolved reactive silica in LRL were unaffected by acidification to pH 4.7. These patterns derive largely from diatom population dynamics. Following an increase under ice cover, silica is rapidly depleted by the spring diatom bloom. A slight increase in midsummer is followed by depletion in autumn by a fall bloom. Annual average concentrations of silica were also not affected at pH 5.6 and 5.1 (Figure 5). However, an apparent (but small) treatment effect was observed at pH 4.7 when treatment-basin values were significantly higher than reference-basin values ($\alpha < 0.05$, Wilcoxon signed rank test). The greatest difference occurred during February 1991 (pH 4.7) when silica in the acidified basin reached 0.24 mg of SiO_2 per liter but was undetectable in the reference basin.

In contrast, during experimental acidification of ELA Lake 223 a decline in soluble silica to ~40% of the long-term mean occurred below pH 5.6 (*29*). This decline was attributed to the appearance in large numbers of *Asterionella ralfsii*, an acidophilic diatom. Prior to acidification, ELA Lake 223 was not dominated by diatoms. Acidification caused a change in algal class, that is, a replacement of acid-sensitive green and blue-green algae with acid-tolerant diatoms. The more common response of acidifying lakes is thought to be a shift in the diatom community from alkaliphilic species to acidophilic ones. For example, *A. ralfsii* is one of the indicators of acidification frequently used in paleoecological studies (*30*).

Possible explanations for the increase in silica in LRL at pH 4.7 include hydrologic differences, presence of a somewhat acid-tolerant diatom community, and the influence of another silica-utilizing community. Overall,

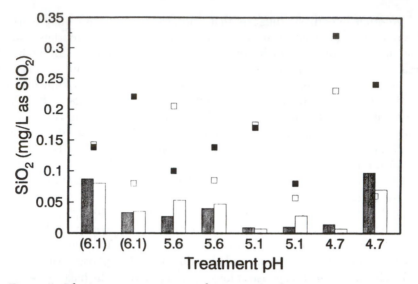

Figure 5. Silica summer averages and winter maxima. Bars are summer averages: shaded, treatment; and unshaded, reference. Squares are winter maxima: black, treatment; and white, reference.

the plot of summer average and winter maximum silica values (Figure 5) suggests that hydrologic factors (groundwater inflow, precipitation, and lake stage) played a much more important role in defining year-to-year variations than did the acid treatment.

During a drought in the region (during both years at pH 5.1 and the first year at pH 4.7; 1987–1990), no groundwater inflow occurred and lake levels dropped nearly 1.0 m. The drought years had the lowest summer averages of dissolved silica for both treatment and reference basins. Although no groundwater inflow occurred through the end of acidification, near-normal precipitation in summer 1990 raised lake levels ~15 cm by fall 1990. Silica concentrations returned to predrought values in both basins in year 2 at pH 4.7.

In years with normal water table levels (prior to 1987), measurable groundwater inflow (high in silica) occurred only near the extreme southeast shoreline of LRL, providing silica to the reference basin only. This inflow does not account for the similarity in treatment and reference basin concentrations prior to the drought, nor does it readily explain the increase observed in both basins in the second year at pH 4.7. These observations suggest that groundwater input per se is not the dominant source of silica to LRL, but that other hydrologic factors such as precipitation and lake stage are more important.

LRL receives nearly all its water in precipitation directly to its surface, but precipitation is generally very low in silica and probably provides little silica to LRL. Nevertheless, silica could be provided to both basins by groundwater interflows (events of short duration in which runoff flows through soils near the shoreline and littoral sediments that contain leachable silica). Presumably, groundwater interflows would be more frequent in years with higher precipitation; however, the contribution of interflows is difficult to quantify.

Lake level could also influence silica levels. For example, the decline in lake level resulted in a large loss of surface area and decreased the amount of contact between lake water and sandy littoral areas that contain weatherable silicate minerals. Rates of weathering are usually enhanced by acidification and would help to explain the interbasin differences observed at pH 4.7. However, weathering rates of LRL sediment are unknown (*see* Sediment Processes section). In addition, although the differences between the basins were significant, they were small (0.01–0.03 mg of SiO_2 per liter) and may be accounted for by small differences in hydrological factors.

A large population of the freshwater sponge, *Spongilla lacustra*, exists in LRL, and field observations indicated that the sponge population declined somewhat with acidification. This decline may have influenced silica concentrations, but the effects are unquantified thus far. Unfortunately, only limited data are available regarding the taxonomic progression of phytoplankton in LRL. However, because the near-surface concentration of silica in lakes is usually intimately linked to diatom population dynamics, it seems likely that acid-tolerant species did not become more common in LRL during acidification.

Internal Alkalinity Generation. *Overview.* One of the major concepts to emerge from controlled acidification experiments is internal alkalinity generation (IAG). The importance of this mechanism in regulating the alkalinity of some acid-sensitive lakes was first recognized in the Experimental Lakes Area (ELA) of northwestern Ontario, Canada. It was found that a substantial fraction of the sulfuric acid added to lower the lake's pH was disappearing and that much more acid was required to maintain a target pH than would be predicted from the lake's volume and initial alkalinity (e.g., *40*). Similar results were obtained from acidification experiments conducted in enclosures and ion balances measured at McCloud Lake (Florida) in the early 1980s (*41, 42*). Investigations by ELA researchers (*40, 43–45*) and LRL researchers (*17, 27, 28, 46*) during the 1980s led to an improved understanding of the mechanisms involved in IAG, the factors affecting its importance, and the development of quantitative models to predict IAG rates by various mechanisms (e.g. *45, 47*).

Several studies have shown that in some cases in-lake processes can generate more alkalinity than watershed processes. By constructing a detailed alkalinity budget for Lake 239, Schindler et al. (43) showed that IAG was 4.5 times more important than terrestrial processes. Cook et al. (40) used the input–output approach on artificially-acidified Lake 223. Their findings were similar to those of Schindler et al. (43) in that IAG neutralized 66–81% of the acid additions. In general, IAG is not important in lakes with large watersheds and short hydraulic retention times. However, it can be important in regulating alkalinity and pH in seepage lakes (groundwater recharge systems), which tend to have long water residence times and receive most of their water as precipitation directly on the lake surface (27, 47).

ANC (or alkalinity) can be defined by the electroneutrality condition as the difference between the strong base cations and the strong inorganic and organic ($RCOO^-$) acid anions (48):

$$ANC = [Ca^{2+}] + [Mg^{2+}] + [Na^+] + [K^+] + [NH_4^+] -$$

$$[SO_4^{2-}] - [Cl^-] - [NO_3^-] - [RCOO^-] \quad (4)$$

In addition, aluminum ions (Al^{3+}, $AlOH^{2+}$, and $AlOH_2^+$) must be considered in some acidic lakes. Soluble Fe^{2+} and Mn^{2+} forms may contribute significantly to the ion balance in anoxic hypolimnia of some lakes.

Four main processes are involved in IAG:

1. sulfate reduction, a microbial process occurring primarily in near-surface sediments below the oxic–anoxic boundary;

2. cation production by ion exchange, in which H^+ replaces base cations on exchange sites of organic and mineral sediments;

3. cation production by weathering reactions of minerals, such as aluminosilicate clays; and

4. nitrogen transformations, such as algal assimilation in the water column and denitrification in surficial sediments.

In addition, reductive dissolution of Fe or Mn (hydr)oxides at the sediment–water interface may be an important but ephemeral (seasonal) source of IAG in the anoxic hypolimnia of lakes. The nature and potential importance of each of these processes is described briefly as follows.

Removal of sulfate from the water column can occur by either assimilatory or dissimilatory reduction. Assimilatory reduction occurs in the water column, whereas uptake by plankton results in the formation of organic S.

$$106CO_2 + 16NO_3^- + HPO_4^{2-} + 122H_2O + 19H^+ + 0.5SO_4^{2-} \longrightarrow$$

$$C_{106}H_{264}O_{110}N_{16}P_1S_{0.5} + 139O_2 \quad (5)$$

Dissimilatory reduction by anaerobic bacteria occurs in the anoxic hypolimnion of stratified lakes and in sediments just below the oxic–anoxic boundary. It produces H_2S,

$$SO_4^{2-} + 2CH_2O \longrightarrow H_2S + 2HCO_3^- \tag{6}$$

which is subsequently incorporated into organic matter or metal sulfides.

The rate of reduction generally has been found to be proportional to $[SO_4^{2-}]$ in the water column (e.g., *46, 47, 49*), even though recent investigation suggests that the process is complicated and not solely diffusion-controlled (*50*). Sulfate reduction and fixation as organic S, metal sulfides, or elemental S in near-surface sediments creates a concentration gradient that promotes diffusion of SO_4^{2-} from lake water into the sediments. Sulfate reduction produces a net alkalinity gain only if the reduced sulfur is permanently incorporated into the sediment (*49, 51, 52*).

The processes of cation production by weathering or ion exchange cannot be differentiated by ion-budget calculations, but they may be distinguished on the basis of kinetics. In early experiments to investigate the role of sediments as buffers, sequential additions of sulfuric acid to well-mixed sediment slurries produced rapid increases in soluble Ca^{2+} that reached stable concentrations within 24 h (*53, 54*). The rapid production of cations suggested that the dominant process was ion exchange

$$nH^+ + MX \longrightarrow M^{n+} + nHX \tag{7}$$

where M^{n+} is a cation with a charge of $+n$ and X is the exchanger surface of a mineral or organic particle. Microcosm and mesocosm acidification studies (*43, 55*) also resulted in increased Ca^{2+} concentrations that were attributed to ion exchange. Laboratory techniques that evaluate soil cation-exchange capacity (to model the response of soil to acidic deposition; e.g., *56*) can be used to quantify the reservoir of exchangeable base cations in surficial lacustrine sediments and model the response to lake acidification.

The incongruent dissolution of aluminosilicates, such as

$$2NaAlSi_3O_{8(s)} + 2H_2CO_3^* + 9H_2O \longrightarrow$$
$$2Na^+ + 2HCO_3^- + Al_2Si_2O_5(OH)_{4(s)} + 4Si(OH)_4 \tag{8}$$

is a relatively slow process, occurring over days to years. Significant advances in the understanding of mineral weathering processes and rates have been made in the past decade (e.g., *57, 58*). Weathering is an important source of alkalinity and acid neutralization in terrestrial systems and in drainage lakes with significant watersheds; over long time scales it is the dominant mechanism. Although weathering-induced IAG in seepage lakes is probably small compared with other IAG processes, Sherman (59) showed that several feldspar minerals found in LRL sediments were undersaturated with respect to pore water at various sediment depths. However, the contribution of aluminosilicate weathering to IAG in LRL has not been quantified.

Congruent dissolution of carbonates, such as

$$CaCO_3 + H^+ \longrightarrow Ca^{2+} + HCO_3^- \tag{9}$$

occurs rapidly in the undersaturated, dilute waters of acid-sensitive (low-ANC) lakes, but few or no carbonate minerals are found in the local geology of such lakes. However, wind-blown dust from farms or prairie land may provide a minor source of calcium and bicarbonate.

Nitrogen transformations such as nitrate assimilation, denitrification, and decomposition (eqs 10–12) contribute to alkalinity by consuming H^+. In contrast, both ammonium assimilation and nitrification (eqs 13 and 14) consume alkalinity via the production of H^+.

$$106CO_2 + 138H_2O + 16NO_3^- \longrightarrow$$
$$C_{106}H_{260}O_{106}N_{16} + 16OH^- + 138O_2 \tag{10}$$
$$5CH_2O + 4NO_3^- + 4H^+ \longrightarrow 5CO_2 + 2N_2 + 7H_2O \tag{11}$$
$$C_{106}H_{260}O_{106}N_{16} + 16H^+ + 106O_2 \longrightarrow$$
$$106CO_2 + 106H_2O + 16NH_4^+ \tag{12}$$
$$106CO_2 + 106H_2O + 16NH_4^+ \longrightarrow$$
$$C_{106}H_{260}O_{106}N_{16} + 16H^+ + 106O_2 \tag{13}$$
$$NH_4^+ + 2O_2 \longrightarrow NO_3^- + H_2O + 2H^+ \tag{14}$$

In regions where atmospheric deposition of HNO_3 is high, nitrogen transformations may be important in the acid–base chemistry of surface waters. When nitrate exceeds the requirements of algae (N:P >> 10:1–20:1), denitrification may become more important than nitrate assimilation (28, 60). Rudd et al. (61) and Kelly et al. (62) proposed that denitrification removes nitrate less efficiently than algal assimilation, and that the dominance of denitrification allows acidification to occur. However, the alkalinity budgets of many low-ANC lakes are not strongly affected by nitrogen transformations because the effects of nitrate and ammonium retention roughly cancel each other. Chemical budgets indicate this to be the case in LRL (17) and several other low-ANC lakes (27, 28, 40, 63). Seasonal changes in alkalinity may be affected by N transformations, however. For example, Garrison et al. (25) attribute a decrease in under-ice ANC in nearby Max Lake to increased nitrification.

Several studies have shown that sulfate reduction and cation production (the sum of ion-exchange and weathering processes) together account for 60–100% of the IAG measured. Reduction of SO_4^{2-} accounted for over half (53%) of the IAG, and production of Ca^{2+} was the second most important mechanism (39%) in the alkalinity budget for Lake 239 (*43*). Findings for artificially-acidified Lake 223 were similar (*40*), but sulfate reduction was more important because the addition of H_2SO_4 increased $[SO_4^{2-}]$ and stimulated the rate of SO_4^{2-} reduction. Similarly, Lin et al. (*63*) calculated ion balances for Vandercook Lake, a dilute seepage system close to LRL, and found that cation production accounted for 46% of the IAG and sulfate reduction accounted for 54%.

Little Rock Lake.   Net IAG in LRL takes place primarily in or near the sediment. It can be evaluated by measurements of pore-water chemistry, comparison of hypolimnetic and epilimnetic chemistry, and calculation of ion budgets. An example of each approach follows.

Numerous measurements of pore-water chemistry have been made in LRL throughout the experiment (*4, 17, 59*). Typical vertical pore-water profiles (Figure 6) indicate that the sediments are acting as sinks for sulfate

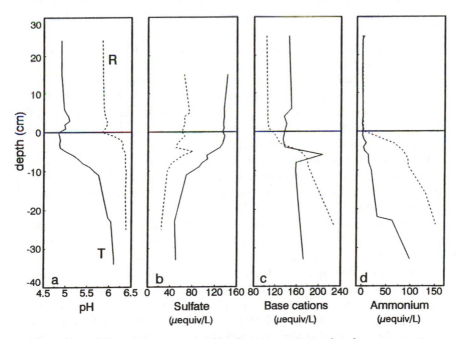

Figure 6. Sediment pore-water profiles for various IAG-related parameters in treatment basin (T, solid line) and reference basin (R, dashed line), 5-m sites, July 1990.

and H^+ from the water column (Figures 6a and 6b) and as sources to the water column for base cations (Ca^{2+}, Mg^{2+}, K^+, and Na^+) and ammonium (Figures 6c and 6d). The contributions of sediments to the cation content of lake water and loss of sulfate from the water column to the sediments can be quantified from the pore-water profiles and extrapolated to annual factors by use of Fick's law (e.g., 46, 59).

The pore-water profiles also indicate a possible treatment effect. In the treatment basin before acidification to pH 4.7, much higher levels of ANC (alkalinity) and higher pH were found in the pore water just 1–2 cm below the sediment–water interface (59). In contrast, pore-water pH profiles obtained in the same site in the summers of 1990 and 1991 show pH < 5.0 in the upper 5–10 cm of sediment. Corresponding profiles for a site in the reference basin did not show such a depression (Figure 6a; ref. 4).

Young et al. (64) proposed that the exchangeable acidity of sediments should be considered when lime doses are calculated in order to adequately neutralize acidic surface waters. Although the exchangeable acidity of LRL sediments was not measured, the low pore-water pH values suggest that acidification of sediment exchange sites occurred to a depth of 5–10 cm in the acidified basin at pH 4.7 (see Sediment Processes section). The increase in exchangeable H^+ in the surficial sediments will slow the recovery of the lake's pH and alkalinity after acid loading is stopped, because the exchange process will have to be reversed and the released H^+ neutralized by other processes.

The alkalinity generated in or near the sediments of the treatment basin accumulates in the hypolimnion during periods of water-column stratification (the reference basin only weakly stratifies). This accumulation results in a substantial increase in alkalinity and pH in the bottom waters of the treatment basin each summer. For example, in August 1990 alkalinity was nearly 900 μequiv/L and pH was 6.15 at 9-m depth (~1 m above the sediment–water interface). Corresponding near-surface alkalinity was −27 μequiv/L, and pH was 4.8. Peaks in 9-m alkalinity increased during the acidification experiment—from 135 μequiv/L in 1986, to 490 in 1988, to ~900 in 1990—primarily because of an increased rate of sulfate reduction with increasing $[SO_4^{2-}]$.

The charge-balance definition of alkalinity (eq 4) allows us to calculate the relative contribution of each ion to the hypolimnetic alkalinity from the difference between the epilimnetic and the hypolimnetic volume-weighted concentrations. Estimates for August 1990 show that alkalinity contributions follow the order: Fe^{2+} > NH_4^+ ~ SO_4^{2-} > Ca^{2+} > Mg^{2+} ~ K^+ ~ Al^{3+} > Na^+ ~ Mn^{2+}. Ferrous ion and ammonium comprised most of the hypolimnetic alkalinity (49 and 18%, respectively). Results for other years were similar, except that Fe^{2+} and NH_4^+ contributions were less than SO_4^{2-} and Ca^{2+} contributions (4, 17). These results reflect the trend of increasing hypolimnetic NH_4^+ during the acidification experiment and a somewhat higher than usual buildup in hypolimnetic Fe^{2+} during 1990.

During August 1990, volume-weighted hypolimnetic alkalinity was ~350 μequiv/L. If all of this alkalinity were available as net acid neutralization capacity, the resulting homogeneous basin alkalinity would be about +2.5 μequiv/L. However, after fall overturn (October 1990) the alkalinity was only −15.5 μequiv/L (still higher than the preoverturn volume-weighted epilimnetic alkalinity of −21 μequiv/L). Clearly, much of hypolimnetic alkalinity was lost by reoxidation of reduced compounds following destratification of the lake and reoxygenation of the bottom waters. For example, virtually all the Fe(II) is reoxidized in the lake water, even though rates of oxidation are slower at lower pH (2, 3, 35). Nearly all of the alkalinity lost between August and October can be accounted for by reoxidation of iron. The extent of reduced sulfur reoxidation is still unknown (49).

If we omit the ions contributing little or no net gain (i.e., Fe^{2+}, NH_4^+, and probably Mn^{2+}) to the alkalinity generated in the hypolimnion during August 1990, the decrease in sulfate between the epilimnion and hypolimnion was responsible for about 50% of the increase in alkalinity; the increase in Ca^{2+} was responsible for 30%. Increases in Mg^{2+} and K^+ contributed 7 and 6%, respectively. The increase in Al^{3+} contributed about 5%, but this cannot be considered to be mitigative because of the toxicity of aluminum to aquatic biota. The relative contributions of each ion to the hypolimnetic alkalinity are similar to the relative contributions to whole-basin alkalinity as determined by ion budgets.

Water budgets are as yet incomplete for the the past 3 years of acidification, and ion budgets thus are also incomplete for these years. The following discussion is based on net annual budgets developed by Tacconi (65) and Weir (17) for the pretreatment period and the first 3 years of acidification. A summary of their budgets for sulfate and alkalinity (Table III) shows some interesting trends. Areal sulfate reduction rates were comparable in the two basins before acidification began. However, the average rate for the treatment basin was twice that of the reference basin for the first 3 years of acid loading because of higher $[SO_4^{2-}]$, which stimulated reduction. Similarly, net areal rates of IAG in the two basins were nearly the same in the preacidification period. However, IAG nearly doubled in the treatment basin during the first 3 years of acid loading. A decrease in net IAG in the reference basin during the first 3 years of acidification reflects problems with budget analysis in one of the years for the reference basin only (the budget did not close during the 1986–1987 treatment year).

On the basis of mass balance calculations through the first 3 years of acid additions (17), only 33% of the added acid resulted in a decrease in lake alkalinity. A second 33% was neutralized by in-lake (IAG) processes, of which sulfate reduction accounted for slightly more than half and cation production for slightly less than half. Approximately 33% of the total sulfate load (wet and dry deposition, and acid additions) was lost via outflow. Therefore, about half of the added acid remained in the water column; two thirds of it was unreacted and one third was neutralized by base cations.

Table III. Sulfate and Alkalinity Budgets for LRL Basins

Treatment	Inputs	Outputs	Acid Additions	Δ Storage	Internal Production
Sulfate Budgets					
1984–1985 (Pretreatment)					
Treatment basin	40	27	0	−14	−27
Reference basin	37	23	0	−12	−26
1985–1988 (Average of first three treatment years)					
Treatment basin	41	48	115	55	−53
Reference basin	37	23	0	−7	−21
Alkalinity Budgets					
1984–1985 (Pretreatment)					
Treatment basin	−10	15	0	21	46
Reference basin	−11	12	0	22	45
1985–1988 (Average of first three treatment years)					
Treatment basin	−14	4	−115	−44	89
Reference basin	−15	17	0	−15	14

NOTE: All values are in milliequivalents per square meter per year.

Sediment Processes. Treatment-basin water-column base cations Ca^{2+}, Mg^{2+}, and K^+ increased steadily with decreasing pH until the total base-cation concentration at pH 4.7 was about 50 μequiv/L higher than background. Ca^{2+} exhibited the largest increase of these ions (37 μequiv/L). In addition, production of base cations accounted for about half of the IAG measured in LRL for the first 3 years of acid additions (17). The most likely sources of the cations include sediment ion exchange, enhanced weathering or decomposition rates, and groundwater inflow. Because groundwater flows only into the reference-basin of LRL, it cannot be responsible for the increases observed in the treatment basin. To expand our knowledge of the biogeochemical cycle of base cations and to infer the sources of cations, we investigated several attributes of LRL sediments. We measured the total, organically bound, and exchangeable concentrations of base cations contained in the sediments and conducted additional analyses to determine whether Ca^{2+}–NH_4^+ exchange can explain the transient pore-water profiles of calcium observed seasonally in LRL (59).

Methods used in these determinations were described in the Experimental Methods section. The measurements were made on sediments collected during the first year of treatment to pH 5.1 from areas (5-m water-column depth) characterized by fine-grained, organic-rich sediments (organic content greater than 40%; i.e., gyttja) with water content greater than 90% (17, 59). The mineral fraction consisted of approximately equal portions of silt and clay-size particles. The silt portion was predominantly quartz (>50%) with the remainder plagioclase (21–33%) and alkali feldspar (11–20%). The clay portion was 21–33% smectite with 11–20% each vermiculite, illite, kaolinite, quartz, and plagioclase. Some alkali feldspar (5–10%) and chlorite (<5%) were also detected. Baker et al. (*66*) estimated that the fine-grained organic sediments, as characterized by the samples in this study, cover approximately 66% of the lake surface area.

H⁺-Cation Exchange.   This section summarizes the results of laboratory analyses performed on LRL sediments to measure the total, organically bound, and exchangeable concentrations of base cations contained therein. We discuss mineral weathering and decomposition of organic matter with respect to the production of cations and estimate the possible contribution of H^+-cation exchange to water-column chemistry and the generation of alkalinity (IAG). Other sediment processes that may influence interpretation of data, such as bioturbation, are also discussed.

The total content of sediment base cations (Ca^{2+}, Mg^{2+}, K^+, and Na^+), as determined by lithium metaborate fusion (*16*), decreased down-core in both basins from 190 ± 38 mequiv/100 g in the 0–2-cm interval to 160 ± 32 mequiv/100 g in the 12–14-cm interval (Figure 7a). For any depth interval the difference between the basins was less than 5%, and the analysis of sediments for total base cations was not precise enough to confirm the differences between basins detected by other analyses. In contrast, the analyses

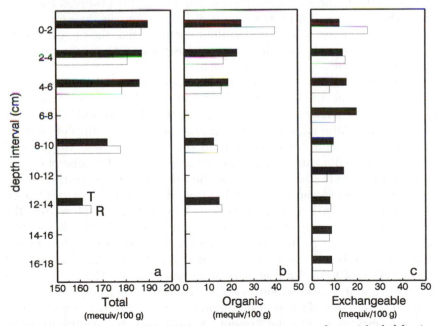

Figure 7. Sediment profiles of base cations in treatment basin (shaded bar) and reference basin (unshaded bar), 5-m sites, October 1987. Key: a, total; b, organic fraction; and c, exchangeable fraction. All are expressed in milli-equivalents per 100 g of wet sediment.

for organically bound and exchangeable cations did show interbasin differences in near-surface sediments. For example, Figure 7b shows the base-cation content obtained by H_2O_2 digestion (15). This portion includes cations that are both exchangeable and more tightly bound in organic matter, but for ease of discussion, we will refer to this fraction as the organic fraction. The 0–2-cm interval of the reference-basin sediments had an organic fraction content ~15 mequiv/100 g higher than that in the treatment basin, which was 25 mequiv/100 g. This fraction decreased down-core in both basins to ~15 mequiv/100 g in the 12–14-cm interval.

The difference between the treatment and reference organic fraction in the 0–2-cm interval can be accounted for by the difference in the exchangeable fraction determined by the ammonium saturation method (18). This fraction represents cations on the exchange sites of mineral and organic particles. The exchangeable fraction in reference-basin sediments decreased from 25 mequiv/100 g in the 0–2-cm interval down to 9 mequiv/100 g in the 16–18-cm interval (Figure 7c). In the 0–2-cm interval of the treatment-basin sediment this fraction was only 12 mequiv/100 g, resulting in an interbasin difference nearly equal to that measured in the organic fraction. In contrast to reference-basin values, treatment-basin values increased with depth to 20 mequiv/100 g in the 6–8-cm interval and then decreased to 9 mequiv/100 g in the 16–18-cm interval. Therefore, the observed trends in the exchangeable and organic fractions of base cations in the near-surface sediments of the two basins support the hypothesis that the increased concentrations of base cations in the treatment-basin water column were derived from the sediments by H^+-cation exchange.

Regarding the other potential explanations for the cation increases in the water column, we are not yet able to provide definitive evidence (pro or con), but the following discourse suggests that they are unlikely sources of the cations. Mineral weathering (i.e., the incongruent dissolution of aluminosilicates) could provide cations to the water column. The mineral content of the sediments contains substantial quantities of weatherable minerals. Several studies have shown that the rates of feldspar weathering are accelerated by increasing $[H^+]$ (58), but this effect generally occurs at much lower pH values (pH < 3) than occur in LRL sediments. Moreover, the production of cations by mineral weathering is accompanied by the production of silica. Although treatment-basin SiO_2 at pH 4.7 was slightly higher than reference-basin values (see Nutrients section), the increase was not nearly sufficient to explain the increase in base cations. The contributions to water-column concentrations and IAG by weathering reactions in LRL sediments, and the effects of acidification on weathering rates are currently under investigation. For the present, we assumed that contributions by weathering were the same in both basins.

Decomposition of organic matter also could provide cations to the water column. In an initial attempt to measure decomposition rates, Tacconi (65) showed that the release of base cations from seston occurred mostly in the first 48 h. This rapid release suggested that they were exchangeable and not fixed in the organic matrix. The decomposition of freshly deposited organic matter thus is an unlikely source of the cations; most organically bound cations are easily and rapidly lost. The rates of organic decomposition were assumed for this discussion to be the same in each basin.

The differences between treatment and reference basins observed in the profiles of exchangeable cations could be explained by:

1. differences in the density and activity of the benthic community (i.e., bioturbation);

2. differences in the factors that control the composition and mass of fluxes to the sediment, resulting in different depositional histories; and

3. treatment effects, in which exchangeable cations are replaced by H^+, thus diminishing the pool of exchangeable cations.

Regarding the first explanation, the benthic macroinvertebrate communities in the treatment and reference basins of LRL are similar in composition and low in density, and treatment caused relatively minor changes (67, 68). Although the effects of benthic activity were not measured at the core sites, there is little reason to believe that bioturbation was important at these depths or different between basins.

It would be an oversimplification to suggest that a pair of cores could be considered duplicates or that information from a single core could be extrapolated to basin-wide processes. However, carefully chosen core sites can provide a basis for reasonable estimates and hypothesis formation. Both basins of LRL have similar morphometric and edaphic conditions (Figure 1), suggesting similar depositional regimes and histories. Core locations were chosen near sites where extensive pore-water measurements had been made (4, 17, 59) and where sediment cores had been dated by ^{210}Pb. Mean mass accumulation rates calculated over the top 4–5 cm were comparable between basins: 105 and 126 g/m^2 per year for the treatment and reference basin, respectively (65, 66).

Because ion exchange is a rapid process, differences in depositional history cannot be inferred solely from profiles of exchangeable cations. The magnitude of the exchangeable fraction does not indicate properties of the solid surface per se. That is to say, it is not a measure of the total number of exchange sites. It is only a measure of those sites occupied by base cations. If corresponding values of exchangeable acidity or total exchange sites were available, more definitive conclusions regarding matrix differences could be made, but these values were not measured in this investigation. Nonetheless, the percent composition of calcium, magnesium, potassium, and sodium in the top 4 cm of sediment shows that the sediments are similar (Figure 8).

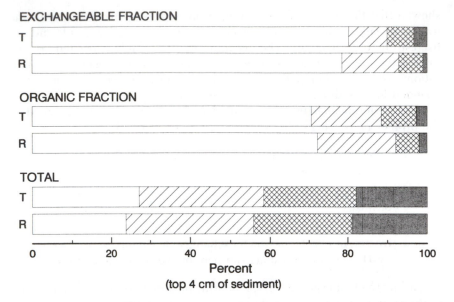

Figure 8. Percent composition of Ca^{2+} (unshaded), Mg^{2+} (diagonally hatched), K^+ (cross-hatched), and Na^+ (shaded) in base-cation pools of the top 4 cm of treatment-basin sediment (T) and reference-basin sediment (R), 5-m sites, October 1987.

The largest interbasin difference in composition of the exchangeable fraction is in the proportion of exchangeable sodium, which was 3 times higher in the treatment-basin sediments than in the reference basin. The apparent enrichment of exchangeable Na^+ in the treatment sediments is consistent with a depletion of exchangeable Ca^{2+}, Mg^{2+}, and K^+, the cations that increased in the water column with decreasing pH. Sodium in the water column showed no trend with decreasing pH, suggesting that no H^+–Na^+ exchange occurred.

Finally, an estimate of the potential increase in the treatment-basin water-column base-cation concentration attributable to ion exchange was obtained by using the following information:

- E_{0-N} is the exchangeable content of base cations in the top N cm of sediment;
- D is the density of sediments, 1.02 g wet/cm^3;
- dry:wet is the dry-to-wet ratio of sediments, 0.07 g/g;
- 0.66SA is 66% of treatment-basin surface area, 6.47×10^4 m^2;
- V_{basin} is the volume of treatment basin, 3.77×10^5 m^3.

The exchangeable content of base cations in the top N cm of sediment, E_{0-N}, is the summation over depth of the exchangeable cation fractions presented graphically in Figure 7c. Values of E_{0-N} are given in Table IV for $N = 2, 4,$ 8, and 12. The increase in the treatment-basin water-column base-cation concentration attributable to complete ion exchange of base cations in the $0-N$-cm interval, ΔBC_{IX}, was calculated by using E_{0-N} determined for both the treatment- and reference-basin sediments (Table IV). The addition of ΔBC_{IX} to the water-column base-cation concentration results in the maximum concentration attainable solely by ion exchange in the near-surface sediments. Although ion exchange will occur to varying degrees at different sediment depths, this calculation provides an estimate of the possible contribution of ion exchange to the water-column chemistry.

For example, the average base-cation concentration in the water column of the treatment basin was 124 μequiv/L in 1987 when the sediment samples were obtained (year 1 at pH 5.1). If we consider the situation in which all the cations in the top 4 cm of treatment-basin sediments (where $E_{0-4} = 27$, giving $\Delta BC_{IX} = 64$) were exchanged for H$^+$ ions, then the highest possible water-column concentration would be 188 μequiv/L (124 + 64). Because the treatment-basin sediments apparently had lost base cations before sampling was done, this calculation probably underestimates base-cation contributions to the lake water by ion exchange. Another calculation may provide an estimate of treatment-basin ΔBC_{IX} for the entire acidification phase if it is assumed that E_{0-N} of the preacidified (pre-1985) treatment basin was similar to E_{0-N} of the reference basin in 1987; ΔBC_{IX} based on this premise is given in Table IV. Prior to acid additions, the average base-cation concentration of the treatment basin was 90 μequiv/L. If we again consider the situation in which all the base cations in the top 4 cm (where $E_{0-4} = 40$, giving $\Delta BC_{IX} = 94$) were exchanged for H$^+$, then the highest possible concentration would be 184 μequiv/L (90 + 94). This result is similar to that obtained by using E_{0-4} measured on the 1987 treatment-basin sediments.

Similarly, the water-column base-cation concentration resulting from complete exchange of the top 2 cm would be 155 μequiv/L (124 + 31,

Table IV. Total Exchangeable Base-Cation Content and Potential Increase with Ion Exchange

Depth Interval (0–N cm)	$E_{0-N}{}^a$ (mequiv/100 g)		$\Delta BC_{IX}{}^b$ (μequiv/L)	
	Treatment-Basin Sediments	Reference-Basin Sediments	Based on Treatment E_{0-N} cm	Based on Reference E_{0-N} cm
0–2	12	25	31	59
0–4	27	40	64	94
0–8	62	59	146	139
0–12	86	74	203	174

^{a}Exchangeable base-cation content of treatment- and reference-basin sediments.
^{b}Potential increase in treatment-basin water-column base-cation content attributable to ion exchange.

based on treatment E_{0-2}) or 149 μequiv/L (90 + 59, based on reference E_{0-2}). The actual base-cation content of the treatment-basin water column was 140 μequiv/L at pH 4.7, which is slightly less than the water-column concentration obtained from complete exchange of only the top 2 cm. The sediment reservoir of exchangeable base cations thus is more than sufficient to account for the observed increase in cations in the water column. This cation-exchange process will have to be reversed and the released H^+ neutralized by other processes, such as sulfate reduction, before complete recovery from acidification can be achieved.

$Ca^{2+}-NH_4^+$ Exchange. Measurements were made to determine the $Ca^{2+}-NH_4^+$ exchange coefficient to ascertain whether $Ca^{2+}-NH_4^+$ exchange can explain the large, rapid changes in pore-water calcium concentrations observed by Sherman (59) in LRL over short periods of time (~1 month). These changes in calcium occurred during periods of ammonium production via decomposition (May–June). The magnitude of the $Ca^{2+}-NH_4^+$ exchange coefficient for LRL sediments would indicate whether NH_4^+ is important in the release of Ca^{2+} from the sediments.

An equation describing the exchange reaction between Ca^{2+} and NH_4^+ is (69):

$$Ca^{2+} + 2NH_4X \longrightarrow 2NH_4^+ + CaX \tag{15}$$

where X is an exchanger surface of a sediment particle. Because this process is reversible, it can be described in terms of an equilibrium constant of the general form:

$$K = \frac{(CaX) \times (NH_4^+)^2}{(NH_4X)^2 \times (Ca^{2+})} \tag{16}$$

where parentheses refer to activities.

Difficulties associated with measuring the activities of ions on a solid phase led many workers to suggest empirical relationships similar to eq 16 in an attempt to define the equilibrium constant, K. A relationship developed by Vanselow (70) assumes that surface activity is proportional to the mole fraction of an ion. For exchange between cations M and N, the surface activity of M is defined by

$$(MX) = \frac{[\overline{M}]}{[\overline{M}] + [\overline{N}]} \tag{17}$$

where $[\overline{M}]$ and $[\overline{N}]$ (in moles per kilogram) are the concentrations of M and N occupying exchange sites. Substituting this assumption into the general form of the equation for the exchange constant results in an exchange constant, K_v:

$$K_v = \frac{[\overline{Ca}] \times \{[\overline{Ca}] + [\overline{NH_4}]\} \times (NH_4^+)^2}{[\overline{NH_4}]^2 \times (Ca^{2+})} \tag{18}$$

Because ion exchange is affected by surface charge, the exchanging cations may be represented in equivalent rather than molar amounts. This convention results in an exchange reaction between calcium and ammonium expressed as follows:

$$\frac{1}{2} Ca^{2+} + NH_4X \longrightarrow NH_4^+ + Ca_{1/2}X \tag{19}$$

where $Ca_{1/2}X$ represents the association of one positive charge from the Ca^{2+} ion with one negative surface charge. This convention was used by Gapon (71), who assumed that surface activity is proportional to the equivalent fraction of an ion. That is, surface activity of cation M^{m+} for exchange between cations M^{m+} and N^{n+} is defined by

$$E_{M1/mX} = \frac{[\overline{M}_{1/m}]}{[\overline{M}_{1/m}] + [\overline{N}_{1/n}]} \tag{20}$$

where $[\overline{M}_{1/m}]$ and $[\overline{N}_{1/n}]$ are the concentrations of M^{m+} and N^{n+} occupying exchange sites and $\{[\overline{M}_{1/m}] + [\overline{N}_{1/n}]\}$ is the total number of sites on the solid phase, both expressed in milliequivalents per 100 g of sediment. The resulting exchange constant (Gapon's coefficient, K_G) for $Ca^{2+}-NH_4^+$ exchange is

$$K_G = \frac{[Ca^{2+}]^{1/2} \times E_{NH_4X}}{[NH_4^+] \times E_{Ca_{1/2}X}} \tag{21}$$

Both the molar and equivalent conventions satisfy the conservation of mass and charge requirements, but neither convention accurately represents reality. The exchange of molar quantities does not consider charge in the computation of surface activity, which is a serious omission when exchanging ions have differing charges. In the equivalent convention, the symbol $M_{1/m}X$ ($m > 1$) has no molecular significance (i.e., there exists no such entity as one-half of a calcium ion). However, Sposito (72) showed that the Vanselow and Gapon equations are interrelated by thermodynamic principles. Goulding (73) stated that none of the empirical constants have been found to be truly constant over a range of exchange conditions, but Gapon's constant (hereafter referred to as Gapon's coefficient) has proven useful in practice.

Gapon's coefficient, K_G, was determined on sediment that was prepared in three different ways: Ca^{2+}-saturated, NH_4^+-saturated, and untreated.

Exchange was measured at typical pore-water concentrations of Ca^{2+} and NH_4^+. For the Ca^{2+}-saturated and NH_4^+-saturated sediments, K_G was calculated by mass balance from measurements of initial and final calcium and ammonium concentrations in solution; the total concentration of exchange sites was known. On the untreated sediment, K_G could be determined only after these simplifying assumptions were made:

1. Ca^{2+} and Mg^{2+} exchange properties are similar and they may be treated as a single ion;

2. NH_4^+ and K^+ exchange properties are similar and they may be treated as a single ion;

3. Na^+ concentrations and effects are negligible.

The low value of the exchange coefficient determined for Ca^{2+}-saturated sediment, 0.05 (± 0.06, $n = 11$), suggested that no replacement of Ca^{2+} by NH_4^+ would occur at typical pore-water concentrations of these ions. This inactivity is not surprising because selectivity is largely a function of ionic charge. The affinity for monovalent ions over the resident divalent ions is much lower. In contrast, similar and higher exchange coefficients were determined for NH_4^+-saturated and untreated sediments: $K_G = 4.6$ (± 1.6, $n = 24$) and 5.5 (± 1.6, $n = 17$), respectively. The Gapon coefficients determined on NH_4^+-saturated and untreated sediments were thus considered to be reasonable and the average of those determinations, $K_G = 5.1$, used in subsequent calculations.

If $Ca^{2+}-NH_4^+$ exchange occurring in LRL sediments is described by the Gapon relationship, the calculated change in NH_4^+ on exchange sites should approximate the change in measured pore-water Ca^{2+}. This relationship was tested for each basin by using pore-water profiles of NH_4^+ and Ca^{2+} obtained in August and October 1988. A range of values of $\Delta E(NH_4^+)$, the volume-weighted change in NH_4^+ occupying exchange sites between times t_1 and t_2 (microequivalents per square centimeter in the top 12 cm of sediment), was calculated from the pore-water profiles for these dates and the following parameters:

- $E'(Ca^{2+})$ is the concentration of Ca^{2+} occupying exchange sites; range of values based on laboratory determination: 7.7–10.0 mequiv/100 g, assumed to be constant down-core;

- K_G is Gapon's coefficient, 5.1, based on laboratory study;

- D is density of sediments, 1.02 g wet/cm^3;

- dry:wet = dry-to-wet ratio of sediments, 0.07 g/g.

The measured change in reference-basin pore-water Ca^{2+} ($\Delta Ca^{2+} = 0.5$ μequiv/cm^2 for the top 12 cm) compares well with the calculated change in NH_4^+ occupying exchange sites ($\Delta E(NH_4^+) = 0.5$–0.64 μequiv/cm^2).

However, ΔCa^{2+} measured for the acidified basin (0.24 μequiv/cm^2 for the top 12 cm) does not agree with the corresponding value of $\Delta E(NH_4{}^+)$ (−1.2 to −1.6 μequiv/cm^2 for the top 12 cm). This result suggests that a process other than $Ca^{2+}-NH_4{}^+$ exchange controls pore-water Ca^{2+} in the treatment basin. For example, the effects of $Ca^{2+}-H^+$ exchange may overwhelm the effects of $Ca^{2+}-NH_4{}^+$ exchange under the increased H^+ loads present in the treatment basin.

Although selectivity of the sediments favors calcium, the results of this study give some support to the idea that $NH_4{}^+$ is an active counterion for Ca^{2+} in sediment pore waters under conditions whereby $Ca^{2+}-H^+$ exchange is not a dominant process. The Gapon coefficient for $Ca^{2+}-NH_4{}^+$ exchange can explain the calcium pore-water profile in the reference basin and may be used to estimate the flux of cations from sediments during periods of high ammonium generation. However, it cannot be used for the same purpose in the treatment basin.

Recovery Predictions.   Chemical budgets allow us to gain an understanding of the processes responsible for generating alkalinity and to determine the response of each to acid loadings. In turn, this information allows us to predict lake response after acid loading stops, that is, to develop recovery models based on the input–output concepts of Vollenweider (74). Although in-lake alkalinity generation has been known for decades (e.g., 75, 76), early lake acidification models considered watershed processes only; the trickle-down model (77) included a zero-order term for IAG. Whole-lake experiments reemphasized the importance of IAG in the regulation of alkalinity and prompted further investigation of biological (e.g., 40, 44, 78) and sediment contributions to IAG (e.g., 45, 53). This research provided the basis for deterministic IAG-based acidification–recovery models (27, 47, 79).

Four forms of the basic IAG model (27, 46) are described here to predict rates of recovery of LRL alkalinity. The models are described in order of increasing complexity and realism, reflecting the inclusion of IAG contributions from more biogeochemical processes (4, 17). As previously discussed, chemical budgets from the first 3 years of acidification indicated that the main processes controlling IAG are sulfate reduction and cation production by ion exchange (in order of importance). Effects of nitrate and ammonium retention roughly cancel each other (in terms of net alkalinity production) (17).

The simplest model, Model 1, was based on the assumption that IAG in the treatment basin after acid additions will be constant (i.e., IAG is a zero-order term) and approximately equal to that measured in the reference basin. The equation for Model 1 is

$$\frac{d[ALK]}{dt} = \frac{1}{V}\{J_{alk} - S_o[ALK] + A(IAG)\} \tag{22}$$

where [ALK] is lake alkalinity (mequiv/m^3); [ALK]$_0$ is −26 mequiv/m^3; V is lake volume (3.6 × 10^5 m^3); J_{alk} is alkalinity loading (−1.4 × 10^6 mequiv/ year); S_o is outflow rate (4.758 × 10^4 m^3/year); A is lake area (9.33 × 10^4 m^2); IAG is reference-basin internal alkalinity generation, average of 4 years (21 mequiv/m^2 per year) (17); and t is 0 at the end of the last summer of acid additions. This model underestimates treatment-basin response because it does not account for the increased rate of SO$_4^{2-}$ reduction resulting from higher [SO$_4^{2-}$] in that basin.

Model 2 was based on the assumption that sulfate reduction is the only important IAG process and that the rate is proportional to [SO$_4^{2-}$] in the treatment basin. Depletion of sulfate is directly correlated to alkalinity production; the rate of IAG decreases with time as [SO$_4^{2-}$] decreases. The sulfate loss coefficient (k_{SO_2}) was estimated to be 0.46 m/year for LRL (17, 47) prior to acid additions, but it may change with severe perturbation. For example, sulfate-reducing bacteria may be inhibited at low pH. No change in k_{SO_4} was evident through the first year at pH 5.1. When the ion budgets at pH 4.7 are complete they will reveal whether a change in k_{SO_4} occurred. Model 2 is described by the following coupled equations:

$$[ALK] = [ALK]_0 - \Delta[SO_4^{2-}] \tag{23}$$

$$\frac{d[SO_4^{2-}]}{dt} = \frac{1}{V}\{J_{SO_4} - S_o[SO_4^{2-}] - k_{SO_4}A[SO_4^{2-}]\} \tag{24}$$

where [SO$_4^{2-}$] is lake sulfate concentration (mequiv/m^3); [SO$_4^{2-}$]$_0$ is 147 mequiv/m^3; J_{SO_4} is sulfate loading (4.0 × 10^6 mequiv/year); and k_{SO_4} is the first-order loss coefficient for sulfate (0.46 m/year). If other IAG processes are important, Model 2 will not yield accurate estimates of the time required for alkalinity recovery. In addition, it may underestimate the time required for recovery in LRL because it does not include loss of alkalinity (and sulfate) by groundwater outflow, an important factor in a seepage system.

Model 3 includes SO$_4^{2-}$ reduction, cation production, and ion losses by outflow. The apparent cation-production term (CP'), was calculated on the basis of the observed increase in water-column base cations. It includes production via both weathering and ion exchange, and was treated as a zero-order (constant) term because a more accurate functional relationship is not available. Model 3 is described by the following coupled equations:

$$\frac{d[ALK]}{dt} = \frac{1}{V}\{J_{alk} - S_o[ALK] + k_{SO_4}A[SO_4^{2-}] + A(CP')\} \tag{25}$$

$$\frac{d[SO_4^{2-}]}{dt} = \frac{1}{V}\{J_{SO_4} - S_o[SO_4^{2-}] - k_{SO_4}A[SO_4^{2-}]\} \tag{26}$$

where CP' is treatment-basin cation production (32 mequiv/m^2 per year). Model 3 is a more reasonable representation of the north-basin response.

It is not yet a complete recovery model, however, and it was developed before we observed that the surficial sediments had become acidified. Because the surface sediments of LRL have become acidified, the cation-exchange process will need to be reversed and the released H^+ neutralized by other processes.

A parameterized model that includes the reversal of cation exchange and subsequent neutralization of released H^+ is under development, but an estimate of the effects of sediment acidification (sediment cation or alkalinity deficit) on rate of recovery can be made by using the equations in Model 3 if the following assumptions are made.

1. Preacidification values of alkalinity and total base cations represent a long-term steady-state condition to which the lake will eventually return.

2. The initial sediment alkalinity deficit (at the end of acidification or the beginning of recovery) is equal to the difference between the preacidification base-cation content of the water column and that measured at the end of acidification. This deficit can be treated as if it were a component of the water-column alkalinity. In other words, total alkalinity at the beginning of recovery, $[ALK]_{T,0}$, is equal to the sum of water-column alkalinity at that time, $[ALK]_0$, and the difference between preacidification base-cation concentration and that at the beginning of recovery $(t = 0)$.

3. The alkalinity produced at the sediment–water interface by sulfate reduction will be distributed between the water column and sediment compartments in proportion to their alkalinity deficit. At the end of acidification in LRL, both the water column and the sediments (expressed in terms of excess cations in the water) had initial alkalinity deficits of ~50 mequiv/m^3, and therefore generated alkalinity was assumed to be distributed equally. In other words, for every 2 mequiv of alkalinity produced, 1 mequiv contributes to water-column alkalinity and 1 mequiv to sediment alkalinity.

4. Apparent cation production, CP', remains constant and represents contributions by mineral weathering or hydrologic input. The model resulting from these assumptions, Model 4, is described by the following equations:

$$\frac{d[ALK]_T}{dt} = \frac{1}{V} \{ J_{alk} - S_o[ALK]_T +$$

$$k_{SO_4} A [SO_4^{2-}] + A(CP') \} \qquad (27)$$

$$\frac{d[SO_4^{2-}]}{dt} = \frac{1}{V}\{J_{SO_4} - S_o[SO_4^{2-}] -$$

$$k_{SO_4}A[SO_4^{2-}]\} \tag{28}$$

$$[ALK] = [ALK]_0 - \frac{\Delta[ALK]_T}{2} \tag{29}$$

where $[ALK]_T$ is total alkalinity as previously defined (mequiv/m³) and $[ALK]_{T,0}$ is −76 mequiv/m³.

Results of the four models (Figure 9a) illustrate the effect on predicted recovery rates of including various alkalinity-generating processes. Models 1 and 2 probably yield upper and lower limits of the time required to recover to the preacidification alkalinity level. Model 3 probably yields an under-estimate of recovery time, in that it does not consider the need to neutralize acidified surficial sediments (and restore base cations on sediment-exchange sites that have been lost during the last years of acid loading). Model 4 probably yields the most accurate estimate of recovery time, but it does not provide a functional relationship for the cation-production term. Based on Model 4, the north basin will reach 50% of the preexperimental alkalinity concentrations in 3–5 years and 90% in ~8 years. Complete recovery is predicted to occur in 12.5–15 years.

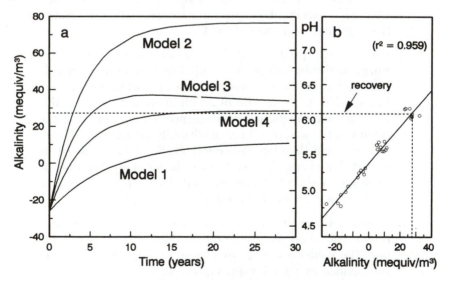

Figure 9. Recovery predictions. Part a: Model 1, reference-basin IAG only; Model 2, sulfate reduction only; Model 3, sulfate reduction, cation production, and outflow; and Model 4, sulfate reduction, cation production, outflow, and sediment neutralization. Part b, Autumn pH–alkalinity correlation: r² = 0.959.

Alkalinity and pH measurements made in LRL during fall turnover show that a reasonable correlation exists between alkalinity and pH (Figure 9b). Calculation of lake-water partial pressure of CO_2, P_{CO_2}, based on turnover values of pH and alkalinity, yielded an average value of $10^{-3.3}$ atm (assuming all but 5 μequiv/L of the measured alkalinity was HCO_3^-). This result indicates that the lake is in equilibrium with atmospheric CO_2 during the fall sampling periods. This relationship allows us to predict the north-basin pH during recovery (Figure 9a, right Y-axis). Based on Model 4 and the pH–alkalinity relationship, the north-basin equilibrium pH will recover to 5.5 in ~3.5 years and to 6.1 in 12.5–15 years.

Of course, the accuracy of these predictions depends on the accuracy of the model assumptions. For more accurate estimates of recovery time, a sediment base-cation recovery component must be added to the model. Some efforts along these lines have been achieved in studies of long-term lake responses to liming. For example, DePinto et al. (79) described a model (ALaRM, the acid lake reacidification model) that included a component encompassing sediment transformations and transport. Further exploration of the functional relationship for the cation-production term is needed to model responses to acid loadings. This investigation is being undertaken during the study of the recovery of LRL from acidification.

Summary and Conclusions

The incremental acidification of LRL over a 6-year period resulted in a gradual change in the lake's major ion chemistry such that the composition at pH 4.7 was substantially different from the pH 6.1 conditions. Sulfate increased by a factor of ~2.8 and replaced bicarbonate as the major anion. Base cations increased in the water column at each treatment pH such that at pH 4.7 total base-cation concentration increased 56% over preacidification values. Increases in Ca^{2+} accounted for about 75% of the increase in total base-cation concentration.

Mass-balance calculations for the first 3 years of acid additions indicate that the principal IAG processes are sulfate reduction and cation production. Specifically, one-third of the total sulfate input (added acid and deposition) was neutralized by in-lake processes. Increased sulfate reduction consumed slightly more than one-sixth and production of cations neutralized somewhat less than one-sixth of the acid added. Of the remaining sulfate, one-third was lost by outflow, and one-third decreased lake alkalinity. Laboratory determinations suggest that sediment-exchange processes occurring in only the top 2 cm of surficial sediments can account for the observed increase in water-column cations. Acidification of the near-surface sediments (with partial loss of exchangeable cations) will slow recovery because of the need to exchange the sediment-bound H^+ and neutralize it by other processes. Reactor-based models that include the primary IAG processes predict that the north basin will achieve 50% recovery in 3–5 years and 90% in ~8 years; complete recovery is predicted to occur in 12.5–15 years.

The results of the LRL experiment suggest that broad generalizations cannot be made regarding the effects of acidification on lacustrine nutrient cycles. Such effects are site-specific and dependent on resident lake biota and seasonal factors such as under-ice pH depressions. Minor disruption of the nitrogen cycle in LRL was found, but this did not include the cessation of nitrification, as reported for other acidified lakes; nitrogen fixation was inhibited, and total nitrogen decreased. Effects of acidification on silica cycling in LRL differed from those observed in other experimental acidifications. Silica concentrations increased slightly relative to the reference during treatment to pH 4.7. This change suggested that the diatom population was somewhat inhibited by the low pH. In spite of laboratory experiments that showed that the release of inorganic P from LRL sediments could be reduced by as much as 90% as pH decreased from 6.0 to 4.5, phosphorus cycling was largely unaffected in LRL. These trends in P agree generally with the findings of other experimental acidification studies reported in the literature.

The LRL experimental acidification verified many of the predictions we made at the outset of the study (5, 6), but some responses of the lake were unanticipated. For example, the increases in base cations occurred as predicted, and the relative importance of in-lake processes in the neutralization of added H_2SO_4 was predicted fairly closely. In contrast, the decrease in total N was contrary to our expectations and the decrease in N_2 fixation was not expected. The combination of field, mesocosm, and laboratory studies at LRL elucidated several biogeochemical processes and enabled us to make mechanistic interpretations of the observed changes.

Acknowledgments

We gratefully acknowledge the cooperation of the Little Rock Lake research team, in particular Tim Kratz and other staff of the University of Wisconsin Trout Lake Limnological Research Station for field sampling and measurement of pH and alkalinity throughout the project; Carl Watras and the Wisconsin Department of Natural Resources for site management; William Rose, U.S. Geological Survey, Madison, WI, for water-budget data; John Eaton, project officer, and the U.S. Environmental Protection Agency, Environmental Research Laboratory at Duluth, MN, which provided primary financial support for the LRL experiment. A large number of undergraduate technicians and graduate students in our laboratory produced the data reported here. Special thanks are given to Naomi Detenbeck, Scott King, Carl Mach, Leslie Sherman, and Janice Tacconi for data gathering and analysis; Larry Baker and Noel Urban for fruitful discussions on IAG processes; and Paul Bloom (Department of Soil Science, University of Minnesota) for helpful discussions on sediment processes. Preparation of this chapter was supported by a grant from the U.S. Geological Survey Water Resources Research Competitive Grants Program. This chapter has not been reviewed by the U.S. EPA for content or policy implications, and no official endorsement should be inferred.

References

1. Watras, C. J.; Frost, T. M. *Arch. Environ. Contam. Toxicol.* **1989**, *18*, 157–165.
2. Mach, C. E.; Brezonik, P. L. *Sci. Total Environ.* **1989**, *87/88*, 260–285.
3. Brezonik, P. L.; Mach, C. E.; Downing, G.; Richardson, N.; Brigham, M. *Environ. Toxicol. Chem.* **1990**, *9*, 871–885.
4. Sampson, C. J. M.S.C.E. Thesis. University of Minnesota, Minneapolis, 1992.
5. Brezonik, P. L.; Baker, L. A.; Detenbeck, N.; Eaton, J. G.; Frost, T. M.; Garrison, P. J.; Johnson, M. D.; Kratz, T. K.; Magnuson, J. J.; McCormick, J. H.; Perry, J. E.; Rose, W. J.; Shephard, B. K.; Swenson, W. A.; Watras, C. J.; Webster, K. E. *Experimental Acidification of Little Rock Lake, Wisconsin: Baseline Studies and Predictions of Lake Responses to Acidification;* Water Research Center, University of Minnesota, St. Paul, MN, Spec. Rep. No. 7, l985.
6. Brezonik, P. L.; Baker, L. A.; Eaton, J. R.; Frost, T. M.; Garrison, P.; Kratz, T. K.; Magnuson, J. J.; Perry, J. E.; Rose, W. J.; Shephard, B. K.; Swenson, W. A.; Watras, C. J.; Webster, K. E. *Water Air Soil Pollut.* **1986**, *31*, 115–121.
7. Brezonik, P. L.; Webster, K. E.; Perry, J. E. *Verh. Int. Ver. Theor. Angew. Limnol.* **1990**, *24*, 445–448.
8. Webster, K. E.; Frost, T. M.; Watras, C. J.; Swenson, W. A.; Gonzalez, M.; Garrison, P. J. *Environ. Pollut.* **1992** *78*, 73–78.
9. Swenson, W. A.; McCormick, J. H.; Simonson, T. D.; Jensen, K. M.; Eaton, J. G. *Arch. Environ. Contam. Toxicol.* **1989**, *18*, 167–174.
10. Solorzano, L. *Limnol. Oceanogr.* **1969**, *5*, 751–754.
11. Langner, C. L.; Hendrix, P. F. *Water Res.* **1982**, *16*, 1451–1454.
12. *Standard Methods for the Examination of Water and Wastewater,* 17th ed.; American Public Health Association: Washington, DC, 1989.
13. Dierberg, F. E.; Brezonik, P. L. *Appl. Environ. Microbiol.* **1981**, *41*, 1413–1418.
14. Flett, R. J.; Hamilton, R. D.; Campbell, N. E. R. *Can. J. Microbiol.* **1976**, *22*, 43–51.
15. Tessier, A.; Campbell, P. G. C.; Bisson, M. *Anal. Chem.* **1979**, *51*, 844–851.
16. Suhr, N. H.; Ingamells, C. O. *Anal. Chem.* **1966**, *38*, 730–734.
17. Weir, E. P. M.S.C.E. Thesis, University of Minnesota, Minneapolis, 1989.
18. Thomas, G. W. In *Methods of Soil Analysis, Part 2*, 2nd ed.; Page, A. L.; Miller, R. H.; Keeney, D. R., Eds.; Am. Soc. Agron. Soil Sci.: Madison, WI, 1982; pp 159–164.
19. Chapman, H. D. In *Methods of Soil Analysis;* Black, C. A., Ed.; American Society of Agronomy and Soil Science: Madison, WI, 1965; pp 891–904.
20. Peech, M. In *Methods of Soil Analysis;* Black, C. A., Ed.; American Society of Agronomy and Soil Science: Madison, WI, 1965; pp 905–913.
21. Baes, A. U.; Bloom, P. R. *Soil Sci.* **1988**, *46*, 6–14.
22. Pagenkopf, G. K. *Environ. Sci. Technol.* **1983**, *17*, 342–347.
23. Campbell, P. G. C.; Stokes, P. M. *Can. J. Fish. Aquat. Sci.* **1985**, *42*, 2034–2049.
24. Rudd, J. W. M.; Kelly, C. A.; Schindler, D. W.; Turner, M. A. *Science (Washington, D.C.)* **1988**, *288*, 1515–1517.
25. Garrison, P. J.; Rose, W. R.; Watras, C. J.; Hurley, J. P. In *Acid Rain Mitigation;* Adams, V. D.; Morgan, E., Eds.; Lewis: Boca Raton, FL, 1993; in press.

26. Kratz, T. K.; Cook, R. B.; Bowser, C. J.; Brezonik, P. L. *Can. J. Fish. Aquat. Sci.* **1987**, *44*, 1082–1088.
27. Baker, L. A.; Brezonik, P. L. *Water Resour. Res.* **1988**, *24*, 65–74.
28. Baker, L. A.; Brezonik, P. L. *Water Resour. Res.* **1988**, *24*, 1828–1830.
29. Schindler, D. W.; Mills, K.; Malley, D.; Findley, D.; Shearer, J.; Davies, I.; Turner, M.; Lindsey, G.; Cruikshank, D. *Science (Washington, D.C.)* **1985**, *228*, 1395–1401.
30. Schindler, D. W.; Frost, T. M.; Gunn, J. M.; Mills, K. H.; Chang, P. S. S.; Davies, I. J.; Findlay, L.; Malley, D. F.; Shearer, J. A.; Garrison, P. J.; Watras, C. J.; Webster, K.; Brezonik, P. L.; Swenson, W. A. *Proc. R. Soc. Edinburgh*, **1991**, *97B*, 193–226.
31. Grahn, O.; Hultberg, H.; Landner, L. *Ambio* **1974**, *3*, 93–94.
32. Ogburn, R. W., III. Ph.D. Thesis, University of Florida, Gainesville, 1984.
33. Ogburn, R. W.; Brezonik, P. L. *Water Air Soil Pollut.* **1986**, *30*, 1001–1006.
34. Curtis, P. J. *Nature (London)* **1989**, *337*, 156–158.
35. Sholkovitz, E. R.; Copeland, D. *Geochim. Cosmochim. Acta* **1982**, *46*, 393–410.
36. Baccini, P. In *Chemical Processes in Lakes*; Stumm, W., Ed.; John Wiley and Sons: New York, 1985; pp 189–205.
37. Detenbeck, N. E. Ph.D. Thesis, University of Minnesota, Minneapolis, 1987.
38. Detenbeck, N. E.; Brezonik, P. L. *Environ. Sci. Technol.* **1990**, *25*, 395–403.
39. Detenbeck, N. E.; Brezonik, P. L. *Environ. Sci. Technol.* **1990**, *25*, 403–409.
40. Cook, R. B.; Kelly, C. A.; Schindler, D. W.; Turner, M. A. *Limnol. Oceanogr.* **1986**, *31*, 134–148.
41. Baker, L. A.; Brezonik, P. L.; Edgerton, E. S. *Water Resour. Res.* **1986**, *22*, 715–722.
42. Baker, L. A.; Perry, T. E.; Brezonik, P. L. In *Lake and Reservoir Management: Practical Applications;* Taggart, J.; Moore, L., Eds.; Proc. 4th Ann. Meeting, North Am. Lake Manag. Soc.: McAfee, NJ, 1984; pp 356–360.
43. Schindler, D. W.; Turner, M. A.; Stainton, M. F.; Linsey, G. A. *Science (Washington, D.C.)* **1986**, *232*, 844–847.
44. Kelly, C. A.; Rudd, J. W. M.; Hesslein, R. H.; Schindler, D. W.; Dillon, P. J.; Driscoll, C. T.; Gherini, S. A.; Hecky, R. E. *Biogeochemistry.* **1987**, *3*, 129–140.
45. Schiff, S. L.; Anderson, R. F. *Can. J. Fish. Aquat. Sci.* **1987**, *44*, 173–187.
46. Brezonik, P. L.; Baker, L. A.; Perry, T. E. In *Chemistry of Aquatic Pollutants;* Hites, R.; Eisenreich, S. J., Eds.; Advances in Chemistry 216; American Chemical Society: Washington, DC, 1987; pp 229–260.
47. Baker, L. A.; Brezonik, P. L.; Pollman, C. D. *Water Air Soil Pollut.* **1986**, *31*, 89–94.
48. Stumm, W.; Morgan, J. J. *Aquatic Chemistry*, 2nd ed.; John Wiley and Sons: New York, 1981; p 188.
49. Baker, L. A.; Brezonik, P. L.; Urban, N. In *Biogenic Sulfur in the Environment;* Saltzman, E. S.; Cooper, W. J., Eds.; ACS Symposium Series 393; American Chemical Society: Washington, DC, 1989; pp 79–100.
50. Urban, N. R. In *Environmental Chemistry of Lakes and Reservoirs;* Baker, L. A., Ed.; Advances in Chemistry 237; American Chemical Society: Washington, DC, 1994; Chapter 10.
51. Giblin, A. E.; Likens, G. E.; White, D.; Howarth, R. W. *Limnol. Oceanogr.* **1990**, *35*, 852–869.
52. Rudd, J. W. M.; Kelly, C. A.; Furutani, A. *Limnol. Oceanogr.* **1986**, *31*, 1281–1291.
53. Baker, L. A.; Brezonik, P. L.; Edgerton, E. S.; Ogburn, R. W., III. *Water Air Soil Pollut.* **1985**, *25*, 215–230.
54. Oliver, B. G.; Kelso, J. R. M. *Water Air Soil Pollut.* **1983**, *20*, 379–389.

55. Perry, T. E. M.S.C.E. Thesis, University of Minnesota, Minneapolis, 1987.
56. Bloom, P. R.; Grigal, D. F. *J. Environ. Qual.* **1985,** *14,* 489–495.
57. Stumm, W.; Wieland, E. In *Aquatic Chemical Kinetics;* Stumm, W., Ed.; John Wiley and Sons: New York, 1990; pp 367–400.
58. Schnoor, J. L. In *Aquatic Chemical Kinetics;* Stumm, W., Ed.; John Wiley and Sons: New York, 1990; pp 475–504.
59. Sherman, L. A. M.S. Thesis, University of Minnesota, Minneapolis, 1988.
60. Rudd, J. W. M.; Kelly, C. A.; Schindler, D. W. *Water Resour. Res.* **1988,** *24,* 1825–1827.
61. Rudd, J. W. M.; Kelly, C. A.; Schindler, D. W.; Turner, M. A. *Limnol. Oceanogr.* **1990,** *35,* 663–679.
62. Kelly, C. A.; Rudd, J. W. M.; Schindler, D. W. *Water Air Soil Pollut.* **1990,** *50,* 49–61.
63. Lin, J. C.; Schnoor, J. L.; Glass, G. E. In *Sources and Fate of Aquatic Contaminants;* Hites, R.; Eisenreich, S. J., Eds.; Advances in Chemistry 216; American Chemical Society: Washington, DC, 1987; pp 229–262.
64. Young, T. C.; Rhea, J. R.; McGlaughlin, G. *Water Air Soil Pollut.* **1986,** *31,* 839–846.
65. Tacconi, J. E. M.S.C.E. Thesis, University of Minnesota, Minneapolis, 1988.
66. Baker, L. A.; Tacconi, J. E.; Brezonik, P. L. *Verh. Int. Ver. Theor. Angew. Limnol.* **1988,** *23,* 346–350.
67. Brezonik, P. L; Webster, K. E.; Perry, J. A. *Verh. Int. Ver. Theor. Angew. Limnol.* **1990,** *24,* 445–448.
68. Brezonik, P. L.; Eaton, J. G.; Frost, T. F.; Garrison, P. J.; Kratz, T. K.; Mach, C. E.; McCormick, J. H.; Perry, J. A.; Rose, W. A.; Sampson, C. J.; Shelley, B. C. L.; Swenson, W. A.; Webster, K. E. *Can. J. Fish. Aquat. Sci.* **1993,** *50,* 598–617.
69. Bohn, H. L.; McNeal, B. L.; O'Connor, G. A. *Soil Chemistry,* 2nd ed.; John Wiley and Sons: New York, 1985.
70. Vanselow, A. P. *Soil Sci.* **1932,** *33,* 95–113.
71. Gapon, Ye. N. J. *Gen. Chem. USSR Engl. Transl.* **1933,** *3,* 144–160.
72. Sposito, G. *Soil Sci. Soc. Am. J.* **1977,** *41,* 1205–1206.
73. Goulding, K. W. T. *Adv. Agron.* **1983,** *36,* 216–257.
74. Vollenweider, R. A. *Schweiz. Z. Hydrol.* **1975,** *37,* 53–84.
75. Hutchinson, G. E. *A Treatise on Limnology;* John Wiley and Sons: New York, 1957; Vol. 1.
76. Hongve, D. *Verh. Int. Ver. Theor. Angew. Limnol.* **1978,** *20,* 743.
77. Schnoor, J. L.; Stumm, W. In *Chemical Processes in Lakes;* Stumm, W., Ed.; John Wiley and Sons: New York, 1985; pp 311–338.
78. Kelly, C. A.; Rudd, J. W. M. *Biogeochemistry* **1984,** *1,* 63.

RECEIVED for review April 1, 1992. ACCEPTED revised manuscript August 6, 1992.

6

Organic Phosphorus in the Hydrosphere

Characterization via ^{31}P Fourier Transform Nuclear Magnetic Resonance Spectroscopy

Mark A. Nanny and Roger A. Minear

Institute for Environmental Studies, University of Illinois at Urbana–Champaign, Urbana, IL 61801

Phosphorus-31 Fourier transform nuclear magnetic resonance (^{31}P FT-NMR) spectra of dissolved organic phosphorus (DOP) species, collected from the epilimnion of a small lake from September 1990 to May 1991, were used to identify and characterize soluble P compounds in lake water. Ultrafiltration and reverse osmosis concentration techniques were used to achieve a 2000-fold DOP concentration factor. The sensitivity of the NMR was further enhanced by the use of the spin-lattice relaxation agent iron ethylenediaminetetraacetate (FeEDTA). These techniques are briefly discussed, in addition to the effects of pH, ionic strength, concentrated humic matrix, and Fe-EDTA on the ^{31}P FT-NMR spectra. Individual DOP species in lake water have not been conclusively identified with ^{31}P FT-NMR spectroscopy. The ^{31}P FT-NMR spectra indicate the presence of mono- and diester phosphates, and the presence of DNA is strongly suggested. ^{31}P FT-NMR spectra show seasonal changes that correlate to seasonal changes in the lake. Varying the sample pH and collecting the subsequent ^{31}P FT-NMR spectra illustrates that not all of the DOP species's signal positions are pH-dependent. This independence indicates possible DOP aggregate or micelle formation.

T HE DISSOLVED PHOSPHORUS FRACTION is the most important aquatic phosphorus compartment in terms of biological growth in an aquatic system because it provides the major source of available phosphorus to phytoplankton. To be biologically useful the dissolved phosphorus compounds must

0065–2393/94/0237–0161$08.75/0
© 1994 American Chemical Society

first be converted into orthophosphate. Therefore, to fully understand the aquatic phosphorus cycle, the identity and the chemical and physical behavior of the dissolved organic phosphorus (DOP) must be known. The removal rates and behavior of DOP in the presence of other dissolved and colloidal material such as humic and fulvic acids; clay colloidal material; and various ions such as Ca^{2+}, Mg^{2+}, and Fe^{2+} or Fe^{3+} must also be addressed.

For these reasons, numerous attempts have been made to identify and characterize DOP, but with little success because it is usually present in very low concentrations. Typical values in lake waters range from 5 to 100 µg of P/L in oligotrophic to eutrophic systems. Colorimetric methods have been used extensively to detect and differentiate between soluble reactive phosphorus (SRP) and soluble unreactive phosphorus (SUP) at concentrations as low as 10 µg of P/L (1). SRP is generally considered to consist of only orthophosphate compounds, whereas SUP is composed of all other phosphorus species, primarily organic phosphorus compounds. The sum of SRP and SUP is equal to the total soluble phosphorus (TSP). These methods were used to study the dynamics of bulk phosphorus fractionation between the sediments, suspended particulate matter, the biota, and the dissolved fraction (2). Despite these studies, very little is known regarding the identity and characteristics of the DOP in the hydrosphere.

Attempts to identify and characterize the DOP fraction have relied on gel chromatography (Sephadex) (3–8), ^{32}P and ^{33}P tracer studies (3–11), bioassays (9–12), enzyme bioassays (13, 14), and high-performance liquid chromatography (HPLC) with a post-column reactor (15, 16). A consistent feature in the gel chromatography studies is the appearance of a high-molecular-weight (HMW) fraction (>30,000–5000 daltons, depending on the exclusion limit of the gel employed), always at the upper limit of the size-exclusion gel used, and a low-molecular-weight fraction that coelutes with orthophosphate. Between the high- and low-molecular-weight fraction peaks, which are distinct and prominent, a continuum of an intermediate-molecular-weight fraction is often present in low concentrations. Sometimes this fraction also is represented by a shoulder in the low-molecular-weight elution region. These studies also indicate that SRP is sometimes present in the HMW fraction.

Incubation of lake water with ^{32}P or ^{33}P as tracers and subsequent gel chromatography reveals that a major pathway exists between dissolved orthophosphate and the particulate phase (3, 5–7). Low-molecular-weight phosphorus forms in the presence of bacteria and algae. SUP is present in the low-molecular-weight fraction and is classified as individual DOP compounds unassociated with particulate or colloidal material. The HMW fraction found in gel chromatography studies is characterized as a colloid that contains phosphorus compounds or incorporates orthophosphate. The colloidal material then releases orthophosphate, replenishing the dissolved phosphorus cycle. In some eutrophic lakes the HMW SRP fraction can make up to 80% of the total soluble phosphorus (6). Studies by Koenings and Hooper (17) and by Franko and Heath (18) demonstrated the binding of orthophosphate by humic material in the presence of ferric ions. This humic–$PO_4{}^{3-}$–Fe(III) complex shows up in the HMW fraction when analyzed with gel chromatography. Upon irradiation with ultraviolet light, the phosphate is released and the ferric ions are reduced to ferrous ions.

Other methods used to characterize and identify DOP involve bioassays with *Chlorella* to study the biological availability and biouptake of the HMW SRP fraction (4, 6). These bioassays indicate that the algal growth responds similarly to HMW SRP and to $PO_4{}^{3-}$. A preference for $PO_4{}^{3-}$ was detected, and not all of the reactive HMW fraction was used. Enzymatic assays used by Herbes et al. (13) tentatively identified inositol hexaphosphate as part of the DOP. Using an anion-exchange HPLC system with a phosphorus-specific post-column reactor, Minear and co-workers (15, 16) possibly have detected inositol hexaphosphate, DNA, and nucleotide fragments in lake waters.

Only a few DOP species have been conclusively identified in natural fresh waters. These species are DNA by Minear (19) and DeFlaun et al. (20), RNA and DNA by Karl and Bailiff (21), and 3′,5′-cyclic adenosine monophosphate by Franko and Wetzel (22). Jefferey (23) detected the presence of phospholipids in sea water. Others have provided circumstantial evidence for inositol hexaphosphate in lake water (24–27) and in aquatic sediments (13).

Despite the variety of analytical methods used and the amount of effort employed, the paucity of information regarding DOP and its importance in aquatic ecosystems indicates a need for new tools if the knowledge concerning DOP is to be expanded. Fourier transform nuclear magnetic resonance (FT-NMR) spectroscopy holds promise, for it is already becoming a highly beneficial tool in the area of environmental analysis. ^{13}C FT-NMR spectroscopy has been used since 1976 (28) to examine humic and fulvic acids. ^{29}Si FT-NMR spectroscopy was recently applied in the detection of polyorganosiloxanes in the environment (29). ^{31}P FT-NMR spectroscopy was recently applied to the characterization of organic phosphorus present in the environment by examining organic phosphorus in soils (30–34) and humic material from soils (35, 36), marine sediments (37), and wastewater-treatment-plant activated sludge (38–41). Mono- and diester and inorganic polyphosphates and occasionally phosphonates were detected in these samples.

Because the application of NMR spectroscopy to environmental samples is relatively new, we focused our studies on the identification and characterization of DOP by ^{31}P FT-NMR spectroscopy. Ultrafiltration and reverse osmosis concentration techniques were employed to increase the dissolved organic phosphorus concentrations to the detection level of ^{31}P FT-NMR techniques (approximately 10–20 mg of P/L). With these concentration methods a DOP concentration factor of up to 2000 is obtainable. This chapter reports the use of ^{31}P FT-NMR spectroscopy in the analysis of DOP. In contrast to soil and sediment ^{31}P FT-NMR studies, we avoided the use of strong base extraction techniques to maintain DOP integrity. Individual DOP compounds have not yet been conclusively identified, but we have detected DOP as mono- and diester phosphates in an engineered lake in Champaign County, Illinois. We also observed temporal changes in the ^{31}P NMR spectra over the period of September through May. From these results, we concluded that ^{31}P FT-NMR spectroscopy is a viable technique with potential to identify and characterize the DOP involved in the aquatic phosphorus cycle.

Background NMR Theory

The basic principle of NMR spectroscopy is that a signal arises from the magnetic dipole of nuclei undergoing a transition between energy levels. A magnetic dipole is created by the intrinsic spinning motion of the nucleus and the electrical charges of its protons (42). The nuclear spin can be described by the spin quantum number, I, which must be equal to or greater than ½ for a magnetic dipole to be present. In addition, the nucleus must have an unpaired spin arising from either an odd number of protons or an odd number of neutrons with an even number of protons present. Many of the most common nuclei examined (such as ^{1}H, ^{13}C, and ^{31}P) have spin quantum numbers equal to ½.

In the presence of an external magnetic field, H_0, the magnetic dipole of the nucleus orients itself in discrete positions relative to H_0, corresponding to specific energy levels. The energy difference between these levels is given by eq 1

$$\Delta E = \frac{\mu B H_0}{I} \tag{1}$$

where ΔE is the energy difference between each pair of levels, μ is the magnetic moment of the nucleus expressed as nuclear magnetons, B is the nuclear magneton constant (5.049×10^{-24} erg/G), and H_0 is the applied external field. As can be seen from eq 1, the applied external magnetic field and the energy difference between the positional levels are directly proportional. In the absence of H_0, there is no difference between the energy levels. For a nucleus with a spin quantum number of $I = $ ½ in an applied external magnetic field, the magnetic dipole has only two discrete positions: it can be aligned with or against H_0.

When energy equivalent to the difference between the energy levels is applied to the system, a transition from the lower to the higher energy level occurs. In NMR spectroscopy, the applied energy that allows this nuclear magnetic dipole transition to occur is a radio-frequency magnetic field, H_1, which is applied perpendicularly to H_0.

The intensity of the signal produced by this transition is proportional to the number of nuclei (n) that change from the lower to the higher energy state. At equilibrium, the population difference between these energy states can be described by the Boltzmann equation

$$\frac{n_{\text{upper}}}{n_{\text{lower}}} = \exp\left(\frac{-\mu B H_0}{IkT}\right) \tag{2}$$

where k is the Boltzmann constant and T is the absolute temperature. At room temperature in an H_0 typical of modern NMR spectrometers, this population difference is very small. Because the population difference varies with the strength of H_0, larger and larger magnets are being developed. However, this development is very difficult and expensive. Thus NMR spectroscopy is limited to concentrated samples, or it requires numerous repeated acquisitions to provide a viable spectrum.

At a specified H_0, the frequency of the H_1 that induces transitions is an intrinsic property of the nucleus. The H_1 magnetic field can be altered by the electron density surrounding the nucleus. Thus, even though H_1 may be at the proper frequency to induce nuclear magnetic dipole transitions, the nucleus may be receiving a magnetic field that is slightly greater or smaller than H_1 as a result of the electron shielding. Because of this shielding, each nucleus of a given element will be excited at a similar but discrete frequency. Thus numerous signals will be present in the spectrum, with positional differences dependent upon the electron density surrounding the nucleus being examined. The signal position, reported as the chemical shift, is given in parts per million (ppm) and defined as

$$\text{chemical shift} = \frac{\text{observed shift (Hz)} \times 10^6}{\text{spectrometer frequency (Hz)}} \tag{3}$$

The observed shift is the difference between the observed signal and a reference signal. ^{31}P NMR spectroscopy often uses the signal produced by 85% H_3PO_4 as the reference signal and assigns it the chemical shift of 0.00 ppm.

Coupling is a phenomenon that affects the signal pattern (singlet, doublet, triplet, etc.). This coupling is caused by the interaction between the magnetic fields from the examined nucleus and electrons surrounding other nearby nuclei. For example, the phosphorus nuclei from mono- and diester phosphates can couple with the protons present in the phosphate group, giving rise to complex signal patterns. To simplify the ^{31}P FT-NMR spectrum, NMR experiments are run decoupled (i.e., the protons are subjected to a separate radio-frequency field that matches the frequency of the proton's resonance). This treatment eliminates the proton's contribution to coupling so that the phosphorus NMR signals are not split, but appear as a single peak.

In a continuous-wave NMR (CW NMR) experiment the H_1 frequency is slowly scanned and each nucleus resonates at its specific frequency, which is a function of the examined nucleus and the electron density that surrounds it. This experiment can be long and tedious because if high resolution is desired, the frequencies must be swept very slowly. If the concentration of the examined nucleus is low, a large number of repeated sweeps may be needed to obtain a detectable signal.

Fourier transform NMR spectroscopy, on the other hand, permits rapid scanning of the sample so that the NMR spectrum can be obtained within a few seconds. FT-NMR experiments are performed by subjecting the sample to a very intense, broad-band, H_1 pulse that causes all of the examined nuclei to undergo transitions. As the excited nuclei relax to their equilibrium state, their relaxation-decay pattern is recorded. A Fourier transform is performed upon this relaxation-decay pattern to provide the NMR spectra. The relaxation-decay pattern, which is in the time domain, is transformed into the typical NMR spectrum, the frequency domain. The time required to apply the H_1 pulse, allow the nuclei to return to equilibrium, and have the computer perform the Fourier transforms on the relaxation-decay pattern often is only a few seconds. Thus, compared to a CW NMR experiment, the time can be reduced by a factor of 1000-fold or more by using the FT-NMR technique.

The major advantage of this dramatic time reduction lies in the ability to scan the sample repeatedly, combine the relaxation-decay patterns collected from each scan, and then perform a Fourier transform upon the final composite relaxation-decay pattern. This technique, in essence, increases the spectral sensitivity by allowing the NMR signals acquired from each scan to be constructively added to each other while the noise cancels itself deconstructively. This approach greatly increases the sensitivity of the instrument and allows NMR experiments to be performed on samples that have low concentrations of the desired nucleus (i.e., for ^{31}P, 20 mg of P/L is a feasible concentration with instrument time of hours to a few days).

It is very important that there be sufficient time between pulses in FT-NMR experiments so that the nuclei can return to the original equilibrium state. If the equilibrium state has not been reached before the next H_1 pulse, the still-excited nuclei will not participate in the transition and thus will produce a decreased signal intensity relative to the previous signal. As the experiment proceeds and more pulses are applied, more nuclei will remain in the exited state until eventually none of the nuclei will be in the lower energy state when pulsed. At this point the sample is saturated and will not produce a signal. The length of time required for the nuclei to relax is called the spin-lattice or T_1 relaxation time.

To reduce the nuclei's T_1 relaxation time, relaxation agents are used. Relaxation agents are usually transition metal complexes, primarily ferric ethylenediaminetetraacetate or chromium acetylacetonate. Transfer of energy from the excited nucleus to the metal ion's lone pair of electrons provides a relaxation pathway and allows the examined nucleus to relax at a much greater rate, thus permitting more scans per unit time. The relaxation agent must be selected carefully. If the relaxation rate is increased too much, signal line broadening will occur, even to the extent that the signal becomes undetectable, because the signal width is inversely proportional to the relaxation time.

The chemical shift associated with ^1H nuclei indicates the type of functional group containing the proton (e.g., alkane, alkene, carboxylic acid, phenolic, and alcohol). Evidence of coupling with nearby protons also provides information about the number of adjacent protons and their identity. ^{13}C NMR spectroscopy provides information regarding the type of carbon atoms present (e.g., aliphatic, aromatic, or carboxylic), as shown by the chemical shift. ^{31}P NMR spectroscopy provides the same information as ^{13}C NMR spectroscopy. However, most of the phosphorus in environmental samples is likely to be in the forms of orthophosphate or its mono- and diesters. Therefore, in the ^{31}P NMR experiment, the effect of the number and type of R groups present in O_x–$PO(OR)_{3-x}$ on the phosphorus nucleus will determine the signal's position. Orthophosphate and monoester phosphate ^{31}P NMR signals appear in the 0–10-ppm region, depending upon the sample pH. Diester phosphate signals appear in the 0- to −5-ppm region, whereas triester phosphates are not likely to be detected in the environment because of concentration limitations related to compound solubility. Polyphosphates appear in the region of −5 to −10 ppm; phosphonates, which are characterized by the C–P bond between the phosphate phosphorus atom and an R group, have signals appearing at 20–25 ppm. All of these chemical shifts are relative to 85% phosphoric acid.

Methodology

To achieve ^{31}P NMR spectra of DOP, ultrafiltration (UF) and reverse osmosis (RO) membrane concentration techniques and T_1 relaxation agents were employed. Concentration factors of 2000 or more are needed to reach the NMR detection threshold for phosphorus, and use of a relaxation agent permits more scans per unit time. ^{31}P FT-NMR studies of soil and sediment do not require such high concentration factors because soil and sediment phosphorus concentrations are much higher and extraction techniques are easily employed. A caveat with soil and sediment extraction methods is that a strong base is required; thus only base-soluble organic phosphorus is isolated. Potential hydrolysis of the organic phosphorus by the strong base is an even greater problem with these extraction methods.

The use of UF and RO membranes as a concentration procedure (compared to other concentration techniques for DOP such as freeze-concentration, freeze-drying–reconstitution, anion-exchange–small-volume elutions, and lanthanide precipitation–sequential dissolution) is preferred. It separates DOP according to molecular size, depending on the nominal molecular-weight cutoff (i.e., pore size) of the membrane used. Our samples are fractionated into a HMW fraction (30,000 daltons), an intermediate- to low-molecular-weight fraction (1000 daltons), and a low-molecular-weight fraction (approximately 300 daltons) (Figure 1). They are purposefully fractionated so as to isolate and concentrate the various DOP fractions observed in

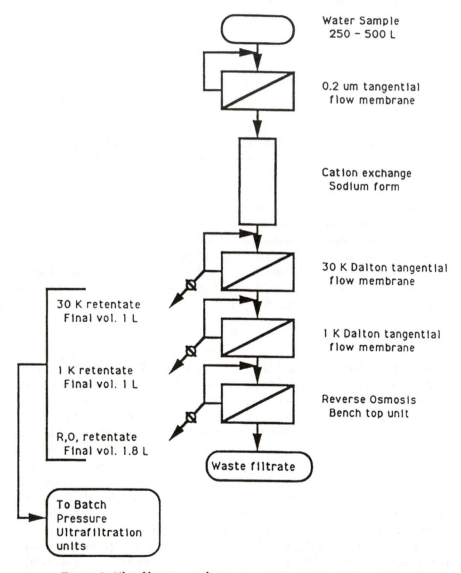

Figure 1. Ultrafiltration and reverse osmosis concentration system.

studies using gel chromatography. At present, only the retentate from the 1000-dalton, tangential-flow UF membrane has a high enough DOP concentration to facilitate ^{31}P FT-NMR spectroscopy. An additional benefit of ultrafiltration and reverse osmosis concentration methods is that the sample does not undergo drastic physical or chemical transformations and thus maintains its integrity.

The T_1 relaxation agent increased the sensitivity of the NMR instrument by decreasing several of the mono- and diester phosphate relaxation times by factors of 2–5. In this way the delay time between scans was decreased (*43*). This change permits an increase in the number of scans observed per unit time. Although the presence of ferric ions creates the potential for precipitation of phosphorus–iron complexes, the addition of a large molar excess of ethylenediaminetetraacetate (EDTA) relative to ferric ions prevented precipitation, even over a large pH region (*44*).

Experimental Section

Sample Collection.   Pelagic lake-water samples were collected at Crystal Lake, an engineered mesotrophic lake in Champaign County, Illinois. It has an average depth of 10 feet with a maximum depth of 13 feet and is fed by groundwater from a 200-foot-deep well. For each sample 250–500 L of water was filtered with a plankton net and stored in 55-L polyethylene (Nalgene) containers until processed in the laboratory approximately 30 min later.

Concentration.   Samples were concentrated by using ultrafiltration and reverse osmosis membranes (Figures 1 and 2). The water was first filtered with a 0.2-μm tangential flow filter to remove algal cells, bacteria, and colloidal and suspended solids. The water then passed through a Na^+ cation-exchange column containing 6.2 dm^3 of 50–100-mesh resin (Dowex 50×8) in the sodium form to remove Ca^{2+} and Mg^{2+} and then onto a second tangential-flow filtration unit consisting of either a 30,000-dalton polysulfone or a 1000-dalton cellulose acetate membrane. The retentate was continuously recycled to the second tangential-flow filter while the filtrate passed to a reverse osmosis bench-top spiral-wound membrane unit (Millipore), in which various membranes were installed. The membranes used were 50% and 80% NaCl rejection reverse osmosis cellulose acetate membranes, a 99% NaCl rejection reverse osmosis polyamide membrane, and a 1000-dalton cellulose acetate ultrafiltration membrane. The retentate from this filtration was recycled back to the spiral-wound membrane. The final volume of the second tangential-flow unit retentate was usually 0.8 L, and for the spiral-wound reverse osmosis unit the final volume was usually 1.5 L. The spiral-wound reverse osmosis retentate was passed through a second smaller cation-exchange column containing the same resin type as the first cation-exchange unit to remove any remaining Ca^{2+} and Mg^{2+}. Samples were further concentrated to a final volume of 8 mL with a sequence of decreasing-volume batch pressure ultrafiltration units (Amicon) containing cellulose acetate filters. Effective pore sizes of 500 or 1000 daltons were used for the low- and intermediate-molecular-weight fractions; 5000- and 10,000-dalton membranes were employed with the HMW fraction.

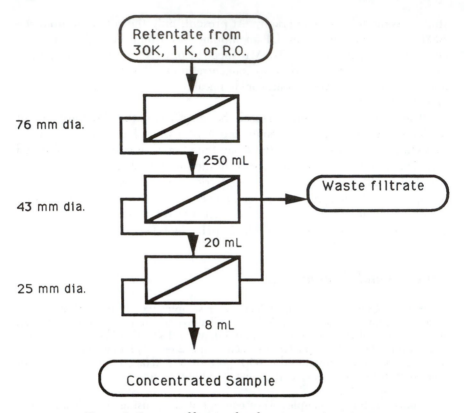

Figure 2. Pressure filtration batch concentration system.

Sample Preparation. Samples for ^{31}P FT-NMR analysis were prepared by adding sodium azide to form a 0.1% solution and to prevent microbial growth, and an appropriate amount of ferric EDTA to provide a phosphorus:iron molecular ratio of 1.5:2.0. Control studies have shown that the addition of sodium azide has no effect upon the ^{31}P FT-NMR spectra. The 0.0103 M ferric EDTA relaxation agent solution had an EDTA:iron ratio of approximately 30 and was made from ferric nitrate (Fischer) and tetrasodium ethylenediaminetetraacetate (EDTA, Sigma) dissolved in deionized water. Total soluble phosphorus and soluble reactive phosphorus concentration measurements were done by the phosphate molybdate test with the oxidation step achieved via potassium persulfate oxidation (1). Samples for pH and ionic strength studies were prepared by making solutions of 150 mg of P/L of each model compound. The model compounds chosen were *myo*-inositol hexaphosphate (IHP, Sigma), inositol monophosphate (IMP, Sigma), DNA from degraded herring sperm (Sigma), choline phosphate (Sigma), serine phosphate (Sigma), and potassium dihydrogen orthophosphate (J. T. Baker). For the pH studies, the sample pH was modified with either dilute HCl or NaOH. The ionic strength was modified by the addition of NaCl. In the ionic strength studies, the pH was held fairly constant by the use of either an acetic acid–sodium acetate buffer or a sodium carbonate–sodium bicarbonate buffer. The contribution of the buffers to the ionic strength was taken into consideration.

Concentrated natural humic matrix was obtained from old, concentrated natural water samples in which the dissolved phosphorus concentration was either too low to be measured or in which it had all hydrolyzed to orthophosphate.

NMR Spectroscopy.   All ^{31}P FT-NMR spectra were collected on one of two instruments (GN 300 narrow bore or GN 300 wide bore) at the University of Illinois, School of Chemical Sciences Molecular Spectroscopy Laboratory. All pH, ionic strength, concentrated humic matrix, and relaxation agent spectra were obtained with the GN 300 narrow-bore spectrometer. Crystal Lake samples were scanned for 24–48 h with the GN 300 wide-bore instrument or over several 12-h periods with the GN 300 narrow-bore NMR, compiling the individual free induction decay patterns (FID) into a single FID. A Fourier transformation was then performed with the composite FID to obtain the final spectrum. The ^{31}P FT-NMR spectra, obtained at 121.648 MHz, were generated by a pulse width of 20–24 μs with a pulse delay of 3–6 s, depending upon the sample. All spectra were proton decoupled, and a spectral width of 10,000 Hz was used. Magnetic shimming and signal phasing were done by computer, and all chemical-shift measurements were measured relative to 85% H_3PO_4. Samples were placed in 10-mm glass NMR tubes (Wilmad Corporation) with inserts containing deuterium oxide (Sigma). Longitudinal (T_1) relaxation measurements were obtained by using the inversion recovery method (i.e., a 180°–t–90° pulse sequence).

Results and Discussion

Matrix Effects.   *pH.* Numerous factors such as sample pH, ionic strength, humic substances, and relaxation agents can modify the NMR spectrum. For example, monoester phosphate chemical shifts are pH-dependent (*44–46*), and we showed (*44*) for several monoester phosphates that as the sample pH is increased, their chemical shifts also increase (Figure 3). This behavior is caused by the ability of monoester phosphates to undergo protonation–deprotonation. Because the monoester phosphate chemical shift is pH-dependent, the curves resulting from plotting pH versus chemical shift are analogous to a titration curve. Thus, monoester phosphate pK_a values can be measured from these pH–chemical shift curves (*44–46*).

According to the theory of Letcher and Van Wazer (*47*), changes in the chemical-shift position depend on three factors: the difference in the electronegativity of the P–X bond, the change in the electron orbital overlap, and the change in bond angles between the atoms attached to phosphorus. Upon deprotonation of a monoester phosphate molecule, the electron density on the deprotonated phosphate oxygen atom changes. After deprotonation, the oxygen atom contains a negative charge, increasing the electron density surrounding the oxygen nucleus. This increase in turn decreases the electron density of the phosphorus nucleus. The phosphorus nucleus, with less electron shielding from the external magnetic field H_0 than it previously had while the oxygen was protonated, resonates at a greater frequency, and its signal appears at a higher chemical-shift value. The other two factors affecting the chemical-shift position are considered to be minor relative to that caused

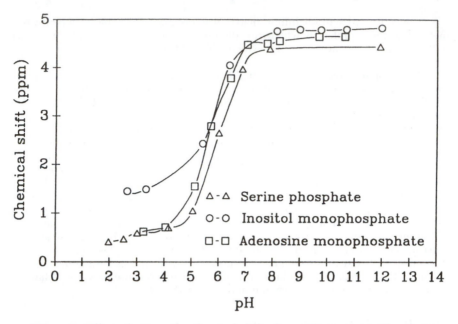

Figure 3. Effect of pH on the chemical shift of several monoester phosphate compounds. Serine phosphate and inositol monophosphate are in a pure water matrix, and adenosine monophosphate is in a concentrated humic–FeEDTA matrix.

by the changes in the P–X bond's electronegativity. The deprotonation–protonation of a monoester phosphate probably does not greatly influence the π-electron orbital density of the P=O bond, and changes in the bond angles are believed to have only very small effects upon the signal position.

Another pH-induced effect is change in the signal pattern of compounds containing multiple monoester phosphate groups. An example of this is inositol hexaphosphate (IHP). At low pH values (2–5) the ^{31}P NMR spectrum of IHP exhibits four individual singlets. The intensity of the two inner peaks is twice the intensity of each of the two outer peaks (Figure 4). Increasing the pH causes this signal pattern to undergo dramatic changes until the original four-peak pattern returns when pH 12 is reached. In the pH 2–5 region the IHP molecule maintains the chair configuration in which the phosphate groups 2,6 and 3,5 (Figure 5), all in the axial position, are involved in hydrogen bonding. These two sets give rise to the two large central peaks; phosphate groups 1 and 4 give rise to the two smaller outer signals (Figure 4). Deprotonation of IHP disrupts the chair configuration because of steric repulsion between the lone electron pairs from the newly deprotonated phosphate groups and the protonated phosphate groups. Changes in the molecular configuration and the chemical shift will continue until all of the phosphorus groups have lost an equivalent number of protons so the molecule can return to the chair configuration.

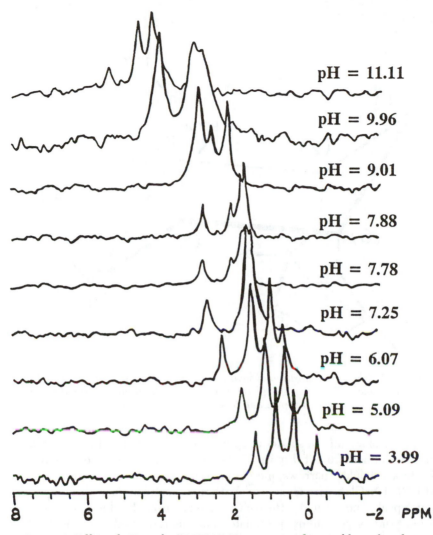

Figure 4. Effect of pH on the ^{31}P FT-NMR spectrum of inositol hexaphosphate in a concentrated humic–FeEDTA matrix.

Ionic Strength. The sample's ionic strength also influences the chemical shift and signal pattern of the NMR spectra. Increasing the NaCl concentration of the solution causes the NMR signals to shift slightly downfield (Figure 6). Downfield shifts of the signal are caused by a decrease in electron charge of the phosphorus nucleus. In turn, the decrease is probably caused by the large number of positive cations clustering around the oxygen atom, allowing it to contain a greater electron density. Increasing the NaCl concentration also affects the signal pattern of compounds containing multiple monoester phosphate groups. At pH 9.8, IHP displays a ^{31}P NMR spectrum

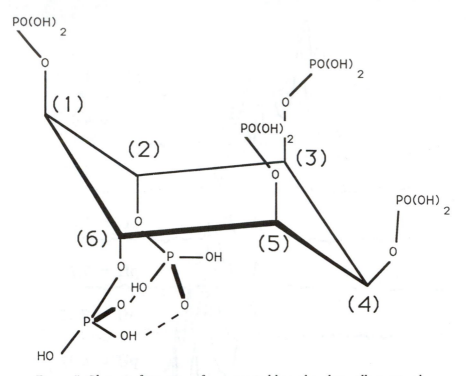

Figure 5. Chair configuration of myo-inositol hexaphosphate, *illustrating hydrogen bonding between phosphate groups 2 and 6. Hydrogen bonding between groups 3 and 5 is not shown for clarity.*

consisting of an intense singlet at approximately 4.1 ppm with a large broad shoulder upfield from it (Figure 7). As the NaCl concentration is increased, this broad shoulder narrows, increasing in intensity, until it becomes a peak. At 1.00 M NaCl the spectrum displays three separate peaks with an emerging fourth peak. These signal pattern changes result from the clustering of cations around the oxygen atoms, protonated and deprotonated. The electrostatic radius is decreased such that steric hindrance between the phosphate groups is reduced and the IHP molecule can begin to return to the chair configuration.

Humic Matrix and T_1 Relaxation Agent. The effects of the concentrated humic matrix and the ferric relaxation agent on the ^{31}P NMR spectrum are similar to that of increasing ionic strength. When orthophosphate is dissolved in a concentrated humic matrix, its signal position is shifted slightly downfield relative to a pure water matrix at an equivalent pH (Figure 8). The addition of FeEDTA to the concentrated humic matrix causes this downfield shift to be even more pronounced. The increase in the downfield shift of signals

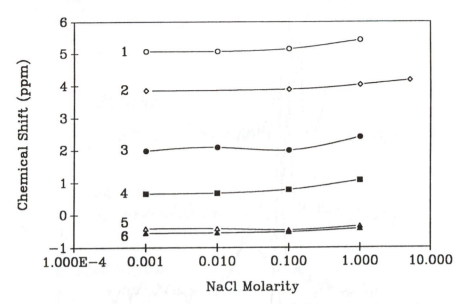

Figure 6. Effect of ionic strength on the chemical shift of inositol monophosphate (IMP), inositol hexaphosphate (IHP), choline phosphate, and DNA at various pH values. Key: 1, IMP, pH 9.9; 2, choline phosphate, pH 10.8; 3, IMP, pH 4.7; 4, IHP, pH 5.0; 5, DNA, pH 9.8; and 6, DNA, pH 4.4.

due to the presence of concentrated humic matrix or the relaxation agent occurs with monoester phosphates such as choline phosphate (Figure 9). The reason for this behavior lies in the ionic-strength-induced downfield shifts. In a concentrated humic matrix, possible intramolecular hydrogen bonding occurs between hydroxyl and carboxylic groups and the phosphates. The ferric relaxation agent, FeEDTA, has been proposed by Elgavish and Granot (48) to form a phosphate–Fe–EDTA complex. If this is the case, there is little doubt that the phosphate oxygen atoms are involved in complexing with FeEDTA, shifting electron density from the phosphorus nucleus.

Lake-Water Samples. Ten lake-water samples were collected from September 1990 to May 1991. The total soluble phosphorus concentration for the concentrated samples ranged from 23.8 to 60.8 mg of P/L, and the soluble reactive phosphorus concentrations ranged from 1.0 to 18.1 mg of P/L (Table I). Dissolved organic carbon concentration values for the concentrated samples ranged from 5000 to 20,000 mg of C/L. The signal-to-noise ratios from 12–14-h runs achieved for the NMR spectra range from 3.0 to 7.0. The pH of the concentrated samples after the addition of FeEDTA fell between the values of 7.00 and 8.00. Addition of the FeEDTA increased the pH by only a few tenths of a pH unit.

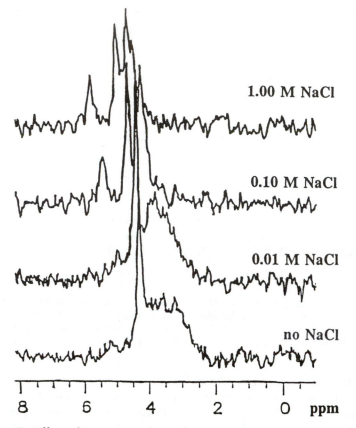

Figure 7. Effect of ionic strength on the ^{31}P FT-NMR spectrum of inositol hexaphosphate at pH 9.8.

Comparison of the ^{31}P NMR spectra reveals several similarities (Figure 10). All spectra contain prominent signals in the monoester region (0.00–5.00 ppm) and less distinct signals between 0.00 and –2.00 ppm, the diester phosphate region. Phosphonates and polyphosphates were not detected in any samples.

A typical spectral pattern of DOP usually consists of four basic components:

- a small broad envelope spanning from approximately 4.00–5.00 ppm.
- an intense single peak in the region from 2.50 to 4.00 ppm.
- a large, intense, broad envelope spanning 0.00–2.00 ppm; often appearing to consist of two or more overlapping peaks.
- several smaller, sometimes overlapping, signals in the region from –2.00 to 0.00 ppm.

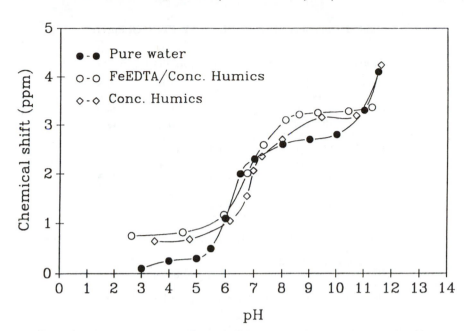

Figure 8. Effect of pH on orthophosphate's chemical shift in various matrices. The sample concentration was 150 mg of P/L. Pure water data are taken from reference 46.

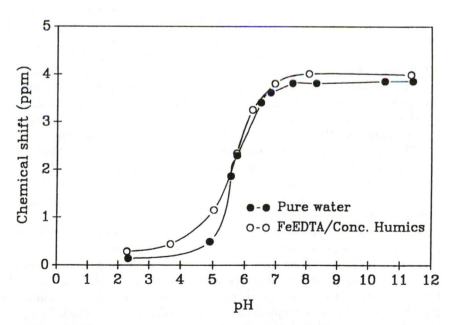

Figure 9. Effect of pH on choline phosphate's chemical shift in various matrices. The sample concentration was 150 mg of P/L.

Table I. Summary of Crystal Lake ^{31}P FT-NMR Spectra

Sample	Date	$S:N^a$	pH	SRP (mg P/L)	TSP (mg P/L)
A[b]	9/5/90	3.2	c	1.0	23.8
B[b]	9/12/90	3.0	c	2.4	25.6
C[d]	9/26/90	3.8	c	3.0	46.0
D[d]	10/24/90	2.4	c	5.2	48.6
E[d]	11/9/90	2.6	7.90	6.0	25.0
F[e]	1/16/91	5.6	7.92	18.1	60.8
G[d]	2/11/91	4.6	7.28	3.4	50.0
H[e]	3/11/91	4.0	7.53	2.8	23.5
I[e]	4/3/91	3.7	7.72	2.5	29.3
J[e]	5/2/91	4.9	7.68	1.8	36.7

[a] Signal-to-noise ratio.
[b] The membrane used was 80% NaCl rejection reverse osmosis cellulose acetate.
[c] No data were collected.
[d] The membrane used was 99% NaCl rejection reverse osmosis polyamide.
[e] The membrane used was 1000-dalton, tangential-flow polysulfone.

Studies of several model monoester phosphate compounds (44), each of which are potentially present in lake water, were performed by examining signal position, and signal pattern if appropriate, over a pH range to obtain an idea of which compounds could be present in the sample. All model compounds were dissolved in a concentrated humic matrix with FeEDTA present. These studies indicate that the signal envelope from 4.00 to 5.00 ppm could result from 3'-adenosine monophosphate, inositol monophosphate, serine phosphate, or choline phosphate (Figures 3 and 9) because ^{31}P NMR signals for all of these compounds appear in this region at pH 7.

The intense singlet that appears between 2.50 and 4.00 ppm is orthophosphate. Correlation of the sample pH and signal position with that of the pH-dependent chemical-shift curve of orthophosphate in a concentrated humic matrix with FeEDTA (Figure 8) confirms this peak's identity.

The identity of the large broad peak spanning from 0.00 to 2.00 ppm is an enigma. ^{31}P NMR spectra obtained at pH 7.00–8.00 suggest that this envelope could be attributed to inositol hexaphosphate. However, spectra obtained at other pH values (which will be described in detail later) show that this envelope has drastically different pH behavior than inositol hexaphosphate. Glycerophosphoryl choline and glycerophosphoryl ethanolamine, even though they are diester phosphates, have been shown by other researchers (49) to produce signals in this region at this pH. Both are degradation products of structural components of cellular membranes (50) and thus would be expected to be fairly ubiquitous in the aquatic environment. But these species are susceptible to further hydrolysis and attack by lipophosphatases, to yield choline phosphate, ethanolamine phosphate, and glyc-

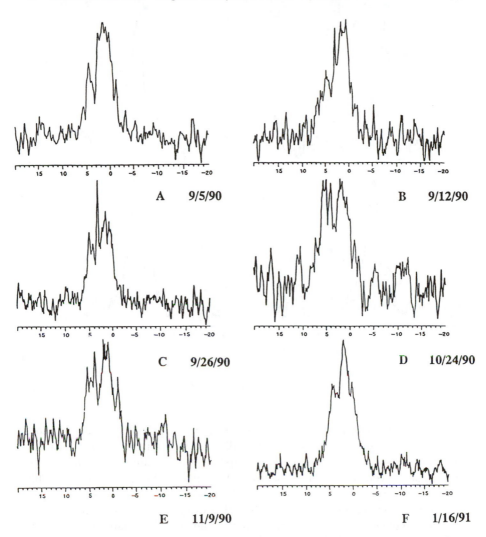

Figure 10. ^{31}P FT-NMR spectra of Crystal Lake samples. Letters correspond to sample designations in Table I. (Continued on next page.)

erol phosphate. Thus it would seem highly unlikely to detect large amounts of these glycerophosphoryl compounds. ^{31}P FT-NMR spectra of soils and sediments (*31–38*), no signals are present between 1.50 and 0.00 ppm.

The final region from −2.00 to 0.00 ppm is attributed to diester phosphates such as RNA, DNA, nucleotide fragments, and phosphatidyl compounds. Signals occur in this region for both DNA and phosphatidyl choline in a concentrated humic matrix with FeEDTA present (Figure 11). ^{31}P FT-NMR evidence supporting the presence of DNA is provided by the spectrum of a sample that has been oxidized by alkaline bromination (Figure 12).

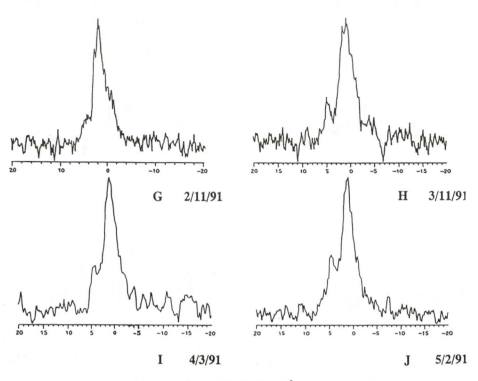

G   2/11/91   H   3/11/91

I   4/3/91   J   5/2/91

Figure 10. Continued.

Alkaline bromination oxidizes organic phosphorus, except for IHP (*51*) and DNA (*52*), to orthophosphate. After alkaline bromination, three distinct signals are present. The signal at 6.5 ppm is presumably caused by orthophosphate, even though at the sample pH of 10.3 orthophosphate would be expected to appear between 3.0 and 4.0 ppm. The second signal at 1.5 ppm is probably remaining unoxidized DOP that originally gave rise to the large signal envelope. The third signal, in the diester phosphate region, is at −1.5 ppm and is probably from DNA.

pH Studies.   In further attempts to characterize and better define the structure of the signal envelopes, the pH of the samples obtained in April and May was varied and ^{31}P NMR spectra were obtained (Figure 13). The pH behavior was similar for both samples. The 4.00−5.00-ppm envelope behaved as a typical monoester phosphate; at pH >7 it was present between 4.00 and 5.00 ppm, but at lower pH values it moved upfield and merged into the large 0.00−2.00-ppm envelope. The large singlet, which appears between 4.00 and 2.50 ppm and is attributed to orthophosphate, was not detected in these samples.

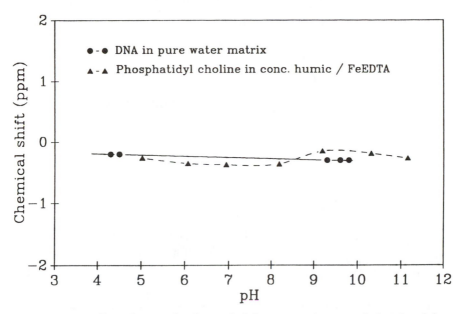

Figure 11. Effect of pH on the chemical-shift position of DNA and phosphatidyl choline. DNA is in a pure water matrix, and phosphatidyl choline is in a concentrated humic–FeEDTA matrix.

The position of the large, broad envelope between 0.00 and 2.00 ppm is only slightly affected by pH changes. This behavior indicates that this signal envelope is not caused by IHP, for if IHP were present two large singlets would appear at 2.3 and 3.0 ppm at pH 9, and at pH 10 the two peaks would be shifted to 4.2 and 3.2 ppm (Figure 4). None of these peaks appear when the sample pH is increased; in fact, no signals are visible in the region between 2.00 and 4.00 ppm. When the pH is decreased to 4, the typical four-peak pattern of inositol hexaphosphate (rather than the broad envelope) would be expected from the ^{31}P FT-NMR spectra of IHP spiked into a concentrated humic matrix with FeEDTA present. On the basis of the sample's low-pH data (pH <4) alone, though, IHP cannot be dismissed; IHP signals may be broad and overlapping because of interactions with the concentrated humic matrix, the relaxation agent, or other cations present. Glycerophosphoryl choline and glycerophosphoryl ethanolamine ^{31}P NMR signals have been reported to appear at 1.5–0.5 ppm (relative to 85% H_3PO_4) and not to be pH-dependent. It seems highly unlikely that these compounds would be present; as mentioned previously, they are fairly unstable in an aqueous environment.

The envelope of signals below 0.00 ppm, although not sufficiently intense to identify any individual peaks, does not exhibit any pH-dependent behavior. This pattern is expected if these signals result from diester phos-

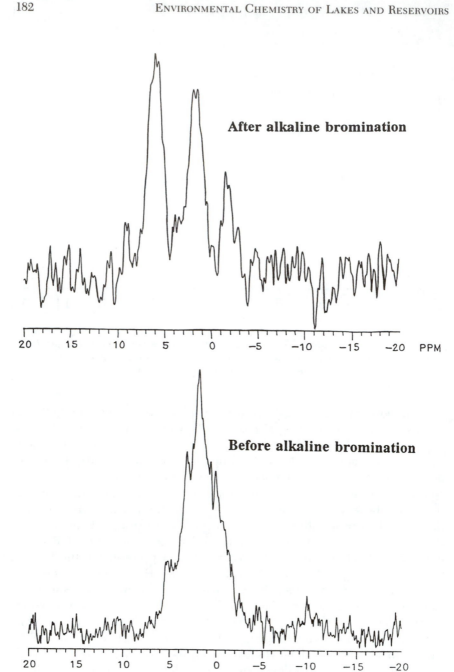

Figure 12. ^{31}P *FT-NMR spectrum of Crystal Lake sample G, 1000-dalton membrane retentate fraction, before and after alkaline bromination. The total soluble phosphorus concentration was 50 mg of P/L.*

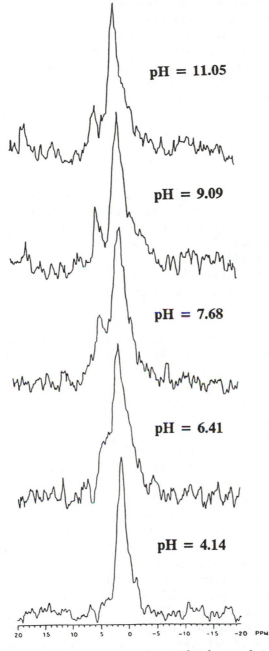

Figure 13. pH behavior of Crystal Lake sample J.

phates, because the model compounds, DNA and phosphatidyl choline, show similar pH-independent behavior (Figure 11).

One hypothesis can be formulated about the identity of the large, broad, 2.00–0.00-ppm envelope. Instead of a unique, individual compound, it probably results from an aggregate or micelle composed of dissolved organic phosphorus compounds or concentrated humic material. The aggregate could possibly arise naturally from the species present in the sample. One such specie is polyethylene glycol, a nonionic surfactant that has been implied in these samples based on ^{13}C NMR. These materials and phosphatidyl compounds, which may also be present, have critical micelle concentrations on the order of 10^{-8} to 10^{-10} M. Aggregate or micelle formation could also be enhanced or caused by the extensive concentration procedures employed. These aggregates or micelles could incorporate dissolved organic phosphorus into their structures. This configuration would shield the phosphorus compounds from pH changes in the surrounding solution and account for the lack of pH-dependent behavior. If several DOP species are present at once in these aggregates, the lack of individual, separate signals could result from signal broadening, which could occur if the DOP is adsorbed to the aggregate surface or incorporated into the micelle's membrane or inner volume.

To test if DOP was being incorporated into an aggregate structure, Crystal Lake sample I was spiked with 15 mg of P/L of phosphatidyl choline (Figure 14). At this concentration, the signal intensity from phosphatidyl choline should be comparable to the intensity of the 0.00–2.00-ppm signal envelope. Instead of the phosphatidyl choline peak appearing at approximately –0.3 ppm, the intensity of the broad envelope (with its maximum at 0.64 ppm) increased its signal-to-noise ratio from 2.5 to 4.6. The further addition of 15 mg of P/L of 3'-adenosine monophosphate, which at pH 6.59 is expected to appear at approximately 4.7 ppm, only increased the intensity of the broad envelope, so that the signal-to-noise ratio became 12.3. This signal is not attributable to orthophosphate, for the sample's SRP concentration was only 5.3 mg of P/L, much too low to produce a signal of such magnitude. Additionally, orthophosphate will appear at 2.2 ppm in a concentrated humic matrix with FeEDTA present. This prediction is based on ^{31}P FT-NMR spectra of orthophosphate spiked into concentrated humic matrix with FeEDTA present (Figure 8). In this final spectrum, a small individual peak appeared with a maximum at 4.0 ppm. It probably arose from free adenosine monophosphate (AMP). Another broad signal, of unknown origin, began to appear at approximately –15.00 ppm.

Incorporation of phosphatidyl choline and AMP signals into the large, broad envelope was unexpected because it had not been observed before with studies involving model compounds in a concentrated humic matrix with FeEDTA present (44). However, these model compound studies were conducted at a phosphorus concentration of 150 mg of P/L, a concentration 10 times greater than the amount that was spiked into sample I. This fact,

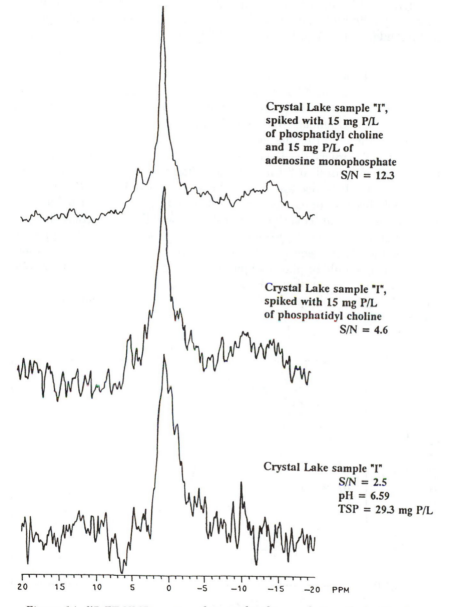

Crystal Lake sample "I",
spiked with 15 mg P/L
of phosphatidyl choline
and 15 mg P/L of
adenosine monophosphate
S/N = 12.3

Crystal Lake sample "I",
spiked with 15 mg P/L
of phosphatidyl choline
S/N = 4.6

Crystal Lake sample "I"
S/N = 2.5
pH = 6.59
TSP = 29.3 mg P/L

*Figure 14. ^{31}P FT-NMR spectra of Crystal Lake sample I, spiked with phos-
phatidyl choline and adenosine monophosphate.*

coupled with the appearance of a small AMP peak at 4.0 ppm in sample I, suggests that fractionation may exist between the free dissolved form and the aggregated form, with the aggregation–micelle fraction able to incorporate only a limited amount of phosphorus and the remaining phosphorus existing independently in the dissolved fraction.

Seasonal Behavior. A seasonal trend appeared in a comparison of the relative intensity of the 4.00–5.00-ppm envelope with that of the 0.00–2.00-ppm envelope over the 10-month sampling period (Figure 15). From early September to mid-January the 4.00–5.00-ppm envelope intensity was approximately three-fourths the intensity of the larger 0.00–2.00-ppm envelope. In mid-February the large 0.00–2.00-ppm signal envelope intensity dramatically increased to 3 times greater than that of the 4.00–5.00-ppm signal intensity. This increase indicated a likely change in the DOP species, presumably because of early spring algal growth. In early January to mid-February the thick ice covering had broken up and the water color changed from a drab olive-brown to a bright green that indicated new algal growth. One possible DOP species contributing to the enhanced 0.00–2.00-ppm signal envelope could be phospholipids incorporated into micelles. Increased concentrations of phospholipids with increased algal growth would be expected because phospholipids are a major cellular membrane component (50).

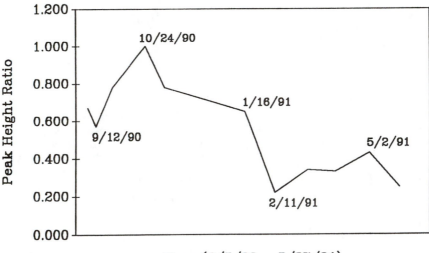

Figure 15. Seasonal changes in dissolved organic phosphorus detected by [31]*P FT-NMR spectroscopy. The peak height ratio was 4.0–5.0 ppm:0.0–2.0 ppm.*

Future Research

^{31}P FT-NMR spectroscopy holds promise of helping to characterize dissolved organic phosphorus, but considerable additional work needs to be done to obtain better sensitivity and resolution. The methods presented in this chapter describe ultrafiltration and reverse osmosis membrane concentration methods with subsequent ^{31}P FT-NMR analysis. With these methods much progress has been made, but further refinement is desired to yield better spectra. One option is to use NMR spectrometers with stronger magnets, but access to these instruments is often limited and they are expensive. Another route pursued extensively by researchers using ^{13}C NMR spectroscopy in studying humic and fulvic acids, coal, and hydrocarbon products is solid-state NMR spectroscopy, usually employing the technique of cross polarization–magic-angle spinning. This may be a potential option in the future, but at present its promise seems dim because of the signal broadening that occurs in solid-state NMR experiments. With the ^{13}C NMR technique signals are often well-separated and appear over a wide spectral range because of the many different types of carbon atoms present (e.g., carboxylic, aromatic, and aliphatic) and slight broadening of the signals will not appreciably decrease resolution. In contrast, with DOP samples, mono- and diester phosphates are presently the only detectable species, and their signals appear only in a very narrow spectral range. Thus any line broadening will cause signal overlap and loss of resolution.

Other pathways to better resolution and enhanced sensitivity may involve methods used in NMR studies of cellular phosphorus dynamics and humic and fulvic materials of water, soil, and sediments. These techniques include a variety of extractions, ion-pairing reagents, adsorption techniques, and lanthanide-shift reagents. For example, an extraction technique that has greatly enhanced the resolution and sensitivity of ^{31}P FT-NMR spectra of Ehrlich ascites tumor cells (53) and HeLa cells (54) is the addition of 35% perchloric acid and removal of acid-insoluble material, followed by filtration and neutralization with K_2CO_3 or NaOH.

Organic solvent extractions may also prove beneficial to enhancing spectral quality. Even though the DOP may not be exclusively isolated, partial removal of the humic matrix, which is potentially responsible for signal broadening, may result. Advantages of isolating DOP into an organic solvent, such as dimethyl sulfoxide, include the reduction and elimination of intra- and intermolecular interactions, which influence signal shape and width. Organic solvents would also allow the use of a wide range of T_1 relaxation agents and lanthanide-shift reagents that are not soluble in an aqueous solution. To enhance DOP isolation by solvent extraction, it may be necessary to employ ion-pairing reagents such as tetrabutylammonium bromide. Besides their ability to increase DOP solubility in nonaqueous solvents, ion-pairing reagents can also reduce intra- and intermolecular interactions in aqueous and nonaqueous solvents. Another route would be to freeze-dry the sample and then perform a strong base–strong acid extraction, such as in the ^{31}P FT-NMR soil and sediment studies. The major disadvantages with this procedure are the isolation of only base- and acid-soluble material and the possible hydrolysis of DOP to orthophosphate.

Other potential DOP isolation methods are based on sorption using anion-exchange resins or nonionic resins. Extensive research in concentrating humic material has employed these methods (55). For isolation and concentration of DOP, these techniques have several disadvantages, such as possible irreversible sorption onto the resins and the concentration of all anionic species present in the sample. Another major drawback is the need to use a strong base to remove the sorbed material from the resin. For sorption to occur with nonionic resins requires the protonation of all weak acids, thus exposing the sample to low pH.

The use of lanthanide-shift reagents (LSRs) might help to resolve overlapping or closely positioned signals in ^{31}P NMR spectra, such as DOP samples that consist of mono- and diester phosphates. Many LSRs are effective with phosphate esters, phosphonates (56–59), and phospholipid micelles (60–63). LSRs with phospholipid micelles have been used to separate the signals of the outer-surface phosphate groups from those of the inner phosphate group. The outer phosphate groups will be the only phosphorus atoms affected by the presence of an LSR and therefore their signal position will change while the inner phosphate group signal position remains constant. Critical micelle constants (CMC) of pure phospholipids have been measured by monitoring changes in the ^{31}P FT-NMR signals versus the phospholipid concentration. Because the CMC is a function of the nonpolar tail of the phospholipid, the length (number of carbons present on the tail) can be determined. If the large signal envelope (0.00–2.00 ppm) of the ^{31}P FT-NMR spectrum of the concentrated lake-water samples is attributed to the formation of a micelle or aggregate composed of several different DOP species, LSRs would aid in the identification of the DOP that was on the micelle surface. Precipitation of the LSR and the concentrated humic matrix possibly may be avoided by the use of macrocyclic ligands, which would have a high enough complexation constant to prevent complexation of the lanthanide ion by the humic material, but still allow interaction between the lanthanide ion and the phosphate group (64–68).

These techniques may further enhance the ability of ^{31}P FT-NMR spectroscopy to be used for the identification of hydrosphere DOP. We have demonstrated already that ^{31}P NMR spectroscopy is a viable technique for DOP identification and characterization. ^{31}P NMR spectroscopy may eventually be useful through several of the techniques mentioned in this chapter for studying interactions of DOP with dissolved humic substances, colloids, and seston and particulate adsorption. Hence, ^{31}P FT-NMR spectroscopy could be applied to studying the fractionation of phosphorus among the aquatic environment's many different components. ^{31}P FT-NMR spectroscopy, coupled with other analytical techniques such as HPLC, gel chromatography, bioassays, and enzyme bioassays, has potential to enhance the understanding of the identity and behavior of DOP in aquatic systems.

Summary

Because NMR spectroscopy is a nuclei-specific technique and has the ability to distinguish between similar compounds, it is an excellent method for identifying similar species in complex matrices. Thus, ^{31}P FT-NMR spectroscopy is ideal for the identification and characterization of the hydrosphere DOP. Even so, NMR spectroscopy is fairly insensitive and requires high sample concentrations. Low DOP concentrations are increased to ^{31}P FT-NMR detection limits by using ultrafiltration and reverse osmosis membranes. Not only is the DOP concentrated, but it is fractionated according to its molecular size. Compared to other concentration and molecular size fractionation techniques for DOP, ultrafiltration and reverse osmosis are relatively rapid and easy.

Much of the information regarding the identity and the characteristics of DOP in lake water has been learned by using ^{31}P FT-NMR spectroscopy. Although the DOP species have not yet been conclusively identified, ^{31}P FT-NMR spectroscopy shows that the DOP is composed of mono- and diester phosphates. The presence of DNA is strongly suggested by alkaline bromination experiments. ^{31}P FT-NMR spectroscopy also indicates that concentrated DOP species form aggregates or micelles, isolating the incorporated DOP from changes in solution pH. These aggregates or micelles are able to incorporate additional organic phosphorus that has been added to the sample. It is still unclear whether the aggregation–micelle phenomena is an artifact of the concentration procedure or is representative of the DOP present in the lake water.

DOP seasonal variations are observed between the relative heights of the monoester phosphate signals. The increase in height of the 0.0–2.0-ppm signal correlates with the breakup of the winter ice and new algal growth. This correlation illustrates that ^{31}P FT-NMR spectroscopy can detect changes in the DOP resulting from temporal changes in the lake.

References

1. *Standard Methods for the Examination of Water and Wastewater*, 17th ed.; Clesceri, L. S.; Greenberg, A. E.; Trussell, R. R., Eds.; American Public Health Association: Washington, DC, 1989; pp 437–452.
2. Wetzel, R. G. *Limnology*, 2nd ed.; Saunders College Publishing: New York, 1983; pp 255–297.

3. Lean, D. R. S. *J. Fish. Res. Board. Can.* **1973**, *30*, 1525–1536.
4. Paerl, H. W.; Downes, M. T. *J. Fish. Res. Board. Can.* **1978**, *35*, 1639–1643.
5. Stainton, M. P. *Can. J. Fish. Aquat. Sci.* **1980**, 37, 472–478.
6. White, E.; Payne, G. *Can. J. Fish. Aquat. Sci.* **1980**, 37, 664–669.
7. Burnison, B. K. *Can. J. Fish. Aquat. Sci.* **1983**, *40*, 1614–1621.
8. Stevens, R. J.; Stewart, B. M. *Water Res.* **1982**, *16*, 1507–1519.
9. Paerl, H. W.; Lean, D. R. S. *J. Fish. Res. Board. Can.* **1976**, *33*, 2805–2813.
10. Nurnberg, G.; Peters, R. H. *Can. J. Fish. Aquat. Sci.* **1984**, *41*, 757–765.
11. Orsett, K.; Karl, D. M. *Limnol. Oceanogr.* **1987**, *32*(2), 383–395.
12. Taft, J. L.; Loftus, M. E.; Taylor, W. R. *Limnol. Oceanogr.* **1977**, *22*(6), 1012–1021.
13. Herbes, S. E.; Allen, H. E.; Mancy, K. H. *Science (Washington, D.C.)* **1975**, *187*, 432–434.
14. Pettersson, K. *Int. Rev. Gesamten. Hydrobiol.* **1979**, *64*, 585–607.
15. Kim, S.; Minear, R. A. *Prepr. Pap.* **1991**, American Chemical Society 201st National Meeting, Atlanta, GA.
16. Gadomski, J.; Minear, R. A. *Prepr. Pap.* **1991**, American Chemical Society 202nd National Meeting, New York.
17. Koenings, J. P.; Hooper, F. F. *Limnol. Oceanogr.* **1976**, *21*(5), 684–696.
18. Franko, D. A.; Heath, R. T. In *Aquatic and Terrestrial Humic Materials*; Christman, R. F.; Gjessing, E. T., Eds.; Ann Arbor Science: Ann Arbor, MI, 1983.
19. Minear, R. A. *Environ. Sci. Technol.* **1972**, *6*, 431–437.
20. DeFlaun, M. F.; Paul, J. H.; Davis, D. *Appl. Environ. Microbiol.* **1986**, 654–659.
21. Karl, D. M.; Bailiff, M. D. *Limnol. Oceanogr.* **1989**, *34*(3), 543–558.
22. Franko, D. A.; Wetzel, R. G. *Limnol. Oceanogr.* **1982**, *27*, 27–38.
23. Jefferey, L. M. *J. Am. Oil Chem. Soc.* **1966**, *43*(4), 211–214.
24. Sommers, L. E.; Harris, R. F.; Williams, J. D. H.; Armstrong, D. E.; Syers, J. K. *Soil Sci. Soc. Am. Proc.* **1972**, *36*, 51.
25. White, R. H.; Miller, S. L. *Science (Washington, D.C.)* **1976**, *193*, 885–886.
26. Weimer, W. C.; Armstrong, D. E. *Anal. Chim. Acta* **1977**, *94*, 35–47.
27. Eisenreich, S. J.; Armstrong, D. E. *Environ. Sci. Technol.* **1977**, *11*, 497–501.
28. Stuermer, D. H.; Payne, J. R. *Geochim. Cosmochim. Acta* **1976**, *40*, 1109–1114.
29. Bellama, J. M.; Meyer, S. R.; Pellenbarg, R. E. *Appl. Organomet. Chem.* **1991**, *5*(2), 107–109.
30. Newman, R. H.; Tate, K. R. *Commun. Soil Sci. Plant Anal.* **1980**, *11*, 835–842.
31. Tate, K. R.; Newman, R. H. *Soil Bio. Biochem.* **1982**, *14*, 191–196.
32. Ogner, A.; Langerud, B. R. *Acta Agric. Scand.* **1983**, *33*, 143–144.
33. Hawkes, A. E.; Powlson, D. S.; Randall, E. W.; Tate, K. R. *J. Soil Sci.* **1984**, *35*, 35–45.
34. Gil-Sotres, F.; Zech, W.; Alt, H. G. *Soil Biol. Biochem.* **1990**, *22*(1), 75–79.
35. Kupka, T.; Pacha, J.; Dziegielewski, J. *Magn. Reson. Chem.* **1989**, *27*(1), 21–26.
36. Ogner, G. *Geoderma* **1983**, *33*(2), 215–19.
37. Ingal, E. D.; Schroeder, P. A.; Berner, R. A. *Geochim. Cosmochim. Acta* **1990**, *54*, 2617–2620.
38. Florentz, M.; Granger, P.; Hartemann, P. *Appl. Environ. Microbiol.* **1984**, 519–525.
39. Hill, W. E.; Benefield, L. D.; Jing, S. R. *Water Res.* **1989**, *23*, 1177–1181.
40. Uhlmann, D.; Roske, I.; Hupfer, M.; Ohms, A. *Water Res.* **1990**, *24*, 1355–1360.
41. Jing, S. R.; Benefield, L. D.; Hill, W. E. *Water Res.* **1992**, *26*(2), 213–223.

42. Derome, A. E. *Modern NMR Techniques for Chemistry Research;* Pergamon: New York, 1987; Chapter 4.
43. Wasylishen, R. E. In *NMR Spectroscopic Techniques;* Dybowski, C.; Lichter, R., Eds., Marcel Dekker: New York, 1987.
44. Nanny, M. A.; Minear, R. A. *Prepr. Pap.* **1991,** American Chemical Society 201st National Meeting, Atlanta, GA.
45. Costello, A. J. R.; Glonek, T.; Myers, T. C. *Carbohydr. Res.* **1976,** *46,* 159–171.
46. O'Neill, I. K.; Sargent, M.; Trimble, M. L. *Anal. Chem.* **1980,** *52,* 1288–1291.
47. Letcher, J. H.; Van Wazer, J. R. *Top. Phosphorus Chem.* **1967,** *5,* 75–266.
48. Elgavish, G. A.; Granot, J. J. *J. Magn. Reson.* **1979,** *36,* 147–150.
49. Wahlgren, M.; Drakenberg, T.; Vogel, H. J.; Dejmek, P. *J. Dairy Res.* **1986,** *53,* 539–545.
50. Gurr, M. I.; Harwood, J. L. *Lipid Biochemistry: An Introduction,* 4th ed.; Chapman and Hall: London, 1991.
51. Cosgrove, D. J. *Nature (London)* **1962,** *194,* 1265.
52. Clarkin, C. M.; Minear, R. A. University of Illinois at Urbana–Champaign, unpublished results.
53. Navon, G.; Ogawa, S.; Schulman, R. G.; Yamane, T. *Proc. Natl. Acad. Sci. U.S.A.* **1977,** *74*(1), 87–91.
54. Evans, F. E.; Kaplan, N. O. *Proc. Natl. Acad. Sci. U.S.A.* **1977,** *74*(1), 4909–4913.
55. Aiken, G. R. In *Humic Substances in Soil, Sediment, and Water;* Aiken, G. R.; McKnight, D. M.; Wershaw, R. L.; MacCarthy, P., Eds.; John Wiley and Sons: New York, 1985.
56. Sanders, J. K. M.; Williams, D. H. *Tetrahedron Lett.* **1971,** *30,* 2813–2816.
57. Mandel, F. S.; Cox, R. H.; Taylor, R. C. *J. Magn. Reson.* **1974,** *14,* 235–240.
58. Taylor, R. C.; Walters, D. B. *Tetrahedron Lett.* **1972,** *1,* 63–66.
59. Berkova, G. A.; Zakharov, V. I.; Smirnov, S. A.; Morkovin, N. V.; Ionin, B. I. *Zh. Obshch. Khim.* **1977,** *47*(6), 1431–1432.
60. Kumar, V. V.; Baumann, W. J. *Biophys. J.* **1991,** *59*(1), 103–107.
61. Yoshida, T.; Miyagai, K.; Taga, K.; Okabayashi, H.; Matsushita, K. *Magn. Reson. Chem.* **1990,** *28*(8), 715–721.
62. Dennis, E. A.; Ribeiro A. A. In *Magnetic Resonance in Colloid and Interface Science;* Resing, H., Ed.; American Chemical Society: Washington, DC, 1976.
63. Chrzeszczyk, A.; Wishnia, A.; Springer, C. S., Jr. In *Magnetic Resonance in Colloid and Interface Science;* Resing, H., Ed.; American Chemical Society: Washington, DC, 1976.
64. Elgavish, G. A.; Reuben, J. *J. Am. Chem. Soc.* **1976,** *98*(16), 4755–4759.
65. Reuben, J. *J. Am. Chem. Soc.* **1976,** *98*(12), 3726–3728.
66. Elgavish, G. A.; Reuben, J. *J. Am. Chem. Soc.* **1977,** *99*(6), 1762–1765.
67. Bryden, C. C.; Reilley, C. N.; Desreux, J. F. *Anal. Chem.* **1981,** *53,* 1418–1425.
68. Peters, J. A.; Vijverberg, C. A. M.; Kieboom, A. P. G.; van Bekkum, H. *Tetrahedron Lett.* **1983,** *24,* 3141–3144.

RECEIVED for review September 26, 1991. ACCEPTED revised manuscript June 9, 1992.

CYCLING AND DISTRIBUTION OF MAJOR ELEMENTS

Installing sediment traps to measure phosphorus sedimentation in Lake Michigan.
Photo courtesy of James P. Hurley.

Although it has been less than fashionable as a research topic in recent years, our understanding of the biogeochemistry of major elements is improving rapidly. Renewed interest in the biogeochemistry of major nutrients has coincided with a re-emerging emphasis on the "traditional" pollutants. Similarly, studies of sulfur biogeochemistry have been driven by a pragmatic need to understand lake acidification. Equally important, the behavior of toxic pollutants is inextricably linked to the behavior of major elements. In many instances, our understanding of trace metals and toxic organic substances depends upon our understanding of the biogeochemistry of the major elements. How could one understand the behavior of cadmium without an intimate understanding of the cycling of sulfur?

Chemistry of Dissolved Organic Matter in Rivers, Lakes, and Reservoirs

J. A. Leenheer

U.S. Geological Survey, Water Resources Division, Denver, CO 80225

Recent investigations provide new insight on the structural chemistry of dissolved organic matter (DOM) in freshwater environments and the role of these structures in contaminant binding. Molecular models of DOM derived from allochthonous and autochthonous sources show that short-chain, branched, and alicyclic structures are terminated by carboxyl or methyl groups in DOM from both sources. Allochthonous DOM, however, had aromatic structures indicative of tannin and lignin residues, whereas the autochthonous DOM was characterized by aliphatic alicyclic structures indicative of lipid hydrocarbons as the source. DOM isolated from different morphoclimatic regions had minor structural differences.

T HE GEOCHEMISTRY OF DISSOLVED ORGANIC MATTER (DOM) in natural waters was recently reviewed in a textbook by Thurman (*1*). In the extensive body of literature he reviewed, average dissolved organic carbon (DOC) concentrations varied as follows: 0.7 mg/L for groundwater, 1.0 mg/L for precipitation, 2.0 mg/L for oligotrophic lakes, 5.0 mg/L for rivers, 10.0 mg/L for eutrophic lakes, 15 mg/L for marshes, and 30 mg/L for bogs. The various sources and relative abundance of DOC in natural waters are depicted in Figure 1.

Only about 20% of the DOM in natural waters consists of identifiable compounds that include carbohydrates, carboxylic acids, amino acids, and hydrocarbons (*1*). The remaining 80% of DOM (generally defined as humic substances) consists of complex, environmentally altered residue of plant, bacterial, and fungal origin, of moderate molecular weight (500–5000 daltons). This predominantly acidic undefined DOM has been classified by

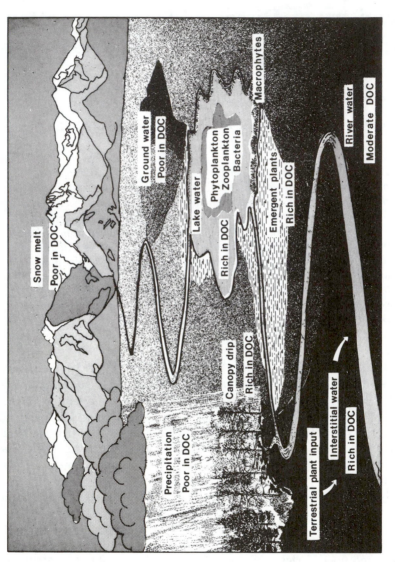

Figure 1. Allochthonous and autochthonous sources of dissolved organic carbon in natural waters. (Reproduced with permission from reference 1. Copyright 1985 Kluwer Academic Publishers.)

adsorption chromatography (2) into hydrophobic and hydrophilic fractions. Hydrophobic acids from natural waters have been operationally defined as dissolved humic substances (3). Acidification of the hydrophobic acid fraction of DOM precipitates humic acid, with fulvic acid remaining in the solution phase. In natural waters about 40% of the DOM is fulvic acid and 10% is humic acid (1). Hydrophilic acids, which account for about 30% of the DOM, are a mixture of undegraded biopolymers and highly degraded humic substances.

Studies of the chemistry of DOM in natural waters are greatly limited by the fact that most of the chemical structures involved are undefined. The complexity of the DOM mixture and the substantial size of the molecules have defeated attempts to characterize about 80% of DOM at the molecular level. Significant information, however, has been obtained on the structural chemistry of DOM since the publication of the book by Thurman (1). This chapter presents both recently published and previously unpublished information about the structural chemistry of DOM in various freshwater environments.

The experimental approach used was application of quantitative organic analytical information to derive an average structural model of certain DOM fractions in a manner analogous to the derivation of the structure of a pure compound. Heterogeneity was minimized for purposes of model synthesis and interpretation by separating DOM into more homogeneous fractions during the isolation procedure and by studies of samples derived from certain "end-member" environments where inputs and diagenetic processes that create and transform DOM are limited. Finally, the structural information obtained from end-member environments was applied to studies of DOM in the integrating environment of the Mississippi River and its associated reservoirs and tributaries.

Analytical Constraints

The complexity of the molecular mixture in various environments has severely limited analytical efforts to resolve DOM into its constituent molecules for structural determination (4). DOM consists of thousands to millions of compounds mixed as solutes in varying aqueous systems. Even if the structures could be determined at the molecular level, DOM chemistry could not be described as the sum of the characteristics of the individual components because of molecular interactions both within the DOM and with other aqueous species that modify component characteristics (4, 5).

Another analytical constraint is the definition of dissolved organic matter. Filters in the 0.1–1.0-μm size range pass colloids that are not truly dissolved. Studies discussed in this chapter will be limited to natural organic solutes that are either isolated by adsorption chromatography or ultrafiltered through 0.005-μm filters. Although neither of these techniques will absolutely exclude colloidal matter, their operational classification represents DOM better than conventional filtration methods do.

In spite of these difficulties with DOM chemistry, environmental chemists are frequently asked what molecular structures within the mixture are responsible for contaminant binding, haloform production, light attenuation, protonation characteristics, and other problems of environmental relevance. The chemist usually hypothesizes that DOM features such as aromaticity, polarity, functional-group content and configuration, molecular interactions, and molecular size can explain the observed phenomena. However, models of DOM (or DOM-fraction) structures must be based on average-mixture analyses to support these hypotheses. Such models represent average properties of thousands to millions of mixed compounds.

Mixture complexity must be minimized before structural studies can begin. One approach is fractionation of the mixture to concentrate and isolate the property of interest (5–7). An alternative is to study DOM found in environmental end-member systems. End-member environments are water bodies for which inputs of organic matter (allochthonous versus autochthonous) and climate (polar versus tropic) are homogeneous compared to those of most water bodies. The following research presents a fractionation of DOM isolated from end-member systems; moderately definitive molecular models were derived.

Fulvic Acid Structures from End-Member Environments

Correlation of Structure with Source.   Allochthonous-derived DOM (8) was isolated from the Suwannee River at its origin in the Okefenokee Swamp in southern Georgia. The fulvic acid fraction, which is responsible for the black coloration of the water, was extensively characterized (9). Several average molecular models based on quantitative analytical data were presented in that report (10) to denote the mixture characteristics of fulvic acid. One model, modified to depict biochemical sources and based on quantitative analytical data (10), is presented in Structure 1. Other models of Suwannee River fulvic acid (based on lignins, terpenoids, tannins, and flavonoid sources) were previously proposed (11).

The aromaticity of the model in Structure 1 reveals the tannin–lignin origin of the Suwannee River fulvic acid. Aliphatic structures, generally short-chain and branched, are commonly terminated by carboxyl groups. Aliphatic alicyclic structures from carbohydrate and lipid residues are also present. Ester linkages, both cyclic and linear, are indicated by a hydrolysis study (12). Hydrogen saturation and drying of the fulvic acid upon isolation probably created lactone ester linkages. Carbohydrate residues appear to be oxidized and joined to the remainder of the structure so that few alcoholic hydroxyl groups remain. Aromatic ketonic groups appear to be points of attachment between aliphatic groups and the aromatic core (13). The bio-

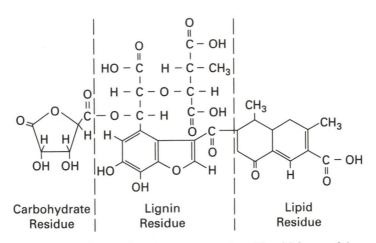

Structure 1. One of several average structural models of fulvic acid from the Suwannee River, Georgia.

chemicals from which fulvic acid is derived appear to be extensively degraded through biological oxidation (*14*). They are coupled by oxidative radical mechanisms involving oxygen, oxygen functional groups such as phenols and carbonyl groups, and unsubstituted positions of carbon double bonds (*15*).

DOM is derived from autochthonous sources such as phytoplankton and photosynthetic bacteria (*16*) at Big Soda Lake near Fallon, Nevada. This lake is alkaline (pH 9.7) and chemically stratified. It contains DOC concentrations as high as 60 mg/L and dissolved salt concentrations as high as 88,000 mg/L (*17*). The DOM in this lake is colorless. The fulvic acid fraction was isolated by adsorption chromatography (Amberlite XAD-8 resin) (*18*) and by zeotrophic distillation of water from *N,N*-dimethylformamide (*19*). Average molecular model synthesis was achieved in a manner similar to that used for fulvic acid from the Suwannee River. The characterization data are presented in Table I and the structural model is presented in Structure 2.

This model of fulvic acid derived from an autochthonous source is notable for its aliphatic alicyclic character. The model was generally derived from the marine lipid model of Harvey et al. (*20*), in which polyunsaturated lipids from phytoplankton are oxidatively coupled and cross-linked by free radicals at unsaturated sites to produce an aliphatic alicyclic structure. All open, straight-chain hydrocarbon structures in this model were degraded (presumably by biological β-oxidation) to carboxyl groups on aliphatic alicyclic rings. The resistance of aliphatic alicyclic structures to biological oxidation is well documented (*14*). The absence of chromophoric structures in the model confers resistance to degradation by ultraviolet radiation.

Structure 2 precisely adheres to the analytical constraints of Table I, but a wide variety of aliphatic alicyclic terpenoids, steroids, and porphyrins may serve as its precursors. The Harvey et al. (*20*) precursor model was favored

Table I. Molecular Data Used for Average Structural Model of Fulvic Acid from Big Soda Lake

Parameter	Method	Value
Number-average molecular weight	Equilibrium ultracentrifugation	550 daltons
	Vapor-pressure osmometry in tetrahydrofuran (four point)	500 daltons
Elemental contents corrected for moisture and ash content:		
Carbon	Combustion; measurement of CO_2	55.9%
Hydrogen	Combustion; measurement of H_2O	5.7%
Oxygen	Reductive pyrolysis; measurement of CO	36.3%
Nitrogen	Dumas method	1.6%
Sulfur	Combustion; measurement of SO_4	0.5%
Average molecular formula	Synthesis of number-average molecular weight and elemental analysis	$C_{26}H_{32}O_{12}N$
Index of hydrogen deficiency (ϕ)	$\phi = [(2C + 2) - H]/2$	11.0 rings or π bonds or both
Carbon distribution by type of carbon	Quantitative ^{13}C-NMR spectrometry in ^{12}C-enriched dimethyl sulfoxide for indicated integration range	26 C total
Aliphatic	0–65 ppm	14 C

H—C—O (alcohol, ether, ester, acetal, ketal)	65–110 ppm	4 C
Aromatic and olefinic	110–165 ppm	3 C
Carboxyl plus ester	165–185 ppm	5 C
Hydrogen distribution:		
Exchangeable hydrogen distribution by type of hydrogen:		32 H total
Carboxyl	Titrimetry, trifluoroethyl alkylation, ^{1}H-NMR spectrometry	4 H
Alcohol	Trifluoroethyl alkylation and ^{1}H-NMR spectrometry	1 H
Nonexchangeable hydrogen distribution by type of hydrogen	Quantitative ^{1}H-NMR spectrometry as sodium salt at pH 8.0. Indicated integration range	27 H nonexchangeable
Isolated aliphatic $H_3C—C=O$, $H_2C—C=O$, $HC—C=O$, $H_3C—C=C$, $H_2C—C=C$, $HC—C=C$	0–1.9 ppm	15 H
	1.9–3.2 ppm	8 H
$H_3C—O$, $H_2C—O$, $HC—O$	3.2–6.5 ppm	3 H
Aromatic	6.5–8.5 ppm	1 H
Oxygen distribution by type of oxygen		12 O total
Carboxyl	Titrimetry	8 O
Ester	Infrared spectrometry of tetrabutylammonium salt	2 O
Alcohol	Trifluoroethyl alkylation and ^{1}H-NMR spectrometry	1 O
Ether	Total C—O and C=O linkages determined by ^{13}C-NMR minus carboxyl, ester, and alcohol C—O and C=O linkages	1 O

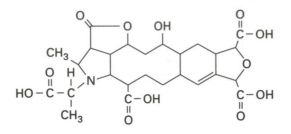

Structure 2. One of several average structural models of fulvic acid from Big Soda Lake, Nevada.

over terpenoid and steroid precursors because quaternary carbon linkages, which are present in terpenoids and steroids, were not detected in the attached proton test ^{13}C NMR spectra. Likewise methyl groups, attached to these quaternary carbons, were not abundant in the ^{1}H NMR spectrum of fulvic acid from Big Soda Lake.

Fulvic acid was isolated in Big Soda Lake above and below the chemocline, which occurs at 34-m depth. Water near the lake surface has moderate salinity and is oxygenated, whereas water below the chemocline is hypersaline and anoxic (*17*). In spite of these environmental differences the chemical character of the fulvic acid from above or below the chemocline did not vary, as determined by elemental analyses and NMR spectrometry.

The ^{14}C age determination of the fulvic acid isolated from water near the lake surface was 2300 years before the present, whereas the ^{14}C age was 4900 years before the present for the fulvic acid isolated from water below the chemocline. These old ages for both fulvic acids from Big Soda Lake are in marked contrast to that reported for fulvic acid from the Suwannee River, less than 30 years before the present (*11*). The refractory nature of this type of fulvic acid derived from phytoplankton and photosynthetic bacteria is significant for carbon-cycling studies.

Methods of Structural Analysis. The most significant differences between structural models 1 and 2 are the prominent aromatic carbon content in model 1 and the aliphatic alicyclic ring content in model 2. Determinations of aromatic carbon content and ring content of fulvic acid might be useful for identifying sources and processes of degradation and fractionation. However, neither of these procedures is simple and straightforward.

Aromatic carbon content cannot be directly determined from ^{13}C NMR spectrometry because it overlaps with olefinic carbon. Aromatic and olefinic hydrogens can be resolved in ^{1}H NMR spectrometry, but the chemical shifts of methine hydrogens on esters of secondary alcohols overlap with chemical shifts of olefinic hydrogen in the ^{1}H NMR spectra of fulvic acids. The ring content (Θ) is a difference determination between the index of hydrogen deficiency (ϕ) minus the carbonyl carbon content minus 0.5 (aromatic plus olefinic carbon content).

$$\Theta = \phi - \text{carbonyl C} - 0.5 \text{ (aromatic plus olefinic C)}$$

As shown by the data and methods presented in Table I, the analytical requirements for determining structural models are prohibitive for geochemical studies that survey many different environments.

A method was devised for estimating aromatic plus olefinic carbon content and total ring content. ^{1}H NMR spectral data are used to estimate aromatic plus olefinic carbon content, and ^{13}C NMR and ^{1}H NMR spectral data are combined with elemental carbon, hydrogen, and ash analyses to determine ring content.

First, peak heights are measured at five points in the ^{1}H NMR spectra (Figure 2). All ^{1}H NMR spectra of fulvic acids described in this study were determined as the sodium salt in D_2O at pH 8 (*21*). Peak heights were used rather than peak areas to minimize overlapping spectral contributions from various proton structures. From structural-model considerations, peak 1 appears to be a combination of methylene and methine protons in aliphatic alicyclic rings and branched methyl groups located beta to carbonyl groups of a carboxylic acid, ester, or ketone. The structural model rules out meth-

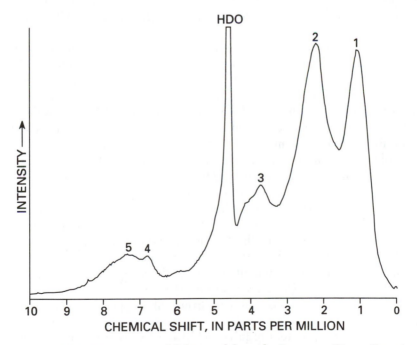

Figure 2. ^{1}H NMR spectrum of fulvic acid from the Suwannee River, Georgia.

ylene protons in open-chain hydrocarbons as significant contributors to this peak. Peak 2 represents methyl, methylene, and methine protons on carbons attached to aromatic rings or on adjacent carbonyl carbons. Peak 3 represents methylene and methine protons on carbons singly bonded to oxygen in alcohol, and ether functional groups. Peak 4 represents protons on aromatic carbons in the ortho-ring position to phenol or phenolic ether or ester moieties; these aromatic protons commonly occur in tannin and lignin structures. Therefore, peak 4 is a surrogate indicator of tannin and lignin content. Peak 5 indicates protons on aromatic rings that are minimally substituted with electron-donating or electron-withdrawing substituents.

Peak 5 is the best surrogate indicator of aromatic carbon content because of its strong peak intensity in the aromatic proton region. In investigations of fulvic acids from a number of end-member environments, peak 5 was much more conservative than peak 4 to changes resulting from degradation and fractionation processes. Similarly, peak 1 was determined to be much more conservative to degradation and fractionation processes than peaks 2 and 3. Therefore, the conservative nature of protons in peaks 5 and 1 makes their peak-height ratio a surrogate indicator of the percentage of aromatic plus olefinic carbon in fulvic acids from diverse environments. This application assumes that the relative proportions of protons to aromatic, olefinic, and aliphatic carbon in peaks 1 and 5 are constant. The peak-height ratios of peaks 2, 3, and 4 to peak 1 can be used to obtain information about degradation and fractionation processes involving structural moieties in these peaks.

Fulvic acid from the Suwannee River was used to calibrate peak-height ratios for aromatic carbon content. The application of this method to fulvic acid samples with known aromatic plus olefinic carbon content from various environments is shown in Table II. Aromatic plus olefinic carbon percentages calculated by the peak-height ratio method using ^{1}H NMR data closely agree with these percentages computed from ^{13}C NMR data, with the exception of the Big Soda Lake samples.

Fulvic acid from Big Soda Lake apparently has a greater percentage of olefinic carbons than fulvic acids from the Suwannee River, Calcasieu River, and Mississippi River, as shown by differences in molecular models 1 and 2. Therefore, the ^{1}H NMR peak-height ratio method underestimates the aromatic plus olefinic carbon content of the fulvic acid from Big Soda Lake. The fulvic acid sample from the Calcasieu River, which appears to be of allochthonous origin because of its aromatic plus olefinic carbon content, had nearly the same ring content as the Big Soda Lake sample, which is of autochthonous origin. Therefore, ring content does not appear to be a useful term in distinguishing allochthonous from autothonous sources for fulvic acids.

The lower ring content of fulvic acid from the Mississippi River might be related to high concentrations of suspended sediment in the river. The

Table II. Data Used To Determine Aromatic Plus Olefinic Carbon Percentage and Ring Content of Fulvic Acids

Determination	Method	Suwannee River above Fargo, GA	Big Soda Lake near Fallon, NV	Calcasieu River near Kinder, LA	Mississippi River near St. Francisville, LA
Aliphatic carbon (%)	Quantitative ^{13}C-NMR spectrometry	27.0	52.3	26.5	38.4
Carbohydrate, ether, alcohol carbon (%)	^{13}C-NMR	17.5	14.1	18.9	15.4
Aromatic plus olefinic carbon (%)	^{13}C-NMR	30.5	11.7	32.9	25.1
Carboxyl, ester, amide carbon (%)	^{13}C-NMR	19.0	19.7	14.6	17.1
Ketonic carbon (%)	^{13}C-NMR	6.0	1.8	6.1	2.8
Peak 5:1 ratio	Peak height at 7.4 ppm divided by peak height at 1.2 ppm in ^{1}H-NMR spectrum	0.160	0.020	0.162	0.132
Aromatic plus olefinic carbon calculated from ^{1}H-NMR data (%)	Peak 5:1 ratio times 190.6[a]	30.5	3.8	30.9	25.1
Number of carbons per 1000-dalton molecule	Percentage of carbon (ash free) times 10	45.0	46.6	47.2	46.0
Number of hydrogens per 1000-dalton molecule	Percentage of hydrogen (ash free) times 0.833	43.3	57.0	46.9	54.9
Index of H deficiency (ϕ) per 1000-dalton molecule	$\phi = [(2C + 2) - H]/2$	24.4	19.1	24.8	19.6
Number of rings (Θ) per 1000-dalton molecule	$\Theta = \phi -$ carbonyl C $- 0.5$ (aromatic plus olefinic carbon)	6.3	7.2	7.3	4.7

[a] 190.6 is derived from aromatic-carbon percentage divided by peak 5:1 ratio of fulvic acid from the Suwannee River.

sediment might fractionate fulvic acid by adsorbing components containing rings and associated functional groups. The most useful means of source identification is aromatic plus olefinic carbon content. ^{1}H NMR data and the peak-height ratio method to determine aromatic plus olefinic carbon content produce results with errors that are not significant for broad interpretive purposes.

Variation among Sampling Sites.   Dissolved humic substance samples from seven end-member environments were isolated for study. Autochthonous inputs to DOM were expected to dominate in Big Soda Lake and in Island Lake, which is a groundwater-sustained eutrophic lake in the sandhills of western Nebraska. Allochthonous inputs to DOM from a swamp environment predominate in the Suwannee River. They also dominate in the Calcasieu River in western Louisiana, but the proportion of swampland is much lower there. The Temi River is a tropical blackwater tributary of the Orinoco River in Venezuela, where allochthonous inputs dominate. The entire Sagavanirktok River basin is located north of the tree line on the North Slope of Alaska; a mixture of allochthonous and autochthonous inputs was expected for the various rivers and lakes in this basin. Lastly, Hidden Lake Creek, which is the outlet of Hidden Lake on the Kenai Peninsula of Alaska, was sampled to determine if nutrient inputs from decaying salmon were contributing to primary production and autochthonous inputs to DOM.

^{1}H NMR data from these seven sites are presented by spectral peak-height ratios in Table III. The sites were listed in order of increasing aromatic plus olefinic carbon percentages. Fulvic acids from all the lake samples are much lower in aromatic plus olefinic carbon content than those from river samples. These results confirm the hypothesis that autothonous inputs result in dissolved humic substances that have a low aromatic plus olefinic carbon content. The lake samples also are lower in the ratios of peak 2 (carboxylated chains and aliphatic ketones), peak 3 (carbohydrates), and peak 4 (phenolic tannins and lignins) to peak 1 (branched methyl groups and alicyclic aliphatics) than are the river samples.

Fulvic acid isolated from Island Lake provided direct spectral evidence of protons attached to aliphatic alicyclic rings. The ^{1}H NMR spectrum of this sample (Figure 3) has a broad medium-intensity peak near 1.8 ppm. The chemical shifts of protons on aliphatic alicyclic rings are commonly split into two broad peaks centered at 1.1 and 1.8 ppm because of their variable boat and chair configurations (23).

The Sagavanirktok River is intermediate in aromatic carbon content. This river drains the bogs on the Arctic tundra (allochthonous inputs) and several lakes (autochthonous inputs). Samples from the Suwannee and Calcasieu rivers are very similar in peak-height ratios with the exception of peak 2:1 ratio, which is much lower for the Calcasieu River sample. The data in Table II indicate that the Calcasieu River fulvic acid has a greater ring content

Table III. ¹H-NMR Spectral Peak-Height Ratios and Aromatic Plus Olefinic Carbon Contents of Dissolved Humic Substances

Sample and Site	Peak 2:1[a]	Peak 3:1[a]	Peak 4:1[a]	Peak 5:1[a]	Peak 4:5[a]	Aromatic Plus Olefinic Carbon (%)[b]
Fulvic acid from Big Soda Lake near Fallon, Nevada[c]	0.643	0.119	0.008	0.020	0.40	3.8
Fulvic acid from Island Lake near Oshkosh, Nebraska[d]	0.565	0.232	0.022	0.050	0.44	9.5
Fulvic acid from Hidden Lake Creek near Soldotna, Alaska[d]	0.596	0.184	0.046	0.056	0.80	10.7
Fulvic acid from Sagavanirktok River above Dead Horse, Alaska[d]	0.739	0.316	0.105	0.102	1.02	19.4
Fulvic acid from Suwannee River above Fargo, Georgia[d]	1.022	0.436	0.150	0.160	0.94	30.5
Fulvic acid from Calcasieu River near Kinder, Louisiana[d]	0.580	0.589	0.155	0.162	0.96	30.9
Fulvic and humic acid from Temi River, blackwater tributary of Orinoco River, Venezuela[e,f]	0.820	HOD—solvent peak interferes	0.250	0.250	1.00	47.7

[a] Peak heights were measured at points noted in Figure 2.
[b] Percentage of aromatic plus olefinic carbon was computed from the peak 5:1 ratio as described in Table II.
[c] Fulvic acid was isolated by the methods of references 18 and 19.
[d] Fulvic acid was isolated by the methods of reference 22.
[e] The Temi River sample was collected by Robert Stallard of the U.S. Geological Survey.
[f] Fulvic acid was isolated by the method of reference 26.

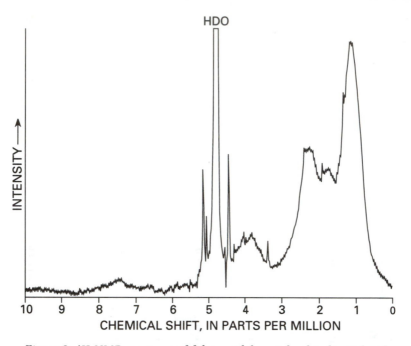

Figure 3. ¹H NMR spectrum of fulvic acid from Island Lake, Nebraska.

than that from the Suwannee River. An oxidative degradation reaction of a terpenoid hydrocarbon that might explain these results is shown in reaction 1.

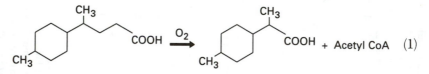

Aliphatic side-chain oxidation (reaction 1) would result in an increase of carboxyl groups at points of chain branching on alicyclic rings. This degradation would result in a 50% decrease in peak 2 and an increase in the ring content. The branched methyl group located beta to the carboxyl group would give the observed peak at 1.1 ppm for peak 1 along with some of the alicyclic methylene protons.

　　The Suwannee River was sampled at its origin at the outlet of the Okefenokee Swamp. This fulvic acid, therefore, is likely to be less degraded than a sample from the Calcasieu River that was taken near its mouth on the estuary during a warm, low-flow period in early summer. Metal-ion solubility controls and sorption on mineral surfaces in upland soils also might fractionate the fulvic acid in the Calcasieu River. In contrast, the Suwannee River mineral–soil solubility controls are less significant.

The Temi River samples had the largest aromatic carbon content because of inclusion of the humic acid fraction. Tropical blackwater rivers are known to contain large percentages (as much as 30%) of humic acid DOM because of the low-conductivity waters and lack of solubility controls associated with the sandy podzols in the tropical rain forest (*24, 25*).

Fulvic Acid Structures from Integrating Environments

Most aquatic environments integrate organic inputs from allochthonous and autochthonous sources and process them through various degradations and fractionations. DOM in the integrating environment of the lower Mississippi River was studied from July 1987 to June 1991. Seven sampling cruises were made during this time in which parcels of water were sampled in Lagrangian manner at 15–17 sites between St. Louis, Missouri, and New Orleans, Louisiana. Figure 4 shows a map of the lower Mississippi River and its major tributaries.

The first three cruises were made during low-discharge conditions in midsummer, late fall, and late spring. The next four cruises were made during a mixture of low- and high-discharge conditions in different reaches of the river system.

Preparation of Samples. During the first two sampling cruises, about 100 L of depth- and flow-weighted representative samples were screened to remove sand. Then the samples were passed through a continuous-flow centrifuge to remove suspended silt larger than to 0.3 μm, a tangential flow ultrafilter (30,000-dalton porosity, cellulose-membrane filter) to remove colloids, and a 10-L methyl methacrylate resin column (Amberlite XAD-8) to isolate organic solutes (*26*). These operations were carried out on the research vessel "Acadiana".

During sampling cruises 3–5, 20 L of sample from the ultrafilter was preserved with chloroform, shipped back to an analytical laboratory, and vacuum-evaporated to 500 mL before being passed through a 500-mL column of XAD-8 resin (*27*). This revised procedure should have increased the recovery of DOM because of a 10-fold increase in the resin:water ratio. In addition, hydrophilic organic solutes that did not adsorb on the XAD-8 resin were isolated by a vacuum-evaporation procedure whereby water is distilled from acetic acid and inorganic salts precipitate in the acetic acid (*28*). Recoveries of DOC by the two procedures are presented in Table IV.

Recoveries of DOC were variable among sampling sites; most of this variability was apparently caused by losses during the laboratory isolation procedure. The many steps in the isolation procedure for dissolved hydrophilic substances might have resulted in loss of as much as 50% of that fraction. The procedure used to concentrate DOC in the dissolved-humic-substances fraction increased recovery, but about one-third of it still

Figure 4. Map of the lower Mississippi River and major tributaries.

passed through the resin column even with a 1:1 resin:water ratio. A discrete break between nonpolar and polar properties appears to distinguish between humic and hydrophilic substances dissolved in water.

Characterization of Samples.   Fifty-five samples of dissolved humic substances were isolated and characterized during the first five sampling cruises. Characterizations of samples obtained during the first two sampling cruises included elemental analyses, molecular-weight distributions, acid–base titrimetry, solution-density determinations, stable-carbon-isotope determinations, and determinations of ^{13}C NMR and ^1H NMR spectra. The

Table IV. DOC Yields (%) in Humic and Hydrophilic Isolates

Sampling Site	Fraction (Dissolved)	Cruise 2, Nov–Dec 1987	Cruise 3, May–June 1988
Illinois River downstream from Meredosia, Illinois	Humic substances	59.2	75.8
	Hydrophilic substances	ND[a]	7.1
Mississippi River downstream from Hickman, Kentucky	Humic substances	50.1	61.1
	Hydrophilic substances	ND	12.4
Mississippi River at Helena, Arkansas	Humic substances	47.8	69.5
	Hydrophilic substances	ND	12.7
Mississippi River downstream from Vicksburg, Mississippi	Humic substances	84	58.4
	Hydrophilic substances	ND	9.1

[a]Not determined.

stable-carbon-isotope data were relatively invariant between sampling sites; overlap of isotopic ratios between phytoplanktonic and terrestrial plant inputs in freshwater systems was documented (29). After the first two sampling cruises it became apparent that ^{1}H NMR spectrometry was the most useful characterization technique for chemical and geochemical studies, and the other characterizations were discontinued.

The first sampling cruise (July–August 1987) was made during low-discharge conditions on the Mississippi River and its tributaries; all water-temperature measurements were about 30 °C. DOC concentrations, DOC loads, and water discharge for the July–August 1987 sampling cruise are presented in Table V. This first sampling cruise was the only cruise in which evidence for instream loss of humic substances was documented by following the downriver movement of water. Maximum loads were detected in the Mississippi River upstream from its junction with the Ohio River. DOC loads did not increase downriver with DOC-load inputs from tributaries, although river discharge nearly doubled. DOC concentrations that were typically 3–4 mg/L near St. Louis decreased to 1–2 mg/L downstream from the junction with the Ohio River.

^{1}H NMR spectral data provided additional evidence that dissolved humic substances were exceptionally degraded during the first sampling cruise as a result of high temperatures and low discharges (Figure 5). The polyethylene glycol indicated by a peak in Figure 5 was a new river contaminant that was isolated along with the dissolved humic substances (27), and formate was a reagent used in the isolation procedure.

Although the peak 5:1 ratio indicating aromatic plus olefinic carbon content appeared typical, the peak 2:1 and 4:5 ratios were the lowest for

Table V. DOC Concentrations, DOC Loads, and River Discharge

Sampling Site	DOC Concentration (mg/L)	DOC Load (metric tons/day)	River Discharge[a] (m³/s)
Mississippi River near Winfield, Missouri	4.2	500	1,370
Mississippi River at Hartford, Illinois	3.2	420	150
Missouri River at Hermann, Missouri	3.8	860	2,600
Mississippi River at St. Louis, Missouri	3.7	1,250	3,900
Mississippi River at Chester, Illinois	3.9	1,440	4,300
Ohio River at Olmsted, Illinois	1.7	300	2,100
Mississippi River downstream from Hickman, Kentucky	1.9	1,030	6,300
Mississippi River at Helena, Arkansas	2.0	1,190	6,900
White River at river mile 11.5	1.1	30	330
Arkansas River at river mile 55.9	1.0	70	790
Mississippi River upstream from Arkansas City, Arkansas	1.9	1,250	7,600
Old River Outflow Channel near Knox Landing, Louisiana	3.1	550	2,100
Mississippi River near St. Francisville, Louisiana	2.3	1,230	6,200
Mississippi River below Belle Chasse, Louisiana	1.7	ND[b]	ND

NOTE: Values were obtained during July–August 1987 sampling cruise.
[a]As reported by Moody and Meade, ref. 30.
[b]ND is not determined.

samples obtained on this first cruise. The most extreme case was the sample from the Ohio River at Olmsted, Illinois. Discharge data indicated that 72% of the discharge at this site originated in the Tennessee and Cumberland rivers, which enter the Ohio River upstream from Olmsted, Illinois. Therefore, the DOM at this site represented discharges from the reservoirs of the Tennessee Valley Authority (TVA) system where long residence times, high water temperatures, and autochthonous inputs altered the nature of dissolved humic substances. The low peak 2:1 ratio, indicative of carboxylated aliphatic chains and aliphatic ketones, did not result in a low carboxylic acid content (4.9 mmol/g for the sample from the Ohio River). Degradation of

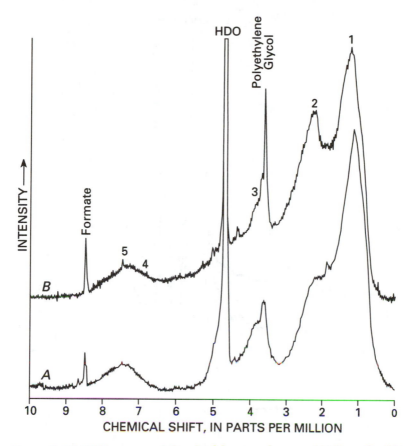

Figure 5. ¹H NMR spectra of dissolved humic substances (A) from the Ohio River at Olmsted, Illinois; and (B) from the Mississippi River near St. Francisville, Louisiana; sampled July–August 1987.

aliphatic chains as hypothesized in reaction 1 probably caused the low peak 2:1 ratio.

The fifth cruise (June 1989) sampled low-discharge conditions in the Mississippi River and its tributaries in the St. Louis, Missouri, area and moderate to high discharges with rising stages of the Ohio, White, Arkansas, Yazoo, and lower Mississippi rivers. The marked effect of discharge conditions on the nature of dissolved humic substances is shown by the three ¹H NMR spectra in Figure 6. Input of allochthonous dissolved organic substances increased the ratios of peak 2 (carboxylated chains and aliphatic ketones), peak 3 (carbohydrates), peak 4 (phenolic aromatics in tannins and lignins), and peak 5 (aromatic protons) to peak 1 in spectra B and C as compared to spectrum A.

During the first five sampling cruises on the Mississippi River near its mouth below Belle Chasse, Louisiana, the correlations among peak spectral ratios, discharge, and suspended sediment concentrations were measured

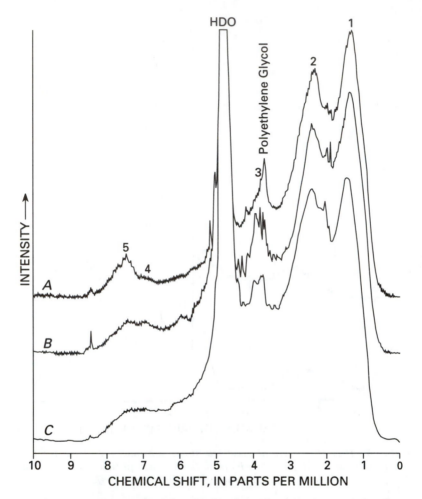

Figure 6. ¹H NMR spectra of dissolved humic substances from (A) the Illinois River at Hardin, Illinois; (B) the Arkansas River at Pendleton, Arkansas; and (C) the Mississippi River downstream from Belle Chasse, Louisiana; sampled June 1989.

as shown in Table VI. The greatest correlations with discharge were seen between peaks 3 and 1 (carbohydrate content) and peaks 4 and 5 (tannin and lignin residues). A "flushing effect" (i.e., increasing DOC concentrations with increasing discharge) has been measured by a number of research studies (31).

The nature of dissolved hydrophilic substances in the three samples whose spectra are shown in Figure 6 is indicated by ¹H NMR spectral patterns in Figure 7. The major peak at 3.8 ppm in Figure 7 indicates that carbohydrates are the major component of all three spectra. The additional peaks at 4.2 ppm and the broad 7.4–8.2-ppm peak in the sample from the

Table VI. Correlations (r²) of Peak Spectral Ratio, Discharge, and Suspended Sediment Concentrations

^{1}H-NMR Peak Spectral Ratio	Discharge (r²)	Sediment Concentration (r²)
Peak 2:1	0.356	0.339
Peak 3:1	0.758	0.714
Peak 4:1	0.797	0.676
Peak 5:1	0.822	0.600
Peak 4:5	0.710	0.622

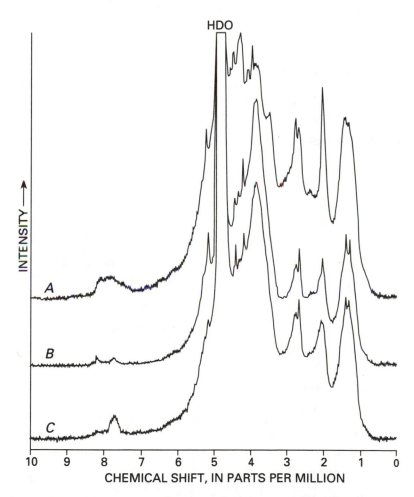

Figure 7. ^{1}H NMR spectra of hydrophilic substances isolated from (A) the Illinois River at Hardin, Illinois; (B) the Arkansas River at Pendleton, Arkansas; and (C) the Mississippi River downstream from Belle Chasse, Louisiana; sampled June 1989.

Illinois River indicate the oxidative degradations shown in reactions 2 and 3.

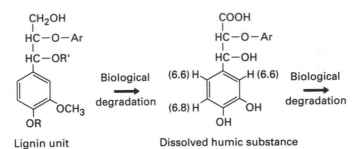

Lignin unit   Dissolved humic substance

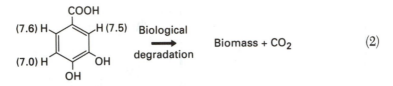

Dissolved hydrophilic substance

$$\text{(2)}$$

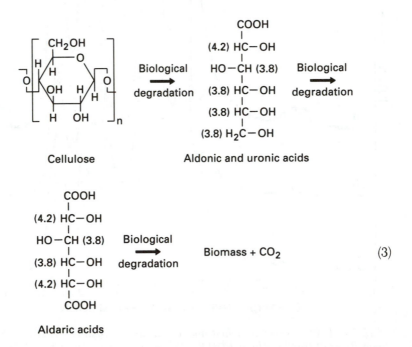

Cellulose   Aldonic and uronic acids

Aldaric acids

$$\text{(3)}$$

In reactions 2 and 3, the numbers in parentheses denote ^{1}H NMR chemical shifts in parts per million.

In reaction 2 lignin is demethylated, fragmented, and oxidized to dissolved humic substances (32). These substances are further oxidized to aromatic phenolic acids (hydrophilic acids) that are completely degraded to biomass and carbon dioxide. The successive changes in ^{1}H NMR chemical shifts of aromatic hydrogens on humic- and hydrophilic-substance isolates are consistent with the sequence shown in reaction 2. Similarly, carbohydrates (such as cellulose) degrade through the sequence shown in reaction 3 (33). In the Figure 7 spectra, the hydrophilic fractions from the three samples appear to be a mixture of carbohydrate acids, with aldaric acids predominating in the sample from the Illinois River.

During high-discharge conditions, which prevailed during the sixth sampling cruise (February–March 1990), DOC concentrations were 1.5–2.0 times those measured during low-discharge conditions. DOC concentrations, DOC loads, and river discharges for the sixth sampling cruise are presented in Table VII. The conservative nature of DOC transport under high-discharge conditions is indicated by the fact that the DOC loads were additive for the Ohio River at Olmsted, Illinois (with inputs from the upper Ohio, Wabash, Cumberland, and Tennessee rivers) and for the Mississippi River downstream from Hickman, Kentucky (with inputs from the upper Mississippi and Ohio rivers). Inputs of DOC from the Arkansas River (not sampled) are detected by the DOC-load increase at the sampling site upstream from Arkansas City, Arkansas. Even during high-discharge conditions, the effect of TVA reservoirs can be seen in the small DOC concentrations in the Cumberland and Tennessee rivers. The DOC load listed in Table VII for the lower Mississippi River downstream from Vicksburg, Mississippi is an order of magnitude greater than that determined during low-flow conditions (Table V).

Summary and Conclusions

Data in this and other reports indicate three sources for dissolved organic substances in rivers, lakes, and reservoirs: lignins, carbohydrates, and lipids. This study and others using degradative approaches (24) clearly established lignin as a precursor for dissolved humic substances. Similarly, carbohydrates are known to be a component of dissolved organic substances in natural waters (1).

The aliphatic alicyclic hydrocarbon precursor is not well recognized as the major aliphatic component in dissolved humic substances, although it was previously postulated to occur (11). This precursor might arise from terpenoid hydrocarbon lipids, but the data presented in this chapter favor polyunsaturated lipid precursors that are oxidatively coupled and cyclized by free-radical mechanisms (20). Degradative studies have not identified this aliphatic component in recognizable fragments. The quantitative, structural-model approach presented here combines the results of ^{13}C NMR, ^{1}H NMR,

Table VII. DOC Concentrations, DOC Loads, and River Discharge

Sampling Site	DOC Concentration (mg/L)	DOC Load (metric tons/day)	River Discharge[a] (m³/s)
Ohio River at Uniontown, Kentucky	3.3	1,890	6,620
Wabash River near New Haven, Illinois	5.0	1,010	2,340
Cumberland River near Smithland, Kentucky	2.9	540	2,170
Tennessee River near Calvert City, Kentucky	2.5	1,420	6,570
Ohio River at Olmsted, Illinois	3.4	4,730	16,100
Mississippi River near Cache, Illinois	5.3	2,080	4,540
Mississippi River downstream from Hickman, Kentucky	3.6	6,530	20,990
Mississippi River downstream from Fulton, Tennessee	3.7	7,290	22,810
Mississippi River at Helena, Arkansas	3.7	7,460	23,340
Mississippi River upstream from Arkansas City, Arkansas	4.2	12,000	33,000
Mississippi River downstream from Vicksburg, Mississippi	4.2	12,400	34,120
Mississippi River near St. Francisville, Louisiana	4.1	9,320	26,320
Mississippi River downstream from Belle Chasse, Louisiana	3.8	8,780	26,730

NOTE: Values were obtained during February–March 1990 sampling cruise.
[a]As reported by Moody and Meade, ref. 34.

elemental analysis, and functional-group analysis to quantify the ring content. The number of rings in dissolved humic substance structure (disregarding aromatic ring structures) rules out the possibility of significant open-chain aliphatic hydrocarbon content. From a biochemical standpoint, the accumulation of ring-carboxylated, aliphatic alicyclic structures (reaction 1) in dissolved humic substances is a logical consequence of the relative resistance of these structures to biodegradation (14).

The clear differentiation of allochthonous from autochthonous sources of DOM in integrating freshwater environments is not presently possible.

Minimum levels of terrigenous organic components can be quantified by aromatic plus olefinic carbon content or as lignin oxidation products. However, the aliphatic alicyclic component can originate from either source, and aliphatic alicyclic humic substances that appear to be of autochthonous origin may be of allochthonous origin with the aromatic component removed by fractionation or degradation processes. For differentiation of sources, stable carbon isotope ratios do not work as well for freshwater systems as for marine systems because of overlapping isotopic ratios between phytoplanktonic and terrestrial plant inputs (29). Perhaps organic nitrogen content and nitrogen isotope ratios could be used to define sources better.

In conclusion, DOM in water is a complex mixture of degraded moderate-sized biomolecules. As degradation proceeds in the aquatic environment, carbohydrates and lignin degradation products disappear within days to a few years. Only the carboxylated aliphatic alicyclic component persists for thousands of years in the marine environment (35) or in isolated lake environments like Big Soda Lake, Nevada.

Acknowledgments

The author thanks Stephen W. Robinson and Yousif K. Kharaka of the U.S. Geological Survey, Menlo Park, California, for initiating the research investigation of the ^{14}C age of DOM in Big Soda Lake, Nevada. The ^{14}C ages were determined by Stephen W. Robinson.

Samples of DOM from the Sagavanirktok River and Hidden Lake Creek in Alaska and from Island Lake, Nebraska, were collected and analyzed with the assistance of Patricia A. Brown, Eric A. Stiles, and Ted I. Noyes of the U.S. Geological Survey, Denver, Colorado. Jeffrey Koenings of the Alaska Fish and Game Department, Soldotna, Alaska, suggested and coordinated the Hidden Lake Creek study. Thomas C. Winter and James W. LaBaugh suggested and coordinated the Island Lake, Nebraska, study.

Robert H. Meade, John A. Moody, and Herbert H. Stevens of the U.S. Geological Survey, Denver, Colorado, deserve credit for designing and conducting the hydrological study of the Mississippi River that was used to compute DOC loads for this report. The author also thanks Patricia A. Brown, Terry I. Brinton, Wesley L. Campbell, John R. Garbarino, Ted I. Noyes, James F. Ranville, Terry F. Rees, Robert F. Stallard, and Howard E. Taylor of the U.S. Geological Survey, Denver, Colorado, for assisting in sample collection and analyses for the Mississippi River study.

Lastly, Robert L. Wershaw and Kevin A. Thorn rendered invaluable assistance in obtaining and interpreting the ^{13}C NMR data presented in this chapter.

References

1. Thurman, E. M. *Organic Geochemistry of Natural Waters;* Martinus Nijhoff/ Junk: Boston, MA, 1985; 497 pp.

2. Leenheer, J. A.; Huffman, E. W. D., Jr. *J. Res. U.S. Geol. Sur.* **1976**, *4*, 737–751.
3. Thurman, E. M.; Malcolm, R. L. *Environ. Sci. Technol.* **1981**, *15*, 463–466.
4. Hayes, M. H. B.; MacCarthy, P.; Malcolm, R. L.; Swift, R. S. In *Humic Substances II: In Search of Structure*; Hayes M. H. B.; MacCarthy, P.; Malcolm, R. L.; Swift, R. S., Eds.; Wiley: New York, 1989; pp 3–34.
5. Leenheer, J. A.; Brown, P. A.; Noyes, T. I. In *Aquatic Humic Substances: Influence on Fate and Treatment of Pollutants*; Suffet, I. H.; MacCarthy, P., Eds.; Advances in Chemistry 219; American Chemical Society: Washington DC, 1989; pp 25–39.
6. Leenheer, J. A. In *Humic Substances in Soil, Sediment, and Water: Geochemistry, Isolation, and Characterization*; Aiken, G. R.; McKnight, D. M.; Wershaw, R. L.; MacCarthy, P., Eds.; Wiley: New York, 1985; pp 409–429.
7. Amy, G. L.; Liu, H. In *Organic Substances and Sediments in Water: Humics and Soils*; Baker, R. A., Ed.; Lewis: Chelsea, MI, 1991; Vol. 1, pp 99–110.
8. Malcolm, R. L.; McKnight, D. M.; Averett, R. C. In *Humic Substances in the Suwannee River, Georgia: Interactions, Properties, and Proposed Structures*; Open-File Report 87–557; Averett, R. C.; Leenheer, J. A.; McKnight, D. M.; Thorn, K. A., Eds.; U.S. Geological Survey: Denver, CO, 1989; pp 5–21.
9. *Humic Substances in the Suwannee River, Georgia: Interactions, Properties, and Proposed Structures*; Open-File Report 87–557; Averett, R. C.; Leenheer, J. A.; McKnight, D. M.; Thorn, K. A., Eds.; U.S. Geological Survey: Denver, CO, 1989; 377 pp.
10. Leenheer, J. A.; McKnight, D. M.; Thurman, E. M.; MacCarthy, P. In *Humic Substances in the Suwannee River, Georgia: Interactions, Properties, and Proposed Structures*; Open File Report 87–557; Averett, R. C.; Leenheer, J. A.; McKnight, D. M.; Thorn, K. A., Eds; U.S. Geological Survey: Denver, CO, 1989; pp 331–360.
11. Thurman, E. M.; Malcolm, R. L. In *Aquatic and Terrestrial Humic Materials*; Christman, R. F.; Gjessing, E. T., Eds.; Ann Arbor Science: Ann Arbor, MI, 1983; pp 1–23.
12. Bowles, E. C.; Antweiler, R. C.; MacCarthy. P. In *Humic Substances in the Suwannee River, Georgia: Interactions, Properties, and Proposed Structures*; Open File Report 87–557; Averett, R. C.; Leenheer, J. A.; McKnight, D. M.; Thorn, K. A., Eds.; U.S. Geological Survey: Denver, CO, 1989; pp 205–230.
13. Leenheer, J. A.; Wilson, M. A.; Malcolm, R. L. *Org. Geochem.* **1989**, *11*, 273–280
14. *Microbial Degradation of Organic Compounds*; Gibson, D. T., Ed.; Marcel Dekker: New York, 1984; 535 pp.
15. Bollag, J. M. In *Aquatic and Terrestrial Humic Materials*; Christman, R. F.; Gjessing, E. J., Eds.; Ann Arbor Science: Ann Arbor, MI, 1983; pp 126–141.
16. Zehr, J. P.; Harvey, R. W.; Oremland, R. S.; Cloern, J. S.; George, L. H; Lane, J. L. *Limnol. Oceanogr.* **1987**, *32*, 781–793.
17. Kharaka, Y. K.; Robinson, S. W.; Law, L. M.; Carothers, W. W. *Geochim. Cosmochim. Acta* **1984**, *48*, 823–835.
18. Leenheer, J. A. *Environ. Sci. Technol.* **1981**, *15*, 578–587.
19. Leenheer, J. A.; Brown, P. A.; Stiles, E. A. *Anal. Chem.* **1987**, *59*, 1313–1319.
20. Harvey, G. R.; Boran, D. A.; Chesal, L. A.; Tokar, J. M. *Mar. Chem.* **1983**, *12*, 119–132.
21. Noyes, T. I.; Leenheer, J. A. In *Humic Substances in the Suwannee River, Georgia: Interactions, Properties, and Proposed Structures*; Open File Report 87–557; Averett, R. C.; Leenheer, J. A.; McKnight, D. M.; Thorn, K. A., Eds.; U.S. Geological Survey: Denver, CO, 1989; pp 231–250.

22. Leenheer, J. A.; Noyes, T. I. Water-Supply Paper 2230; U.S. Geological Survey: Reston, VA, 1984; 16 pp.
23. *The Sadtler Handbook of Proton NMR Spectra;* Simons, W. W., Ed.; Sadtler Research Laboratories: Philadelphia, PA, 1978; 1254 pp.
24. Ertel, J. R.; Hedges, J. I.; Devol, A. H.; Richey, J. E.; Ribeiro, M. N. G. *Limnol. Oceanogr.* **1986,** *31,* 739–754.
25. Leenheer, J. A. *Acta Amazonica* **1980,** *10,* 513–526.
26. Leenheer, J. A.; Meade, R. H.; Taylor, H. E.; Pereira, W. E. In *U.S. Geological Survey Toxic Substances Hydrology Program: Proceedings of the Technical Meeting* (Phoenix, AZ, Sept 26–30, 1988); Water Resources Investigations Report 88–4220; Mallard, G. E.; Ragone, S. E., Eds.; U.S. Geological Survey: Reston, VA, 1989; pp 501–512.
27. Leenheer, J. A.; Wershaw, R. L.; Brown, P. A.; Noyes, T. I. *Environ. Sci. Technol.* **1991,** *25,* 161–168.
28. Leenheer, J. A. *Proc. Am. Soc. Limnol. Oceanogr.* June 12–16, Boulder, CO, 1988; p 139.
29. Mook, W. G.; Tan, F. C. In *Biogeochemistry of Major World Rivers, SCOPE 42;* Degens, E. T.; Kempe, S.; Richey, J. E., Eds.; Wiley: New York, 1991; pp 245–264.
30. Moody, J. A; Meade, R. H. Open File Report 91–485; U.S. Geological Survey: Denver, CO, 1992; 143 pp.
31. Spitzy, A.; Leenheer, J. A. In *Biogeochemistry of Major World Rivers, SCOPE 42;* Degens, E. T.; Kempe, S.; Richey, J. E., Eds.; Wiley: New York, 1991; pp 213–232.
32. *Lignin Biodegradation: Microbiology, Chemistry, and Potential Applications;* Kirk, T. K.; Higuchi, T.; Chang, H., Eds.; CRC Press: Boca Raton, FL, 1980; Vol. 1, 241 pp.
33. White, A.; Handler, P.; Smith, E. L. *Principles of Biochemistry,* 3rd ed.; McGraw-Hill: New York, 1964; pp 22–42.
34. Moody, J. A.; Meade, R. H. Open-File Report 92–651; U.S. Geological Survey: Denver, CO, 1993, 227 pp.

RECEIVED for review September 26, 1991. ACCEPTED revised manuscript August 7, 1992.

Long-Term Changes in Watershed Retention of Nitrogen

Its Causes and Aquatic Consequences

John L. Stoddard

ManTech Environmental Technology, Inc., U.S. Environmental Protection Agency, Environmental Research Laboratory, Corvallis, OR 97333

Nitrogen saturation occurs when the supply of nitrogenous compounds from the atmosphere exceeds the demand for these compounds on the part of watershed plants and soil microbes. Several factors predispose forested watersheds to N saturation, including chronically high rates of N deposition, advanced stand age, and large pools of soil N. Many watersheds in the eastern United States exhibit symptoms of N saturation. A sequence of recognizable stages produces characteristic long-term and seasonal patterns of lake-water and stream-water NO_3^- concentrations that reflect the changes in rates and relative importance of N transformations as these watersheds become more N sufficient. The early stages of N saturation are marked by increases in the severity and frequency of NO_3^- episodes. The later stages of N saturation are marked by elevated baseflow concentrations of NO_3^- from groundwater. The most advanced symptoms of N saturation usually occur in regions with the most elevated rates of N deposition. Long-term increases in surface-water NO_3^- have important implications for surface-water acidification, but probably will not lead to freshwater eutrophication.

HISTORICALLY, NITROGEN DEPOSITION has not been considered a serious threat to the integrity of aquatic systems. Most terrestrial systems have been assumed to retain N strongly. In such cases there is a small probability that deposited N would make its way to the surface waters that drain these terrestrial systems. Nitrogen within aquatic ecosystems can arise from a variety of sources, including point-source and non-point-source pollution, biological fixation of gaseous N, and deposition of nitrogen oxides and ammonium. When N was known to be affecting aquatic systems, some source other than deposition was assumed to be responsible. The amounts of N supplied to aquatic systems by these other sources often outweigh by a large margin the amount of N potentially supplied by atmospheric deposition. During the past decade, however, our understanding of the transformations that N undergoes within watersheds has increased greatly. In areas of the United States where nonatmospheric sources of N are small, we can begin to infer cases in which N deposition is having an impact on aquatic systems.

This chapter will establish how watersheds and surface waters are likely to change as the impacts of N deposition become more severe. Aber et al. (1) described the changes that the terrestrial components of undisturbed watersheds undergo as the effects of N deposition increase. I will present the aquatic equivalents of the stages described by Aber et al. (1) and outline the key characteristics of these stages as they influence seasonal and long-term aquatic N dynamics. The second purpose of this chapter is to present accumulated evidence, primarily from the eastern United States, that otherwise-undisturbed watersheds show impacts from elevated N deposition and that the severity of these impacts corresponds to the historical levels of N deposition that the various sites have experienced. The focus on watersheds that are undisturbed, other than by elevated rates of deposition, is important. Other disturbances (especially tree harvesting) can have significant and long-lived effects on N cycling. Elimination of other potential sources of disruption to the N cycle allows us to narrow the field of stresses that might be contributing to the effects we observe.

Estimation of the effects of N deposition on aquatic systems is made difficult by the large variety of forms of N found in air, deposition, watersheds, and surface waters, as well as by the myriad pathways through which N can be cycled in terrestrial and aquatic ecosystems. These complexities separate N deposition from its effects and reduce our ability to attribute known aquatic effects to known rates of N deposition. The organization of this chapter reflects this complexity. Because an understanding of the ways that N is cycled through watersheds is critical to our understanding of N effects, I begin with a brief description of the N cycle and of the transformations of N that may occur in watersheds. I then discuss the two most likely effects of N deposition (acidification and eutrophication).

Nitrogen Inputs

Watersheds are generally several orders of magnitude larger than the surface waters that drain them. Thus most of the atmospheric deposition that may potentially enter aquatic systems falls first on some portion of the watershed. Nitrogen may be deposited to the watershed or directly to water surfaces in a variety of forms, including NO_3^-, NH_4^+, and organic N in wet and dry deposition. In addition, plants may absorb gaseous N (as NO, NO_2, or nitric acid vapor) (2, 3). N thus absorbed may subsequently enter the watershed N budget as litterfall through the death of plant biomass (4, 5).

Concentrations of NO_3^- and NH_4^+ in precipitation vary widely throughout North America; they depend largely on the proximity of sampling sites to sources of emissions. Galloway et al. (6) report mean concentrations of 2.4 μequiv/L for NO_3^- and 2.8 μequiv/L for NH_4^+ for a site in central Alaska sampled in 1980–1981. In the Sierra Nevada Mountains of California, mean concentrations of NO_3^- and NH_4^+ for the period 1985–1987 were 5.0 and 5.4 μequiv/L, respectively (7). In a comparison of N deposition at lake and watershed monitoring sites in northern United States and southern Canada, Linsey et al. (8) found NO_3^- concentrations ranging from 15 to 40 μequiv/L and NH_4^+ concentrations from 10 to 50 μequiv/L for 1970–1982 in an area that was considered remote but may be influenced by prairie dust and long-range acidic deposition; neither ion dominated over the other. In some areas closer to anthropogenic N sources (e.g., in the northeastern United States and southeastern Canada) volume-weighted mean NO_3^- concentrations range from 30 μequiv/L (e.g., in the Adirondack and Catskill mountains of New York) to 50 μequiv/L (e.g., in the eastern Great Lakes region). Mean NH_4^+ concentrations range from 10 to 20 μequiv/L in the same areas (9). Ammonium concentrations are highest (~40 μequiv/L) in the agricultural areas of midwestern United States.

Some uncertainty exists in all estimates of NH_4^+ deposition derived from measurements made in the National Atmospheric Deposition Program/ National Trends Network (NADP/NTN) because of the method of sample collection used by cooperators in this program. Samples are collected weekly from buckets that are covered at all times except during active precipitation. Precipitation that falls early in the week may sit for several days before being collected and filtered for analysis. A high probability exists that some NH_4^+ collected in NADP collectors will be biologically assimilated (transformed into organic forms of N) before samples are filtered. It is not currently known what the magnitude of this problem may be.

Deposition of N depends on its concentration in precipitation, the volume of water falling as precipitation, and the amount of N in dry deposition (2). The last of these values (dry deposition) is difficult to measure and is often estimated as a fraction (e.g., 30–40%) of wet deposition (10). Sisterson et al. (11) discussed direct measurements of dry deposition at regionally representative sites. They concluded that deposition of N species ranges from 40–50% of wet deposition in the northeast to ~80% of wet deposition in the southeastern United States. Given the range of concentrations previously mentioned and the volumes of precipitation falling in different regions of North America, estimates of N deposition rates range from less than 12 equiv/ha per year in Alaska to more than 800 equiv/ha per year in northeastern United States (Table I).

Table I. Rates of Nitrogen Deposition in Several Areas of North America

Area	NO_3^-	NH_4^+	Total	Source
Alaska[a] (Poker Flat)	6.9	4.8	11.7	6
Sierra Nevada, CA[b] (Emerald Lake)	79	85	164	7
Ontario, Canada[c] (Experimental Lakes Area)	125	140	265	8
British Columbia, Canada[c]	260	130	390	14
Upper Midwest[d]	300	210	510	155
Southeastern U.S.[e] (Walker Branch, TN)	540	180	720	2
New Hampshire[c]	464	200	664	154
Catskills[a]	580	292	874	79
Adirondacks[d]	590	190	780	155

NOTE: All deposition values are in microequivalents per liter.
[a]Dry deposition was estimated as 35% of total deposition.
[b]Dry deposition was sampled as part of snowpack; no correction was made for dry deposition.
[c]Bulk precipitation measurements; no correction was made for dry deposition.
[d]Values were corrected for dry deposition based on ratios of Hicks (201).
[e]Includes estimates for dry deposition and gaseous uptake of N.

Generally, NO_3^- dominates over NH_4^+ at sites close to emission sources (8, 12). Dissolved organic nitrogen (DON) concentrations are highly variable in precipitation but often amount to 25–50% of inorganic N deposition (8, 12–14). In some areas, DON can occur in greater concentrations than the inorganic species (15).

In addition to wet and dry deposition, many high-elevation sites may receive substantial inputs of N from clouds or fog (15–17). Few quantitative estimates of cloud deposition are available, but results from one site on Whiteface Mountain in the Adirondacks indicate that clouds and fog can contribute up to 40% of total deposition (18). Rates of wet and dry deposition at Whiteface Mountain were comparable to the Adirondack values given in Table I (~740 equiv/ha), but total deposition rates (including cloud and fog deposition) averaged 1170 equiv/ha.

The Nitrogen Cycle

Atmospheric N can enter aquatic systems either as direct deposition to water surfaces or as N deposition to the terrestrial portions of a watershed. Nitrogen deposited to the watershed is routed and transformed by watershed processes. It may eventually reach aquatic systems in forms only indirectly related to the original deposition. The transformations that N undergoes within the watershed (e.g., in soils, by microbial action, and in plants) play a major role in determining what forms and amounts of N eventually reach surface waters. Much of the challenge of determining when N deposition is affecting aquatic systems depends on our ability to identify which N transformations are occurring and which are not. A large part of the following discussion is therefore focused on terrestrial processes that alter the forms and rates of N supply. Most of these processes also occur within lakes and streams, and their strengths can determine whether the effects of N deposition will be felt immediately (e.g., through the eutrophication of headwater lakes) or downstream (e.g., in estuaries and coastal waters). In a very real sense N cycling within the terrestrial ecosystems controls whether N deposition will reach aquatic systems (and in what concentrations), whereas N cycling within lakes and streams controls whether the N will have any measurable effect. Nitrogen assimilation, mineralization, nitrification, denitrification, and nitrogen fixation are important processes that affect the fate of N from atmospheric deposition.

Nitrogen Assimilation.   Nitrogen assimilation is the uptake and metabolic use of N by plants and soil microbes (Figure 1). Assimilation by the terrestrial ecosystem controls the form of N eventually released into surface waters, as well as affecting the acid–base status of soil and surface waters.

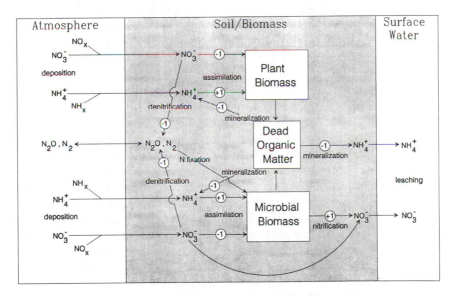

Figure 1. A simplified watershed nitrogen cycle, with major pathways (arrows) and their effects on the watershed hydrogen budget (numbers in circles) shown. Circled numbers represent the number of hydrogen ions transferred to the soil solution or surface water (+1) or from the soil solution or surface water (−1) for every molecule of NO_3^- or NH_4^+ that follows a given pathway. For example, nitrification follows the pathway for NH_4^+ assimilation into microbial biomass (+1) and is leached out as NO_3^- (+1), for a total hydrogen ion production of +2 for every molecule of NO_3^- produced.

Terrestrial assimilation is a major form of N removal in watersheds and may be sufficient to prevent all atmospherically derived N from reaching surface waters (19).

Nitrogen is the most commonly limiting nutrient in North American forest ecosystems (20, 21). The form of N used by terrestrial ecosystems strongly affects the acidifying potential of N deposition (Figure 1). Ammonium uptake is an acidifying process (i.e., uptake of NH_4^+ releases 1 mole of H per mole of N assimilated).

$$NH_4^+ + R \cdot OH \longrightarrow R \cdot NH_2 + H_2O + H^+ \tag{1}$$

The biological uptake of NO_3^-, on the other hand, is an alkalinizing process (i.e., uptake of NO_3^- consumes 1 mole of H per mole of N assimilated).

$$R \cdot OH + NO_3^- + H^+ \longrightarrow R \cdot NH_2 + 2O_2 \tag{2}$$

Most forested watersheds undergo a dormant period, or at least a period of much reduced growth, during the winter months. Because assimilation of N is also reduced during this season, watersheds have a much lower ability to retain N during the winter and early spring. This seasonality is responsible for the commonly observed pattern of higher surface-water NO_3^- concentrations in winter and spring than in summer and fall (discussed under Stage 0 N loss). If spring snowmelt occurs before substantial forest growth begins in the spring, snowmelt NO_3^- concentrations can be substantial, even in areas of moderate N deposition. Concentrations of NH_4^+ in surface waters, on the other hand, are rarely elevated at any season because soil cation exchange; low mobility; and competition among vegetation, mycorrhizal roots, and nitrifiers all contribute to watershed NH_4^+ retention.

Muller and Bormann (22) proposed that spring ephemeral plants in the understories of forested watersheds may act as "vernal dams" controlling the loss of N from watersheds during the spring. Spring ephemeral plants may exploit the light, water, and nutrients available before forest growth and canopy closure make conditions less optimal in later spring and summer. Zak et al. (23) used radioactive N tracers to corroborate that spring ephemeral communities are associated with N retention during spring. However, they found that most of the retention was attributable to assimilation by soil microbes, not plants. There has been a tendency to think of N assimilation as primarily a plant-mediated process, but assimilation by microbes may be a very important mechanism minimizing N losses from watersheds throughout the growing season. Most forest soils are characterized by a very large, relatively inert, pool of organic N (24), which suggests that microbial uptake and immobilization of N are the dominant processes in most watersheds. Before watershed N demand (from both forest and soils) can be met or exceeded, uptake requirements of both vegetation and soil microbial communities must be met. If one accepts the common wisdom that forest vegetation is a superior long-term competitor for N (*24, 25*), then the evolution of a N-sufficient watershed can be seen to consist of two stages: fulfillment of vegetation N demand, followed by fulfillment of soil microbial demand.

Assimilation by aquatic plants, a key process in the potential eutrophication of surface waters by N, may also play a role in their acid–base status. Uptake of NO_3^- in lakes is an alkalinizing process (Figure 1) that may be stoichiometrically important in some lakes (*26*). Aquatic plants generally favor the uptake of NH_4^+ over the uptake of NO_3^-; NH_4^+ uptake is energetically favorable because NO_3^- must first be reduced before it is physiologically available to algae (*27*). McCarthy (*28*) summarized several studies that consistently show that potential (saturated) NH_4^+ uptake rates are greatly enhanced in N-deficient cells. This relationship is now used, along with various other indices, as a basis for identifying the degree of N limitation in phytoplankton (*29, 30*).

A crucial difference between aquatic and terrestrial ecosystems is that N additions do not commonly stimulate growth in aquatic systems, as seems to be the case in many terrestrial systems. N limitation may be the exception in aquatic systems rather than the rule. The question of whether N limitation is a common occurrence in surface waters will play a large role in determining whether N deposition affects the trophic state of aquatic ecosystems.

The effects of N supply on uptake and growth rates in phytoplankton and periphyton are the subject of volumes of literature, a summary of which is beyond the scope of this chapter. However, certain aspects of the limitation of algal growth by the supply of N and other nutrients will be discussed later in the section on eutrophication by N deposition. Other details on algal nutrition can be found in reviews by Goldman and Glibert (*31*), Button (*32*), Kilham and Hecky (*33*), and Hecky and Kilham (*34*).

Mineralization.   Mineralization is the bacterial decomposition of organic matter; it releases NH_4^+ that can subsequently be nitrified to NO_3^-. Mineralization is an important process in watersheds, as it recycles N that would otherwise be tied up in soil organic matter following the death of plants, or as leaflitter (Figure 1). In a comparative study of mineralization in soils, Nadelhoffer et al. (*35*) found N mineralization rates ranging from 6000 to 9600 equiv/ha per year under deciduous tree species and from 2800 to 5800 equiv/ha per year under coniferous species. These rates should be compared to N deposition rates of 600–900 equiv/ha per year for high-deposition areas of the northeast (Table I). Nadelhoffer et al. (*35*) also reported estimated rates of N uptake that were 5–20% higher than rates of mineralization. These rates suggest that internal cycling sources of N far outweigh external sources such as deposition under most conditions.

The effect of mineralization on the acid–base status of draining waters depends on the form of N produced. The conversion of organic N (e.g., from leaflitter) to NH_4^+ consumes 1 mole of H per mole of N produced (Figure 1); it can be thought of as the reverse of the reaction in eq 1. Organic N that is mineralized and subsequently oxidized (nitrified) to NO_3^- (eq 3) produces a net of 1 mole of H per mole of NO_3^- produced. Because the production of organic N (i.e., assimilation) can either produce or consume hydrogen (depending on whether NO_3^- or NH_4^+ is assimilated), the net (ecosystem) effect of mineralization depends on both the species entering the watershed and the species leaving the watershed (Figure 1).

Mineralization often has the initial effect (e.g., immediately after leaffall) of immobilizing N (36). In ecosystems where plant growth is limited by the availability of N, mineralization is also limited by N in the sense that addition of N to the leaflitter speeds decay and increases the rate at which N is immobilized by decomposers (37, 38). This initial immobilization period is marked by a net increase in the N content of leaflitter. Nitrogen limitation of decomposition follows in part from the low N content typical of litter, which arises from the translocation of N out of leaves during senescence. The immobilization phase of mineralization is followed by a period of slow release of inorganic N from the soil microbial pool (36).

Nitrification.   Nitrification, the oxidation of NH_4^+ to NO_3^-, is mediated by bacteria and fungi in both the terrestrial and aquatic portions of watersheds. It is an important process in controlling the form of N released to surface waters by watersheds, as well as in controlling the acid–base status of surface waters (Figure 1). Nitrification is a strongly acidifying process, producing 2 moles of H for each mole of N (NH_4^+) nitrified.

$$NH_4^+ + 2O_2 \longrightarrow NO_3^- + 2H^+ + H_2O \qquad (3)$$

Because nitrification in forest soils transforms NH_4^+ into NO_3^-, the acidifying potential of deposition (attributable to N) is often defined as the sum of NH_4^+ and NO_3^- on the assumption that all N will leave the watershed as NO_3^- (39). Unless NO_3^- leaving the watershed is accompanied by base cations (e.g., Ca^{2+}), it will acidify lakes and streams by lowering their acid-neutralizing capacity (ANC).

Nitrification is limited in most soils by the supply rate of NH_4^+ (40, 41). Competition exists between nitrifiers and vegetation, which may both be limited by the availability of NH_4^+. This microbial demand for NH_4^+, coupled with the high cation-exchange capacity of most temperate forest soils, leads to surface-water NH_4^+ concentrations that are usually undetectable. Nitrification rates may also be limited by inadequate microbial populations, lack of water, allelopathic effects (toxic effects produced by inhibitors manufactured by vegetation), or by low soil pH.

Among these potential limiting factors, soil pH plays an obviously vital role in any discussion of the acidification of surface waters by N deposition. Nitrification has traditionally been thought of as an acid-sensitive process (*1*, *42*), but high rates of nitrification have been reported recently from very acid soils (i.e., pH < 4.0) in the northeastern United States (*41*, *43*, *44*) and in Europe (*45*, *46*). In the southeastern United States, Montagnini et al. (*47*) were unable to find any pH effect on nitrification or to stimulate nitrification by buffering acid soils. In a survey of sites across the northeastern United States, McNulty et al. (*48*) found no correlation between nitrification rates and soil pH, but a strong association ($r^2 = 0.77$) with rates of N deposition. The weight of evidence suggests that nitrification will occur at low soil pH values as long as the supply of NH_4^+ is sufficient and that any inhibition of nitrification by low soil pH can be overcome by excess N availability.

Rates of nitrification in lakes and streams may also be limited by low concentrations of NH_4^+. Supply rates of NH_4^+ from watersheds are often low (except in cases of point-source pollution), and nitrifying organisms have little substrate with which to work. Two exceptions to this generality are cases in which NH_4^+ deposition is extremely high (such as near agricultural areas) and in which NH_4^+ is produced within the aquatic system. Experiments on whole lakes and in mesocosms have confirmed the acidifying potential of ammonium additions from deposition to surface waters (*49*, *50*). Ammonium deposition is especially deceptive because in the atmosphere it can combine as a neutral salt with SO_4^{2-} to produce precipitation with near-neutral pH values, as seen in the Netherlands (*51*). Once deposited, however, the ammonium can be assimilated (leaving an equivalent amount of hydrogen) or nitrified (leaving twice the amount of hydrogen). Canadian whole-lake experiments generated conflicting evidence that nitrification in lakes is an acid-sensitive process. Rudd et al. (*52*) presented data indicating that nitrification was blocked at pH values less than 5.4 in an experimentally acidified lake and that this situation led to a progressive accumulation of NH_4^+ in the water column. Schindler et al. (*49*) reported nitrification in an experimentally fertilized lake at pH 4.6. Rudd et al. (*52*) hypothesized that nitrification will occur in low-pH lakes only when winter pH values are sufficiently high (pH > 5.4) to allow nitrifiers to become well established before the low pH values resulting from nitrification develop. This situation existed in the experimental lake described by Schindler et al. (*49*).

High NH_4^+ concentrations may also result through decomposition in lakes whose deeper waters become anoxic during periods of stratification (usually late winter or late summer). Nitrification of this NH_4^+ occurs when lakes mix during spring or fall. The mixing process supplies the oxygen necessary for nitrifying organisms to survive and metabolize (*53*). In this case, the main influence of nitrification is to recycle N within the system and to supply NO_3^- either to denitrifiers or to N-deficient algae.

Denitrification.   Denitrification is the biological reduction of NO_3^- to produce gaseous forms of reduced nitrogen (N_2, NO, or N_2O) (54). It is an anaerobic process (i.e., it occurs only in environments where oxygen is absent) whose end product is lost to the atmosphere (Figure 1). In terrestrial ecosystems, denitrification occurs in anaerobic soils, especially boggy, poorly drained soils, or in anaerobic microsites. It has traditionally been considered a relatively unimportant process outside of wetlands (55). However, denitrification could be an episodic process occurring after such events as spring snowmelt and heavy rainstorms, when soil oxygen tension is reduced (56). No single equation can describe the denitrification reaction because several end products are possible. However, denitrification is always an alkalinizing process that consumes 1 mole of H for every mole of N denitrified (Figure 1).

Denitrification can be involved in the production or consumption of N_2O, a product that may have considerable significance as a greenhouse gas (57, 58). In a review of the effects of acidic deposition on denitrification in forest soils, Klemedtsson and Svensson (59) concluded that denitrification rates are often limited by the availability of oxygen and may therefore be relatively insensitive to increases in N deposition. Acidification of soils has an uncertain effect on overall rates of denitrification, but it strongly affects the end product of denitrification and favors the production of N_2O over N_2 (56, 60). Even under extreme conditions, denitrification is unlikely to be a significant sink for watershed N. Nonetheless, it may be significant in the global atmospheric budget of N_2O (60, 61).

Denitrification plays a much larger role in N dynamics in aquatic ecosystems than it does in terrestrial ones. In streams, rivers, and lakes, bottom sediments are the main sites for denitrification (62), although open-water denitrification has also been reported (63). In lake and stream sediments NO_3^- is potentially available from the water column, but it is produced mainly when organic matter is broken down within the sediments and the resulting NH_4^+ is subsequently oxidized (62). Denitrification is an especially important process in large rivers, for which Seitzinger (62) reported denitrification rates that were 7–35% of total N inputs. Denitrification in aquatic ecosystems is an alkalinizing process, consuming 1 mole of H for every mole of NO_3^- denitrified.

Estimates of denitrification rates range from 54 to 345 $\mu mol/m^2$ per hour in streams with high rates of organic matter deposition, 12 to 56 $\mu mol/m^2$ per hour in nutrient-poor oligotrophic lakes, and 42 to 171 $\mu mol/m^2$ per hour in eutrophic lakes (62). Rudd et al. (64) reported an increase in the rate of denitrification from less than 0.1 to over 20 $\mu mol/m^2$ per hour in an oligotrophic lake when nitric acid was added in a whole-lake experimental acidification. This result suggests that freshwater denitrification may be limited by NO_3^- availability. In deep muds of slow-flowing streams, the process can effectively reduce NO_3^- concentrations in the water column by as much as 200 μequiv/L over a 2-km length of stream (65, 66). This depletion amounts to 75% of the daily NO_3^- input during a growing season. Thus denitrification can be considered as a method for NO_3^- removal in the management of some slow-moving streams with a deep organic substrate (67).

Nitrogen Fixation.   Gaseous atmospheric nitrogen (N_2) can be fixed to produce NH_4^+ by a wide range of single-celled organisms, including blue-green algae (*Cyanobacteria*), and various aerobic and anaerobic bacteria. Symbiotic nitrogen-fixing nodules are present on the roots of some early successional forest species (68). In headwater streams, nodules on rooting structures of riparian vegetation (e.g., *Alnus* sp.) can also be important N fixers (69). Ordinarily, N fixation has no direct effect on the acid–base status of soil or surface waters.

$$N_2 + H_2O + 2R\cdot OH \longrightarrow 2R\cdot NH_2 + \frac{3}{2} O_2 \qquad (4)$$

Nitrogen fixation in excess of biological demand, however, can lead to nitrification or mineralization of organic N and ultimately to acidification of soil or surface waters (70, 71).

Nitrogen fixation counteracts denitrification losses of N from surface waters and is fundamental to replenishing fixed forms of N in all aquatic ecosystems. It is thought to be the main process responsible for maintaining surplus inorganic N in lakes and streams. It is therefore basic to the concept that primary production in most lakes and streams is limited by phosphorus (72).

Rates of N fixation are generally related to trophic status in fresh water. Howarth et al. (73) showed that fixation in low-, medium- and high-nutrient lakes is generally <0.02, 0.9–6.7, and 14.3–656.9 mmol/m^2 of N per year, respectively. Fixation is also closely correlated with the abundance of blue-green algae (53). This relationship suggests that the algae, rather than bacteria, dominate N fixation in lakes. Although N fixation does occur in sediments, that source is of minor importance compared to N fixation in the water column. Only in very nutrient-poor lakes, where N loading from all other sources is small, is N fixation in sediments of overall significance (e.g., 32 and 6% of total inputs in Lake Tahoe and Mirror Lake, respectively) (73).

Nitrogen Saturation

Much of the debate over whether aquatic systems are being affected by N deposition centers on the concept of nitrogen saturation of forested watersheds. Nitrogen saturation can be defined as a situation in which the supply of nitrogenous compounds from the atmosphere exceeds the demand for these compounds on the part of watershed plants and microbes (1, 74). Under conditions of N saturation, forested watersheds that previously retained nearly all of N inputs due to a high demand for N by plants and microbes begin to have higher loss rates of N. These losses may be in the form of leaching to surface waters or to the atmosphere through denitrification. These two potential loss pathways have profoundly different impacts on the acid–base status of watersheds and surface waters (see following discussion). Their relative importance in advanced stages of N saturation will be a decisive characteristic determining the severity of the impact of N saturation.

Progressive Stages.   Aber et al. (1) proposed a hypothetical time course for a watershed response to chronic N additions (Figure 2), describing both the changes in N cycling that are proposed to occur and the plant responses to changing levels of N availability. Aber et al. include in their hypothetical time course a trajectory for the loss of N to surface-water runoff, which suggests a simple response (N leaching) in the later stages of N saturation. This chapter seeks to establish whether stages equivalent to those shown in Figure 2 can be described for surface waters and to determine whether the response of surface waters to advanced stages of N saturation is as simple as is suggested in Figure 2.

Stage 0 of the Aber et al. (1) conceptual model is the pretreatment condition; inputs of N from deposition are at background levels and watershed losses of N are negligible (Figure 2). Increased deposition occurs in Stage 1, but effects on the terrestrial ecosystem are not evident. For a limiting nutrient such as N, a fertilization effect might result in increased ecosystem production and tree vigor at Stage 1. Retention of N is very efficient and little or no N would be lost annually to surface waters that drain Stage 1 watersheds. Many forested watersheds in the United States would be considered to exist at this stage.

Negative effects occur in Stage 2 of the Aber et al. (1) hypothetical time course. However, these effects are subtle, nonvisual, and require long time scales for detection. Only in Stage 3 do the effects on the forests become visible and result in major environmental impacts. Aber et al. (1) emphasize that different species and environmental conditions could alter the timing of the effects illustrated in Figure 2.

A number of factors may contribute to a watershed's progression through the stages of N loss, including elevated N deposition, stand age, and high soil-N pools. High rates of N deposition play a clear role, as the ability of forest biomass to accumulate N must be finite. At very high long-term rates of N deposition, the ability of forests and soils to accumulate N will be exceeded. The only remaining pathway for loss of N (other than runoff) is denitrification. Although high rates of N deposition may favor increased rates

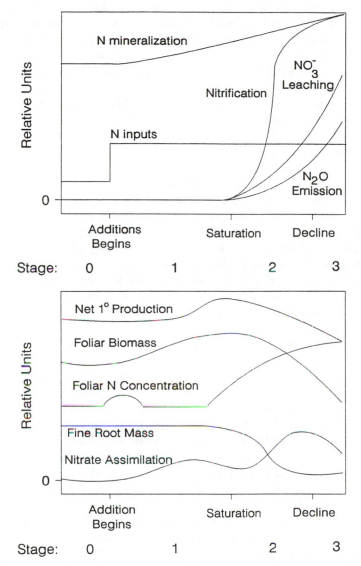

Figure 2. Hypothetical time course of forest ecosystem response to chronic nitrogen additions. Top: relative changes in rates of nitrogen cycling and nitrogen loss. Bottom: relative changes in plant condition (e.g., foliar biomass and nitrogen content, fine root biomass) and function (e.g, net primary productivity and nitrate assimilation) in response to changing levels of nitrogen availability. (Redrawn with permission from reference 1. Copyright 1989 American Institute of Biological Sciences.)

of denitrification, many watersheds lack the conditions necessary for substantial denitrification (e.g., low oxygen tension, high soil moisture, and temperature). Another important factor in N loss from watersheds is the age of the forest stands. A loss in the ability to retain N is a natural outcome of forest maturation, as demand for N on the part of more slowly growing tree species may plateau in later stages of forest development or decline as forests achieve a "shifting-mosaic steady state" (75). Uptake rates of N into vegetation are generally maximal around the time of canopy closure for conifers and somewhat later (at higher rates) in deciduous forests because of the annual replacement of canopy foliage in these ecosystems (76).

Finally, soil-N status may also affect N loss rates. Where large soil-N pools exist, they imply that soil microbial processes that are ordinarily limited by the availability of N are instead limited by some other factor (e.g., availability of labile organic carbon) and contribute to the likelihood that watersheds will leach NO_3^- (24, 77). N saturation occurs in a sequence beginning with the fulfillment of vegetation N demand, which is followed by the fulfillment of soil microbial N demand. The existence of large soil-N pools suggests that the second of these requirements may be easily met. The possible importance of all three factors (deposition, stand age, and soil N) in shifting watersheds from one stage of N loss to another will be discussed later in the context of surface water evidence of watershed N saturation.

Watershed Nitrogen Saturation.   The loss of N from watersheds occurs in stages that correspond to the stages of terrestrial N saturation described by Aber et al. (1). The most obvious characteristics of these stages of N loss are changes in the seasonal and long-term patterns of surface-water NO_3^- concentrations, which reflect the changes in N cycling that are occurring in the watershed. The N cycle at Stage 0 is dominated by forest and microbial uptake, and the demand for N has a strong influence on the seasonal NO_3^- pattern of receiving waters (Figure 3). The normal seasonal NO_3^- pattern in a stream draining a watershed at Stage 0 would include very low, or immeasurable, concentrations during most of the year and measurable concentrations only during snowmelt (in areas where snowpacks accumulate over the winter months) or during spring rainstorms. The small loss of NO_3^- during the dormant season is a transient phenomenon. It results because snowmelt and spring rains commonly occur in these environments before substantial forest and microbial growth begin in the spring. (Winter mineralization of soil organic N may be an exception to this inactivity; see ref. 78.) Thus some of the N stored in soils and snowpack may pass through the watershed during extreme hydrological events and generate a pulse of elevated NO_3^- concentration (Figure 3b). The key surface-water characteristics of Stage 0 watersheds are very low NO_3^- concentrations during most of the year and maximum spring NO_3^- concentrations that are smaller than the concentrations typical of deposition.

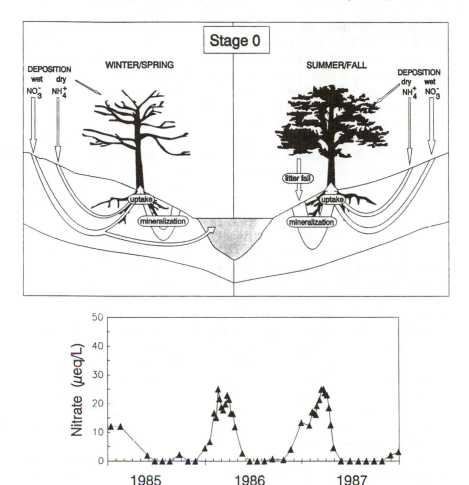

Figure 3. Top panel: Schematic representation of nitrogen cycle in watershed at Stage 0 of watershed nitrogen loss. Sizes of arrows indicate the magnitude of process or transformation. Differences between winter–spring and summer–fall seasons are shown on opposite sides. At Stage 0, nitrogen transformations are dominated by plant and microbial assimilation (uptake), with little or no NO_3^- leakage from the watershed during the growing season. Bottom panel: Small amounts of NO_3^- may run off during snowmelt, producing the typical Stage 0 seasonal NO_3^- pattern. Data in lower panel are from Black Pond, Adirondack Mountains. (Data are taken from reference 157.)

At Stage 1 the seasonal pattern typical of Stage 0 watersheds is amplified. This amplification of the seasonal NO_3^- signal may be the first sign that watersheds are proceeding toward the chronic stages (i.e., Stages 2 and 3 in Figure 2) of N saturation (42, 79). This suggestion is consistent with the changes in N cycling that are thought to occur at Stage 1. A conceptual understanding of these changes derives from the most common definition of nutrient limitation. Implicit in the definition of nutrient limitation is the idea that "the current supply rate [of a nutrient] prevents the vegetation from achieving maximum growth rates *attainable within other environmental constraints*" (ref. 80; emphasis added). During the cold season these environmental constraints can be severe, and maximum attainable growth rates are clearly much lower than in the warm months. Although much of this discussion is couched in terms of forest trees, the same arguments apply to soil microbial communities (e.g., decomposers, nitrifiers) that may be as important as vegetation in controlling N loss from watersheds (80).

Overall limitation of forest growth in the early stages of N saturation is characterized by a seasonal cycle of limitation by physical factors (e.g., cold and diminished light) during late fall and winter and by nutrients (primarily N) during the growing season. The effect of increasing the N supply (e.g., from deposition) is to postpone the seasonal switch from physical to nutrient limitation during the breaking of dormancy in the spring (Figure 4) and to prolong the seasonal N saturation that is characteristic of watersheds at this stage. At Stage 1, this switch is delayed enough that substantial NO_3^- may leave the watershed during extreme hydrological events in the spring. Watershed loss of N at Stage 1 is still a seasonal phenomenon, and the annual N cycle is still dominated by uptake (Figure 5), but NO_3^- leaching is less transient than at Stage 0. The key characteristics of Stage 1 watersheds are episodes of surface-water NO_3^- that exceed concentrations typical of deposition (Figure 5b). Elevated NO_3^- during these episodes may result from preferential elution of anions from melting snow (81, 82) or from the contribution of N mineralization to the soil pool of NO_3^- that may be flushed during high-flow periods (43, 83).

In Stage 2 of watershed N loss the seasonal onset of N limitation is even further delayed. As a consequence, biological demand exerts no control over winter and spring N concentrations, and the period of N limitation during the growing season is much reduced (Figure 6). The annual N cycle, which was dominated by uptake at Stages 0 and 1, is instead dominated by N loss through leaching and denitrification at Stage 2; sources of N (deposition and mineralization) outweigh N sinks (uptake). The same mechanisms that produce episodes of high NO_3^- during extreme hydrological events at Stage 1 also operate at Stage 2. But more importantly, NO_3^- leaching can also occur at Stage 2 during periods when the hydrological cycle is characterized by deeper percolation (Figure 6). If biological demand is sufficiently depressed during the growing season, N begins to percolate below the rooting zone and elevated groundwater NO_3^- concentrations result. Nitrification becomes an important process at Stage 2 (Figure 2; also *see* ref. 1). Lowered biological demand leads to a buildup of NH_4^+ in soils and nitrification may be stimulated. This change is pivotal in the N cycle because nitrification is such a

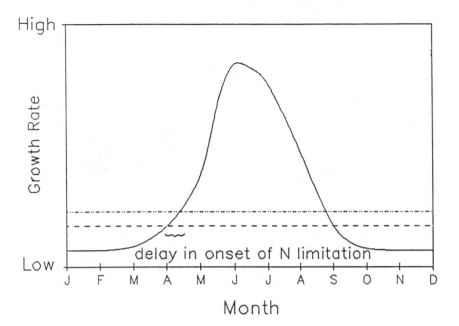

Figure 4. Conceptual model of mechanism responsible for increased incidence and severity of NO_3^- episodes in watersheds in early stage of nitrogen saturation (Stage 1). Solid line is potential growth rate of forest if all nutrients are sufficient and only environmental variables (e.g., temperature, light, and water) limit growth. Dashed line is potential growth rate if environmental variables are optimal, but N is in short supply. Points where dashed and solid lines cross are transitions from N-unlimited to N-limited growth and vice versa. Dotted line is potential growth rate if environmental variables are optimal and N supply is moderate. Shift in onset of N limitation in spring will result from increase in nitrogen supply and is likely to coincide with timing of snowmelt in northern forested watersheds.

strongly acidifying process (Figure 1). The key characteristics of Stage 2 watersheds are elevated baseflow concentrations of NO_3^-, which result from high groundwater concentrations (Figure 6b). Episodic NO_3^- concentrations are as high as Stage 1, but the seasonal pattern at Stage 2 is damped by an increase in baseflow concentrations to levels as high as those found in deposition.

In Stage 3 the watershed becomes a net source of N rather than a sink (Figure 7). Nitrogen-retention mechanisms (e.g., uptake by vegetation and microbes) are much reduced. Mineralization of stored N may add substantially to N leaving the watershed through leaching or in gaseous forms. As in Stage 2, nitrification rates are substantial. The combined inputs of N from deposition, mineralization, and nitrification can produce concentrations of NO_3^- in surface waters that exceed inputs from deposition alone. The key

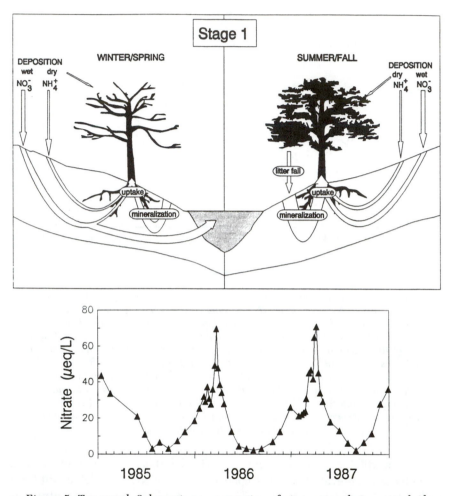

Figure 5. Top panel: Schematic representation of nitrogen cycle in watershed at Stage 1 of watershed nitrogen loss. Size of arrows indicates the magnitude of process or transformation. Differences between winter–spring and summer–fall seasons are shown on opposite sides. As in Stage 0, uptake dominates the nitrogen cycle during the growing season at Stage 1 and little or no NO_3^- leaks from the watershed during the summer and fall. The primary difference between Stage 0 and Stage 1 is the delay in the onset of N limitation during the spring season (see text and Figure 4). Bottom panel: Large runoff events (e.g., snowmelt or rainstorms) during the dormant season can produce episodic pulses of high NO_3^- concentrations, as shown in the typical Stage 1 seasonal NO_3^- cycle. Data in bottom panel are from Constable Pond in the Adirondack Mountains. (Data are taken from reference 157.)

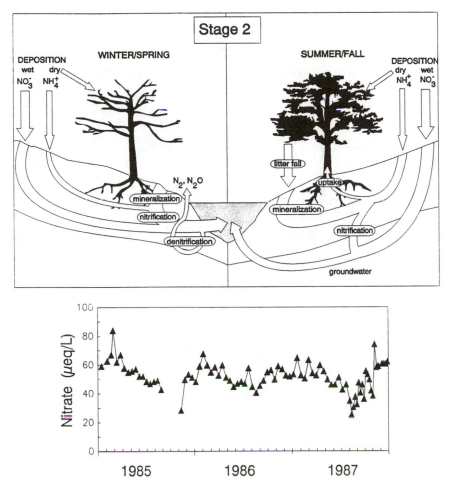

Figure 6. Top panel: Schematic representation of nitrogen cycle in watershed at Stage 2 of watershed nitrogen loss. Size of arrows indicates the magnitude of process or transformation. Differences between winter–spring and summer–fall seasons are shown on opposite sides. Uptake of nitrogen by forest plants and microbes is much reduced at Stage 2, resulting in loss of NO_3^- to streams during winter and spring and to groundwater during the growing season. Loss of gaseous forms of nitrogen through denitrification may also be elevated at Stage 2 if conditions necessary for denitrification are present (see text). Although episodes of higher NO_3^- concentrations continue to occur during high-flow events such as spring snowmelt, the primary difference between Stage 1 and Stage 2 is the presence of elevated NO_3^- concentrations in groundwater. Bottom panel: The typical seasonal NO_3^- pattern at Stage 2 includes both high episodic concentrations and high base-flow concentrations. Data in bottom panel are from Fernow Experimental Forest, Control Watershed No. 4, West Virginia.

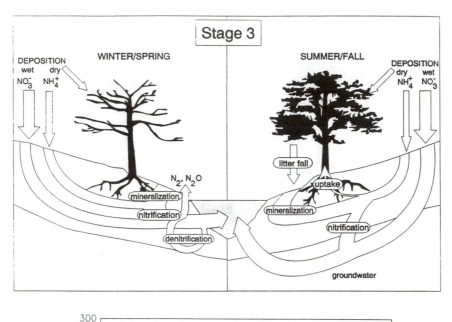

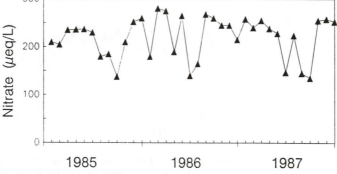

Figure 7. Top panel: Schematic representation of nitrogen cycle in watershed at Stage 3 of watershed nitrogen loss. Size of arrows indicates the magnitude of process or transformation. Differences between winter–spring and summer–fall seasons are shown on opposite sides. At Stage 3 no sinks for nitrogen exist in the watershed and all inputs, as well as mineralized nitrogen, are lost from the system either through denitrification or in runoff water. Because mineralization supplies nitrogen in excess of deposition, concentrations of NO_3^- in runoff may exceed those in deposition. Bottom panel: Typical seasonal NO_3^- pattern at Stage 3 includes concentrations at all seasons in excess of concentrations attributable to deposition and evapotranspiration, as at Dicke Bramke in Germany; data are from reference 190.

characteristics of Stage 3 watersheds are these extremely high NO_3^- concentrations and the lack of any coherent seasonal pattern in NO_3^- concentrations (Figure 7b).

Conceptually, the stages of watershed N loss can be thought of as occurring sequentially, as a single watershed progresses from being strongly N deficient to strongly N sufficient. This sequence is consistent with the conceptual model presented by Aber et al. (*1*; also Figure 2) and can be supported by two lines of evidence that are presented in the following sections of this chapter. The first line of evidence comes from space-for-time substitutions (*84*), where the occurrence of various stages across a gradient of current N deposition is used as a surrogate for the temporal sequence that a single site might undergo if it were exposed to chronically elevated levels of N deposition. This technique is commonly applied to current environmental problems for which a good historical record is not available (*85*). The second line of evidence comes from long-term temporal trends at single sites, where increases in N efflux from watersheds (observable as increasing trends in NO_3^- concentration) and changes in the seasonal pattern of NO_3^- concentration can be directly attributed to the combined effects of chronic N deposition and other factors (e.g., forest maturation). The few cases in which individual sites have progressed from Stage 0 to Stage 1 or Stage 2 of watershed N loss are especially useful in establishing that N saturation occurs as a temporal sequence in areas of high N deposition. These lines of evidence are discussed in the following sections.

The Consequences of Nitrogen Loss from Watersheds

Two possible consequences of watershed N loss are surface-water acidification and eutrophication. The acidification processes in lakes and streams are conventionally separated into chronic (long-term) and episodic (event-based) effects. Surface waters are generally considered acidic if their acid-neutralizing capacity (ANC) is less than zero. The ANC of a lake or stream is a measure of the water's capacity to buffer acidic inputs. It results from the presence of carbonate or bicarbonate (i.e., alkalinity), aluminum, and organic acids in the water (*86*). In the past decade a great deal of emphasis was placed on chronic acidification in general, and on chronic acidification by SO_4^{2-} in particular (*87, 88*). This emphasis on SO_4^{2-} has resulted largely because sulfur deposition rates are often higher than those for N (S deposition rates are approximately twice the N deposition rates in the Northeast; ref. *9*) and because NO_3^- appears to be of negligible importance in surface waters sampled during summer and fall index periods (*89*). As mentioned previously, summer and fall are seasons when watershed demand for N is normally very high. This timing creates a low probability that N, in any form, will be leached into soil and surface waters unless the watersheds are in stages 2 and 3 of N loss. Under the usual conditions (i.e., in Stages 0 and 1), N leaking from terrestrial ecosystems is more likely to be a transient (or seasonal) phenomenon than a chronic one.

Eutrophy generally refers to a state of nutrient enrichment (53), but the term is commonly used to refer to conditions of increased algal biomass and productivity, the presence of nuisance algal populations, and a decrease in oxygen availability for heterotrophic organisms. Eutrophication is the process whereby lakes, estuaries, and marine systems progress toward a state of eutrophy. In lakes, eutrophication is often considered a natural process, progressing gradually over their long-term evolution. The process can be significantly accelerated by the additional input of nutrients from anthropogenic sources. The subject of eutrophication was extensively reviewed by Hutchinson (90), the National Academy of Sciences (91), and Likens (92).

Chronic Acidification.   The most comprehensive assessment of chronic acidification of lakes and streams in the United States comes from the National Surface Water Survey (NSWS). It was conducted as part of the National Acid Precipitation Assessment Program (NAPAP). The NSWS surveyed the acid–base chemistry of both lakes and streams by using an index period concept. The goal of the index period concept was to identify a single season of the year that exhibited low temporal and spatial variability and that, when sampled, would allow the general condition of surface waters to be assessed (89). For lakes, the index period selected was autumn overturn, the period when most lakes are mixed uniformly from top to bottom. For streams the chosen index period was spring baseflow, the period after spring snowmelt and before leafout (93). Because of the strong seasonality of the N cycle in forested watersheds (described earlier), the choice of index period plays a very large role in the assessment of whether N is an important component of chronic acidification.

The Eastern Lake Survey (89) was based on a probability sampling of lakes during fall overturn, and the National Stream Survey (NSS; ref. 94) was based on a probability sampling of streams during a spring baseflow period. The results suggest that N compounds make only a small contribution to chronic acidification in North America. Henriksen (95) proposed that the ratio of $NO_3^-:(NO_3^- + SO_4^{2-})$ in surface waters be used as an index of the influence of NO_3^- on chronic acidification status. This index assesses the importance of NO_3^- relative to the importance of SO_4^{2-}, which is usually considered more important in chronic acidification. A value greater than 0.5 indicates that NO_3^- has a greater influence on the chronic acid–base status of surface waters than does SO_4^{2-}

Median values of $NO_3^-:(NO_3^- + SO_4^{2-})$ ratios for acid-sensitive regions of the United States sampled in the NSWS are shown in Table II. Data are taken mainly from refs. 89 and 96, with data from additional regional surveys included in Table II for comparison (97–99). In all regions the median values of $NO_3^-:(NO_3^- + SO_4^{2-})$ are all less than 0.5, and most are less than 0.2.

Table II. Concentrations and Ratio of NO$_3^-$ to (NO$_3^-$ + SO$_4^{2-}$) in Lakes and Streams of Acid-Sensitive Regions of the United States

Region[a]	Year	pH	NO$_3^-$[b]	SO$_4^{2-}$[b]	NO$_3^-$:(NO$_3^-$ + SO$_4^{2-}$)
Eastern Lake Survey[c]					
Southern Blue Ridge	1985	6.98	3.1	31.8	0.09
Florida	1985	6.56	1.0	93.7	0.01
Upper Midwest	1985	7.09	0.7	57.1	0.01
Maine	1985	6.91	0.2	74.6	0.00
Southern New England	1985	6.81	0.8	141.1	0.01
Central New England	1985	6.77	0.3	101.2	0.00
Adirondack Mountains	1985	6.71	0.6	118.7	0.01
Catskill–Poconos	1985	7.02	0.7	159.3	0.00
National Stream Survey[d]					
Poconos–Catskills	1986	6.96	6	169	0.03
Northern Appalachians	1986	6.60	30	171	0.14
Valley and Ridge	1986	7.05	10	154	0.09
Mid-Atlantic Coastal Plain[e]	1986	5.98	—	—	—
Southern Blue Ridge	1986	6.99	8	17	0.28
Piedmont	1986	6.80	2	48	0.03
Southern Appalachians	1986	7.33	16	58	0.32
Ozarks–Ouachitas	1986	6.62	1	59	0.02
Florida	1986	5.48	5	22	0.19
Catskill Regional Survey[f]	1984–86	6.60	29	138	0.17
Great Smoky Mountains Stream Survey[g]	1984–86	5.58	61.1	76.1	0.44
Virginia Trout Stream Survey[h]	1987	6.70	0	72	0.00

[a] Values are weighted medians for each region.
[b] Values are given in microequivalents per liter.
[c] Data are from reference 89.
[d] Values for pH are for entire region (94); medians for NO$_3^-$, SO$_4^{2-}$ and the NO$_3^-$:(NO$_3^-$ + SO$_4^{2-}$) ratio exclude sites with potential agricultural or other land-use impacts (96).
[e] The influence of agricultural and land-use practices could not be ruled out for any of the sites in the mid-Atlantic coastal plain (96).
[f] Median value for repeated measurements at 51 streams (data are from reference 79).
[g] Mean value for repeated measurements at five streams (data are from reference 99).
[h] Median value of 341 streams (data are from reference 98).

These values suggest that NO_3^- does not contribute significantly to chronic acidification in most regions of the United States. Several southeastern regions exhibit ratios in the 0.2–0.4 range, primarily because their current SO_4^{2-} concentrations are relatively low. The Southern Blue Ridge, in particular, has the lowest SO_4^{2-} concentrations found in the NSS, and the relatively high $NO_3^-:(NO_3^- + SO_4^{2-})$ ratios in this region could be considered misleading. Taken in total, regional survey data suggest that the Catskills, Northern Appalachians, Valley and Ridge Province, and Southern Appalachians all show some potential for chronic acidification attributable to NO_3^-. However, in all of the regions shown in Table II chronic acidification is more closely tied to SO_4^{2-} than to NO_3^-.

The highest $NO_3^-:(NO_3^- + SO_4^{2-})$ ratios and the highest mean NO_3^- concentrations were recorded in streams of the Great Smoky Mountains, where baseflow NO_3^- concentrations were as high as, or in some cases higher than, baseflow SO_4^{2-} concentrations (99). Cook et al. (99) stressed that this study included a small number of streams, that the region is known for geological sources of SO_4^{2-} (the Anakeesta geological formation), and that high SO_4^{2-} and NO_3^- concentrations tend to occur simultaneously. The results are entirely consistent, however, with soil lysimeter studies carried out in the Great Smoky Mountains. According to these studies, NO_3^- concentrations in deep soil lysimeters are higher than input concentrations from deposition and throughfall (172). Because deep soil-water concentrations can be assumed to approximate groundwater and baseflow stream-water concentrations, these streams and watersheds in the Great Smoky Mountains are the only known examples in the United States of watersheds at Stage 3 of watershed N loss, as described earlier.

Chronic acidification resulting from N deposition is much more common in Europe than in North America (39). Many European sites show chronic increases in N export from their watersheds (101, 102), and at sites with the highest stream-water NO_3^- concentrations (i.e., Lange Bramke and Dicke Bramke in West Germany) NO_3^- concentrations no longer show the seasonality that is expected from normal watershed processes (39). Henriksen and Brakke (101) reported regional chronic increases in surface-water NO_3^- in Scandinavia in the past decade. These increases in NO_3^- concentration are associated with increasing concentrations of aluminum, which is toxic to many fish species (103, 104). Some evidence (105) suggests that NO_3^- has a greater ability to mobilize toxic aluminum from soils than does SO_4^{2-}. Chronic acidification attributable to ammonium deposition has also been demonstrated in the Netherlands (51, 106). As described earlier, ammonium in deposition can be nitrified to produce both NO_3^- and hydrogen ions, which are subsequently leaked into surface waters. Rates of NO_3^- and NH_4^+ deposition are much higher in Europe (in some places deposition is >2000 equiv/ha per year; ref. 107) than in the United States (Table I). Chronic N acidification may be more evident in Europe than in North America because N saturation is further developed in Europe.

Episodic Acidification. Except in extreme cases, N loss from watersheds is more likely to be an episodic or seasonal process than a chronic one. Therefore, data used to assess the contribution of NO_3^- to acidification must be collected on an intensive schedule so short-term effects of NO_3^- increases can be characterized. This situation places severe limitations on the type of data that can be used in this assessment. Data from regional surveys, although they provide excellent spatial coverage, are hampered by the need to collect samples during a stable index period. In most areas the spring period when NO_3^- effects are most likely to be observable would not meet the requirements described earlier for a stable index period. As a result the regional importance and severity of episodic acidification have not been quantified; that is, the regional information on chronic acidification that was gained from the NSWS has no parallel in episodic acidification, and all conclusions must be based on site-specific data. Even within a given area, such as the Adirondack Mountains, major differences can be evident in the occurrence, nature, location (lakes or streams), and timing of episodes at different sites.

It has been estimated that 1.4–7.4 times as many streams in the eastern United States undergo episodic acidification than are chronically acidic (*108*). Similarly, the number of episodically acidic Adirondack lakes is estimated to be 3 times higher than the number of chronically acidic lakes (*108*). Wigington et al. (*109*) reported that acidic episodes occur in a wide range of geographic locations in the northeastern, southeastern, and western United States, as well as in Scandinavia, Europe, and Canada.

A number of processes contribute to the timing and severity of acidic episodes (*42*). During high-discharge periods the most important of these processes are

- dilution of base cations (*110*);
- increases in organic acid concentrations (*111*);
- increases in SO_4^{2-} concentrations (*112*); and
- increases in NO_3^- concentrations (*110, 113, 114*).

In addition to these factors, which produce the chemical conditions characteristic of episodic events, the likelihood of an acidic episode is also influenced by the chemical conditions prevailing before the episode begins. Episodes are more likely to be acidic, for example, if the baseflow ANC of the stream or lake is low. In this way, acidic anions, especially SO_4^{2-}, can contribute to the severity of an acidic episode by lowering the baseflow ANC of the stream or lake, even though they do not increase during the event (*97*).

In many cases all of these processes will contribute to episodes in a single aquatic system. Dilution, for example, probably plays a role in all episodic decreases in ANC and pH in all regions of the United States (*109*).

Dilution results from the increased rate of runoff and the channeling of runoff through shallower soil layers that occurs during storms or snowmelt. The shorter contact time produces runoff with a chemical composition closer to that of atmospheric deposition than is typical of baseflow conditions (115–117). Because precipitation is usually more dilute than stream or lake water, storm runoff produces surface waters that are more dilute than during nonrunoff periods. In a sense, dilution sets the baseline condition to which is added the effects of organic acids and atmospherically derived SO_4^{2-} and NO_3^-.

Little information exists about the effects of changes in organic acids during episodes. Driscoll et al. (118) and Eshleman and Hemond (119) concluded that organic acids did not contribute to snowmelt episodes in the Adirondacks or in Massachusetts, respectively. At Harp Lake in Canada, organic acidity is believed to remain constant (120) or decrease (121) during snowmelt episodes. Haines (122) and McAvoy (123) documented increases in organic acidity during rain-caused episodes in coastal Maine and in Massachusetts.

Storage of SO_4^{2-} in watersheds and its subsequent release $_4^{2-}$ during episodic events are well documented in many parts of Europe (109), but the process has not been found commonly in the United States. Sulfate episodes have been described for the Leading Ridge area of Pennsylvania (124) and at Filsen Creek in Minnesota (125), but they are not widespread. Sulfate does contribute to episodic acidity, however, in the sense that concentrations may remain high during events and contribute to a lower baseline ANC. The effects of other factors, such as increased NO_3^-, are supplemental to any constant effect of SO_4^{2-} in lowering the baseline ANC (97).

In the discussion of the contribution of NO_3^- to chronic acidification, the importance of NO_3^- was related to the importance of SO_4^{2-}, the primary acidic anion in anthropogenically acidified lakes and streams. For episodic acidification, the importance of short-term increases in NO_3^- concentration must be assessed in relation to the other processes that contribute to acidic episodes. In the Adirondacks, for example, strong NO_3^- pulses in both lakes (110, 113) and streams (118) are apparently the primary factor contributing to depressed ANC and pH during snowmelt. Schaefer et al. (126) examined intensive monitoring data from 11 Adirondack lakes and concluded that the magnitude of the episodes experienced by lakes depends strongly on their base-cation concentration. They concluded that lakes with high base-cation concentrations (and therefore high ANC values) undergo episodes that are largely the result of dilution by snowmelt. Low-ANC lakes, on the other hand, undergo episodes that result largely from increases in NO_3^- concentrations. At intermediate ANC levels, lakes are affected by both base-cation dilution and NO_3^- increases. Therefore these lakes may undergo the greatest increases in acidity during snowmelt episodes (Figure 8). Murdoch and Stoddard (127) reported similar results for streams in the Catskill Mountains,

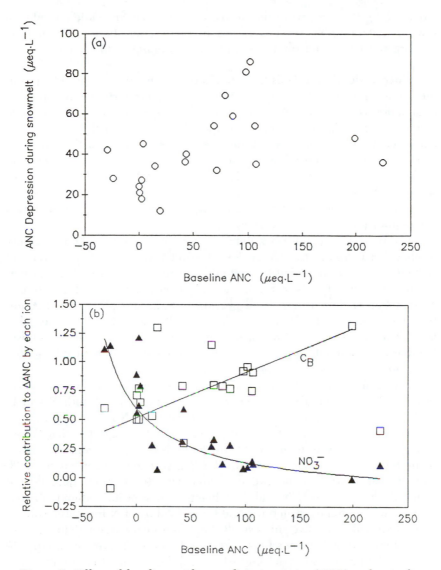

Figure 8. Effect of baseline acid-neutralizing capacity (ANC) and episodic conditions in Adirondack lakes. a, Relationship between baseline ANC and the springtime depression in ANC (baseline ANC—minimum ANC) for 11 lakes sampled in 1986 and 1987. b, The relative contributions of base cations (C_B) and nitrate (NO_3^-) to the springtime ANC depressions in Adirondack lakes. Lakes at intermediate ANC values undergo the largest springtime depressions in ANC. Lakes with lower baseline ANC are affected more by NO_3^- pulses, and lakes with higher baseline ANC are affected more by base-cation dilution. Solid lines represent best-fit relationships. (Redrawn with permission from reference 126. Copyright 1990 American Geophysical Union.)

where NO_3^- increases were the primary determinant of acidic episodes in low- to moderate-ANC streams during the spring. Both base-cation dilution and organic acidity contributed significantly to episodes in the fall.

Eutrophication. Thus far N has been discussed in terms of its prominence as an acidic anion (i.e., as NO_3^-). As in terrestrial ecosystems, inorganic forms of N also act as nutrients in aquatic systems, and a possible consequence of chronic N loss from watersheds is the fertilization of lakes and streams. Establishing a link between N deposition and the eutrophication of aquatic systems depends on a determination that the productivity of the system is limited by N availability and that N deposition is a major source of N to the system. In many cases the supply of N from deposition is minor when compared to other anthropogenic sources, such as pollution from either point or nonpoint sources.

The productivity of fresh waters is generally limited by the availability of phosphorus, rather than N (reviewed in ref. 34). Although conditions of N limitation occur in freshwater systems, they are often either transitory or the result of high inputs of P from anthropogenic sources such as sewage. N limitation is often a short-lived phenomenon because N-deficient conditions favor the growth of blue-green algae (128), many of which are capable of N fixation. In lakes N fixation may be considered a natural mechanism that has contributed to the long-term evolution and ubiquity of P limitation by ensuring an adequate supply of N (72).

Nitrogen limitation can occur naturally (i.e., in the absence of anthropogenic P inputs) in lakes with very low concentrations of both N and P; such low-nutrient lakes are common in the West and Northeast (30). Suttle and Harrison (30) and Stockner and Shortreed (129) suggested that P concentrations in these systems are too low to allow blue-green algae to thrive; blue-green algae are poor competitors for P at very low concentrations (128, 130). In these systems the two nutrients are often closely coupled, and constant shifts between N and P deficiency may occur without obvious changes in community structure. Additional loading of N from atmospheric deposition is likely to have only a small effect on primary productivity because the system quickly becomes P-limited. In a literature survey of 62 separate nutrient-limitation studies in lakes, Elser et al. (131) found that simultaneous additions of N and P produced the largest growth response in 82% of the experiments. These results underline the likelihood that a lake limited by one nutrient may quickly become limited by another if the lake becomes enriched with the original limiting nutrient.

Estimations of nutrient limitation in lake ecosystems follow three major lines of reasoning:

1. evidence from ambient nutrient concentrations and the nutritional needs of algae,

2. evidence from bioassay experiments at various scales, and

3. evidence from nutrient dynamics and input–output studies (*34, 132*).

Bioassay experiments provide the most direct evidence of nutrient limitation because they involve the experimental addition of selected nutrients (singly or in combination) to assemblages of algae under controlled conditions. Analyses of ambient nutrient conditions are less direct indicators of limitation because the biotic response (i.e., biostimulation) is not measured, but is instead inferred from geochemical principles. In this sense the nutrient ratio approach measures potential nutrient limitation rather than actual limitation, but it often shows results consistent with bioassay results. The ambient nutrient ratio approach has been criticized widely because it ignores several factors known to be important to algal growth. The use of only inorganic nutrient species in the ratios, for example, has been criticized because many algal species are known to make use of organic forms, especially of P (*132*). Algal growth may also be more dependent on the supply rates of nutrients than on their ambient concentrations (*31, 133*). Many species of algae may therefore not be limited by nutrients whose ambient concentrations are so low as to be undetectable.

Much of the acceptance of the idea that freshwater lakes are primarily P-limited stems from the close correlations between P concentrations and lake productivity or algal biomass (usually measured as chlorophyll concentration) that have been observed in a large number of lake studies (reviewed in refs. 27 and 134). More recently, researchers have begun to question the ubiquity of the phosphorus:chlorophyll relationship and to identify some of the factors that lead to the large variability observed in this relationship in nature (*128, 135–138*). Recent reexaminations of the data suggest that the phosphorus:chlorophyll relationship is best described as sigmoidal (*139*) and that the slope of the relationship is significantly affected by N concentrations, particularly at high concentrations of P (>1 μmol/L) that are likely to be caused by anthropogenic inputs. McCauley et al. (*139*) found that N had little effect on the phosphorus:chlorophyll relationship at low concentrations of P. This effect is expected in nutrient-poor lakes where the primary effect of N additions would be to push lakes into a phosphorus-deficient condition.

Ambient Nutrient Ratios. Arguments based on ambient nutrient concentrations stem from the early work of Redfield (*140*), who examined particulate concentrations of nutrients from samples of nutrient-sufficient algae taken from marine systems worldwide and found surprisingly consistent results for the ratios of C:N:P concentrations (106:16:1). Deviations from these ratios are considered evidence that one nutrient or another is limiting to algal growth (e.g., N:P ratio values below 16:1 suggest N limitation; values above 16:1 suggest P limitation). Because the relative supply rates of P and N determine whether one or the other nutrient is in short supply, various researchers have extended the interpretation of the Redfield ratio to include ambient nutrient concentrations in water (Redfield's original work was with intracellular concentrations) and applied the nutrient ratio criteria to lakes to predict their likely limiting conditions (135, 138, 141). This method has the potential to illustrate regional patterns and has gained some support from the results of bioassay experiments. This idea has been refined to exclude from the ratio those forms of N and P that are not biologically available, especially organic forms of N. As a result, good predictions of nutrient limitation can now be made from ratios of total dissolved inorganic nitrogen (DIN) to total phosphorus (TP) (142).

Morris and Lewis (142) conducted nutrient-addition bioassays on natural assemblages of phytoplankton from many lakes and compared their results to DIN:TP values measured in the lakes at the same time as the experiments were conducted. They found that lakes with DIN:TP values less than 9 (using molar concentrations) could be limited by either N or P; often additions of both nutrients were required to stimulate growth. Lakes with DIN:TP values less than 2 were always limited by N. The discrepancy between the 16:1 Redfield ratio and the 2:1 ratio suggested by Morris and Lewis (142) results from a number of differences between the Redfield measurements and the method used here. First and foremost, the DIN:TP ratio compares dissolved concentrations of inorganic N to total concentrations of P in ambient water; the Redfield ratio is based on cellular or particulate organic concentrations. A large amount of variability also results from the range of threshold N:P ratios for individual algal species. Suttle and Harrison (30), for example, reported N limitation at ratios ranging from 7:1 to 45:1 for single species. Such a wide range for single species suggests that critical ratios may also be quite variable in nature. Finally, the 2:1 ratio is a conservative estimate of the nutrient-limitation threshold, in the sense that it marks the boundary beyond which all experimental results indicate N limitation (142).

If a DIN:TP value of 2 is used as the threshold below which lakes are considered to be N-limited and is applied to lake data from the NSWS (89, 143), it is possible to estimate the number of N-limited lakes in most of the mountainous regions of the United States (Table III). The NSWS was designed to survey the acid–base chemistry of lakes by using a probability design. Because the selection of lakes is a stratified random sample of the lakes in each region, results can be extrapolated to the target population of lakes in each region. As an acidification survey, the NSWS excludes many lakes in highly disturbed areas (e.g., urban and agricultural areas). The results in Table III should therefore be interpreted in the context of the target population they represent: nonurban, nonindustrial, nonagricultural lakes above a certain size in each region (size cutoffs were >4 ha in the Eastern Lake Survey and >1 ha in the Western Lake Survey). Lakes with total P concentrations greater than 0.65 μmol/L (20 μg/L) have been ex-

Table III. Nitrogen-Limited Lakes in Subregions of the United States Sampled by NSWS

Region	Number of Lakes in Subregion	Estimated Number[a] of N-Limited Lakes	N-Limited Proportion of Population (%)	"Excess P"[b] (μmol/L)
Eastern Lake Survey[c]				
Adirondacks (1A)	1290	41	3	0.3
Poconos–Catskills (1B)	1506	177	12	0.6
Central New England (1C)	1494	72	5	0.3
Southern New England (1D)	1325	257	19	0.7
Northern New England (1E)	1542	77	5	0.2
Northeastern Minnesota (2A)	1499	300	20	0.5
Upper Peninsula, Michigan (2B)	1050	322	31	0.5
Northcentral Wisconsin (2C)	1511	209	14	0.7
Upper Great Lakes Area (2D)	4515	1530	34	0.6
Southern Blue Ridge (3A)	286	10	4	0.3
Florida (3B)	2138	234	11	0.7
Total	18,155	3228	18	0.6
Western Lake Survey[d]				
California (4A)	2406	573	24	0.6
Pacific Northwest (4B)	1706	622	36	0.3
Northern Rockies (4C)	2379	857	36	0.4
Central Rockies (4D)	2299	931	41	0.6
Southern Rockies (4E)	1609	754	47	1.0
Total	10,398	3736	36	0.5

[a]Estimates are based on ratios of dissolved inorganic nitrogen contrations ($NO_3^- + NH_4^+$) to total phosphorus concentrations.
[b]"Excess phosphorus; calculated as the amount of added DIN necessary to increase the DIN:TP ratio to 2.0.
[c]Data are from Kanciruk et al. (202); excluding lakes with total phosphorus >0.65 μ mol/L (20 μg/L).
[d]Data are from reference 203, excluding lakes with total phosphorus >0.65 mol/L (20 μg/L).

cluded from this analysis because many of them may have experienced anthropogenic inputs of P (53, 144). This test is therefore conservative for N limitation, both because the DIN:TP value chosen (<2) is a conservative measure of N limitation (142) and because some lakes with naturally high concentrations of P may be excluded. These lakes are more likely to be N-limited than lakes with low P concentrations.

Nitrogen Limitation by Region. The proportions of lakes that can be considered N-limited vary widely from region to region, with the greatest number being found, as expected, in the West. The highest proportions were found in the Rocky Mountains (40% of lakes in the area exhibited low DIN:TP ratios), but all subregions of the West contained substantial numbers (>20%) of potentially N-limited lakes. The smallest proportions of N-limited lakes were found in the Northeast and Southeast (9 and 10%, respectively, of the lakes in these regions exhibited low DIN:TP ratios). One surprise in this analysis is the number of lakes in the upper Midwest that appear to be N-limited. Taken as a whole, 28% of this region's lakes had DIN:TP ratios less than 2.

A more direct indication of nutrient limitation than is available from nutrient ratios can be gained from bioassay experiments. In this procedure a small volume of natural lake water is enclosed and various known concentrations of potentially limiting nutrients are added (145–147). A growth response (usually measured as an increase in biomass) in treatments containing an added nutrient constitutes evidence of limitation by that nutrient. The results of such experiments are available for only a few selected nutrient-poor lakes, however. They indicate a variety of responses, including strong P limitation (148), limitation by P and iron (147), simultaneous N and P limitation in which the two nutrients are so closely balanced that addition of one alone simply leads to limitation by the other (147), and limitation primarily by N (142, 149). No clear pattern of N or P limitation develops from an examination of these few studies.

The potential for N deposition to contribute to the eutrophication of freshwater lakes is probably quite limited. Eutrophication by atmospheric inputs of N is a concern only in lakes that are chronically N-limited. This condition occurs in some lakes that receive substantial inputs of anthropogenic P and in many lakes where both P and N are found in low concentrations (e.g., Table III). In the former case the primary dysfunction of the lakes is an excess supply of P, and controlling N deposition would be an ineffective method of water-quality improvement. In the latter case the potential for eutrophication by N addition (e.g., from deposition) is limited by low P concentrations; additions of N to these systems would soon lead to N-sufficient, and phosphorus-deficient, conditions. The results of the NSWS shown in Table III, for example, can be used to calculate the increase in N concentration that would be required to push N-limited lakes into P limitation (assuming total P concentrations do not change). An increase of only

0.5 μmol/L in N concentration would be sufficient to induce P limitation in half of the N-limited lakes in the NSWS target population. In the eastern United States the largest N increases (0.7 μmol/L) would be required in Florida and Southern New England, where anthropogenic inputs of P are most likely. The largest increases of all (1 μmol/L) would be required in the Southern Rocky Mountains, which has the highest P concentrations of any of the western subregions (median = 0.26 μmol/L). This P concentration probably results from natural sources of P (e.g., volcanic bedrock). Increases in N deposition to some of the regions in Table III would probably lead to measurable increases in algal biomass in those lakes with low DIN:TP ratios *and* substantial total P concentrations, but the number of lakes that meet these criteria is likely to be quite small.

Regional Analysis of the Stages of Watershed Nitrogen Loss

If elevated rates of N deposition contribute to watershed N loss, then patterns of N loss might be expected to mimic patterns of N deposition. In the following discussion, the distribution of monitoring sites in the eastern United States with NO_3^- patterns suggestive of the varying stages of watershed N loss is presented as occurring across a gradient in N deposition from relatively low to very high. This space-for-time substitution is a useful tool in assessing whether N saturation is progressing in high-deposition areas. Although important exceptions do occur (e.g., *see* discussion of Pennsylvania streams in the next section), lakes and streams with Stage 0 patterns tend to occur in low- to moderate-deposition areas; sites with NO_3^- patterns typical of later stages of N loss tend to occur in high-deposition areas. Results are presented as a series of maps, and the discussion progresses from relatively low-deposition areas in the northeastern corner of the country (Maine) to higher deposition areas in western New England and New York, and moves south into the very-high-deposition areas of southern Pennsylvania and West Virginia. Additionally, results from the few monitoring sites in the western United States, where deposition rates are uniformly low, are presented separately.

In general, sites are included in this analysis if published data of more than 2 years' duration in the 1980s are available and if at least a seasonal sampling schedule was used to collect the data. Each site has been assigned to one of six classes. In addition to four classes representing the four stages of watershed N loss, many of the Stage 1 and Stage 2 sites have data of long enough duration to determine whether they are experiencing upward trends in NO_3^- concentrations; Stage 1 and Stage 2 sites with upward NO_3^- trends are included as separate classes on the maps for the following sections. As mentioned earlier, temporal trends in NO_3^- provide important verification that the stages of N loss occur as a sequence in sites experiencing chronically elevated rates of N deposition.

The Northeastern United States.   For approximately 100 sites in the Northeast, the data are intensive enough to determine their status with respect to watershed N loss (Figure 9). From the standpoint of N, the Northeast is the most data-intensive region of North America. Seasonal data are available from a spatially extensive network of sites, so that geographic patterns in watershed N loss can be inferred. Importantly, these geographic patterns in watershed N loss follow the geographic pattern in N deposition, with the most severe effects observable in the Adirondack and Catskill mountains, where deposition rates are high, and little N loss in Maine, where rates of N deposition are roughly 50% lower (42).

Figure 9. Location of acid-sensitive lakes and streams in the northeastern United States where the importance of NO_3^- to seasonal water chemistry can be determined. Only data from undisturbed watersheds are included. (Data are from references 109, 114, 118, 119, 151b, 154, 156–159, 162, 164, 191–194.)

Data from sites in Maine, Vermont, and the Adirondack and Catskill mountains are available from the U.S. Environmental Protection Agency's (EPA) Long-Term Monitoring (LTM) Project (*150*). In most cases the records from these sites are sufficiently long (7–10 years) to detect trends where they are present. Most of the other sites in Figure 9 have data of shorter duration, and it is not known whether they exhibit long-term trends.

All of the sites in the state of Maine show some seasonality in NO_3^- concentrations, with peak concentrations of less than 20 μequiv/L during spring snowmelt and negligible concentrations during all other seasons; the concentrations suggest typical Stage 0 NO_3^- patterns. Data from five LTM lakes in Maine exhibit no long-term trends in NO_3^- concentrations (*151a*).

Seasonal data are available from two sites in New Hampshire, where deposition rates are slightly higher than in Maine. Baird et al. (*151b*) studied episodic acidification during snowmelt at Cone Pond, New Hampshire, and were unable to detect any NO_3^- in inlet water. Researchers at the Hubbard Brook Experimental Forest in New Hampshire have been studying the watershed processes and their effects on stream-water chemistry since 1963 (*152*). In reference Watershed No. 6, stream-water NO_3^- concentrations undergo strong seasonal cycles with peak concentrations as high as 85 μequiv/L during snowmelt, similar to Stage 1 watershed N loss. Both NO_3^- and hydrogen ion concentrations increase during snowmelt at Hubbard Brook, although SO_4^{2-} concentrations decrease slightly (*153, 154*). No long-term trend in stream-water NO_3^- concentration is detectable for the 23-year period of record at Hubbard Brook (*155*).

In the state of Vermont, 6 of 24 LTM lakes exhibit strong seasonal NO_3^- cycles, and in these lakes seasonal NO_3^- increases are the most important mechanism contributing to spring ANC minima (*156*). These six watersheds exhibit the characteristics of Stage 1 of N loss, and the remaining 18 lakes exhibit Stage 0 patterns. None of the sites exhibit trends in NO_3^- concentrations for the period 1981–1989.

Driscoll and Van Dreason (*157*) reported data from 16 Adirondack LTM lakes collected between 1982 and 1990. The patterns in acid–base chemistry at Constable Pond (Figure 10), one of the Adirondack LTM lakes, can be considered typical of surface waters in the Northeast with episodically elevated concentrations of NO_3^- (e.g., at Stages 1 and 2). Constable Pond becomes acidic only during seasons of high runoff, as a result of base-cation dilution and episodic NO_3^- increases. Sulfate concentrations are relatively invariant. In Constable Pond, as in other Adirondack lakes and streams, short-term changes in NO_3^- were highly correlated, and chemically consistent, with changes in the concentrations of acidic cations (hydrogen and aluminum) (*155*). As mentioned earlier, although dilution of base cations and increases in NO_3^- appear to be the primary causes of episodic acidification in Constable Pond, these episodes are excursions from an already low baseline ANC, which can be attributed largely to high SO_4^{2-} concentrations.

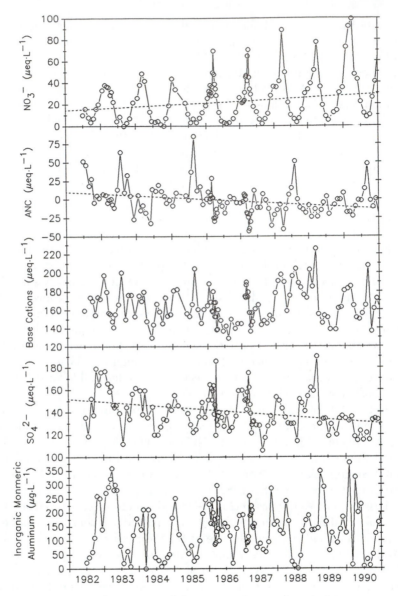

Figure 10. Temporal patterns in lake-water NO_3^-, acid-neutralizing capacity (ANC), base cations $(Ca^{2+} + Mg^{2+} + Na^+ + K^+)$, SO_4^{2-}, and inorganic monomeric aluminum (Al_i) at Constable Pond, a long-term monitoring site in the Adirondack Mountains. Trend lines are shown for variables with significant trends ($p < 0.10$ in seasonal Kendall tau test). Seasonal pattern is typical of Adirondack lakes, with seasonal minima in ANC coincident with seasonal maxima in NO_3^- and Al_i. Many Adirondack lakes exhibited upward trends in NO_3^- in the 1980s; the primary increase was in episodic NO_3^- concentrations. (Data are from reference 157.)

The period of record is long enough for the Adirondack LTM lakes that long-term trends in NO_3^- concentrations can be estimated (Table IV; ref. 157). Nine out of sixteen Adirondack lakes exhibited significant increases in NO_3^- concentrations, with typical slopes of trends being $+1$ μequiv/L per year. Data from the seven remaining lakes all suggested upward trends in NO_3^- that were not significant (Table IV). Additionally, plots of temporal NO_3^- patterns indicate that the primary change in these lakes is in their spring values (Figure 10) and that baseflow NO_3^- concentrations are relatively unchanged. This analysis suggests that these lakes are in Stage 1 of watershed N loss and that their condition is worsening.

In low-ANC streams of the Catskills increases in NO_3^-, base-cation dilution, and high-baseline SO_4^{2-} concentrations all contribute to acidic episodes (97). In Biscuit Brook, an intensively studied headwater stream in the Catskills, concentrations of NO_3^- approach or exceed those of SO_4^{2-} during episodes (Figure 11; ref. 127). Values for the ratio of $NO_3^-:(NO_3^- + SO_4^{2-})$, as presented in Table II, illustrate both the general importance of NO_3^- to the acid–base dynamics of this stream and the increase in importance of NO_3^- during high-flow events (Figure 11). Nitrate concen-

Table IV. Trends in NO_3^- Concentrations for Adirondack Long-Term Monitoring Lakes

Lake Name	n^a	Change in NO_3^{-b} (μequiv/L per year)	p
Arbutus Lake	90	**+1.06**	**<0.01**
Big Moose Lake	98	+0.92	0.20
Black Lake	99	+0.00	0.60
Bubb Lake	99	+0.32	0.43
Cascade Lake	98	+0.15	0.65
Clear Pond	97	**+0.48**	**0.02**
Constable Pond	99	**+1.65**	**0.03**
Dart Lake	99	**+0.90**	**0.08**
Heart Lake	98	**+0.83**	**0.01**
Lake Rondaxe	99	**+0.72**	**0.09**
Little Echo Pond	97	+0.00	0.23
Moss Lake	99	+0.42	0.30
Otter Pond	94	**+1.33**	**0.02**
Squash Pond	93	**+1.80**	**0.03**
West Pond	99	**+0.41**	**0.07**
Windfall Lake	99	+0.46	0.40

NOTE: Slopes are calculated from seasonal Kendall tau test (*204, 205*). Trends with slopes significantly different from 0 ($p < 0.10$) are shown in bold print.
[a]Number of individual observations; the period of record for most sites is from June 1982 to December 1990.
[b]Slopes are medians of the differences between all possible pairs of sequential observations within the same month (*204*).
SOURCE: Data are taken from reference 157.

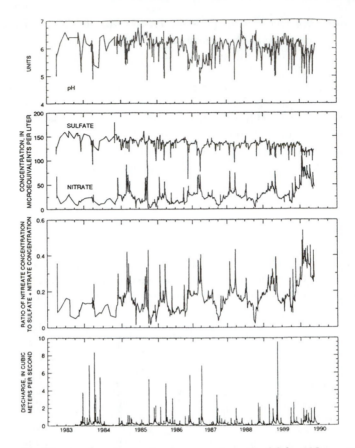

Figure 11. Temporal patterns in stream-water pH, SO_4^{2-}, NO_3^-, ratio of $NO_3^-:(NO_3^- + SO_4^{2-})$, and stream discharge at Biscuit Brook in the Catskill Mountains, 1983–1990. All chemical variables undergo strong seasonality, with strong dependence on stream discharge. As in Adirondack lakes, minima in ANC are coincident with maxima in NO_3^-. Values for the ratio of $NO_3^-:(NO_3^- + SO_4^{2-})$ approach or exceed 0.5 during episodes and indicate that NO_3^- can have as important an acidifying influence as SO_4^{2-} during high-flow events. Significant increases ($p < 0.05$ in seasonal Kendall tau trend test) in NO_3^- concentration ($+2$ µequiv/L per year) and $NO_3^-:(NO_3^- + SO_4^{2-})$ ($+0.014$) were recorded during the 1980s, indicating the increasing importance of NO_3^- both in absolute terms and in comparison to SO_4^{2-}. (Reproduced with permission from reference 127. Copyright 1992 American Geophysical Union.)

trations remain elevated (>20 µequiv/L) throughout all seasons, except for a short period during the autumn. This seasonal pattern has two important implications. First, it suggests that the watershed of this headwater stream (as well as those of seven other headwater LTM streams; 158) has reached Stage 2 of watershed N loss, as indicated by elevated groundwater and stream baseflow NO_3^- concentrations. Second, it suggests that the N demand on the part of the watershed exceeds the N supply only during a short period immediately following leaffall in the autumn, when the major demand for N is on the part of decomposers rather than plants (127).

Historical data are available from 19 large streams (third and fourth order) in the Catskill Mountains, some of which have been monitored since early in this century (79, 97). Trend analyses indicate that NO_3^- concentrations have increased in all of the streams (Table V), with most of the increase occurring in the past 2 decades (97, 127). These increases are not attributable to watershed anthropogenic sources of N (i.e., point and nonpoint sources of pollution) and are similar to trends observed in eight headwater streams monitored in the 1980s as part of the U.S. EPA LTM project (Table V) (97, 127, 158).

However, the seasonal NO_3^- pattern of the large streams differs from that of Biscuit Brook and the LTM streams. In the large streams, the primary change in NO_3^- has been in episodic concentrations. At four historical Catskill sites where stream discharge data are available, for example, the relationships between NO_3^- concentration and discharge were steeper in the 1980s than in the past (Figure 12). Most of the increase in NO_3^- has occurred at high flows. Apparently these streams have progressed from Stage 0 in the early part of the record to Stage 1 in the 1970s and 1980s. Thus far only small increases in baseflow NO_3^- concentrations have occurred. If these increases continue, however, as they apparently have at Biscuit Brook and the Catskill LTM streams, then important documentation of the progression from Stage 0 to Stage 1 to Stage 2 would exist; further monitoring should verify whether such a sequence is taking place. Taken in all, monitoring data from Catskill streams suggest that the region currently contains a mixture of Stage 1 and Stage 2 watersheds, and that N loss from all watersheds is increasing with time.

Additional evidence indicates increasing NO_3^- concentrations in the Northeast. Kramer et al. (159) reported on NO_3^- trends for four northeastern streams in the U.S. Geological Survey (USGS) Bench-Mark Stream network. Excluding data from one stream that drains a disturbed watershed, two of the three streams (Esopus Creek in the Catskill Mountains and Young Woman's Creek in Pennsylvania) exhibit increasing trends of about $+1$ μequiv/L of NO_3^- per year for the period 1967 to 1985. Smith et al. (160) examined trends in NO_3^- data from 383 stream locations in the United States collected between 1974 and 1981. They reported increases at 167 sites, especially east of the 100th meridian. Many of the increasing trends could be attributed to increased use of fertilizers in agricultural areas, particularly in the Midwest. In addition to agricultural runoff, Smith et al. identified atmospheric deposition as a major source of NO_3^- in surface waters, particularly in forested basins of the East (e.g., New England and the mid-Atlantic) and Upper Midwest. Despite widespread use of fertilizers in most of the regions covered by the Smith et al. study, they found a high degree of

Table V. Changes in NO_3^- Concentration in Historical Monitoring Streams and U.S. EPA Long-Term Monitoring Streams in the Catskill Mountains of New York

Site	Before 1945	1945–1970	After 1970
Historical sites			
Batavia Kill[a]	+0.24	+0.21	+0.28
Bear Kill above Grand Gorge[b]	—	—	+0.70
Bear Kill above Hardenbergh Falls	+0.34	—	—
Beaver Kill[a]	+0.05	+0.10	+1.76
Birch Creek above Pine Hill	—	+0.60	+2.68
Birch Creek at Pine Hill	−0.01	+0.68	+0.73
Biscuit Brook			
Bush Kill	+0.11	+0.00	+2.28
Bushnellville Creek[a]	+0.04	+0.25	+1.57
East Branch Neversink, Headwaters			
East Branch Neversink, Midlength			
Esopus Creek above Big Indian	+0.08	—	—
Esopus Creek below Big Indian	−0.16	−0.01	+1.98
Esopus Creek at Coldbrook	+0.24	−0.08	+2.00
High Falls Brook			
Hollow Tree Brook			
Little Beaver Kill[a]	+0.00	+0.01	+0.85
Manor Kill	−0.12	−0.55	+0.97
Neversink River	—	+0.33	+1.28
Rondout Creek	—	+0.00	+1.79
Schoharie Creek at Prattsville	+0.64	−0.13	+1.93
Stony Clove Creek[a]	−0.00	+0.08	+3.77
West Kill	+0.19	—	—
Woodland Creek[a]	+0.02	+0.08	+3.95
Long-term monitoring sites (current)		*Since 1983*	
Beaver Kill		n.s.	
Biscuit Brook		+2.03	
East Branch Neversink, Headwaters		n.s.	
East Branch Neversink, Midlength		+1.28	
High Falls Brook		+1.90	
Hollow Tree Brook		+2.20	
Rondout Creek		+2.93	
Woodland Creek		n.s.	

NOTE: — indicates insufficient data for analysis; n.s. indicates trend is not significant.
[a]Data for these sites are available only for periods before 1945 and from 1977 to 1979. Trends reported for the periods of missing data are based on regression lines for the entire data set; median values cannot be listed.
[b]Data are available for fewer than 2 years in one or more time periods at this site. Trends were not calculated during these time periods at this site, but median values and sample sizes are listed.
SOURCE: Reproduced from reference 127. Copyright 1992 American Geophysical Union.

correlation between stream basin yield of NO_3^- and rates of N deposition. The significance of this correlation should be strongly questioned, however, given the overwhelming importance of fertilizer use in this study.

A cautionary note in the interpretation of long-term N trends is introduced by examination of long-term data from streams at the Hubbard Brook

Experimental Forest (HBEF). Data from control Watershed No. 6 for 1963–1977 suggested a strongly increasing trend in NO_3^- (*161*). These data have been used to suggest that the HBEF watersheds are undergoing N saturation (*25*). Examination of the entire 23-year record (1965–1983) from Watershed No. 6, however, shows no long-term trend (*154, 155*). This result emphasizes the importance of examining N processes in a truly long-term context. Although some of the data reported here for the Catskill Mountains can be considered truly long-term (up to 65 years of record), data for the Adirondack Mountains (*157*) and other areas of the United States (*160*) span only 1–2 decades and should be interpreted with caution.

Finally, no significant effect of NO_3^- on episodic acidification has been observed in some areas of the Northeast. Morgan and Good (*162*) reported data on 10 streams in the New Jersey Pine Barrens and found mean annual NO_3^- greater than 1 μequiv/L only in residential and agricultural watersheds. Importantly, Swistock et al. (*163*) and Sharpe et al. (*164–166*) reported data on episodic acidification of several streams in the Laurel Hill area of southwestern Pennsylvania. They found that NO_3^- played only a minor role in stream acidification and fish kills. Rates of N deposition in this region of Pennsylvania are among the highest in the United States, yet these watersheds appear to remain at Stage 0 of watershed N loss. As mentioned earlier, the three characteristics that predispose watersheds to N saturation are elevated N deposition, advanced stand age, and high soil-N pools. Important information could be gained from an examination of the latter two factors at these seemingly anomalous sites in southwestern Pennsylvania.

The Southeastern United States. There are far fewer sites with sufficiently intensive data to determine their watershed N loss status in the Southeast than in the Northeast (Figure 13). The spatial coverage of sites in the Southeast with seasonal data is poor, and few sites have the long-term data needed to detect trends in NO_3^- concentrations.

One important exception is Control Watershed No. 4 at the Fernow Experimental Watershed near Parsons, West Virginia. Edwards and Helvey (*167*) reported data on stream chemistry in Watershed No. 4 for 1971–1987. Their data indicate that there has been a substantial increase in the loss of NO_3^- from this watershed over the past 2 decades (Figure 14b) and that mean annual NO_3^- concentrations have increased at the rate of approximately +3 μequiv/L per year. The apparent decrease in mean annual NO_3^- in the late 1980s is not significant and is primarily driven by hydrological changes. Important analytical changes occurred during the period of record at this site [NO_3^- concentrations were measured by the Hach method (*168*) prior to 1981 and by ion chromatography thereafter (*167*)], however, and the trend data should be interpreted with caution. Nonetheless, intensive data measured with consistent methods during the 1980s suggest that baseflow NO_3^- concentrations are currently as high as or slightly higher than those in deposition and that episodic concentrations are somewhat higher yet (Figure

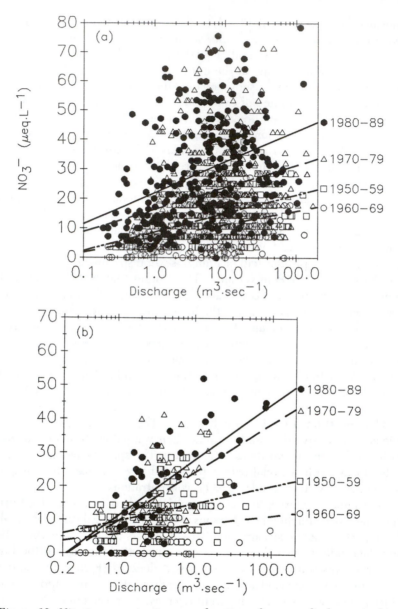

Figure 12. Nitrate concentration as a function of stream discharge in four Catskill streams during 4 decades of data collection, 1950–1989: a, Schoharie Creek at Prattsville; b, Neversink River at Claryville. Regression lines for each decade are from least-squares regression of concentration on the log of stream discharge. All slopes are significantly different from 0 (p < 0.01) except for Rondout Creek in the 1960s. All sites indicate that NO₃⁻ concentrations at high discharges were higher in the 1970s and 1980s than in previous decades. Little change is evident in baseflow NO₃⁻ concentrations. The stability suggests that these sites are at Stage 1 of watershed N loss but that episodic conditions are worsening. (Reproduced with permission from reference 127. Copyright 1992 American Geophysical Union.)

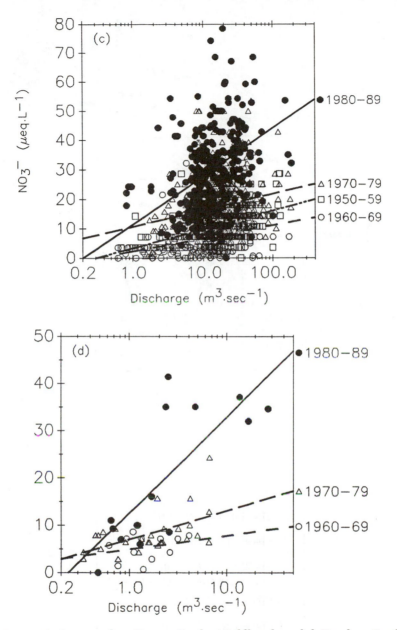

Figure 12. Continued. c, Esopus Creek at Coldbrook; and d, Rondout Creek at Lowes Corners.

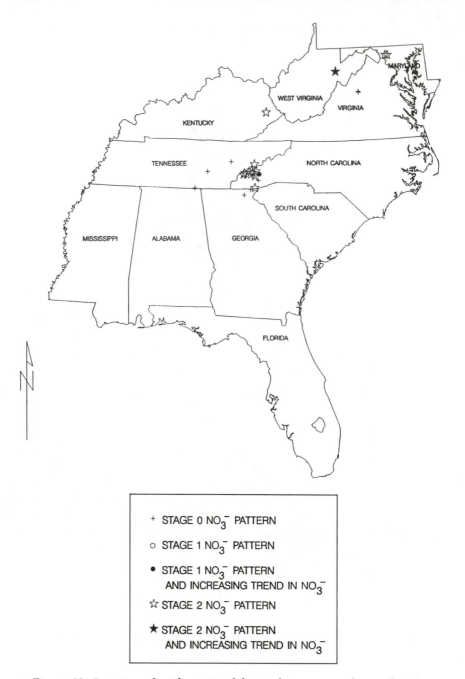

Figure 13. Location of acid-sensitive lakes and streams in the southeastern United States where the importance of NO₃⁻ to seasonal water chemistry can be determined. Only data from undisturbed watersheds are included. (Data are from references 99, 109, 159, 167, 170–172, 174–176, 195, 196.)

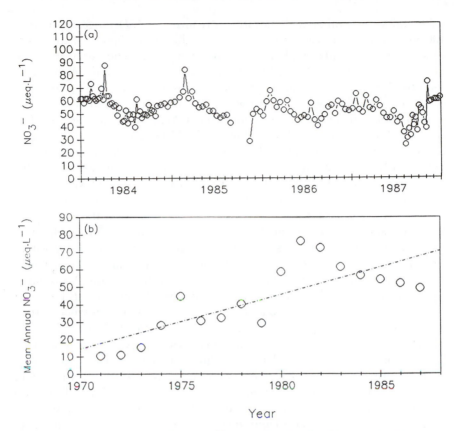

Figure 14. Stream-water chemistry for control watershed No. 4 at the Fernow Experimental Forest in West Virginia. a, Seasonal pattern in NO_3^- indicates both elevated baseflow concentrations and episodic increases during high-flow events in the spring and thus suggests that this watershed has reached Stage 2 of watershed N loss. b, The long-term trend suggests that substantial changes in NO_3^- export have occurred over the past 2 decades. (Data are from reference 167.)

14a). The Fernow site may be the best available example of a watershed at Stage 2 of watershed N loss. This suggestion is strengthened by the results of fertilizer experiments that were carried out in the 1970s at a nearby Fernow watershed. Edwards et al. (*169*) reported that an application of ammonium nitrate (at 336 kg/ha of N) resulted in large increases in stream-water exports, but that the largest portion of N loss was measured in deep seepage. This experimental result confirms that groundwater N losses can be substantial in Stage 2 watersheds.

The highest recorded NO_3^- concentrations in streams draining undisturbed watersheds in the United States come from the Great Smoky Mountains in Tennessee and North Carolina. Nitrate concentrations in Raven Fork

(170) and several other high-elevation streams (99, 171) range from 50 to 100 μequiv/L. In all cases they are comparable to, or higher than, SO_4^{2-} concentrations (see also Table II). In a survey of stream chemistry at a large number of sites in the Great Smoky Mountains, Silsbee and Larson (172) reported baseflow NO_3^- concentrations ranging from 0.2 to 90 μequiv/L. NO_3^- concentrations were highest at higher elevations and in areas of old-growth spruce–fir forest that have never been logged. In many cases NO_3^- concentrations in streams of the Great Smoky Mountains are higher than N concentrations in deposition. This comparison implies both that rates of biological N uptake are low and that mineralization rates are high (173). This pattern also indicates watersheds that have reached Stage 3 of watershed N loss. At this point watersheds may become sources, rather than sinks, for N. As mentioned earlier, the suggestion that watersheds in this area are at Stage 3 is supported by results from soil lysimeter studies, in which NO_3^- fluxes in deep soil lysimeters were higher than N inputs from deposition and throughfall (100). Unfortunately, few data are available to suggest the original source of the N now being mineralized in this region. The data of Silsbee and Larson (172) suggest strongly that forest maturation is linked to the process of NO_3^- leakage from Great Smoky Mountain watersheds. Mineralization of soil N appears to be high only in old-growth forests (171).

Small increases in NO_3^- concentrations during hydrological events have been recorded at sites in a few other areas of the Southeast, including northeastern Georgia (174), where maximum concentrations were ~12 μequiv/L. Cosby et al. (175) examined 7 years' worth of data from two streams in Virginia and found no evidence of NO_3^- episodes; NO_3^- concentrations are always less than 15 μequiv/L in these streams. Swank and Waide (176) reported data from seven undisturbed watersheds at the Coweeta Hydrologic Laboratory in North Carolina, where the volume-weighted mean concentrations of NO_3^- were less than 1.5 μequiv/L. In all of these cases, undisturbed watersheds exhibit Stage 0 patterns of watershed N loss.

The Western United States. Several studies have reported the existence of NO_3^- episodes in western United States, including the North Cascades (177, 178) and Sierra Nevada Mountains (179) (Figure 15). In general, the maximum NO_3^- concentrations observed in the West are less than 15 μequiv/L, substantially lower than in most of the eastern United States. Lakes in the mountainous West, however, tend to be much more dilute and therefore more sensitive to acidic deposition than lakes in the East. Of lakes in the Sierra Nevada Mountains, for example, 39% have ANC values less than 50 μequiv/L, as do 26% of the lakes in the Oregon Cascades and 17% of the lakes in the North Cascades (143).

Rates of N deposition are also much lower in the West (Table I), and episodic NO_3^- concentrations of 15 μequiv/L should be placed in the context of mean concentrations of NO_3^- and NH_4^+ in deposition of 5.0 and 5.4

Figure 15. Location of acid-sensitive lakes and streams in the western United States where the importance of NO_3^- to seasonal water chemistry can be determined. Only data from undisturbed watersheds are included. (Data are from references 109, 117, 159, 177, 179, 197–199.)

μequiv/L, respectively (7). Combined with base-cation dilution and small concentrations of SO_4^{2-}, the NO_3^- increases observed during episodes at Emerald Lake in the Sierra Nevada Mountains have been sufficient to drive ANC to 0 on two occasions in the past decade (7, *180*). Data from the outflow at Emerald Lake in 1986 and 1987 (Figure 16) indicate that minimum ANC values are coincident with maximum concentrations of NO_3^- and diluted base-cation concentrations resulting from high discharge. However, at no

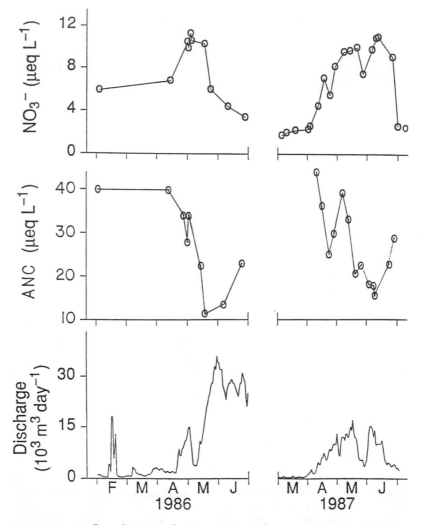

Figure 16. Outflow chemistry from two snowmelt seasons (1986 and 1987) at Emerald Lake, a high-elevation lake in the Sierra Nevada Mountains of California. Maximum NO_3^- concentrations are coincident with ANC minima during the early stages of snowmelt in 1986 and with a rain-on-snow event in 1987. Nitrate episodes are smaller in magnitude than at sites in the eastern United States, but western lakes may be more susceptible to episodic acidification because they have a lower baseline acid-neutralizing capacity than most eastern lakes. (Reproduced with permission from reference 180. Copyright 1991 American Geophysical Union.)

time has the pH of Emerald Lake fallen below 5.5, a level commonly considered the threshold for injury to fish populations. ANC values of 0 can be caused by base-cation dilution alone (a natural process).

The state of episodic acidification in the Sierra Nevada Mountains (and the rest of the West) therefore remains uncertain because few data exist and the data that are available indicate ANC depressions to a value of 0 μequiv/L, but not below. For the time being, most western sites would be classified as Stage 0 watersheds. However, in systems as acid-sensitive as those at high elevations in the West, episodic acidification resulting from NO_3^- may be possible in the watersheds that are only seasonally N-saturated.

Links Between Nitrogen Loss and Nitrogen Deposition

The presence of elevated NO_3^- concentrations in streams draining undisturbed forested watersheds does not necessarily implicate N deposition as the source of the N being exported. Currently, little direct evidence links N deposition with either elevated baseflow NO_3^- concentrations or elevated episodic NO_3^- concentrations and acidic episodes. This lack of evidence stems at least partially from a lack of appropriate data to link deposition to stream-water NO_3^- concentrations. High concentrations of NO_3^- during snowmelt, for example, may result when NO_3^- stored in the snowpack during the winter months is released while the forest is still dormant. The reduced biological activity typical of the winter months creates less demand for N, and snowpack NO_3^- may simply run off without entering the N cycle of the forest or watershed. Several mechanisms, however, amplify the signal produced by atmospheric deposition of N to snowpacks. These mechanisms, which will be discussed in more detail, include

- preferential elution of NO_3^- during the early stages of snowmelt;
- nitrification of NH_4^+ to NO_3^- within soils or the snowpack;
- dry deposition of N to the snowpack before melting; and
- mineralization of soil-N pools over the winter.

In areas with large snowpacks (e.g., much of the Northeast and all of the mountainous West), ions have been shown to drain from the pack in the early stages of snowmelt. This process leads to concentrations that are much higher than the average concentration of the snowpack itself (82). Differential elution of acid anions (like NO_3^-) during the initial stages of snowmelt has been shown to be responsible for the elevated NO_3^- concentrations observed in parts of Scandinavia (81), Canada (82), the Adirondack Mountains (181), the Midwest (182), and the Sierra Nevada Mountains (180). Ammonium deposited to the snowpack (either wet or dry deposition) can subsequently be nitrified to NO_3^- in soils or while still in the snowpack. This reaction would produce NO_3^- concentrations elevated over those calculated from NO_3^- deposition alone (83, 110, 114, 183). Dry deposition of N compounds to the snowpack may also be an important source of NO_3^- in snowmelt water (110, 183). Jeffries (82) presented a review of snowpack storage and release of pollutants during snowmelt.

Some evidence does exist that mechanisms other than atmospheric deposition contribute to NO_3^- episodes, at least on a small scale. Rascher et al. (43), for example, showed that mineralization of organic matter in the soil during the winter months and subsequent nitrification contribute substantially to snowmelt NO_3^- concentrations at one site in the Adirondacks. Schaefer and Driscoll (83) suggested that a similar phenomenon contributes to NO_3^- pulses during snowmelt at 11 Adirondack lakes and that the contribution from mineralization is greater in low-ANC and acidic lakes. Murdoch and Stoddard (127) presented similar results for streams in the Catskill Mountains. Stottlemyer and Toczydlowski (184) also reported that mineralization contributes to snowmelt NO_3^- at a site on the upper peninsula of Michigan. It is not currently known how widespread this phenomenon is.

Some question remains of whether N produced from mineralization is indirectly supplied by atmospheric deposition. Mineralization recycles N from leaflitter, a portion of which undoubtedly originates as atmospheric deposition. Friedland et al. (18) calculated that roughly 30% of the annual N demand of spruce–fir forests in the Adirondack Mountains can be met by deposition alone, setting an upper limit to the proportion of mineralized N that may come indirectly from deposition on an annual basis. In addition, chronic N deposition may also influence the loss of mineralized N by altering the timing of N cycle and providing an opportunity for N to leave the watershed during snowmelt (see discussion of Stage 1), by contributing to the pool of organic N (through fertilization) available for mineralization, or by increasing the rate of mineralization through fertilization (the so-called "priming effect"; 185).

Many of these data suggest that NO_3^- episodes are more severe now than they were in the past, and in some cases that baseflow NO_3^- concentrations have increased. These surface-water N increases have occurred at a time when N deposition has been relatively unchanged in the eastern United States (186–188). If we accept the idea that increases in lake and stream-water NO_3^- concentrations are evidence that N saturation of watersheds is progressing, then current data suggest that present levels of N deposition (350–700 equiv/ha per year) are sufficient to drive this progression in the Adirondack Mountains, the Catskill Mountains, the mountains of West Virginia, and the Great Smoky Mountains. Data from the National Stream Survey (96) suggest a strong correlation between concentrations of streamwater N (NO_3^- + NH_4^+) at spring baseflow and levels of wet N deposition (NO_3^- + NH_4^+) in each of the NSS regions (Figure 17a). The only exception

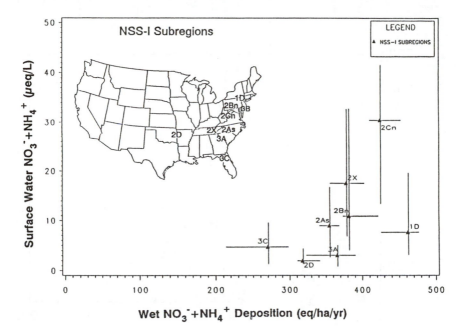

Figure 17a. Relationship between median wet deposition of N (NO_3^- + NH_4^+) and median surface-water N (NO_3^- + NH_4^+) concentrations for physiographical districts within the National Stream Survey that have minimal agricultural activity. [Subregions are Poconos–Catskills (1D), Southern Blue Ridge Province (2As), Valley and Ridge Province (2Bn), Northern Appalachians (2Cn), Ozarks–Ouachitas (2D), Southern Appalachians (2X), Piedmont (3A), mid-Atlantic Coastal Plain (3B), and Florida (3C)]. (Panel a is reproduced with permission from reference 96. Copyright 1991 American Geophysical Union.)

to this relationship is the Pocono–Catskill region, where N deposition is highest (450 equiv/ha per year) but where stream-water N concentrations fall below what is expected on the basis of results from the other regions. The median stream-water NO_3^- value for the Catskills alone (Table II) is 29 μequiv/L (79). This value fits the usual relationship much more closely, suggesting that watersheds in the southern portion of this region (the Poconos) are retaining N more strongly than the northern portion. This interpretation is also consistent with the low loss rates of N from watersheds in southwestern Pennsylvania that were discussed earlier.

Driscoll et al. (155) summarized the relationship between N export and N deposition that is indicated by input–output budget data from a large number of watersheds in the United States and Canada; these data are augmented in Figure 17b with results from recently published reports. Driscoll et al. (155) stressed that the data illustrated in Figure 17b were collected by using widely differing methods and various time scales (from 1 year to

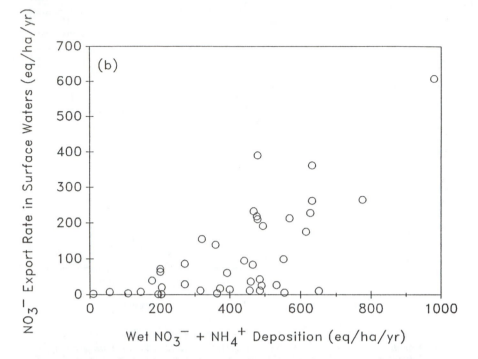

Figure 17b. Relationship between wet deposition of N (NO_3^- + NH_4^+) and rate of N export for watershed studies throughout North America. Sites with significant internal sources of N (e.g., from alder trees) have been excluded. (Original data in panel b are from reference 155; additional data are from references 117, 119, 127, 164, 167, 171, 174, 175, 179, 192, 193, 195.)

several decades). However, both relationships illustrated in Figure 17 indicate that watersheds in many regions of North America are retaining less than 75% of the N that enters them and that the amount of N being leaked from these watersheds is higher in areas where N deposition is highest. This pattern is consistent with what we would expect if large areas of the eastern United States were experiencing the early stages of N saturation, but is not proof of this situation. Both analyses suggest a threshold value of wet deposition of N (~300 equiv/ha per year), above which substantial watershed losses of N might begin to occur. Because dry deposition is not included in these analyses, we can only speculate that threshold values for total deposition (wet + dry) would be substantially higher.

Conclusions

Considerable evidence indicates that high rates of N deposition are contributing to the degradation of water quality in several regions of the United States. The primary effect of deposition appears to be the episodic acidification of surface waters. There is no hard evidence of chronic acidification by NO_3^- in the United States (streams in the Great Smoky Mountains may be an exception, but few high-quality data have been published for this area), and no conceptual mechanism exists for deposition to contribute to the long-term eutrophication of lakes.

When seasonal data on stream and lake chemistry are available, it is possible to place sites into a progressive scheme of watershed N loss according to the stages of N saturation described for terrestrial ecosystems by Aber et al. (*1*). Sites that would be classified at later stages of watershed N loss tend to occur in areas of high N deposition, but substantial variability exists. Apparently high rates of N deposition are necessary, but not sufficient, conditions leading to high rates of watershed N loss. Stand age may also contribute to long-term changes in the ability of watersheds to retain N; watersheds with old-growth forests leak more NO_3^- in both the Northeast (*19*) and Southeast (*171*). Another possible source is high levels of soil N (but few data are available).

Sites from three subregions discussed in detail in this chapter bear further examination with respect to stand age and soil N status. Many lakes in the Adirondack Mountains exhibit Stage 1 patterns of watershed N loss (i.e., Figures 5 and 10); forests in the same area were last harvested in the early 1900s (*189*). Streams in the Catskill region exhibit both Stage 1 and Stage 2 patterns (Figures 11 and 12); forests in the Catskill Mountains were also last harvested in the early 1900s. Control watershed No. 4 at Fernow, West Virginia, exhibits the dramatically elevated baseflow NO_3^- concentrations typical of Stage 2 watersheds (Figure 14); forests at Fernow were selectively logged at the turn of the century (*167*). The forests in the watersheds of all these sites were last harvested at more or less the same time, roughly 90 years ago. Yet the streams and lakes that drain them exhibit different stages of watershed N loss.

The dominant spatial trend evident in this analysis is for sites further south, and sites with higher deposition rates, to exhibit more advanced stages of watershed N loss. The Fernow site may be closer to forest maturation because it experiences a longer growing season than sites to the north; it also is in the region of the country that receives the highest rates of N deposition. Forests at the Adirondack sites might be expected to take longer to reach maturation because of their shorter growing seasons, and they also receive lower rates of N deposition (Table I). Detailed examination of these three sites suggests that both N deposition and stand age play roles in determining what stage of watershed N loss a site has reached. This analysis, however, ignores information from sites in southwestern Pennsylvania where forests are roughly the same age as in these three subregions, but where streams exhibit Stage 0 patterns of seasonal NO_3^- concentrations despite very high rates of N deposition. Further examination of these apparently anomalous sites, especially with regard to their soil-N status, might yield important information about the long-term ability of forested watersheds to retain atmospheric N.

Finally, the most convincing evidence for United States watersheds showing signs of N saturation comes from sites where long-term trends and changes in the seasonal patterns of NO_3^- concentrations can be identified. Data from numerous site-specific studies in the Adirondack, Catskill, and mid-Appalachian mountains show both that long-term increases in N loss from forested watersheds have occurred (primarily in the past 2 decades) and that sites progress through stages like those described here. Early signs of watershed N loss are episodic increases in NO_3^-, especially during the snowmelt season. Later stages of watershed N loss are characterized by increases in groundwater NO_3^- concentrations, which create elevated concentrations in streams at baseflow and damped seasonal fluctuations in NO_3^-.

Acknowledgments

I thank John Aber, Helga Van Miegroet, Charles Driscoll, M. Robbins Church, and Larry Baker for their reviews of various iterations of this chapter; their comments made important contributions to the overall quality. I also thank Mary Beth Adams (Fernow Experimental Forest, Northeastern Forest Experiment Station), Charles Driscoll (Syracuse University), and Peter Murdoch (U.S. Geological Survey, Albany, NY) for the use of their data. The preparation of this document was funded by the U.S. Environmental Protection Agency under Contract No. 68–C8–0006 with ManTech Environmental Technology, Inc. It has been subjected to the Agency's peer and administrative review and approved for publication.

References

1. Aber, J. D.; Nadelhoffer, K. J.; Steudler, P.; Melillo, J. M. *BioScience* **1989**, *39*, 378–386.
2. Lindberg, S. E.; Lovett, G. M.; Richter, D. D.; Johnson, D. W. *Science (Washington, D.C.)* **1986**, *231*, 141–144.
3. Rowland, A.; Murray, A. J. S.; Wellburn, A. R. *Rev. Environ. Health Vol. V* **1985**, *4*, 495–343.
4. Parker, G. G. *Adv. Ecol. Res.* **1983**, *13*, 57–133.
5. Olson, R. K.; Reiners, W. A.; Lovett, G. M. *Biogeochemistry* **1985**, *1*, 361–373.
6. Galloway, J. N.; Likens, G. E.; Keene, W. C.; Miller, J. M. *J. Geophys. Res.* **1982**, *87*, 8771–8786.
7. Williams, M. W.; Melack, J. M. *Water Resour. Res.* **1991**, *27*, 1563–1574.
8. Linsey, G. A.; Schindler, D. W.; Stainton, M. P. *Can. J. Fish. Aquat. Sci.* **1987**, *44*, 206–214.
9. Stensland, G. J.; Whelpdale, D. M.; Oehlert, G. In *Acid Deposition: Long Term Trends*; National Academy Press: Washington, DC, 1986.
10. Baker, L. A. In *Acidic Deposition and Aquatic Ecosystems: Regional Case Studies*; Charles, D. F., Ed.; Springer-Verlag: New York, 1991; pp 645–652.

11. Sisterson, D. L.; Bowersox, V. C.; Olsen, A. R.; Meyers, T. P.; Vong, R. L. *Deposition Monitoring: Methods and Results;* State of Science/Technology Report No. 6; National Acid Precipitation Assessment Program: Washington, DC, 1990.

12. Altwicker, E. R.; Shanaghan, P. E.; Johannes, A. H. *Water Air Soil Pollut.* **1986,** *28,* 71–88.

13. Manny, B. A.; Owens, R. W. *J. Great Lakes Res.* **1983,** *9,* 403–420.

14. Feller, M. C. In *Forest Hydrology and Watershed Management;* IAHS–AISH: Vancouver, British Columbia, Canada, 1987; pp 33–47.

15. Moore, I. D.; Nuckols, J. R. *J. Hydrol.* **1984,** *74,* 81–103.

16. Lovett, G. M.; Reiners, W. A.; Olsen, R. K. *Science (Washington, D.C.)* **1982,** *218,* 1303–1304.

17. Weathers, K. C.; Likens, G. E.; Bormann, F. H.; Eaton, J. S.; Bowden, W. B.; Anderson, J. L.; Cass, D. A.; Galloway, J. N.; Keene, W. C.; Kimball, K. D.; Huth, P.; Smiley, D. *Nature (London)* **1986,** *319,* 657–658.

18. Friedland, A. J.; Miller, E. K.; Battles, J. J.; Thorne, J. F. *Biogeochemistry* **1991,** *14,* 31–55.

19. Vitousek, P. M. *Ecol. Monogr.* **1977,** 47, 65–87.

20. Cole, D. W.; Rapp, M. In *Dynamic Properties of Forest Ecosystems;* Reichle, D. E., Ed.; Cambridge University Press: New York, 1981; pp 341–409.

21. Vitousek, P. M.; Howarth, R. W. *Biogeochemistry* **1991,** *13,* 87–115.

22. Muller, R. N.; Bormann, F. H. *Science (Washington, D.C.)* **1976,** *193,* 1126–1128.

23. Zak, D.; Groffman, P. R.; Pregitzer, K. S.; Christensen, S.; Tiedje, J. M. *Ecology* **1990,** *71,* 651–656.

24. Johnson, D. W. *J. Environ. Qual.* **1992,** *21,* 1–12.

25. Agren, G. I.; Bosatta, E. *Environ. Pollut.* **1988,** *54,* 185–197.

26. Kelly, C. A.; Rudd, J. W. M.; Schindler, D. W. *Water Air Soil Pollut.* **1990,** *50,* 49–61.

27. Reynolds, C. S. *The Ecology of Freshwater Phytoplankton;* Cambridge University Press: Cambridge, England, 1984.

28. McCarthy, J. J. *Can. Bull. Fish. Aquat. Sci.* **1981,** *210,* 211–233.

29. Vincent, W. F. *J. Plankton Res.* **1981,** *3,* 685–697.

30. Suttle, C. A.; Harrison, P. J. *Limnol. Oceanogr.* **1988,** *33,* 186–202.

31. Goldman, J. C.; Glibert, P. M. *Limnol. Oceanogr.* **1982,** 27, 814–827.

32. Button, D. K. *Microbiol. Rev.* **1985,** 49, 270–297.

33. Kilham, P.; Hecky, R. E. *Limnol. Oceanogr.* **1988,** *33,* 776–795.

34. Hecky, R. E.; Kilham, P. *Limnol. Oceanogr.* **1988,** *33,* 796–822.

35. Nadelhoffer, K. J.; Aber, J. D.; Melillo, J. M. *Ecology* **1985,** *66,* 1377–1390.

36. Aber, J. D.; Melillo, J. M.; Nadelhoffer, K. J.; Pastor, J.; Boone, R. D. *Ecol. Appl.* **1991,** *1,* 303–315.

37. Melillo, J. M.; Naiman, R. J.; Aber, J. D.; Lindins, A. E. *Bull. Mar. Sci.* **1984,** *35,* 341–356.

38. Taylor, B. R.; Parkinson, D.; Parsons, W. F. J. *Ecology* **1989,** *70,* 97–104.

39. Hauhs, M.; Rost-Siebert, K.; Raben, G.; Paces, T.; Vigerust, B. In *The Role of Nitrogen in the Acidification of Soils and Surface Waters;* Malanchuk, J. L.; Nilsson, J., Eds.; Nordic Council of Ministers: Copenhagen, Denmark, 1989; Vol. 10, pp 5-1–5-37.

40. Likens, G. E.; Bormann, F. H.; Johnson, N. M.; Fisher, D. W.; Pierce, R. S. *Ecol. Monogr.* **1970,** *40,* 23–47.

41. Vitousek, P. M.; Grosz, J. R.; Grier, C. C.; Melillo, J. M.; Reiners, W. A.; Todd, R. L. *Science (Washington, D.C.)* **1979,** *204,* 469–474.

42. Driscoll, C. T.; Schaefer, D. A. In *The Role of Nitrogen in the Acidification of Soils and Surface Waters;* Malanchuk, J. L.; Nilsson, J., Eds.; Nordic Council of Ministers: Copenhagen, Denmark, 1989; Vol. 10, pp 4.1–4.12.
43. Rascher, C. M.; Driscoll, C. T.; Peters, N. E. *Biogeochemistry* **1987**, *3*, 209–224.
44. Novick, N. J.; Klein, T. M.; Alexander, M. *Water Air Soil Pollut.* **1984**, *23*, 317–330.
45. Van Breemen, N.; Burrough, P. A.; Velthort, E. J.; van Dobben, H. F.; de Wit, T.; Ridder, T. B.; Reigners, H. F. R. *Nature (London)* **1982**, *299*, 548–550.
46. Stams, A. J. M.; Booltink, H. W. G.; Jutke-Schipholt, I. J.; Beemsterboer, B.; Woittiez, J. R. W.; Van Breemen, N. *Biogeochemistry* **1991**, *13*, 241–255.
47. Montagnini, F.; Haines, B. L.; Swank, W. T.; Waide, J. B. *Can. J. For. Res.* **1989**, *19*, 1226–1234.
48. McNulty, S. G.; Aber, J. D.; McLellan, T. M.; Katt, S. M. *Ambio* **1990**, *19*, 38–40.
49. Schindler, D. W.; Turner, M. A.; Hesslein, R. H. *Biogeochemistry* **1985**, *1*, 117–133.
50. Schiff, S. L.; Anderson, R. F. *Can. J. Fish. Aquat. Sci.* **1987**, *44*, 173–187.
51. Van Breemen, M.; van Dijk, H. F. G. *Environ. Pollut.* **1988**, *54*, 249–274.
52. Rudd, J. W. M.; Kelly, C. A.; Schindler, D. W.; Turner, M. A. *Science (Washington, D.C.)* **1988**, *240*, 1515–1517.
53. Wetzel, R. G. *Limnology;* W. B. Sanders: Philadelphia, PA, 1983.
54. Payne, W. J. *Denitrification;* Wiley-Interscience: New York, 1981.
55. Post, W. M.; Pastor, J.; Zinke, P. J.; Stangenberger, A. G. *Nature (London)* **1985**, *317*, 613–616.
56. Melillo, J. M.; Aber, J. D.; Steudler, P. A.; Schimel, J. P. In *Environmental Biogeochemistry, Ecol. Bull. (Stockholm);* Hallberg, R., Ed.; Stockholm, Sweden, 1983; Vol. 35, pp 217–228.
57. Matson, P. A.; Vitousek, P. M. *BioScience* **1990**, *40*, 667–672.
58. Hahn, J.; Crutzen, P. J. *Philos. Trans. R. Soc. London Ser. B* **1982**, *296*, 521–541.
59. Klemedtsson, L.; Svensson, B. H. In *Critical Loads for Sulphur and Nitrogen;* Nilsson, J.; Grennfelt, P., Eds.; Nordic Council of Ministers: Copenhagen, Denmark, 1988; Vol. 15, pp 343–362.
60. Bowden, W. B.; Bormann, F. H. *Science (Washington, D.C.)* **1986**, *233*, 867–869.
61. Bowden, W. B. *Biogeochemistry* **1986**, *2*, 249–279.
62. Seitzinger, S. P. *Limnol. Oceanogr.* **1988**, *33*, 702–724.
63. Keeney, D. R.; Chen, R. L.; Graetz, D. A. *Nature (London)* **1971**, *233*, 66–67.
64. Rudd, J. W. M.; Kelly, C. A.; Schindler, D. W.; Turner, M. A. *Limnol. Oceanogr.* **1990**, *35*, 663–679.
65. Kaushik, N. K.; Robinson, J. B.; Sain, P.; Whitely, H. R.; Stammers, W. N. In *Canadian Symposium on Water Pollution Research;* Canadian Board of Fisheries and Aquatic Sciences: Toronto, Canada, 1975; pp 110–117.
66. Chatarpaul, L.; Robinson, R. B. In *Methodology for Biomass Determinations and Microbial Activities in Sediments;* ASTM STP 673; Litchfield, C. D.; Seyfried, P. I., Eds.; American Society for Testing and Materials: Washington, DC, 1979; pp 119–127.
67. Robinson, J. B.; Whitely, H. R.; Stammers, W. N.; Kaushik, N. K.; Sain, P. In *Best Management Practices for Agriculture and Silviculture;* Lohr, R. C., Ed.; Ann Arbor Science: Ann Arbor, MI, 1979.
68. Boring, L. R.; Swank, W. T.; Waide, J. B.; Henderson, G. S. *Biogeochemistry* **1988**, *6*, 119–159.

69. Binkley, D. *Forest Nutrition Management;* Wiley-Interscience: New York, 1986; p 290.
70. Franklin, J. F.; Dryness, C. T.; Moore, D. G.; Tarrant, R. F. In *Biology of Alder;* Franklin, J. F.; Trappe, J. M.; Tarrant, R. F.; Hansen, G. M., Eds.; U.S. Forest Service: Corvallis, OR, 1968; pp 157–172.
71. Van Miegroet, J.; Cole, D. W. *Soil Sci. Soc. Am. J.* **1985**, *49*, 1274–1279.
72. Schindler, D. W. *Science (Washington, D.C.)* **1977**, *195*, 260–262.
73. Howarth, R. W.; Marino, R.; Lane, J.; Cole, J. J. *Limnol. Oceanogr.* **1988**, *33*, 669–687.
74. Skeffington, R. A.; Wilson, E. *J. Environ. Pollut.* **1988**, *54*, 159–184.
75. Bormann, F. H.; Likens, G. E. *Am. Sci.* **1979**, *67*, 660–669.
76. Turner, R. S.; Cook, R. B.; Van Miegroet, H.; Johnson, D. W.; Elwood, J. W.; Bricker, O. P.; Lindberg, S. E.; Hornberger, G. M. *Watershed and Lake Processes Affecting Surface Water Acid-Base Chemistry;* Acidic Deposition: State of Science and Technology Report 10; National Acid Precipitation Assessment Program: Washington, DC, 1990.
77. Joslin, J. D.; Kelly, J. M.; Van Miegroet, H. *J. Environ. Qual.* **1992**, *21*, 12–30.
78. Foster, N. W.; Nicolson, J. A.; Hazlett, P. W. *J. Environ. Qual.* **1989**, *18*, 238–244.
79. Stoddard, J. L.; Murdoch, P. S. In *Acidic Deposition and Aquatic Ecosystems: Regional Case Studies;* Charles, D. F., Ed.; Springer-Verlag: New York, 1991; pp 237–271.
80. Binkley, D.; Driscoll, C. T.; Allen, H. L.; Schoeneberger, P.; McAvoy, D. *Acidic Deposition and Forest Soils: Context and Case Studies of the Southeastern United States;* Ecological Studies; Springer-Verlag: New York, 1989; Vol. 72, 149 pp.
81. Johannessen, M.; Henriksen, A. *Water Res.* **1978**, *14*, 615–619.
82. Jeffries, D. S. In *Acidic Precipitation: Soils, Aquatic Processes, and Lake Acidification;* Norton, S. A.; Lindberg, S. E.; Page, A. L., Eds.; Springer-Verlag: New York, 1990; Vol. 4, pp 107–132.
83. Schaefer, D. A.; Driscoll, C. T. *Water Air Soil Pollut.*, in press.
84. Pickett, S. T. A. In *Long-Term Studies in Ecology, Approaches and Alternatives;* Likens, G. E., Ed.; Springer-Verlag: New York, 1989; pp 110–135.
85. Sullivan, T. J. In *Acidic Deposition and Aquatic Ecosystems: Regional Case Studies;* Charles, D. F., Ed.; Springer-Verlag: New York, 1991; pp 615–639.
86. Sullivan, T. J.; Driscoll, C. T.; Gherini, S. A.; Munson, R. K.; Cook, R. B.; Charles, D. F.; Yatsko, C. P. *Nature (London)* **1989**, *338*, 408–410.
87. Sullivan, T. J.; Eilers, J. M.; Church, M. R.; Blick, D. J.; Eshleman, K. N.; Landers, D. H.; DeHaan, M. S. *Nature (London)* **1988**, *331*, 607–609.
88. Brakke, D. F.; Henriksen, A.; Norton, S. A. *Water Resour. Bull.* **1989**, *25*, 247–253.
89. Linthurst, R. A.; Landers, D. H.; Eilers, J. M.; Brakke, D. F.; Overton, W. S.; Meier, E. P.; Crowe, R. E. *Characteristics of Lakes in the Eastern United States: Population Descriptions and Physico–Chemical Relationships;* EPA–600/4–86/007a; U.S. Environmental Protection Agency: Washington, DC, 1986; Vol. 1.
90. Hutchinson, G. E. *Am. Sci.* **1973**, *61*, 269–279.
91. *National Academy of Sciences; Eutrophication: Causes, Consequences, Correctives;* National Academy Press: Washington, DC, 1969.
92. Likens, G. E., Ed. *Nutrients and Eutrophication: The Limiting-Nutrient Controversy;* American Society of Limnology and Oceanography: Lawrence, KS, 1972; Special Symposia Vol. 1.
93. Messer, J. J.; Ariss, C. W.; Baker, J. R.; Drouse, S. K.; Eshleman, K. N.;

Kinney, A. J.; Overton, W. S.; Sale, M. J.; Schonbrod, R. D. *Water Resour. Bull.* **1988**, *24*, 821–829.

94. Kaufmann, P. R.; Herlihy, A. T.; Elwood, J. W.; Mitch, M. E.; Overton, W. S.; Sale, M. J.; Messer, J. J.; Cougar, K. A.; Peck, D. V.; Reckhow, K. H.; Kinney, A. J.; Christie, S. J.; Brown, D. D.; Hagley, C. A.; Jager, H. I. *Chemical Characteristics of Streams in the Mid-Atlantic and Southeastern United States: Population Descriptions and Physico–Chemical Relationships;* EPA–600/3–88/021a; U.S. Environmental Protection Agency: Washington, DC, 1988; Vol. 1.

95. Henriksen, A. In *Critical Loads for Sulphur and Nitrogen;* Nilsson, J.; Grennfelt, P., Eds.; Nordic Council of Ministers: Copenhagen, Denmark, 1988; Vol. 15, pp 385–412.

96. Kaufmann, P. R.; Herlihy, A. T.; Mitch, M. E.; Messer, J. J.; Overton, W. S. *Water Resour. Res.* **1991**, *27*, 611–627.

97. Stoddard, J. L. *Water Resour. Res.* **1991**, *27*, 2855–2864.

98. Webb, J. R.; Cosby, B. J.; Galloway, J. N.; Hornberger, G. M. *Water Resour. Res.* **1989**, *25*, 1367–1377.

99. Cook, R. B.; Elwood, J. W.; Turner, R. R.; Bogle, M. A.; Mulholland, P. J.; Palumbo, A. V. *Water Air Soil Pollut.*, in press.

100. Johnson, D. W.; Van Miegroet, H.; Lindberg, S. E.; Todd, D. E.; Harrison, R. B. *Can. J. For. Res.* **1991**, *21*, 769–787.

101. Henriksen, A.; Brakke, D. F. *Water Air Soil Pollut.* **1988**, *42*, 183–201.

102. Hauhs, M. In *Acidic Precipitation;* Adriano, D. C.; Hauhs, M., Ed.; Springer-Verlag: New York, 1989; pp 275–305.

103. Henriksen, A.; Lien, L.; Truaen, T. S.; Sevaldrud, I. S.; Brakke, D. F. *Ambio* **1988**, *17*, 259–266.

104. Brown, D. J. A. *Environ. Pollut.* **1988**, *54*, 275–284.

105. James, B. R.; Riha, S. J. *Soil Sci. Soc. Am. J.* **1989**, *53*, 259–264.

106. Schuurkes, J. A. A. R. *Experientia* **1986**, *42*, 351–357.

107. Rosen, K. In *Critical Loads for Sulphur and Nitrogen;* Nilsson, J.; Grennfelt, P., Eds.; Nordic Council of Ministers: Copenhagen, Denmark, 1988; Vol. 15, pp 269–294.

108. Eshleman, K. N. *Water Resour. Res.* **1988**, *24*, 1118–1126.

109. Wigington, P. J.; Davies, T. D.; Tranter, M.; Eshleman, K. *Episodic Acidification of Surface Waters Due to Acidic Deposition;* Acidic Deposition: State of Science and Technology Report 12; National Acid Precipitation Assessment Program: Washington, DC, 1990.

110. Galloway, J. N.; Schofield, C. L.; Hendrey, G. R.; Peters, N. E.; Johannes, A. H. In *Ecological Impact of Acid Precipitation;* SNSF-project: Oslo, Norway, 1980; pp 264–265.

111. Sullivan, T. J.; Christopherson, N.; Muniz, I. P.; Seip, H. M.; Sullivan, P. D. *Nature (London)* **1986**, *323*, 324–327.

112. Johannessen, M.; Skartveit, A.; Wright, R. F. In *Ecological Impact of Acid Precipitation;* SNSF-project: Oslo, Norway, 1980; pp 224–225.

113. Driscoll, C. T.; Schafran, G. D. *Nature (London)* **1984**, *310*, 308–310.

114. Schofield, C. L.; Galloway, J. N.; Hendrey, G. R. *Water Air Soil Pollut.* **1985**, *26*, 403–423.

115. Driscoll, C. T.; Newton, R. M. *Environ. Sci. Technol.* **1985**, *19*, 1018–1024.

116. Peters, N. E.; Murdoch, P. S. *Water Air Soil Pollut.* **1985**, *26*, 387–402.

117. Stoddard, J. L. *Limnol. Oceanogr.* **1987**, *32*, 825–839.

118. Driscoll, C. T.; Wyskowski, B. J.; Cosentini, C. C.; Smith, W. E. *Biogeochemistry* **1987**, *3*, 225–241.

119. Eshleman, K. N.; Hemond, H. F. *Water Resour. Res.* **1985**, *21*, 1503–1510.

120. Servos, M. R.; Mackie, G. L. *Can. J. Zool.* **1986**, *64*, 1690–1695.
121. LaZerte, B. D.; Dillon, P. J. *Can. J. Fish. Aquat. Sci.* **1984**, *41*, 1664–1677.
122. Haines, T. A. *Water Air Soil Pollut.* **1987**, *35*, 37–48.
123. McAvoy, D. C. *Water Resour. Res.* **1989**, *25*, 233–240.
124. Lynch, J. A.; Hanna, C. M.; Corbett, E. S. *Water Resour. Res.* **1986**, *22*, 905–912.
125. Schnoor, J. L.; Palmer, W. D., Jr.; Glass, G. E. In *Modeling of Total Acid Precipitation Impacts;* Schnoor, J. L., Ed.; Butterworth: Toronto, Canada, 1984; pp 155–173.
126. Schaefer, D. A.; Driscoll, C. T.; Van Dreason, R.; Yatsko, C. P. *Water Resour. Res.* **1990**, *26*, 1639–1647.
127. Murdoch, P. S.; Stoddard, J. L. *Water Resour. Res.* **1992**, *10*, 2707–2720.
128. Smith, V. H. *Limnol. Oceanogr.* **1982**, *27*, 1101–1112.
129. Stockner, J. G.; Shortreed, K. S. *Limnol. Oceanogr.* **1988**, *33*, 1348–1361.
130. Schindler, D. W.; Paerl, H. W.; Keller, P. E.; Lean, D. R. S. In *Hypereutrophic Ecosystems;* Junk: Dordrecht, Netherlands, 1980; Vol. 2, pp 221–229.
131. Elser, J. J.; Marzolf, E. R.; Goldman, C. R. *Can. J. Fish. Aquat. Sci.* **1990**, 47, 1468–1477.
132. Howarth, R. W. *Ann. Rev. Ecol.* **1988**, *19*, 89–110.
133. Healey, F. P. *Crit. Rev. Microbiol.* **1973**, *September 1973*, 69–113.
134. Peters, R. H. *Limnol. Oceanogr.* **1986**, *31*, 1143–1159.
135. Smith, V. H.; Shapiro, J. *Environ. Sci. Technol.* **1981**, *15*, 444–451.
136. Pace, M. L. *Can. J. Fish. Aquat. Sci.* **1984**, *41*, 1089–1096.
137. Hoyer, M. W.; Jones, J. R. *Can. J. Fish. Aquat. Sci.* **1983**, *40*, 192–199.
138. Prairie, Y. T.; Duarte, C. M.; Kalff, J. *Can. J. Fish. Aquat. Sci.* **1989**, *46*, 1176–1182.
139. McCauley, E.; Downing, J. A.; Watson, S. *Can. J. Fish. Aquat. Sci.* **1989**, *46*, 1171–1175.
140. Redfield, A. C. In *James Johnstone Memorial Volume;* Liverpool University Press: Liverpool, England, 1934; pp 176–192.
141. Schindler, D. W. *Limnol. Oceanogr.* **1978**, *23*, 478–486.
142. Morris, D. P.; Lewis, W. M., Jr. *Freshwater Biol.* **1988**, *20*, 315–327.
143. Landers, D. H.; Eilers, J. M.; Brakke, D. F.; Overton, W. S.; Kellar, P. E.; Silverstein, W. E.; Schonbrod, R. D.; Crowe, R. E.; Linthurst, R. A.; Omernik, J. M.; Teague, S. A.; Meier, E. P. *Characteristics of Lakes in the Western United States: Population Descriptions and Physico–Chemical Relationships;* EPA–600/3–86/54a; U.S. Environmental Protection Agency: Washington, DC, 1987; Vol. 1.
144. Vollenweider, R. A. *Water Management Research;* (Water management research Vol. DAS/CSI/68.27); OECD: Paris, France, 1968.
145. Melack, J. M.; Kilham, P.; Fisher, T. R. *Oecologia* **1982**, *52*, 321–326.
146. Setaro, F. V.; Melack, J. M. *Limnol. Oceanogr.* **1984**, *29*, 972–984.
147. Stoddard, J. L. *Hydrobiologia* **1987**, *154*, 103–111.
148. Melack, J. M.; Cooper, S. D.; Holmes, R. W.; Sickman, J. O.; Kratz, K.; Hopkins, P.; Hardenbergh, H.; Thiem, M.; Meeker, L. *Chemical and Biological Survey of Lakes and Streams Located in the Emerald Lake Watershed, Sequoia National Park;* Final Report A3–096–32; California Air Resources Board: Sacramento, CA, 1987.
149. Goldman, C. R. *Limnol. Oceanogr.* **1988**, *33*, 1321–1333.
150. Newell, A. D.; Powers, C. F.; Christie, S. J. *Analysis of Data from Long-term Monitoring of Lakes;* EPA600/4–87/014; U.S. Environmental Protection Agency: Corvallis, OR, 1987.

151a. Kahl, J. S.; Norton, S. A.; Haines, T. A.; Davis, R. B. *Water Air Soil Pollut.*, in press.
151b. Baird, S. F.; Buso, D. C.; Hornbeck, J. W. *Water Air Soil Pollut.* **1987**, *34*, 325–338.
152. Likens, G. E.; Bormann, F. H.; Pierce, R. S.; Eaton, J. S.; Johnson, N. M. *Biogeochemistry of a Forested Ecosystem;* Springer-Verlag: New York, 1977.
153. Johnson, N. M.; Driscoll, C. T.; Eaton, J. S.; Likens, G. E.; McDowell, W. H. *Geochim. Cosmochim. Acta* **1981**, *45*, 1421–1437.
154. *An Ecosystem Approach to Aquatic Ecology: Mirror Lake and Its Environment;* Likens, G. E., Ed.; Springer-Verlag: New York, 1985.
155. Driscoll, C. T.; Schaefer, D. A.; Molot, L. A.; Dillon, P. J. In *The Role of Nitrogen in the Acidification of Soils and Surface Waters;* Malanchuk, J. L.; Nilsson, J., Eds.; Nordic Council of Ministers: Copenhagen, Denmark, 1989; Vol. 10, pp 6.1–6.45.
156. Stoddard, J. L.; Kellogg, J. H. *Water Air Soil Pollut.*, in press.
157. Driscoll, C. T.; Van Dreason, R. *Water Air Soil Pollut.*, in press.
158. Murdoch, P. S.; Stoddard, J. L. *Water Air Soil Pollut.*, in press.
159. Kramer, J. R.; Andren, A. W.; Smith, R. A.; Johnson, A. H.; Alexander, R. B.; Oehlert, G. In *Acid Deposition, Long-Term Trends;* National Academy Press: Washington, DC, 1986; pp 231–299.
160. Smith, R. A.; Alexander, R. B.; Wolman, M. G. *Science (Washington, D.C.)* **1987**, *235*, 1607–1615.
161. Schindler, D. W. *Can. J. Fish. Aquat. Sci. Suppl. 1* **1987**, *44*, 6–25.
162. Morgan, M. D.; Good, R. E. *Water Resour. Res.* **1988**, *24*, 1091–1100.
163. Swistock, B. R.; DeWalle, D. R.; Sharpe, W. E. *Water Resour. Res.* **1989**, *25*, 2139–2147.
164. Sharpe, W. E.; DeWalle, D. R.; Leibfried, R. T.; Dinicola, R. S.; Kimmel, W. G.; Sherwin, L. S. *J. Environ. Qual.* **1984**, *13*, 619–631.
165. Sharpe, W. E.; Leibfried, V. G.; Kimmel, W. G.; DeWalle, D. R. *Water Resour. Bull.* **1987**, *23*, 37–46.
166. Sharpe, W. E.; DeWalle, D. R.; Swistock, B. R. In *Headwaters Hydrology;* American Water Resources Association: Washington, DC, 1989; pp 517–525.
167. Edwards, P. M.; Helvey, J. D. *J. Environ. Qual.* **1991**, *20*, 250–255.
168. *Drinking Water Analysis Handbook;* Hach Chemical Company: Ames, IA, 1977.
169. Edwards, P. M.; Kochenderfer, J. N.; Seegrist, D. W. *Water Resour. Bull.* **1991**, *27*, 265–274.
170. Jones, H. C.; Noggle, J. C.; Young, R. C.; Kelly, J. M.; Olem, H.; Ruane, R. J.; Pasch, R. W.; Hyfantis, G. J.; Parkhurst, W. J. *Investigations of the Cause of Fishkills in Fish-rearing Facilities in Raven Fork Watershed;* TVA/ONR/WR–83/9; Tennessee Valley Authority, Office of Natural Resources: Knoxville, TN, 1983.
171. Elwood, J. W.; Sale, M. J.; Kaufmann, P. R.; Cada, G. F. In *Acidic Deposition and Aquatic Ecosystems: Regional Case Studies;* Charles, D. F., Ed.; Springer-Verlag: New York, 1991; pp 319–364.
172. Silsbee, D. G.; Larson, G. L. *Hydrobiologia* **1982**, *89*, 97–115.
173. Joslin, J. D.; Mays, P. A.; Wolfe, M. H.; Kelly, J. M.; Garber, R. W.; Brewer, P. F. *J. Environ. Qual.* **1987**, *16*, 152–160.
174. Buell, G. R.; Peters, N. E. *Water Air Soil Pollut.* **1988**, *39*, 275–291.
175. Cosby, B. J.; Ryan, P. F.; Webb, R.; Hornberger, G. M.; Galloway, J. N. In *Acidic Deposition and Aquatic Ecosystems: Regional Case Studies;* Charles, D. F., Ed.; Springer-Verlag: New York, 1991; pp 297–318.
176. Swank, W. T.; Waide, J. B. *Ecol. Stud.* **1988**, *66*, 57–79.

177. Loranger, T. J.; Brakke, D. F.; Bonoff, M. B.; Gall, B. F. *Water Air Soil Pollut.* **1986,** *31,* 123–129.
178. Loranger, T. J.; Brakke, D. F. *Water Resour. Res.* **1988,** *24,* 723–726.
179. Melack, J. M.; Stoddard, J. L. In *Acidic Deposition and Aquatic Ecosystems: Regional Case Studies;* Charles, D. F., Ed.; Springer-Verlag: New York, 1991; pp 503–530.
180. Williams, M. W.; Melack, J. M. *Water Resour. Res.* **1991,** 27, 1575–1588.
181. Mollitor, A. V.; Raynal, D. J. J. *Air Pollut. Control Assoc.* **1983,** *33,* 1032–1036.
182. Cadle, S. H.; Dasch, J. M.; Grossnickle, N. E. *Water Air Soil Pollut.* **1984,** 22, 303–319.
183. Cadle, S. H.; Dasch, J. M.; Kopple, R. V. *Environ. Sci. Technol.* **1987,** *21,* 295–299.
184. Stottlemyer, R.; Toczydlowski, D. *Can. J. Fish. Aquat. Sci.* **1990,** 47, 290–300.
185. Haynes, R. J. *Mineral Nitrogen in the Plant-Soil System;* Academic: Orlando, FL, 1986; p 483.
186. Husar, R. B. In *Acid Deposition: Long-Term Trends;* National Academy Press: Washington, DC, 1986; pp 48–92.
187. Simpson, J. C.; Olsen, A. R. *1987 Wet Deposition Temporal and Spatial Satterns in North America;* PNL–7208; Pacific Northwest Laboratory: Richland, WA, 1990.
188. Bowersox, V. C.; Sisterson, D. L.; Olsen, A. R. *Int. J. Environ. Stud.* **1990,** *36,* 83–101.
189. Driscoll, C. T.; Newton, R. M.; Gubala, C. P.; Baker, J. P.; Christensen, S. W. In *Acidic Deposition and Aquatic Ecosystems: Regional Case Studies;* Charles, D. F., Ed.; Springer-Verlag: New York, 1991; pp 133–202.
190. Wright, R. G.; Hauhs, M. *Reversibility of Acidification: Soils and Surface Waters;* Report 23/1990; Norwegian Institute for Water Research: Oslo, Norway, 1990.
191. Kahl, J. S.; Norton, S. A.; Cronan, C. A.; Fernandez, I. J.; Bacon, L. C.; Haines, T. A. In *Acidic Deposition and Aquatic Ecosystems: Regional Case Studies;* Charles, D. F., Ed.; Springer-Verlag: New York, 1991; pp 203–235.
192. DeWalle, D. R.; Sharpe, W. E.; Edwards, P. J. *Water Air Soil Pollut.* **1988,** *40,* 143–156.
193. Barker, J. L.; Witt, E. C. *Water Resour. Invest. U.S. Geol. Surv.* **1990,** 89–4113,
194. Phillips, R. A.; Stewart, K. M. *Water Resour. Bull.* **1990,** *26,* 489–498.
195. Katz, B. G.; Bricker, O. P.; Kennedy, M. M. *Am. J. Sci.* **1985,** *285,* 931–962.
196. Weller, D. E.; Peterjohn, W. T.; Goff, N. M.; Correll, D. L. In *Watershed Research Perspectives;* Correll, D. L., Ed.; Smithsonian Institution Press: Washington, DC, 1986; pp 329–421.
197. Welch, E. B.; Spyridakis, D. E.; Smayda, T. *Water Air Soil Pollut.* **1986,** *31,* 35–44.
198. Eilers, J. M.; Sullivan, T. J.; Hurley, K. C. *Hydrobiologia* **1990,** *199,* 1–6.
199. Gilbert, D. A.; Sagraves, T. H.; Lang, M. M.; Munson, R. K.; Gherini, S. A. *R&D Lake Acidification Assessment Project: Blue Lake Acidification Study;* Vol. Report 009.5–89.2; Pacific Gas and Electric: San Ramon, CA, 1989.
200. Hicks, B. B. In *The Role of Nitrogen in the Acidification of Soils and Surface Waters;* Malanchuk, J. L.; Nilsson, J., Eds.; Nordic Council of Ministers: Copenhagen, Denmark, 1989; Vol. 10, pp 3-1–3-21.
201. Kanciruk, P.; Eilers, J. M.; McCord, R. A.; Landers, D. H.; Brakke, D. F.; Linthurst, R. A. *Characteristics of Lakes in the Eastern United States: Data*

Compendium of Site Characteristics and Chemical Variables; EPA/600/4–86/007c; Environmental Protection Agency: Washington, DC, 1986; Vol. 3.

202. Eilers, J. M.; Kanciruk, P.; McCord, R. A.; Overton, W. S.; Hook, L.; Blick, D. J.; Brakke, D. F.; Kellar, P. E.; DeHaan, M. S.; Silverstein, M. E.; Landers, D. H. *Characteristics of Lakes in the Western United States: Data Compendium for Selected Physical and Chemical Variables;* EPA/600/4–86/054b; Environmental Protection Agency: Washington, DC, 1987; Vol. 2.

203. Hirsch, R. M.; Slack, J. R.; Smith, R. A. *Water Resour. Res.* **1982,** *18,* 107–121.

204. Hirsch, R. M.; Slack, J. R.; *Water Resour. Res.* **1984,** *20,* 727–732.

RECEIVED for review September 26, 1991. ACCEPTED revised manuscript August 10, 1992.

Mass Fluxes and Recycling of Phosphorus in Lake Michigan

Role of Major Particle Phases in Regulating the Annual Cycle

Martin M. Shafer and David E. Armstrong

Water Chemistry Program, Water Science and Engineering Laboratory, University of Wisconsin, Madison, WI 53706

Biogeochemical cycling of phosphorus in a water column in southern Lake Michigan was examined, and the significance of major particle phases to the annual mass flux and recycling of phosphorus was assessed. Comparison of 1982–1983 total P and total filtrable P concentrations with data from 1990–1991 and other published data showed little change. The measured annual primary flux of P to the sediment surface reflected rapid sedimentation of both allochthonous particles and spring diatom production. Diatoms were the dominant vector of P to the sediment surface. Terrigenous phases and autochthonous calcite were also significant. More than half of the mixed-period diatom P demand was provided by colloidal and particulate P, nearly 60% of diatom-associated P was recycled within the water column, and 55–58% of total primary P flux was recycled at the sediment surface. Ultimately 2.2% of C and 2.7% of P became incorporated into recent sediments. The amount of P supplied by resuspension was relatively small compared with water-column standing pools and major flux vectors. With its relatively long residence, the response time for P changes with respect to loading should be on the order of 5–15 years.

T HE PHOSPHORUS STATUS OF THE GREAT LAKES is a continuing concern because of the linkages between P, primary production, and water quality

0065–2393/94/0237–0285$10.50/0
© 1994 American Chemical Society

(*1–4*). External loading plays a major role in regulating P concentrations (*1, 5, 6*). However, P status and primary production are also controlled by the recycling of P within the lake (*7, 8*). Input–output models provide a basis for assessing relations between external loading and in-lake P concentrations (*1, 5, 6*). However, quantitative evaluation of P recycling by mass-balance models is limited by the complexity and lake-specific nature of internal recycling processes. Consequently, the importance of internal recycling in controlling the P status of Lake Michigan is uncertain.

Internal recycling of P within the water column is linked to production, removal, and recycling of P-containing natural particles (*9, 10*). Incorporation of P into natural particles through primary and secondary production and chemical adsorption results in a downward flux of P associated with settling particles. Allochthonous inputs and sediment resuspension also contribute particulate P to the water column. Phosphorus recycling from particulate to dissolved forms occurs through particle dissolution–decomposition and P desorption. Thus, recycling of P can be assessed by determining the input and fate of natural particles moving through the water column. We used this approach to evaluate the role of particle-mediated processes in regulating the concentration of P in Lake Michigan.

Materials and Methods

Location.   The station investigated was located at (42° 40′ N, 87° 00′ W), approximately in the center of the southern basin of Lake Michigan, at a water-column depth of 160 m. The location was selected as representative of midlake conditions in the southern basin. Data from an intensive 1-year sampling program (1982–1983; average sampling interval 3 weeks) and additional data sets from cruises in 1978–1980 and 1989–1991 provided the foundation for this chapter.

Field Procedures.   *Sedimenting Particles.*   A sediment trap array, moored at the sampling station, was retrieved and redeployed 15 times over the course of the main study period (early April 1982 to late March 1983). Deployment intervals averaged 3 weeks. The cylindrical acrylic traps employed were similar in design to those described by Wahlgren and Nelson (*11*). The standard trap used had an aspect ratio of 4.0 (16-cm diameter) and an open area of 162 cm^2 with a baffle and 198 cm^2 without a baffle.

Paired traps (one poisoned with sodium azide, the other unpoisoned) were deployed at eight depths on a 3/4-inch polyester braided line. Traps (upper surface) were located at 4.2, 7.3, 13.0, 20.5, 49.4, 87.7, 116.7, and 131.1 m above the bottom. The upper trap (28.9 m below the water surface) was below the depth of the summer thermocline.

Retrieval was accomplished with an acoustic release (Helle Engineering) located between the bottom trap and a concrete anchor. All trap surfaces were acid-washed and thoroughly rinsed before use. Trapped particles, including those retained on the trap funnel, were kept at 4 °C in the dark for transport to the laboratory for processing.

Standing-Crop Particles.   Standing-crop, noncolloidal, particulate matter was sampled by using two techniques: 1, niskin casts followed by filtration onto track-etched filters and 2, serial sieve fractionation in-line with continuous-flow centrifugation, which enabled the collection of gram quantities of suspended particles.

1. A Teflon-coated sampler (General Oceanics, 10-L Go-Flo) on a stainless steel hydrowire was used to obtain whole-water samples and samples for filtration from 16 points in the water column on each cruise. Samples were pressure filtered onboard ship, within a Class-100 laminar flow enclosure, in an all-Teflon column–filter holder (Savilex) through tared 0.4-μm track-etched filters (Nuclepore). Samples, both total P and total filtrable P, were acidified with high-purity HCl (Ultrex) and immediately frozen.

2. The in-line-sieving continuous-flow centrifugation system consisted of a deck-mounted pumping system coupled to a plastic serial-sieving unit, which in turn was coupled to two high-capacity continuous-flow tubular bowl centrifuges (Sharples Model AS–12V). The sieving unit isolated four size fractions of larger particles (>508, 508–212, 212–114, and 111–63 μm), and the centrifuges retained particles between 63 and 1 μm. All centrifuge surfaces that came into contact with lake water were either lined with or manufactured of poly(vinylidene fluoride) (Kynar) to reduce the possibility of contamination. An acid-leached polyethylene terephthalate (Mylar) sheet was placed inside the circumference of the centrifuge bowl to enable rapid recovery of trapped particles. Particles and lake water (4 L), retained in the centrifuge bowl after the self-sealing device isolated the bowl, were recovered and kept at 4 °C for transport to the laboratory. Time delay from the collection to laboratory processing varied from 24 to 36 h. At typical lake-water particle concentrations of 1 mg/L, more than 0.5 g of suspended particles could be collected in 1 h. On-station time limited the number of large-volume depths sampled per cruise to six or seven. The depths chosen were selected to be representative of distinct layers of the water column and also to correspond to depths at which sediment traps were deployed.

Sampling and processing techniques designed to isolate and concentrate submicrometer-sized (colloidal) particles and associated elements were employed on 1989–1991 cruises. Three distinct approaches were exploited.

1. Filtration (12) of whole lake water from niskin casts and centrifugate from continuous-flow centrifuges was accomplished with track-etched filters (Nuclepore) of various pore sizes. For P, the fractionation protocol employed filters rated at 1.0, 0.4, 0.05, and 0.03 μm. Additional 0.2- and 0.1-μm filters were added for colloid mass–size distribution measurements.

2. For continuous-flow ultracentrifugation, large volumes (200–250 L) of centrifugate from the ship-based centrifuges were collected and transported to the laboratory. There a subsample (~175 L) was processed through a continuous-flow ultracentrifuge operating at 104,000 × g. The efficiency of the ship-based centrifuges is 40–60% for 1-μm particles and nearly 100% for 3-μm particles. The ultracentrifuge efficiently collects particles larger then 30 nm (density-dependent). Therefore, the material isolated by the ultracentrifuge spans an approximate size range of 30 nm to 3 μm.

3. Tangential-flow ultrafiltration was used with approximately 150-L samples of centrifugate from the ship-based centrifuges. These samples were processed through a large tangential-flow cell (regenerated cellulose membrane, Millipore Pellicon) while on station.

Particles in the ultraconcentrates were examined either in suspension or after removal of solution by filtration.

Sediment Core. Three replicate subcores were taken from a 10- × 10- × 4-inch box core retrieved from the 160-m station. Each subcore was sectioned at 0.5-cm intervals from 0.0 to 3.0 cm and at 1.0-cm intervals from 3 to 15 cm. Sectioned sediment samples for chemical analyses were immediately frozen and later freeze-dried before acid digestion.

Laboratory Procedures. *Size Fractionation.* Particles collected in the centrifuge (<63 μm) were further size-fractionated in the laboratory by serial filtration of the suspension through 8-inch-diameter plastic sieves with 19- and 8.2-μm mesh openings. Particle loading was kept low (<15 mg per sieve) to avoid clogging, and filtered centrifuge water was used when needed to aid in the processing. Particles that passed the 8.2-μm sieves, which in most samples was >40% of the total sample mass, were either collected on a variety of tared filters or isolated by centrifugation. For mass determinations, 1.0- and 0.4-μm Nuclepore filters were employed. Loading on one of the 1.0-μm filters was kept low (<0.3 mg) to estimate the fraction of mass between 0.4–1.0 and 1.0–8.2 μm. Frozen samples and controls were dried by lyophilization. Filters were reweighed on a microbalance (Perkin-Elmer AD–2) and vials on an analytical balance (Mettler H51AR). After drying the samples were stored at −10 °C in desiccators.

Trap particles were size-fractionated in the laboratory in a manner similar to that described for the standing-crop particles. Four additional mesh sizes corresponding to those used in the pumping-system sieve unit were used to give complete correspondence with standing-crop size fractions.

In summary, for each trap or standing-crop sampling point, eight particle-size fractions were created with nominal cutoffs of 508, 212, 114, 63, 19, 8.2, 1.0, and 0.4 μm. The 600 standing-crop and 1000 trap-mass fractions were chemically analyzed for phosphorus and major and trace elemental composition. Major particle types were identified and enumerated by optical microscopy.

Chemical Analyses.   Subsamples (<20 mg) of standing-crop–trap particulate matter were solubilized by acid digestion in sealed all-Teflon bombs (Bombco Inc.) in a procedure modified from Eggiman and Betzer (*13*). National Institute of Standards and Technology (NIST) standard reference materials River Sediment (SRM 1645), Urban Particulate Matter (SRM 1648), and mixed-element liquid spikes were used to check bomb performance.

Analysis of major elements (except Si) and total phosphorus on bomb-digested samples was accomplished by inductively coupled plasma emission spectrometry (ICP, ARL model 34,000). Silicon was analyzed colorimetrically (*14*). Phosphorus in total digests was also determined colorimetrically by the method of Murphy and Riley (*15*), as modified by Erickson (*16*). To avoid interference from fluoride ion used in the digestion technique, sample volumes were restricted to <1.5 mL in the standard P analytical protocol.

Biogenic silicon (BSI) was determined, with minor modifications, by the method of DeMaster (*17*). As adapted, the technique involved time-course leaching of ≤20-mg samples of particulate matter in 30 mL of 1.0% Na_2CO_3 in a water bath at 85 °C. Silica in leachates was quantified either colorimetrically (Technicon autoanalyzer procedure) or by nitrous oxide flame atomic absorption. A high-temperature catalytic-combustion technique (Perkin Elmer 240C) was used for particulate organic carbon determinations. Particulate inorganic (carbonate) carbon was measured on the same instrument by CO_2 evolution after treatment of the particles with phosphoric acid.

Whole-water (total P) and filtrable phosphorus (total filtrable P, 0.4 μm) analyses were performed by the ascorbic acid technique (*18*) after digestion with persulfate.

Component Model.   Major particle phases can be described by application of a component model. The component model is based on chemical analyses of particle mass fractions, aided in part by optical microscopic techniques. It determines the weight percentage of major components comprising the total particulate matter pool. Details of the full eight-component model were described by Shafer (19). Major components were discussed in Shafer and Armstrong (20).

Phosphorus Concentrations in Lake Michigan

Total, Particulate, and Filtrable P. Water-column-integrated, total-phosphorus concentration changes over the annual cycle were relatively modest (Figure 1). Total P concentration in the 160-m water column ranged from 734 to 897 mg/m^2 with a mean of 824 mg/m^2 (5.2 μg/L), a range of approximately 10% about the mean. Total filtrable P concentrations, nepheloid P regeneration excluded, ranged from 351 to 521 mg/m^2 (mean was 418 mg/m^2 or 2.6 μg/L). Filtrable reactive-P measurements on selected samples from spring cruises averaged 0.9 μg/L, approximately one-third of the total filtrable P values. Particulate P concentrations ranged from 241 to 410 mg/m^2 (mean was 333 mg/m^2 or 2.1 μg/L), representing on average about 55% of the total P burden (Figure 1).

When P in the nepheloid zone was excluded from the integrations, areal total P concentrations increased from 733 mg/m^2 in mid-April to 845 mg/m^2 in mid-July, and then declined to 630 mg/m^2 by late November.

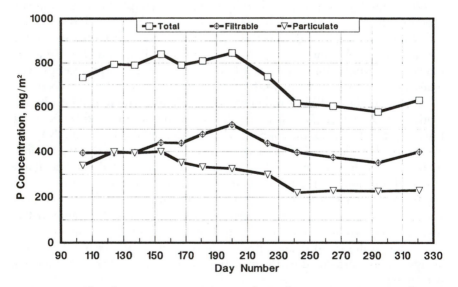

Figure 1. Phosphorus concentrations in Lake Michigan, 1982. Data are from a station in the center of the southern basin. Nepheloid regenerated phosphorus is not included.

However, P accumulated in the nepheloid zone (bottom 20 m) between June and November in an amount approximately equal to that lost from the water column above it (Figure 2). Loss from the water column occurred mostly in the particulate P form, but filtrable P was the main form appearing in the nepheloid zone. A significant increase in filtrable phosphorus levels was also noted in the midsummer hypolimnion (50–125 m), a change that reflects diagenetic loss of phosphorus during settling of the diatom front.

Concentrations in 1982–1983 were similar to concentrations we measured at less frequent intervals through 1991. Total water-column integrated P concentration measured in April 1991 was almost identical to that of April 1982 (760 versus 746 mg/m^2 and 4.8 versus 4.7 μg/L). A slightly higher percentage of the April 1982 total P was particulate than the April 1991 total P (47% versus 40%). One year earlier in 1990, the total-P concentration in late May (646 mg/m^2 or 4.0 μg/L, 41% particulate) was 20% lower than that measured in late May and early June of 1982 (~800 mg/m^2 or 5.0

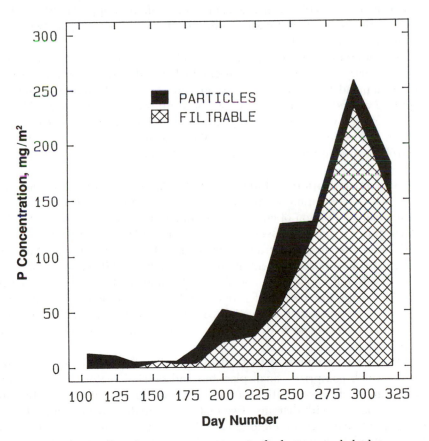

Figure 2. Phosphorus concentrations in the bottom nepheloid zone.

μg/L). The difference resulted from lower particulate-P levels (filtrable-P concentrations were comparable: 380 mg/m^2 or 2.4 μg/L in 1991 and 390–400 mg/m^2 in 1982). Seasonal trends in P levels were, however, similar in 1990 and 1982. Both filtrable and total P increased from May levels to maximal values in July. In 1982 total P rose from 800 to 900 mg/m^2 (5.0 to 5.6 μg/L) and filtrable P rose from 400 to 540 mg/m^2 (2.5 to 3.3 μg/L). In 1990 total P increased from 650 to 959 mg/m^2 (4.1 to 6.0 μg/L) and filtrable P rose from 380 to 600 mg/m^2 (2.4 to 3.7 μg/L). Similarly, our 1982–1983 total-P concentrations were in close agreement with 1983–1986 values published by Lesht (6). These results indicated that total-P concentrations have changed little over the past decade. However, changes in the food web may have influenced relationships between total-P levels and primary production (21, 22).

Even though the concentration changes are relatively small (for a 40-mg/m^2 change over the 160-m water column, ΔP is 0.25 μg/L), the actual quantities of P are large. A concentration change of 100 mg/m^2 corresponds to 5354 metric tons of P over the main lake, an amount approximately equal to the estimated annual total-P loading in the mid 1970s (5) or twice the loading in the mid 1980s (6).

The interannual P patterns suggest a relatively nondynamic P cycle. Yet major primary-production cycles occur in Lake Michigan. Either the biological demand for P is small in comparison to the internal pool and the supply provided by external loading, or P utilization by phytoplankton is largely offset by subsequent recycling to dissolved P or by release from bottom and resuspended sediments. We evaluate the role of P cycling in controlling the supply of P for primary production.

Size Distribution of Particulate P.   Although interannual changes in total P concentration are relatively small, variations in particulate P are substantial. Evaluation of changes according to particle size range provides information on the processes influencing particulate P concentration. In 1982, from the onset of stratification to late July, integrated epilimnetic particulate phosphorus concentrations rose from 65 to 100 mg/m^2. However, these levels fell rapidly to <50 mg/m^2 by mid-September. The change coincided with the period of active $CaCO_3$ precipitation and deposition (Figure 3). Scavia and Fahnenstiel (23) also noted that both primary production and grazing were dramatically reduced during episodes of $CaCO_3$ precipitation.

Most of the epilimnetic particulate P was found in the 1.0–8.2-μm size range, which comprises small plankton and detrital material. During the mixing and early stratification periods, a major fraction of total particulate P was associated with diatoms. Two diatom-dominated size fractions, 63–19 and 211–114 μm, accounted for 20 and 12%, respectively, of the integrated total particulate P burden at peak levels in spring. Total diatom-P contri-

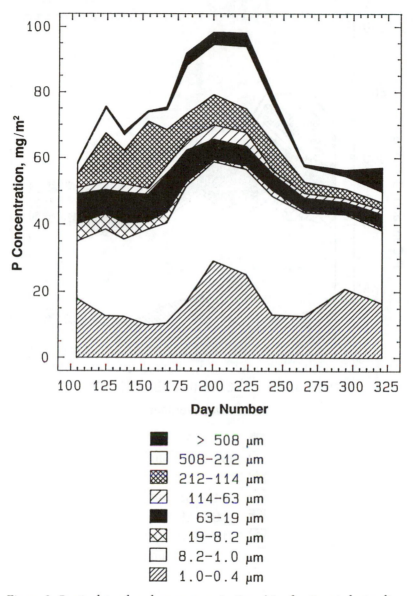

Figure 3. *Particulate phosphorus concentrations (size-fractionated–standing-crop) integrated over the epilimnion.*

bution ranged from 30 to 40% in spring. Zooplankton P (>212 μm) was a significant fraction (21–26%) of the epilimnetic particulate P in midsummer.

When concentrations were integrated over the entire water column (Figure 4), the importance of diatoms and small particles became evident. Throughout the year, more than 50% of the particulate P was contained in

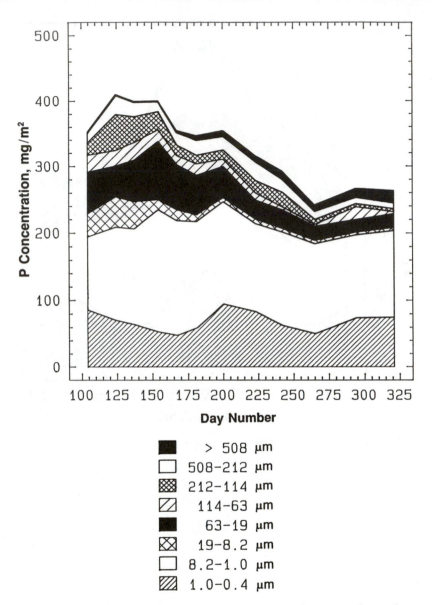

Figure 4. Particulate phosphorus concentrations (size-fractionated–standing-crop) integrated over the entire water column.

<8.2-μm particles. In late summer and fall, when diatoms and larger alloch-thonous aggregates had settled from the system, nearly 80% of particulate P was associated with <8.2-μm particles. Presentation of P concentrations as integrated values belies the complexity of a given profile. Particulate-P profiles especially show considerable detail. They correlate strongly with

algal layering, particularly in the deep chlorophyll maximum. A typical profile sequence is shown in Figure 5.

Phosphorus Transport to the Sediment Surface

Total-P Flux.   Two approaches were combined to calculate removal by deposition. During the stratified period deposition was determined from the mass flux, measured by using sediment traps suspended at eight levels in the water column, and the P concentration in the sediment trap material. Positioning of traps at key depths allowed primary and resuspended flux components to be deconvoluted.

Deposition during the mixed period (up to day 165) was calculated from a mass balance on water-column Si and the Si:P ratio in sediment trap material, because sediment traps overestimate the net particle deposition flux in a mixing water column (*19*). Our calculations assumed that losses of dissolved reactive Si resulting from diatom uptake that are not accounted for by increases in particulate biogenic Si are caused by Si deposition. The estimate of mixed-period P deposition was conservative because we assumed that nondiatom particulate P was removed at a rate similar to diatom P. We also assumed that loss of P in traps resulting from diagenesis–dissolution was negligible. The use of short collection periods (2–3 weeks) and a poison should minimize loss.

The annual total P flux to the sediment surface was about 185 mg/m^2 (Figure 6), corresponding to about 25% of the P in the water column in mid-April. Deposition of P during the mixed period (April to mid-June) was almost 30% of the annual flux. In comparison, Eadie et al. (*24*) estimated a somewhat higher primary flux of about 250 mg of P/m^2. The larger value probably resulted in part from a higher average P concentration in the primary flux (1.2 versus 0.95 mg/g) and less frequent upper-water-column summer deployments. Still, the concordance between independent measurements was quite good.

Nearly 30% of total P deposition occurred in the 1-month period initiated at time of full thermal stratification (mid-June). By early August 66% of the total P had reached the sediment. Rapid sedimentation of both allochthonous particles and spring diatom production in the now-stable water column were the source of these large percentages. Thus mid-water-column net P fluxes were at their annual maximum during this period, with rates in the range of 1200–1800 μg/m^2 per day (Figure 7). Upper hypolimnetic P fluxes in midsummer ranged from 300 to 400 μg/m^2 per day and declined to 200 μg/m^2 per day by November (Figure 7).

Calcium carbonate precipitation and sedimentation from mid-August through October (ref. 20; *see also* the following discussion of components) significantly enhanced P fluxes. The impact of calcite precipitation on P removal from the epilimnion is shown in Figure 8. Phosphorus flux into the

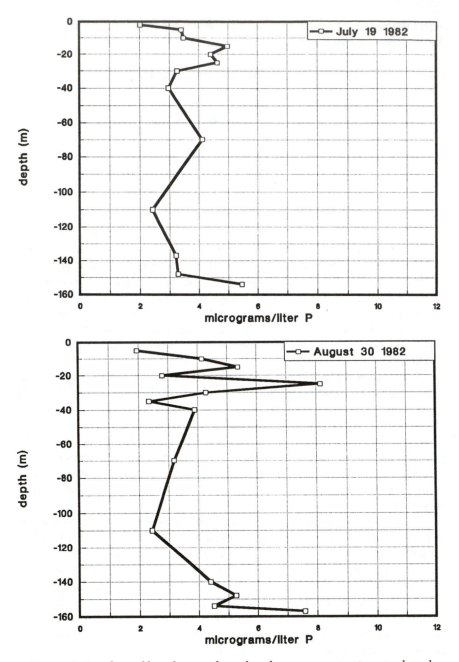

Figure 5. Depth profiles of particulate phosphorus concentration at selected times in 1982. Continued on next page.

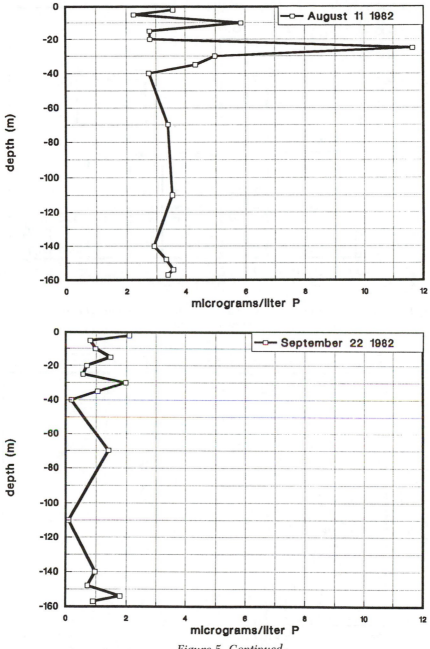

Figure 5. Continued.

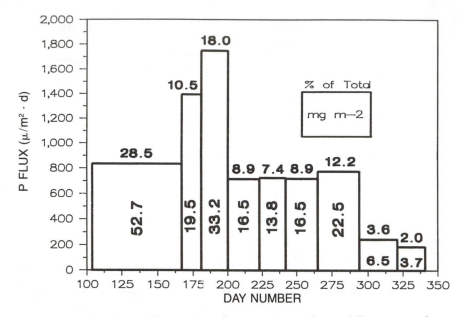

Figure 6. Phosphorus flux to the sediment surface. The total flux over study period was 185 mg/m².

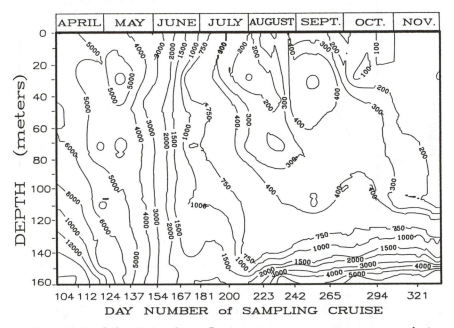

Figure 7. Isopleths of phosphorus flux (micrograms per square meter per day) in the water column, measured by using sediment traps. The values are the sum of size fractions.

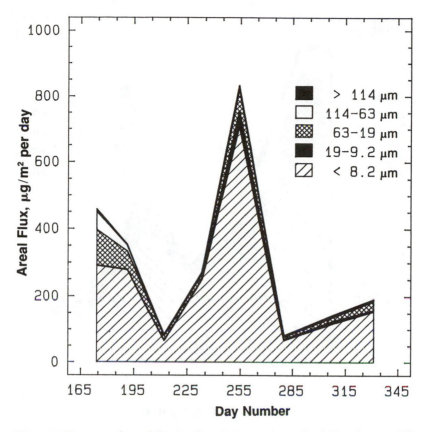

Figure 8. Downward areal flux (sediment-trap-measured) of phosphorus at 29 m during the period of thermal stratification.

uppermost trap (29 m) just before and after the major precipitation episode was about 75 μg/m^2 per day. However, P fluxes of 800 μg/m^2 per day were observed during active precipitation.

Trap and standing-crop data show the front of carbonate-associated P as it settled through the water column, reaching the bottom sediment between days 250 and 260. Most of the primary P flux reaching the sediment during days 265 to 294 (22.5 mg^2) was carbonate-associated. Sediment-trap-based P fluxes in the nepheloid region typically ranged from 1500 μg/m^2 per day (20 m off the bottom) to over 5000 μg/m^2 per day (4 m off the bottom) and showed steady development as the season progressed (Figure 7).

The size distribution of P in sedimenting material at a mid-water-column depth (111 m) is shown in Figure 9. The dominance of small particle sizes (19–8.2 and <8.2 μm) is immediately apparent, representing about 70% on a seasonally averaged basis. The presence of smaller fractions was enhanced

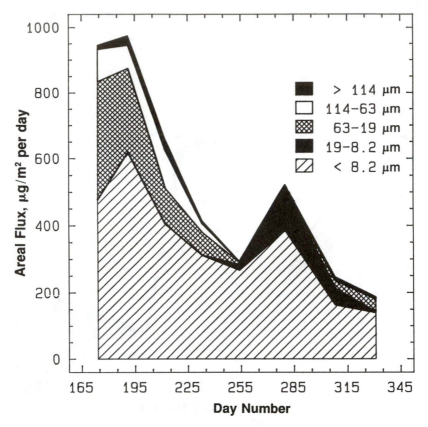

Figure 9. Downward areal flux (sediment-trap-measured) of phosphorus at 111 m during the period of thermal stratification.

after mid-August by the introduction of massive amounts of calcite particles into the two smallest size classes. Only a very small portion of the <8.2-μm fraction from the traps was <2 μm in dimension. In contrast, the standing-crop collections typically contain a large <2-μm component.

The significance of small fractions on an annual basis was notably reduced when expressed as a flux because of a substantial contribution from large particles during periods of high total mass flux. Also conspicuous was the impact of the spring diatom bloom; the growth and subsequent decline of P in the 63–19- and 114–63-μm size fractions was a direct reflection of the spring bloom. At its peak these two fractions contained 55% of the total P flux. Large particles (>114 μm) contributed less than 6% of the P flux. Near-bottom traps contained a higher percentage of medium-sized particles as a result of aggregation in the dynamic nepheloid region.

Phosphorus settling rates were calculated as the ratio of P fluxes to standing-crop concentrations (Figure 10). Observed rates ranged from <0.1

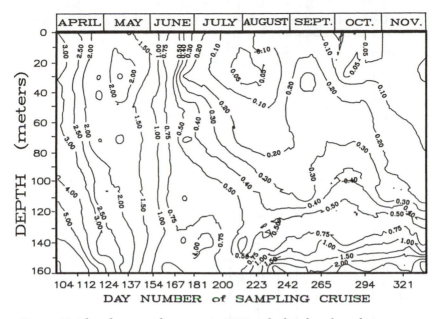

Figure 10. Phosphorus settling rates in 1982, calculated as the sediment-trap-measured depositional flux divided by the total particulate P concentration (meters/day).

to >3 m/day and reflected variation in particle vectors, hydrodynamic environments, and P regeneration. Rates >1.0 m/day probably were not true net bulk settling velocities. Typical values were 0.5 m/day for diatom-associated P and 0.1 m/day for nonsilicious organic matter. Carbonate precipitation enhanced settling rates by a factor of 2–3.

The time-sequence pattern of P deposition to the sediment surface differed markedly from that observed with silicon (25). Biogenic silicon underwent relatively little dissolution in the water column as compared with phosphorus, which resulted in a bell-shaped deposition pattern (Figure 11). Rapid regeneration of P during sedimentation resulted in a slower net settling rate for P than for biogenic silicon.

Particulate Matter Components Controlling P Deposition.   To assess the role of various particle events in elemental cycling, major particle types in size-fractionated standing-crop and trap collections were identified though microscopic examination and chemical analysis. Subsequently, a particle components model was developed to quantify major particle types on the basis of the elemental composition of size-fractionated particulate matter (19, 20). The components model facilitates quantitative assessment of the cycles of natural particles and associated elements.

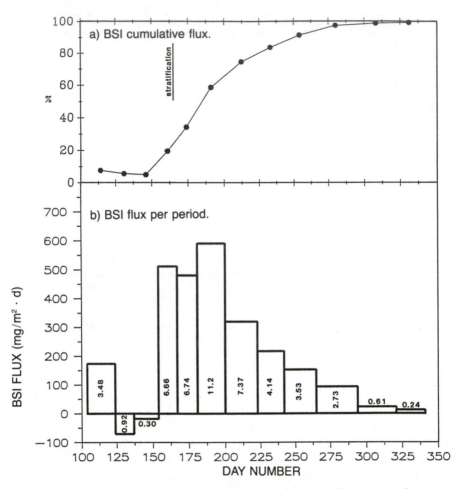

Figure 11. Primary flux of biogenic silica to the sediment surface: a, cumulative flux; b, flux and flux per period (milligrams/square meter).

The major particle components detected varied dramatically over the annual cycle (Figure 12), as well as between trap and standing-crop collections. Biogenic silica (diatoms) was the dominant component of the particle flux between April and June. Calcite, produced by in-lake precipitation, became a major component from early August to November. Proportions of other components were less, but became significant at certain times of the year. When integrated over the year, fluxes of components to the sediment surface ranged from 5 g/m² for fecal pellets to 106 g/m² for diatoms (Table I). In comparison to the settling material (Figure 12a), the suspended particulate matter (Figure 12b) contained more organic matter and less terrigenous mineral material (*19*).

Association of P with major components of the particle flux was investigated to identify and quantify the processes controlling P removal and recycling in the water column. The approach involved coupling fluxes of particle components, derived from the components model, with the P concentration of the component obtained from chemical analysis of pure particle fractions. The isolation of clean samples of individual particle types was achieved through the combination of size fractionation and sampling over a range of times and depths.

The calculated flux of diatoms to the sediment surface was about 105 g/m^2. Based on a diatom-P concentration (diatoms captured in near-bottom traps) ranging from 0.8 to 1.2 mg/g, diatom P deposition amounted to 85–125 mg/m^2. Calcite deposition during August and September was 18–26 g/m^2, calculated from trap flux measurements and the component model estimate of calcite in trap material. From a measured P concentration in calcite captured in mid-water-column traps of 1.2 mg/g, a calcite-P flux of 21–32 mg/m^2 was obtained. The annual allochthonous (terrigenous) material flux to the sediments was estimated to be 33 g/m^2. Combining the flux with the measured P concentration of 1.25–1.5 mg/g (terrigenous-dominated trap collections and buried bottom sediment) gave the terrigenous P deposition flux of 41–50 mg/m^2. Deposition of fecal pellets amounted to about 5 g/m^2, for an estimated annual fecal pellet-P flux of 10–20 mg/m^2. The P content of fecal pellets (2–4 mg/g) was estimated from carbon values. Flux estimates of major particle vectors are those reaching bottom sediment, thereby accounting for phase loss in the water column.

Comparison of the depositional fluxes shows that diatoms were the most important particle component transporting P to the sediment surface, accounting for 50–55% of the flux (Table II). Terrigenous material and calcite were also important transport vectors. Deposition varied markedly with season, as shown by the time series plot of the major particle components (Figure 13). The total P flux calculated by using the particle components model agreed with the flux measured by sediment traps (157–227 versus 185 mg/m^2). The close agreement indicated that the major particle vectors were represented and associated P concentrations were accurately quantified.

Phosphorus Cycling and Regeneration

Uptake and Release by Diatoms.   Diatoms represent a major sink and transport vector for P in Lake Michigan. The importance of diatom production is shown by the changes in dissolved reactive Si levels that resulted from incorporation into diatoms. During the mixed period (late March to mid-June) dissolved reactive Si decreased by 30.7 g/m^2, representing 36% of the spring concentration. An additional 11.4 g/m^2 was re-

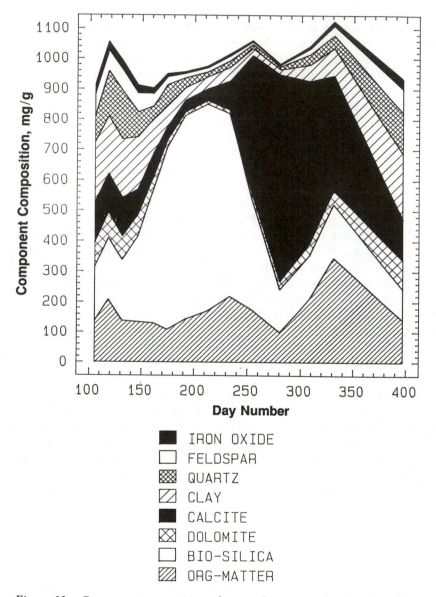

Figure 12a. Component composition of particulate material collected in sediment traps at 111 m depth. The values are the sum of size fractions.

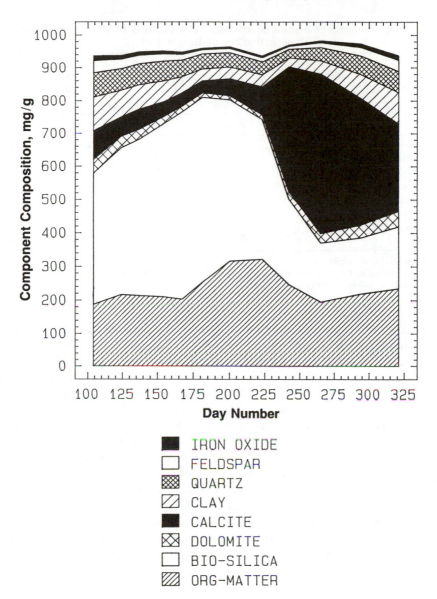

Figure 12b. Suspended particulate matter integrated over the water column.

moved in the subsequent stratified period. We calculated the corresponding P demand resulting from diatom production on the basis of measurements of P and Si concentrations in diatoms obtained by the size-fractionation procedure (Table III). Total diatom demand for P of about 280 mg/m² represented about 37% of the spring total P concentration, somewhat less than the corresponding Si demand of almost 50%. Over 70% of the P demand

Table I. Component Composition of Suspended Particulate Matter and Material Deposited at the Sediment Surface

	Typical Concentration in Suspended Material (mg/g)		Deposited Material, Annual Flux	
Component	Spring	Fall	g/m^2	%
Diatoms	600	230	105.6	61
Calcite[a]	40	450	30.4	17
Terrigenous	260	130	33.2	19
Organic matter	(215)[b]	(205)[b]	25.9	(15)[b]
total nondiatom	55	150	4.8	3
Fecal pellets			5	3[b]
Biogenic silica	(440)[b]	(175)[b]	84.5	(49)[b]
Total	955	960	174[c]	100

[a]Autochthonous calcite.
[b]Not included in total because phase is contained within other components.
[c]Diatoms + calcite + terrigenous + nondiatom organic matter.

occurred prior to stratification in mid-June. The measured diatom P flux to the sediment surface was 85–125 mg/m^2, showing that about 60% of the P incorporated into diatoms was recycled within the water column. The amount estimated to be recycled agreed with measured increases in P concentration in the hypolimnion during the stratified period (Figure 2).

The changes in filtrable P concentrations during the spring period were small in comparison to the magnitude of the P demand generated by primary production and the size of the filtrable P pool. If all P utilized by diatoms came from the spring filtrable P pool, the concentration would be reduced by over 70%. Furthermore, calculated biological P demand from diatoms represents a small fraction of the total demand. We expect nonsiliceous algae to dominate total primary production (see the following discussion of carbon–phosphorus cycling). The filtrable P pool was largely organic and colloidal P; dissolved reactive P concentrations represented only about 30% of the filtrable P in spring. Apparently a major portion of the filtrable P pool was unavailable for rapid utilization by diatoms.

Table II. Component-Specific Annual Fluxes of P

Component	Amount (mg/m^2)	Percent of Total
Diatom-P	85–125	50–55
Calcite-P	21–32	12–15
Terrigenous-P	41–50	22–26
Fecal Pellet-P	10–20	6–8
Total	157–227	

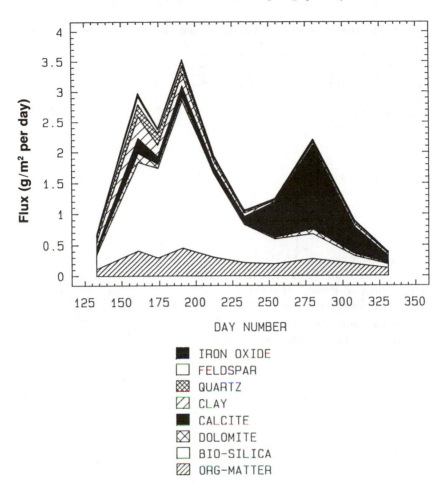

Figure 13. Component flux (sediment-trap-measured) of particulate material deposited at the lake bottom.

The diatom demand occurred mainly during the prestratification period. It is unlikely that external loading provided a major part of the 200 mg of P/m^2 utilized during this 2-month period. Thus, the probable sources were either rapid recycling of diatom-P, other components of the particulate P pool, or bottom sediments.

Evidence that algae utilize P from the particulate pool was found in the large and rapid changes in P content of the <8.2-μm standing-crop fraction and colloidal phases in early spring (Figure 14). These fractions were dominated throughout the mixed period by allochthonous phases, and little dilutional input of new biogenic silica occurred until day 170. Over the period of highest diatom demand, a net removal of 1.5 mg/g of P (5.0 → 3.5 mg/g), or 70 mg/m^2 occurred from this particle pool. The loss of P was also

Table III. Diatom-Associated Uptake, Regeneration, and Deposition of P
in Lake Michigan

Flux Component	Explanation	P Flux (mg/m²)
Diatom phosphorus demand	Diatom P = 2.0 mg/g Si:P = 175	
Mixed period	Si Demand = 30.7 g/m²	180–225
Stratified period	Si Demand = 11.4 g/m²	70–85
Total		250–310
Diatom P flux to sediment surface	Diatom flux = 104 g/m² Diatom P = 0.8–1.2 mg/g	85–125
Diatom P recycled in water column	Demand–deposition	125–225
Increase in midhypolimnion P	Measured changes	85–125

shown by changes in the Si:P ratio in the <8.2 fraction (~1–8.2 μm) (Figure 15) increasing from 40 to 70 over the mixed period. We do not have concurrent colloidal-P data. However, in mid-April 1991 we measured a colloidal-P pool of ~170 mg/m² (9.5 mg of P/g, 110 mg of colloidal mass/m³). If the proportions utilized by diatoms from the colloidal and <8.2-μm fractions were similar, an additional supply of 50 mg/m² was possible. Thus,

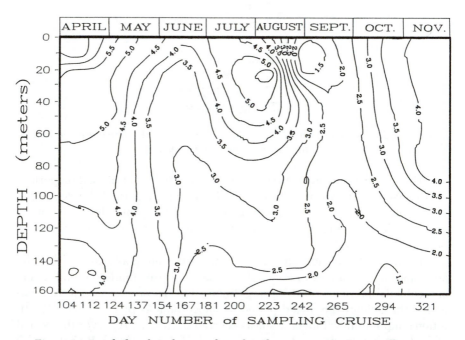

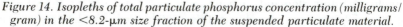

Figure 14. Isopleths of total particulate phosphorus concentration (milligrams/gram) in the <8.2-μm size fraction of the suspended particulate material.

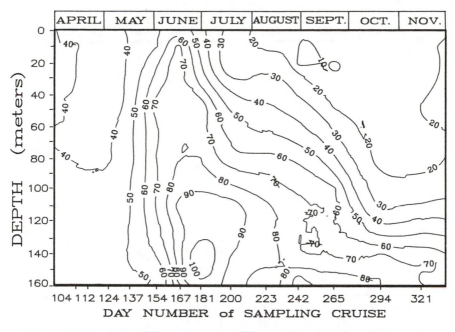

Figure 15. Isopleths of the molar ratio of biogenic Si to P in the <8.2-μm size fraction of the suspended particulate material.

over half of the mixed-period diatom demand could be provided by colloidal and particulate P.

Rapid recycling of diatom P is apparently significant, as revealed by trends in the Si:P ratios of diatom-dominated size fractions of the standing-crop and sediment trap material. A doubling (100 → 200) of the Si:P ratio in the 70–20-μm trap fraction (Figure 16a) was evident. In addition, a three-to fourfold increase (100 → 300–400) occurred in the 111–70-μm trap fraction (Figure 16b), and the ratio in the 70–20-μm standing-crop fraction changed by over a factor of 2 (80 → 200) from late April through mid-June. These particle fractions represented "older" diatoms and clearly pointed to significant loss of P over the mixed period. Additional P was lost from diatoms after stratification developed. This loss resulted in a buildup of filtrable P in the hypolimnion, as discussed previously.

Transport by Calcite. The annual flux of calcite into the uppermost trap (29 m below the lake surface) was estimated to be 35 g/m², and the mean measured P content of this phase was estimated at 1.05 mg/g. A comparison of the calculated 29-m P flux, 37 mg/m², with estimates of deposition to bottom sediment, indicated that 5–16 mg/m² (13–43% of upper-water-column flux) was returned to the water column. This relatively small regeneration flux was not detected in profiles of meta- and hypolimnetic

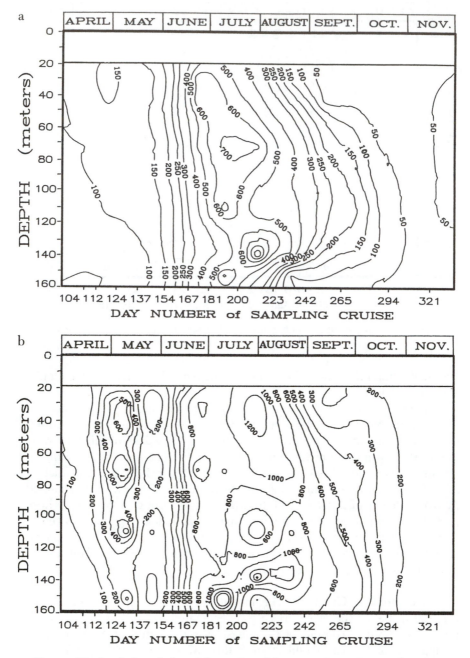

Figure 16. Isopleths of the molar ratio of biogenic Si to P in sediment-trap-collected material: a, 63–19-μm size fraction; and b, 114–63-μm size fraction.

filtrable phosphorus. Estimates of calcite production based on changes in epilimnetic profiles of filtrable calcium are consistent with upper-trap flux. Loss of phosphorus was not proportional to calcite dissolution, as indicated by an increase in P content from 1.05 to 1.2 mg/g.

Uptake and Release by Other Components.   Our approach did not allow quantitative assessment of P release from other particle components during settling because we did not have an estimate of the input term (i.e., flux of P into terrigenous material) and fecal pellets. Though P sorbed to terrigenous material may become available in response to a water-column demand (e.g., primary production), it is unlikely that significant loss of P from terrigenous phases would occur during settling in these nonlabile, previously equilibrated phases. Upper-water-column fecal pellet fluxes were estimated to range from 10 to 20 g/m^2 (*20*). Ferrante and Parker (*26*) reported an average settling distance of 60 m before pellet breakup. Thus 5–15 g/m^2 of fecal pellet mass (15–45 mg of P/m^2), was estimated to be released into the hypolimnion.

Regeneration at the Sediment–Water Interface.   Phosphorus recycling from surficial sediments was calculated by comparing the depositional flux, on the basis of sediment-trap measurements, with the rate of P burial in accumulating sediments. The estimated depositional flux was 185 mg of P/m^2 per year (*see* Figure 6). Estimates derived from the summation of components fall in the range of 161–232 mg/m^2 per year. This range is in good agreement with measured trap flux, indicating that major sedimentation vectors and their associated phosphorus flux are represented.

The mass sedimentation rate (accumulation rate) at the 160-m station is about 150 g/m^2 per year, as measured by ^{210}Pb dating (*27*). The average sedimentation rate for the southern basin is about 69 g/m^2 per year (*28*), corresponding to a sediment focusing factor of 2.17 at our station. The P accumulation flux at our station is about 170 mg/m^2 per year, as calculated from a mean P concentration in surface sediments of 1.135 mg/g and a mass sedimentation rate of 150 g/m^2 per year. The corresponding focusing-corrected accumulation, 78.5 mg of P/m^2 per year, represents 58% of the deposited P (Table IV). We assume that the P concentration in surface sediments at our station is representative of sediments accumulating in other depositional areas of the lake.

The surface-sediment P concentration at our station is actually higher than the mass-weighted mean concentration in the primary depositional flux (1.135 versus 0.950 mg of P/g), a result of less efficient recycling of P than of major components of the depositional flux (biogenic silica, calcite, and organic matter). In total, 63% of the primary mass flux is recycled annually at the sediment surface. This change would result in a hypothetical recycling-corrected P concentration in the primary flux of 2.57 mg/g. A comparison

Table IV. Phosphorus Recycling at Surficial Sediments in Lake Michigan

Flux Component	Explanation	P Flux $(mg/m^2\ y^{-1})$
Depositional flux	Sediment trap measurements	185
Burial flux	Sedimentation rate × sediment P concentration, focusing corrected	78.5
Recycling flux	Deposition–burial	106.5

of this value with surface-sediment concentration indicates an annual recycling percentage of 56%.

As expected for a system near steady state on an annual time scale, the burial rate approximately equals the estimated external loading. From 1979 to 1982, estimated total-P loadings to Lake Michigan decreased from about 100 to 50 mg of P/m^2 per year (5, 6). Sediment mixing results in accumulation rates reflecting conditions averaged over several years. Although it represents 42% of the P deposited annually at the sediment surface, annual burial corresponds to only 10% of the P in the water column in the spring (mid-April) or 28% of the P incorporated into diatoms annually.

Nepheloid Zone.   Detailed profiles of filtrable and total P document the progressive buildup of a nepheloid P pool (Figure 2). Most nepheloid P is in the filtrable (dissolved plus colloidal) fraction and reflects regeneration of the current year's sedimentation. The magnitude of the nepheloid pool accumulating annually is somewhat larger than our independent estimate of annual recycling. Thus, lateral inputs of phosphorus to this region may be significant.

The nepheloid P pool is approximately 50% particulate (>0.4 μm) through mid-August, but shifts rapidly to filtrable forms thereafter. The timing reflects the interplay of diagenesis and supply of phosphorus vectors to the sediment surface. A mean regeneration rate of 1.85 mg of P/m^2 per day is calculated over the period from thermal stratification to maximum areal nepheloid concentration (late November). Rates ranged from <0.2 (late spring) to >4.0 mg of P/m^2 per day (October–November). Only a small fraction of these rates–pools can be attributed to pore-water diffusion of P into the overlying nepheloid region. Pore-water fluxes range from about 0.05 to 0.2 mg of P/m^2 per day for midlake and near-shore sediments, respectively (8, 29). Thus, only 10% of the observed nepheloid filtrable P pool can be attributed to pore-water diffusion.

The nepheloid P pools observed in 1989 and 1990 were similar in magnitude to those measured in 1982. In late August 1989, a nepheloid excess P pool of 241 mg/m^2 (94% filtrable) was measured. In mid July of 1990, the pool contained 167 mg/m^2 (96% filtrable). Concentrations of filtrable P, 2 m off-bottom, were 12.9 μg/L in July 1990 and 23.0 μg/L in August 1989, levels 4 and 5 times the corresponding upper hypolimnetic concentrations.

As illustrated in Figure 17, a large component of nepheloid filtrable P is associated with colloidal-sized particles. The availability of colloid-bound P will depend on whether the association results from repartitioning (i.e., sorption of orthophosphate to different phases after regeneration) or release in colloidal form as carrier particles undergo rapid diagenesis.

In fall the nepheloid zone contains about 20–30 mg/m² of particulate P, compared with a filtrable pool of 150–250 mg/m² that is available for advection at turnover. Roughly one-third of the filtrable pool is colloidal. If uniformly resuspended throughout a 160-m water column, nepheloid colloidal P would represent between 0.3 and 0.5 mg of P/m³ (μg/L).

Epilimnetic Cycling of Phosphorus and Carbon.   Information on carbon cycling is useful in assessing P uptake and regeneration. The few available estimates of annual primary production (*30, 31*) range from 100 to 180 g of C/m². The diatom carbon demand of 12–15 g of C/m² would represent only 8–10% of total carbon fixed if a value of 140 g/m² is used for primary production. Therefore, nonsiliceous algae must account for the remaining demand or large excretory carbon losses must be invoked. If 125 g of C/m² of total carbon is fixed by nonsiliceous autotrophs, the additional phosphorus demand would be ~3000 mg/m² (by using a C:P ratio of 125:1). If nonsiliceous production occurs in a 25-m photic zone over a 6-month

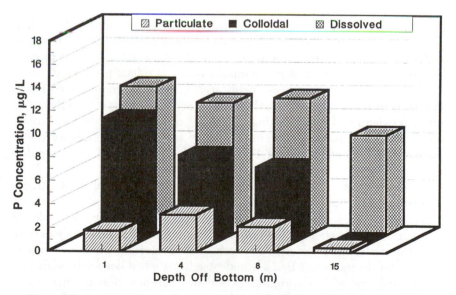

Figure 17. Concentrations of particulate P, colloidal P, and dissolved P in the nepheloid zone in August 1989.

period, and particulate P is 2 μg/L, the residence time of particulate P is only 3.0 days (0.33 per day) (i.e., the pool is turned over 60 times during the 6-month period).

Several estimates of turnover by zooplankton are available. Fluxes of 0.07–0.18 μg of P/L per day were estimated by Scavia et al. (7). Busch and Brooks (32) determined zooplankton excretion rates of 0.4–1.4 μg of P/L per day for summer epilimnetic populations. Equating of these turnover rates to a 6-month period and a 25-m photic zone results in areal fluxes in the range of 500–5000 mg of P/m^2.

In comparison, our measurements of epilimnetic zooplankton biomass over the study period (3540 mg/m^2), combined with Scavia's (7) weight-specific zooplankton regeneration rate of 1.7 μg of P/mg per day, give a flux of ~800 mg of P/m^2. These estimates of phosphorus excretion rates are similar to our carbon-based phosphorus demand.

The fish community also contributes to P regeneration. Kraft (33) presented a mass-balance analysis of P regeneration by alewife in the epilimnion of Lake Michigan that was based on a near-shore mid-1970s scenario. In the late summer months, daily phosphorus regeneration rates obtained were on the order of 1.0 μg of P/L per day (i.e., similar to cited zooplankton rates). Additionally, rates of P incorporation into alewife biomass were about 0.25 μg of P/L per day. Dramatic shifts subsequently occurred in the fish population structure. However, despite the difficulties in extrapolation to off-shore mid-1980s conditions, the potential exists for significant regeneration and accumulation of P via fish.

Resuspension.   Sediment resuspension can play a key role in cycling of P and other particle-associated elements. Resuspension can resupply the water column with both nutrient–contaminant and carrier–sorbent phases. However, measurement of resuspension fluxes is difficult. Although sediment traps have been used, problems arise in applying trap-derived fluxes to resuspension calculations. The behavior of sediment traps at high Reynolds number (winter and spring mixing conditions) is not predictable based solely on standing-crop particle concentrations and settling rates. Thus we calculated the resuspended pool of P by using detailed profiles of suspended material.

This approach is based on the premise that Al can be used as a tracer for bottom sediment material and that the concentration of Al in resuspendable surface sediment is fairly uniform basinwide. Detailed profiles of size-fractionated particulate aluminum concentrations spaced closely in time over the unstratified period show vertical concentration profiles at nearly uniform levels, indicating that a pseudosteady state had been achieved. The mean areal pool of Al during this period was designated as the net resuspended pool (80–90% settles from the water column by September), and the quantity of surface sediment required to supply this pool was calculated.

Assuming a bulk density of 1.05 g/cm^3 and a dry weight fraction of 0.1 for the interface sediment, 0.38 mm of sediment would supply the observed 160-m water-column burden of resuspended phases, approximately half the basinwide average annual linear sedimentation. The corresponding amount of sediment was consistent with the mass of allochthonous components in the water column during the March–May spring mixing period (200–300 mg/m^3). The quantity of resuspended P was calculated as the product of mass of resuspended sediment (g/m^2) and phosphorus concentration in surface sediment (mg of P/g). For a 160-m water column, the amount was 48 mg of P/m^2 (25 mg of P/m^2 for the mean water-column depth of 85 m). The resuspended P flux (25 mg of P/m^2) was also obtained from the product of resuspended Al (mg/m^2) and the P:Al ratio in bottom sediment.

The amount of P supplied by resuspension was relatively small compared with water-column standing pools and major flux vectors. Thus, resuspension of bottom sediments may not be a major mode of phosphorus resupply. The pool of resuspendable P is finite. The deposition–resuspension cycle will not increase the amount of P in this pool unless P is added from another source (e.g., by diffusion of P from lower sediment levels). However, the diffusive flux would be relatively small. The resuspendable particulate P can be recycled during spring mixing by repeated deposition and resuspension, but this cycle does not increase the amount of P in the resuspendable pool. Eadie et al. (24) reported a resuspended P flux (sediment-trap-based) of 3200 mg of P/m^2, 66 times our estimate here. However, this large P flux would require the resuspension of over 2.0 cm of surface sediment and much higher suspended Al levels than were measured in the water column.

Carbon and Phosphorus Burial Efficiencies. The estimate of diatom carbon demand (12–15 g/m^2 per year) is consistent with the flux of carbon to the sediment surface. With sediment-trap fluxes corrected for resuspension, we measured a total annual deposition flux of 12.5 g of C/m^2. In comparison, Eadie et al. (24) obtained 23 g of C/m^2 for a 100-m station, based on three midsummer metalimnion deployments. Of our total, 83% of the carbon was associated with diatoms, and the primary diatom carbon flux was 10.3 g of C/m^2. Thus, about 15–30% of the diatom carbon was regenerated in the water column during sedimentation. Approximately 10% of the diatom flux reached the sediment surface encapsulated in copepod fecal pellets; the remaining 90% was unpackaged.

The estimated primary depositional flux of carbon indicated that only 9% of fixed carbon (total primary production) reached the sediment surface and that diatoms delivered carbon with an average efficiency of 77%, whereas nondiatom carbon was delivered to sediments with an average efficiency of only 2.0%. Carbon is accumulating in recent sediments at a rate of 3.1 g of C/m^2. This rate indicates that 9.4 g of C/m^2 per year (75%) of the primary flux was lost at or near the sediment surface. Therefore, about 2.2% of the fixed carbon ultimately became incorporated into recent sediments. Similarly, about 2.7% of the annual P demand for primary production was removed by sedimentary burial.

Colloidal Phosphorus.   Our data, though necessarily limited in temporal detail, indicate that the colloidal fraction (0.05–1.0 μm) is one of the larger phosphorus pools and probably plays an important role in supplying phosphorus and buffering concentrations in the water column. The magnitude of the colloidal-P association is illustrated in Figures 18 and 19, with selected data from four recent cruises. Data from 1990–1991 are representative of epilimnetic levels, whereas 1989 data apply strictly to nepheloid conditions. Colloidal phosphorus was enriched by at least 50% over the next largest size fraction on all four dates except late May 1990. The extreme enrichment (14-fold) in the nepheloid region indicated diagenesis and reassociation, as well as dilution of the <8.2-μm fraction with carbonate phases. In early spring, colloidal-P concentrations are relatively high and probably serve as a source of P for algal production in the water column during the mixed period. Supporting evidence was found in the May data, showing a depletion of P in colloidal-size fractions.

Selective diagenesis and recycling of P during the summer months resulted in the replenishment of the colloidal fraction to enrichment levels above spring values (*see* mid-July data). However, colloid mass concentrations were lower in midsummer than in spring. The size of the colloidal-P pool actually declined in the main water column during midsummer. The colloidal-P pool represents a major component of the P cycle. However,

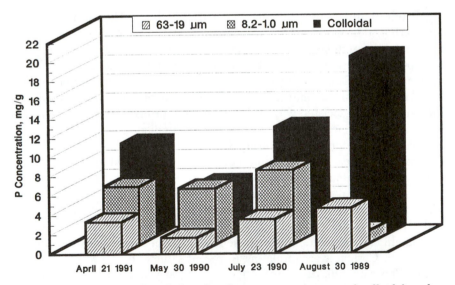

Figure 18. Comparison of the phosphorus concentrations of colloidal and particulate phases on selected dates in 1989–1991.

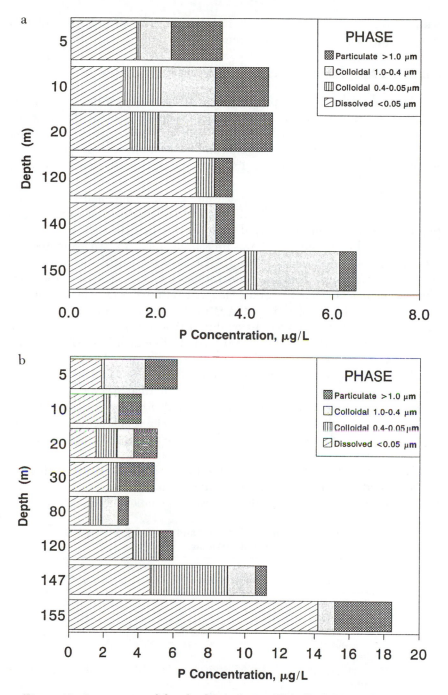

Figure 19. Comparison of dissolved P (<0.05 μm), colloidal P (0.05–0.4 and 0.4–1.0 μm), and particulate P (>1.0 μm) concentrations in Lake Michigan (southern basin): a, May 1990; b, July 1990.

pathways of formation and removal, and the role of colloidal P in regulating P availability and residence times remain uncertain.

Annual Phosphorus Budget and Residence Times

Annual Budget.   Phosphorus fluxes and storage over the annual cycle in Lake Michigan are summarized in Table V and Figure 20. The analysis derives from the detailed data collected in 1982 (early spring to late fall), supplemented by our more recent data and results from other investigations. As discussed previously, the results are representative of conditions over the past decade.

In early spring (April), the 160-m water column contained 750–800 mg of P/m^2, corresponding to 400–425 mg of P/m^2 normalized to the mean depth of 85 m. During the mixed period extending through mid-June, a change in total P of $+49$ mg/m^2 was measured (Table V). We estimated, on the basis of observed silicon cycling, that 53 mg/m^2 of P was lost to the sediment surface over the spring unstratified interval. This calculation implied an estimated mixed period P loading (internal and external) of 102 mg/m^2. A net change of $+66$ mg of P/m^2 between April and November, coupled with measured sediment-trap flux of 185 mg of P/m^2, gave an annual load of 251 mg/m^2. A stratified period load of 149 mg/m^2 was calculated by difference (Table V).

A net loss of approximately 230 mg of P/m^2 occurred in the water column (nepheloid zone excluded) from the onset of thermal stratification to late October. Measured trap fluxes of 122 mg of P/m^2 over this period were almost identical with the measured water-column loss of particulate P (Figure 1). Loadings and fluxes are balanced by regeneration of filtrable P in the nepheloid region.

Loss rates of total P from the stratified water column, based on changes in integrated standing-crop concentrations, ranged from a high of 5.5 mg/m^2 per day (mid-July to late August) to a much lower rate of 0.74

Table V. Summary of Seasonal and Annual P Fluxes in Lake Michigan

Source of P	Spring	Stratifed Period	Total
Change in Storage			
Total water column	49		66
Above nepheloid		-230	
Nepheloid		$+200$	
Output			
Sedimentation	53	132	185
Input			
Internal + external	102	149[a]	251

NOTE: All values are in milligrams per square meter.
[a]Calculated as total input − spring input.

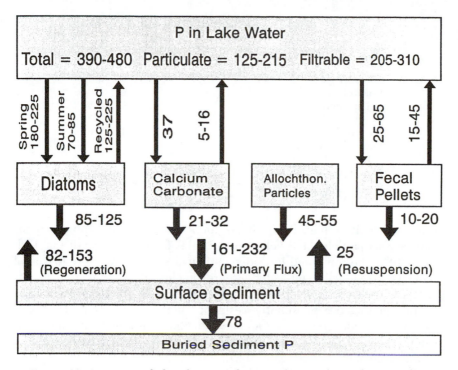

Figure 20. Summary of phosphorus cycle in southern Lake Michigan. Values in boxes are areal concentrations (milligrams/square meter); values on arrows are fluxes (milligrams/square meter per year). Areal concentrations are based on the mean depth (85 m) of the water column.

mg/m^2 per day (late August through late October). An average rate for the entire stratified period of 3.0 mg/m^2 per day was obtained by using a gross P flux of 330 mg/m^2.

Residence Times.   Phosphorus residence times with respect to major depositional processes (*see* Tables II and IV) are summarized in Table VI. In comparison, the total-P residence time based on external loading is about 4.5 years. Residence times were calculated for a mean water-column depth of 85 m, and steady state was assumed. Although transport of P to the sediment surface by the combination of diatoms, calcite, and terrigenous material is relatively rapid, the low burial efficiency results in a relatively long residence time for total P (about 5 years). In comparison, the residence time for Pb is about 0.6 years (*20*). Thus, the response time for P changes with respect to loading should be on the order of 5–15 years.

The effectiveness of a partition coefficient (K_d) approach in predicting trace metal residence times in Lake Michigan was demonstrated by Shafer and Armstrong (*20*). Nearly 99% of the variation in residence time among five trace elements was modeled on the basis of K_d. The same concepts can

**Table VI. Phosphorus Residence Times with Respect
to Various Depositional Processes**

Depositional Process	Total P	Filtrable P
Diatoms	3.4–5.0	1.7–2.5
Calcite	13–20	6.4–10
Terrigenous	8–10	4.0–4.8
Fecal pellet	21–43	11–21
Diatoms + calcite	2.7–4.0	1.4–2.0
Gross deposition (sediment traps)	2.3	1.2
Net deposition (burial)	5.4	2.7

NOTE: All values are given in years.

be applied to P cycling. A K_d (L/kg) for P of ~9×10^5, computed for diatomaceous particles, fell between estimates for Zn (4.4×10^5) and Pb (1.2×10^6). The computed P residence time (3.9 years) was close to that of Zn (3.6 years) and fell slightly off the regression line of residence time versus partition coefficient developed from the five metals (20). (The regression equation predicted a phosphorus residence time of 1.7 years.) The close similarity in residence time and K_d between phosphorus and zinc suggested a similar biological activity and binding mechanism. It may be possible to use these elements as surrogates for each other.

Phosphorus fluxes in Lake Michigan are relatively large (Figure 20). Depositional fluxes, mainly carried by diatoms, calcium carbonate, and allochthonous particles, transport almost 50% of the water column P pool (400–425 mg/m²) to the sediment surface. After stratification, deposition (150 mg of P/m²) decreases the supply of P in the epilimnion available for primary production. Clearly, recycling plays a major role in regulating the fluxes of P in Lake Michigan. The combination of recycling from diatoms in the water column and at the sediment–water interface (about 300 mg/m² per year) substantially exceeds the external loading (about 100 mg/m² per year). The contribution from resuspension is relatively small. Annual burial removes almost 20% of the spring water-column P from the lake. Although the amount of P contained in the water column is linked to the input from external sources, internal regeneration fluxes are large in comparison to the external loading. Thus, changes in external loading have a relatively small influence on the amount of P available to support primary production. However, in the long term, declines in P levels are expected in response to loading reductions (6).

Acknowledgments

This work was funded by the University of Wisconsin Sea Grant College Program under grants from the Office of Sea Grant, National Oceanic and

Atmospheric Administration, U.S. Department of Commerce, and from the State of Wisconsin (Federal Grants: NA80AA–D–00086, Project R/MW–24; NA84AA–D–0065, Project R/MW–37; and NA90AA–D–SG469, Project R/MW–44). The support provided by the Center for Great Lakes Studies (University of Wisconsin–Milwaukee) and captain and crew of the R/V *Neeskay* is especially acknowledged. Phil Emmling provided invaluable assistance during field sampling. In the laboratory, the dedicated efforts of Rae Mindock, Paul Klebs, and James Dowse ensured that the analytical work was of high standards. Mike Fullerton provided the primary carbon data set.

References

1. Chapra, S. C.; Wicke, H. D.; Heidtke, T. M. *J. Water Pollut. Control Fed.* **1983,** *55,* 81–91.
2. Edgington, D. N. In *The Future of Great Lakes Resources;* WIS-SG–84–145; University of Wisconsin Sea Grant Institute: Madison, WI, 1984; pp 25–33.
3. DePinto, J. V.; Young, T. C.; McIlroy, L. M. *Environ. Sci. Technol.* **1986,** *20,* 752–759.
4. Scavia, D.; Fahnenstiel, G. L.; Evans, M. S.; Jude, D. J.; Lehman, J. T. *Can. J. Fish. Aquat. Sci.* **1986,** *43,* 435–443.
5. Chapra, S. C.; Sonzogni, W. C. *J. Water Pollut. Control Fed.* **1979,** *51,* 2524–2533.
6. Lesht, B. M.; Fontaine, T. D., III; Dolan, D. M. *J. Great Lakes Res.* **1991,** *17,* 3–17.
7. Scavia, D.; Lang, G. A.; Kitchell, J. F. *Can. J. Fish. Aquat. Sci.* **1988,** *45,* 165–177.
8. Conley, D. J.; Quigley, M. A.; Schelske, C. L. *Can. J. Fish. Aquat. Sci.* **1988,** *45,* 1030–1035.
9. Scavia, D. *J. Fish Res. Board Can.* **1979,** *36,* 1336–1346.
10. Armstrong, D. E.; Hurley, J. P.; Swackhamer, D. L.; Shafer, M. M. In *Sources and Fates of Aquatic Pollutants;* Hites, R. A.; Eisenreich S. J., Eds.; ACS Symposium Series 216; American Chemical Society: Washington, DC, 1987; pp 491–518.
11. Wahlgren, M. A.; Nelson, D. M. Argonne National Laboratory Annual Report, Part III; ANL–76–88, 1976; pp 103–106.
12. Leppard, G. G.; De Vitre, R. R.; Perret, E.; Buffle, J. *Sci. Total Environ.* **1989,** *87/88,* 345–354.
13. Eggiman, D. W.; Betzer, P. R. *Anal. Chem.* **1976,** *48,* 886–890.
14. Strickland, J. D. H.; Parsons, T. R. *A Practical Handbook of Seawater Analysis,* 2nd ed.; *Bull. Fish. Res. B. Can.* **1972,** No 167, 310 pp.
15. Murphy, J.; Riley, J. P. *Anal. Chim. Acta.* **1962,** *27,* 31–36.
16. Erickson, R. J. "Horizontal Distribution and Flux of Phosphorus in Fish Lake, Wisconsin." Ph.D. Thesis, University of Wisconsin—Madison, Madison, WI, 1980.
17. DeMaster, D. J. *Geochim. Cosmochim. Acta.* **1981,** *45,* 1715–1732.
18. Stauffer, R. E. *Anal. Chem.* **1983,** *55,* 1205–1210.
19. Shafer, M. M. "Biogeochemistry and Cycling of Water Column Particulates in Southern Lake Michigan." Ph.D. Thesis, University of Wisconsin—Madison, Madison, WI, 1988.

20. Shafer, M. M.; Armstrong, D. E. In *Organic Substances and Sediments in Water: Processes and Analytical;* Baker, R. A., Ed.; Lewis: Chelsea, MI, 1991; Vol. 2, pp 15–47.
21. Kitchell, J. F.; Crowder, L. B. *Environ. Biol. Fish.* **1986,** *16,* 205–211.
22. Kitchell, J. F.; Evans, M. S.; Scavia, D.; Crowder, L. B. *J. Great Lakes Res.* **1988,** *14,* 109–114.
23. Scavia, D.; Fahnenstiel, G. L. *J. Great Lakes Res.* **1987,** *13,* 103–120.
24. Eadie, B. J.; Chambers, R. L.; Gardner, W. S.; Bell, G. L. *J. Great Lakes Res.* **1984,** *10,* 307–321.
25. Shafer, M. M.; Armstrong, D. E. *Trans. Am. Geophys. Union* **1988,** *69,* 1143.
26. Ferrante, J. G.; Parker, J. I. *Limnol. Oceanogr.* **1977,** *22,* 92–98.
27. Armstrong, D. E.; Swackhamer, D. L. In *Physical Behavior of PCBs in the Great Lakes;* Mackay, D.; Patterson, S.; Eisenreich, S.; Simmons, M. S., Eds.; Ann Arbor Science: Ann Arbor, MI, 1983; pp 229–244.
28. Robbins, J. A.; Edgington, D. N. *Geochim. Cosmochim, Acta.* **1975,** *39,* 285–304.
29. Quigley, M. A.; Robbins, J. A. *Can. J. Fish. Aquat. Sci.* **1986,** *43,* 1201–1207.
30. Fee, E. J. *J. Fish. Res. Board Can.* **1973,** *30,* 1469–1473.
31. Fahnenstiel, G. L.; Scavia, D. *Can. J. Fish. Aquat. Sci.* **1987,** *44,* 499–508.
32. Busch, J. L.; Brooks, A. S. *Verh. Int. Ver. Theor. Angew. Limnol.* **1988,** *23,* 366–375.
33. Kraft, C. E. "Estimation of Phosphorus Cycling by Fish Using a Bioenergetics Modeling Approach." Ph.D. Thesis. University of Wisconsin—Madison, Madison, WI, 1991.

Received for review October 23, 1991. Accepted revised manuscript August 4, 1992.

Retention of Sulfur in Lake Sediments

N. R. Urban

Lake Research Laboratory, Swiss Federal Institute for Water Resources
and Water Pollution Control (EAWAG/ETH), CH–6047 Kastanienbaum,
Switzerland

*Measurements of S cycling in Little Rock Lake, Wisconsin, and Lake
Sempach, Switzerland, are used together with literature data to show
the major factors regulating S retention and speciation in sediments.
Retention of S in sediments is controlled by rates of seston (planktonic
S) deposition, sulfate diffusion, and S recycling. Data from 80 lakes
suggest that seston deposition is the major source of sedimentary S
for approximately 50% of the lakes; sulfate diffusion and subsequent
reduction dominate in the remainder. Concentrations of sulfate in
lake water and carbon deposition rates are important controls on
diffusive fluxes. Diffusive fluxes are much lower than rates of sulfate
reduction, however. Rates of sulfate reduction in many lakes appear
to be limited by rates of sulfide oxidation. Much sulfide oxidation
occurs anaerobically, but the pathways and electron acceptors remain
unknown. The intrasediment cycle of sulfate reduction and sulfide
oxidation is rapid relative to rates of S accumulation in sediments.
Concentrations and speciation of sulfur in sediments are shown to
be sensitive indicators of paleolimnological conditions of salinity,
aeration, and eutrophication.*

SULFUR FULFILLS MANY DIVERSE ROLES in lakes. As the sixth most abundant element in biomass, it is required as a major nutrient by all organisms. For most algae, S is abundant in the form of sulfate in the water column; however, in dilute glacial lakes in Alaska (*1*) and in some central African lakes (*2*) low concentrations of sulfate may limit primary production. Sulfur also serves the dual role of electron acceptor for respiration and, in reduced forms, source of energy for chemolithotrophic secondary production. Net sulfate reduction can account for 10–80% of anaerobic carbon oxidation in lakes (*3–5*), and hence this process is important in carbon and energy flow. Sulfate reduction, whether associated with uptake of sulfate and incorpo-

0065–2393/94/0237–0323$12.75/0

ration into biomass (assimilatory reduction) or with anaerobic respiration (dissimilatory reduction), generates alkalinity. In many dilute lakes this source of alkalinity is important as a means of neutralizing acid deposition (e.g., 6, 7). Sulfide produced through dissimilatory sulfate reduction can alter the cycling and availability of nutrients, trace metals, and radioisotopes (e.g., 8–10) as well as be toxic to organisms. In addition to direct and indirect effects of sulfur, additional processes are mediated by sulfur-utilizing bacteria. Sulfate-reducing bacteria may be capable of methylating trace metals such as mercury and tin, oxidizing methane, and limiting methane production by competition for electron donors (e.g., 11–14). The numerous possible chemical and biological reactions of sulfur together with the effects of these reactions on numerous other elements render the cycling of S in lakes both important and complex.

Sediment profiles of the forms and rates of accumulation of sulfur have been used in both freshwater and marine systems as indicators of past conditions. Over geologic times, sulfate reduction has played a major role in governing the carbon and oxygen balance of the atmosphere and in regulating global climate (e.g., 15–19). The record of these changing conditions is preserved in the pyrite deposits in marine sediments. Changing conditions of salinity, primary productivity, and oxygenation of bottom waters are similarly recorded in sediment profiles of sulfur species and isotopic composition (e.g., 20–22). In lake sediments, changes in sulfur content and speciation have been interpreted as indicative of changes in trophic status (e.g., 23–25), inputs of acid deposition (e.g., 26–30), and inputs of salt water (21).

Interpretation of sediment stratigraphy can be complicated by diagenetic reactions that alter the original profile. For instance, reactions of sulfur with organic matter have been shown to alter stratigraphic records of organic biomolecules (e.g., 31–33). Such diagenetic alterations can lead to erroneous interpretations of sedimentary profiles. Diffusion of sulfate into sediments can lead to zones of sulfur incorporation that are not contemporaneous with sediment deposition in the same zones (e.g., 21, 28, 34). Organic sulfur from plankton can be transformed into inorganic forms (e.g., 35–38). Clearly, a thorough understanding of the processes controlling retention of sulfur in sediments as well as an understanding of diagenetic transformations within sediments are prerequisites for interpretation of stratigraphic records of sulfur in lake sediments. This chapter uses numerous measurements of sulfur speciation in lake sediments and recent research to clarify the factors that regulate retention of sulfur in lake sediments, the possible diagenetic alterations, and the areas of uncertainty that continue to impede our ability to interpret sediment stratigraphy.

Forms, Abundance, and Patterns of S in Sediments

Sulfur typically is enriched in lake sediments (300–64,000 $\mu g/g$) relative to crustal materials (30–2700 $\mu g/g$; 39, 40) and surface soils (50–2000 $\mu g/g$;

41–45). In some lakes S content and S:C ratios can exceed values typical of marine sediments (*46, 47*), although this abundance tends to be the exception rather than the rule (*9, 20*). In 53 of 72 lakes for which data are available, concentrations of S are higher in surface (recent) sediments than in deeper (background) sediments. This contrast has been variously attributed to eutrophication, air pollution, and diagenesis (e.g., *7, 23–27, 29, 30, 48, 49*). Differences in sedimentary S content among lakes have been attributed to differences in lake-water sulfate concentrations, sediment iron content, and sediment organic carbon content (*24, 26, 30, 50, 51*). However, none of these factors individually can reliably predict the S content of lake sediments (Figure 1).

In lake sediments sulfur exists in both organic and inorganic forms. Inorganic species include H_2S, iron monosulfides (i.e., those with Fe:S ratios close to 1, including amorphous FeS, (hydro)troilite, mackinawite, pyrrhotite, and greigite), pyrite or marcasite, adsorbed or dissolved SO_4^{2-}, elemental sulfur (S^0), polythionates, and a variety of soluble ionic species (SO_3^{2-}, $S_2O_3^{2-}$, S_n^-, and $S_4O_6^{2-}$). In saline lakes gypsum or anhydrite may be present. The dominant species typically are the monosulfides and pyrite. Monosulfides are identified analytically by decomposition with nonoxidizing acid and frequently are called acid-volatile sulfides (AVS). Pyrite and S^0 frequently are analyzed together by the chromium-reduction technique of Zhabina and Volkov (*52*) and referred to jointly as chromium-reducible sulfur (CRS). Organic S compounds are found among proteins, polysaccharides, and lipids; they include thiols, disulfides, thiophenes, thiolanes, sulfones, and sulfoxides. Several volatile species exist [dimethyl sulfide (DMS), methanethiol (MSH), and carbonyl sulfide (COS)], and many species remain to be identified. Analytically, organic S frequently is divided into two groups: carbon-bonded S and ester sulfates. Carbon-bonded S may be analyzed by desulfurization with Raney nickel, but this includes a variety of functional groups. Questions remain about the specificity and comparability of hydriodic acid reduction and acid hydrolysis, the two techniques used to analyze ester sulfates (*41, 53, 54*). Because of a lack of suitable analytical techniques, most of the organic S in sediments has not been characterized.

A wide variation is observed in relative and absolute abundances of different S species in lake sediments. The fraction of sedimentary S in inorganic forms ranges from 0 to 99%. The relative importance of AVS and CRS, the dominant inorganic species, also varies widely (ratios of AVS:CRS range from 0.01 to 100). Concentrations of free H_2S greater than 100 μmol/L occur only in pore waters in lakes with extensive or permanently anoxic hypolimnia (e.g., *55, 56*). Elemental S occurs in low concentrations; in organic-rich sediments the concentrations typically are less than 5 μmol/g (*25, 27, 30*). In more inorganic sediments or in sediments under layers of S-oxidizing bacteria, concentrations of S^0 can reach 280 μmol/g (*27, 56, 57*). Thiosulfate, polythionates, and polysulfides are reactive intermediates present in low concentrations (*57, 58*). Ester sulfates typically

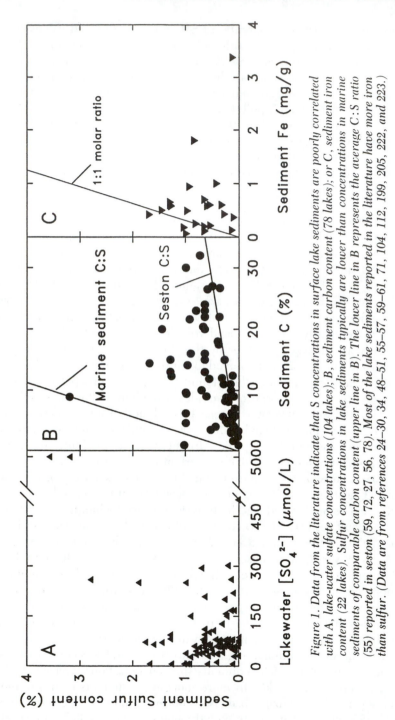

Figure 1. Data from the literature indicate that S concentrations in surface lake sediments are poorly correlated with A, lake-water sulfate concentrations (104 lakes); B, sediment carbon content (78 lakes); or C, sediment iron content (22 lakes). Sulfur concentrations in lake sediments typically are lower than concentrations in marine sediments of comparable carbon content (upper line in B). The lower line in B represents the average C:S ratio (55) reported in seston (59, 72, 27, 56, 78). Most of the lake sediments reported in the literature have more iron than sulfur. (Data are from references 24–30, 34, 48–51, 55–57, 59–61, 71, 104, 112, 199, 205, 222, and 223.)

represent 30–60% of the total organic S (*51, 59–61*). This chapter seeks to identify common patterns of S speciation and the environmental variables and processes responsible for them.

Processes Causing Observed Patterns

Storage of sulfur in lake sediments results from cycling of S in the lake water and postdepositional diagenetic processes in sediments. The major features of S cycling in lakes are well known. Sulfur is a macronutrient; uptake by phytoplankton and subsequent burial of organic S in sediments occur in all lakes (*23, 62, 63*). Putrefaction of organic S releases H_2S (as well as trace amounts of other S gases; *64–66*) that is either fixed in sediments as iron sulfides, lost to the atmosphere via diffusion or ebullition, or oxidized.

Apart from deposition of organic matter, sulfate reduction is the other major source of S found in sediments. Dissimilatory sulfate-reducing bacteria use SO_4^{2-} as an electron acceptor to produce H_2S. As for H_2S produced in algal decomposition, this reduced S may be either fixed in sediments as iron sulfides or organic compounds, or it may be reoxidized by oxygen, metal oxides, or bacteria. A large number of bacterial types (green, purple, colorless, and blue-green) that can oxidize H_2S have been recognized for some time (e.g., *67–69*). The amounts and patterns of S forms in sediments will depend, therefore, on relative rates of seston deposition, sulfate reduction, putrefaction, recycling to the lake water, and diagenetic interconversions of S species.

Deposition, Putrefaction, and Recycling of Seston Sulfur.   Uptake by algae and subsequent sedimentation is a major flux of S within lakes. Sulfur is required by organisms for formation of proteins, structural polysaccharides, sulfolipids, osmotic regulators, and coenzymes. Concentrations of S in algae vary considerably (<1–20 mg/g corresponds to C:S ratios (mass basis) of 22–272; *70, 71*), but do not appear to be influenced by concentrations of SO_4^{2-} in lake water (*1, 72*). A significant fraction of algal S occurs in nonprotein forms such as sulfate esters (*59, 73–75*), but there has been no systematic study of the structure and function of these compounds in freshwater algae. Sulfur generally is taken up by algae as sulfate, although bacteria and some blue-green algae can utilize hydrogen sulfide (*76*). In most lakes sulfate is not a limiting nutrient, and concentrations of sulfate in the water are not greatly affected by algal growth. Quantities taken up by plankton generally are less than 6% of the sulfate pool in the water column (*73, 77*).

Some fraction of planktonic S is released in the water column as H_2S or other volatile S compounds (*65, 66*), and the remainder settles to the sediments, where it is called seston. King and Klug (*73*) calculated that particulate matter (seston) collected in sediment traps had lost 60–75% of the protein S present in algae. Although no change in S content or composition was noted among seston samples collected from traps at 5-, 8-, and 15-m depths in South Lake (59), Losher (56) measured a large progressive decrease in total S content from plankton (8 mg/g) to seston collected at 40-m depth (6 mg/g) to seston collected at 150-m depth (4 mg/g) in Lake Zurich. Relative proportions of carbon-bonded S declined in this sequence, and contributions from inorganic and ester S increased. Existing measurements of S in seston collected in sediment traps (herein termed seston deposition) show a much smaller range of S concentrations than in algae (0.4–14 mg/g; 27, 59, 62, 72, 78), but a large range in C:S ratios (9–105 mass basis). Fluxes of S in seston deposition (31–266 mmol/m² per year; 59, 72, 73) thus are controlled by the magnitude of primary production, recycling of S within the water column, and the unknown factors regulating S content of seston.

Decomposition of seston and recycling of sulfur from sediments remain poorly understood. Assuming that there is no source except seston for the organic S in sediments, King and Klug (73) calculated that 75% of protein S, 42% of ester S, and 46% of total S was mineralized in sediments of Wintergreen Lake. On the basis of similar mass-balance calculations, 43% of ester S and 26% of total S was estimated to be mineralized in South Lake (59) and 60% of total S in Lake Mendota (25). The assumption that all organic S in sediments is derived from seston is not valid in many lakes because much sediment S is derived from the reaction of sulfide with organic matter (72, 79, 80); thus the foregoing estimates probably underestimate the extent of recycling. Correcting for other sources of organic S, Baker et al. (72) estimated that 80% of total seston S is remineralized in Little Rock Lake. No estimates of hydrolysis of sulfate esters are corrected for in situ formation (cf. 79).

Direct measurement of putrefaction is problematic. In laboratory microcosms in which radiolabeled (^{35}S) algae were allowed to settle and decay on top of lake sediments, a net release of less than 5% of the ^{35}S to the water column was observed, and all release occurred within the first 2 weeks (38). However, ongoing microbial uptake of sulfate from the water column may have obscured further release. Maximal potential rates of cystine degradation were estimated by Jones et al. (81) to range from 0.001 to 50 μmol/L per day in Blelham Tarn sediments and by Dunnette (82) to range from 28 to 47 μmol/L per day in sediments from two lakes. Similar measurements of potential rates of hydrolysis of sulfate esters (83) tremendously overestimated the rates calculated by mass balance to occur in sediments of Wintergreen Lake (73). A better understanding of putrefaction is needed to predict retention and concentrations of S in sediments.

Sulfate Reduction.   Dissimilatory sulfate reduction, anaerobic respiration with sulfate as the terminal electron acceptor, is performed by relatively few genera of bacteria (84). Many bacteria and algae are able to reduce sulfate for purposes of assimilation, and several genera are able to reduce elemental S, $S_2O_2^{2-}$, and SO_3^{2-}. Sulfate-reducing bacteria may use a variety of electron donors, including sugars, fatty acids, and hydrogen (*4, 13, 84–88*). Quantitatively, acetate and hydrogen are the most important of these donors (*4, 13, 87, 89, 90*).

Sulfate reduction yields less energy than respiration with oxygen, nitrate, manganese, or iron as the terminal electron acceptor. Hence, thermodynamics predict that it should occur only after alternate electron acceptors have been reduced to low concentrations but above the zone of methanogenesis (*91*). Recently, however, aerobic sulfate reduction was reported (*92, 93*). More typically, oxygen and nitrate are consumed within sediments above 0.5–10 cm, the depth at which sulfate is consumed. Low availability of solid-phase iron and manganese oxides can result in reduction of these species in the same zone as sulfate reduction (e.g., *94, 95*). Kinetics of substrate uptake, microbial growth, and mass transfer as well as thermodynamics determine whether sulfate reduction will occur together with other modes of anaerobic respiration. High affinity for both carbon substrates and electron acceptors allows iron-reducing bacteria to competitively exclude sulfate reducers (*96, 97*) and also allows sulfate reducers to outcompete methanogens (*4, 13, 87*). These competitive interactions are important not only because they promote extensive anaerobic degradation of organic matter (*84, 89*), but also because they determine the diffusive flux of sulfate into sediments by regulating the depth at which sulfate reduction occurs (*4, 90*).

Occurrence and Rates of Sulfate Reduction.   Sulfate reduction is widespread in lakes, as evidenced by depletion of sulfate in sediment pore waters. Pore-water profiles showing depletion of sulfate have been published for more than 35 lakes (e.g., *98, 99*). An absence of sulfate depletion in pore waters does not indicate an absence of sulfate reduction. Sulfate depletion was not evident in pore waters of McNearney Lake, but stable isotope measurements indicated that low rates of sulfate reduction must occur (*1*). Sulfate depletion was noted in 15 of 17 lakes in northeastern North America and Norway (*80, 98*), but uptake of $^{35}SO_4^{2-}$ occurred even in two lakes in which no sulfate depletion was observed. Sulfate production and reduction can occur concurrently, and the former may exceed the latter.

Sulfate reduction occurs in both littoral and pelagic sediments. Cook et al. (*100*) calculated by mass balance that S retention in epilimnetic sediments exceeded that in hypolimnetic sediments. Rudd et al. (*98*) also found higher rates of uptake of $^{35}SO_4^{2-}$ in littoral sediments. In Little Rock Lake, rates of sulfate reduction in intact cores were lower in sandy littoral sediments than in organic-rich pelagic sediments, but rates in shallow bays receiving high leaflitter inputs were comparable to rates in pelagic sediments (*101*).

Reported rates of sulfate reduction range over nearly 3 orders of magnitude (Table I). All measurements reported in Table I are actual, not po-

Table I. Measurements of Sulfate Reduction Rates in Lake Sediments

System	Sulfate Reduction Rate		References
	$nmol/cm^3$ per day	$mmol/m^2$ per year	
Meromictic lakes			
Faro	5–42	36–1200	55
Eutrophic lakes			
Mendota	50–600		85
Wintergreen	458	5580	78
		2560	73
Third Sister	1.6–100		82
Sempach		900–13,000	this study
Maggiore	6–190	2000	104
Lugano	38–69	1200	104
Braband	8500		58
Mesotrophic lakes			
Washington	1.73	43	90
Oligotrophic lakes			
Lawrence	71		4
Little Rock	0–1680	0–5480	101
Estuaries, shallow seas, continental shelf	1–10,000	73–160,000	103

NOTE: All reported measurements are actual, not potential, rates measured with ^{35}S. Blank spaces mean that no data are available.

tential, rates; carrier-free $^{35}SO_4^{2-}$ was used in all cases without addition of unlabeled SO_4^{2-}. Most measurements reported in Table I are from a single date, and some of the variability may reflect seasonal or temperature differences among sites. However, no seasonal variation was observed in Little Rock Lake (101), Wintergreen Lake (78) or Lake Mendota (85). Comparable methodology was used by all investigators reported in Table I; with the exception of Jorgensen (58), all investigators measured reduced ^{35}S only in AVS rather than in CRS. Results of Fossing and Jorgensen (102) and measurements in Lake Sempach (Urban, unpublished data) and Little Rock Lake (101) indicate that very little ^{35}S is incorporated into pyrite during short incubations (<1 h) of freshwater sediments; hence all of the reported rates should be comparable. Rates measured in freshwater lake sediments (2–8500 $nmol/cm^3$ per day) are comparable to rates in estuarine and coastal marine sediments (1–10,000 $nmol/cm^3$ per day; 103). Rates do not increase with increasing trophic status; rates in oligotrophic lakes (71–1680 $nmol/cm^3$ per day; 13, 101) are comparable to rates in eutrophic lakes (6–4000 $nmol/cm^3$ per day; 73, 82, 85, 104). Rates do not vary systematically with concentrations of sulfate in the lake water (Figure 2). Rates also do not vary in proportion to carbon sedimentation rates or sediment carbon content (Figure 3). Clearly, more measurements are needed to further test these conclusions, but the few existing measurements made with comparable techniques contradict several accepted paradigms.

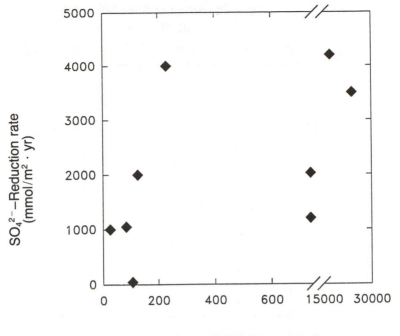

Figure 2. Rates of sulfate reduction in lake sediments reported in the literature range over 3 orders of magnitude and are not correlated with lake sulfate concentrations. All measurements were made with ^{35}S in intact cores or core sections. References are given in Table I.

Factors Controlling Rates of Sulfate Reduction. Factors typically cited as controlling sulfate reduction include temperature, sulfate concentration, and availability of carbon substrates. Although sulfate-reducing bacteria typically exhibit steep responses to temperature (rates increase 2.4- to 3.7-fold per increase of 10 °C; *85, 101, 105*), neither differences between deep and shallow lakes (Table I) nor seasonal variation have been observed in rates of sulfate reduction (*78, 85, 101*). This apparent lack of response of sulfate reduction rates to changes in temperature may indicate that rates are limited by other factors.

Much confusion exists over the effects of sulfate concentration and carbon availability on rates of sulfate reduction (cf. *1, 4, 5, 72, 85, 106*). Sulfate reducers in lake sediments exhibit low half-saturation constants for sulfate (10–70 μmol/L; *4, 72, 78, 85*) as well as for acetate and hydrogen (*4, 13, 87*). The low half-saturation constants allow them to outcompete methanogens for these substrates until sulfate is largely consumed within pore waters (*4, 90*). Low concentrations of sulfate in lakes confine the zone of sulfate reduction to within a few centimeters of the sediment surface (e.g., *4, 90, 98*). The comparability of rates of sulfate reduction in freshwater and marine

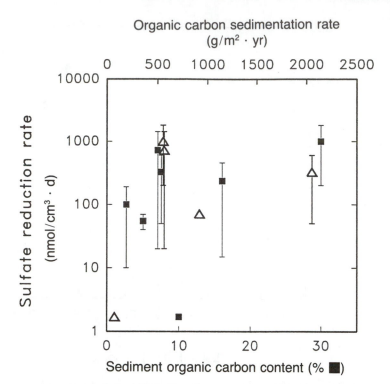

Figure 3. Rates of sulfate reduction (all measured with ^{35}*S) reported in the literature (references in Table I) show no obvious relationship to either sediment carbon content or carbon sedimentation rates (measured with sediment traps). The lowest reported rate of sulfate reduction occurs in the lake with the lowest carbon sedimentation rate, but there is no evidence of carbon limitation among the other lakes. Error bars indicate the range of reported sulfate reduction rates.*

sediments (Table I) and the independence of reduction rates from sulfate concentrations in lake waters (Figure 2) suggest either that sulfate concentration is a poor measure of sulfate availability or that sulfate reduction is not limited by sulfate availability.

Relative rates of sulfate reduction and methanogenesis in lakes of varying trophic status are claimed to indicate that sulfate reduction rates are limited by the supply of sulfate (4, 5, 13). According to this hypothesis, at high rates of carbon sedimentation, rates of sulfate reduction are limited by rates of sulfate diffusion into sediments, and methanogenesis exceeds sulfate reduction. In less productive lakes, rates of sulfate diffusion should more nearly equal rates of formation of low-molecular-weight substrates, and sulfate reduction should account for a larger proportion of anaerobic carbon oxidation. Field data do not support this hypothesis (Table II). There is no relationship between trophic status, an index of carbon availability, and rates of anaerobic

Table II. Relative Importance of Methanogenesis and Sulfate Reduction

Lake	Anaerobic C Oxidation (mmol/m² per day)	% of Anaerobic C Oxidation		References
		SO_4^{2-}	CH_4	
Eutrophic lakes				
Blelham Tarn		5	95	240
Wintergreen[a]	8.8	13	87	4
	108	30	71	78
Mendota[a]		20	80	3
Sempach	18	50–70	30–50	this study
Mesotrophic lakes				
Washington		37	63	90
Oligotrophic lakes				
Lawrence	3	50	50	13
223[a]	9.5	20	72	107
226N[a]	10	16	75	107
227[a]	12	16	82	107
Little Rock	11	83	17	101

NOTE: Blank spaces mean that no data are available.
[a]Denotes lakes with anoxic hypolimnia.

carbon oxidation. There also is no relationship between rates of anaerobic carbon oxidation and the relative contribution of sulfate reduction. In lakes with anaerobic hypolimnia in Table II, sulfate reduction accounts for only 10–30% of anaerobic carbon oxidation, regardless of trophic status. This low percentage may be an artifact of the method of measurement of sulfate reduction. In the three lakes in the Experimental Lakes Area (ELA) sulfate reduction was calculated by mass balance (*107*). Sulfide oxidation may occur at the metalimnion, and gross rates of sulfate reduction may be greater than the calculated net rates. Alternatively, whereas rates of anaerobic carbon oxidation in the ELA lakes are comparable to rates in eutrophic lakes, factors other than carbon availability may affect gross and relative rates of sulfate reduction. In any event, these few existing data do not support the hypothesis that sulfate reduction is limited by diffusion of sulfate into lake sediments.

Existing data lend mixed support to the hypothesis that sulfate reduction is limited by availability of electron donors. Laboratory studies have shown that sulfate reduction in sediments can be stimulated by addition of carbon substrates or hydrogen (e.g., *85, 86*). Increases in storage of reduced sulfur in sediments caused by or associated with addition of organic matter (*108, 109*) also have been interpreted as an indication that sulfate reduction is carbon-limited. Addition of nutrients to Lake 227 in the Experimental Lakes Area resulted in increased primary production and increased storage of sulfur in sediments (*110, 111*). Natural eutrophication has been observed to cause the same effect (*23, 24, 112*). Small or negligible decreases in sulfate concentrations in pore waters of ultra-oligotrophic lakes have been interpreted to indicate that sulfate reduction does not occur in such lakes because of the shortage of available carbon (*1, 113*). With the exception of the laboratory studies of Ingvorsen et al. (*85*) and Smith and Klug (*78*), all of these studies used net storage of sulfur in sediments to infer sulfate reduction rates. Direct measurements of sulfate reduction rates under in situ conditions show no relationship between carbon sedimentation rates or sediment carbon content and rates of sulfate reduction (Figure 3). It is possible that neither of these parameters is an adequate index of carbon availability. An alternative hypothesis is that rates of sulfate reduction are limited by rates of sulfate regeneration.

Rates of sulfate regeneration have not been measured in any lake sediments, but indirect evidence suggests that this is an important process. The relatively low contribution of sulfate reduction to anaerobic carbon oxidation in oligotrophic lakes with anaerobic hypolimnia (Table II) may result from inhibition of sulfide oxidation. High rates of sulfate reduction in eutrophic Lake Sempach relative to Lakes Lugano and Maggiore (Table I) may result from artificial aeration of this lake; aeration has promoted the growth on the sediment surface of a thick layer of *Beggiatoa* that oxidizes sulfide and regenerates sulfate for further reduction. The relatively low rates in meromictic Lake Faro despite high sulfate concentrations may reflect an absence of sulfate regeneration in the sediments. Furthermore, rates of sulfate reduction are orders of magnitude higher than rates of S accumulation (Table III). No studies have examined the source of sulfate or mechanism of sulfate regeneration that sustains the high rates of sulfate reduction.

Possible sources of sulfate include diffusion from the water column, hydrolysis of sulfate esters, and oxidation of reduced sulfur. Diffusion of SO_4^{2-} into sediments cannot supply sulfate at the measured rates of sulfate reduction. Rates of sulfate diffusion into sediments generally are 2 orders of

Table III. Comparison of Rates of Sulfate Reduction, Diffusion, and Retention in Sediments

Lake	Sulfate Reduction[a]	Diffusive Flux	Rate of S Accumulation	References
302N			25	193
302S		91	23–90	193
223		78	5–132	111, 80
South		31	14.2	59
Big Moose		73	13–50	34, 80
Little Rock	1000	16	10–17	72
Wintergreen	2560		122	73
Mendota	2200		34	85
Sempach	8700	45	66–240	this study

NOTE: All values are given in millimoles per square meter per year. Blank spaces mean that no data are available.
[a]All reported measurements were made by injection of $^{35}SO_4^{2-}$ into sediments.

magnitude lower than measured rates of sulfate reduction (Table III). Diffusion could not supply the sulfate required for reduction unless the entire concentration change observed in pore waters occurred within a depth increment of <1 mm. Such situations do exist in microbial mats in saline lakes (e.g., *84*, *114*). Although the limited depth resolution obtained with pore-water equilibrators may cause underestimation of sulfate gradients near the sediment surface, it is unlikely that the gradient is confined to 1 mm. Observed gradients of sulfate in pore waters commonly extend over several centimeters (e.g., *80*, *90*, *99*). Profiles of sulfate reduction also extend several centimeters into sediments (Figure 4; *see also 73, 78, 82, 90, 101*). Typically, more than 50% of sulfate reduction occurs below a depth of 2 cm in sediments (*73*, *78*, *85*, *101*); in Lake Sempach, sulfate gradients (and hence diffusive fluxes) are negligible at this depth (Figure 4).

Hydrolysis of sulfate esters also cannot supply the quantity of sulfate required for sulfate reduction. Hydrolysis of sulfate esters has not been measured directly in any lakes (cf. *73*, *83*), but the annual supply of sulfate esters is less than annual rates of sulfate reduction. In Wintergreen Lake the annual supply of ester sulfate to the sediments is only 4% of annual sulfate reduction (*73*). Similarly, in Little Rock Lake the supply of ester sulfate is less than 1% of the rate of sulfate reduction (*72*). In both lakes, hydrolysis of sulfate esters is estimated to be less than half of the rate of supply to the sediments.

Abiotic oxidation of sulfide by oxygen cannot supply sulfate at rates comparable to rates of sulfate reduction. Unless high concentrations of sulfide develop and the zone of oxidation is much greater than 1 cm, rates of chemical oxidation of sulfide by oxygen will be much less than 1 mmol/m^2 per day (calculated from rates laws found in refs. 115–118). Such conditions can exist in stratified water columns; in the Black Sea water column chemical oxidation rates may be as high as 10 mmol/m^2 per day (*84*). However, in lakes in which sulfide is undetectable in the water column and oxygen disappears within millimeters of the sediment–water interface (e.g., *113*), chemical oxidation of sulfide by oxygen is unlikely to be important.

The studies cited do not clarify what factors determine rates of sulfate reduction in lake sediments. The absence of seasonal trends in reduction rates suggests that temperature is not a limiting factor. Rates of sulfate reduction are not proportional to such crude estimates of carbon availability as sediment carbon content or carbon sedimentation rate, although net reduction and storage of reduced sulfur in sediments often does increase with increasing sediment carbon content. Measured rates of sulfate reduction are not proportional to lake sulfate concentrations, and the relative rates of sulfate reduction and methanogenesis in a variety of lakes do not indicate that sulfate diffusion becomes limiting in eutrophic lakes. Direct comparison of diffusion and reduction rates indicates that diffusion of sulfate into sediments cannot supply sulfate at the rates at which it is reduced. Neither hydrolysis of sulfate

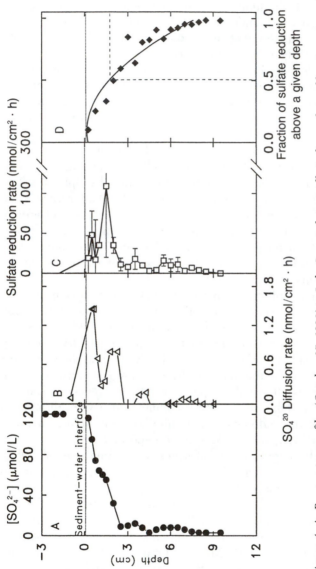

Figure 4. A, Pore-water profiles (October 15, 1990) in Lake Sempach typically indicate that sulfate is consumed within the upper 3 cm. B, Diffusion rates for sulfate calculated from the profile in panel A (T = 5 °C) indicate that the rate is maximal just below the interface, but all rates are less than 2 nmol/cm² per hour. C, Sulfate reduction rates measured with ³⁵S in intact cores on the same date are 2 orders of magnitude greater and do not exhibit the same depth profile as diffusion rates. Error bars indicate the standard deviation among 10–15 replicates. The sulfate profile in panel A was measured by centrifuging pore water from cores identical to those in which sulfate reduction was measured. D, The ³⁵S measurements indicate that 50% of the areal sulfate reduction occurs below 2-cm depth and 25% occurs below 5-cm depth. The pore-water profile indicates that negligible sulfate is supplied by diffusion to this depth.

esters nor oxidation of sulfide by oxygen are capable of supplying sulfate at rates comparable to the rates of sulfate reduction. Two other processes that may limit sulfate reduction include microbial oxidation of sulfide (as suggested by Jorgensen; *119*) and abiotic oxidation by metal oxides. Both of these processes might provide the sulfate required to sustain high rates of sulfate reduction and may explain discrepancies between sulfide production and S accumulation rates (Table III).

Oxidation of Reduced S.   Indirect evidence suggests that microbial oxidation of sulfide is important in sediments. If it is assumed that loss of organic S from sediments occurs via formation of H_2S and subsequent oxidation of sulfide to sulfate (with the exception of pyrite, no intermediate oxidation states accumulate in sediments; cf. *120, 121*), the stated estimates of organic S mineralization suggest that sulfide production and oxidation rates of 3.6–124 mmol/m^2 per year occur in lake sediments. Similar estimates were made by Cook and Schindler (1.5 mmol/m^2 per year; *122*) and Nriagu (11 mmol/m^2 per year; *25*). A comparison of sulfate reduction rates (Table I) and rates of reduced S accumulation in sediments (Table III) indicates that most sulfide produced by sulfate reduction also must be reoxidized but at rates of 716–8700 mmol/m^2 per year. Comparison of abiotic and microbial oxidation rates suggests that such high rates of sulfide oxidation are possible only via microbial mediation.

A variety of bacteria are known to oxidize sulfide. The most well-known examples include the photosynthetic bacteria [*Cyanobacteria*, purple sulfur bacteria (*Chromatiaceae*), green sulfur bacteria (*Chlorobiaceae*), purple nonsulfur bacteria (*Rhodospirillaceae*)], and the "colorless sulfur bacteria" comprised of the three families *Thiobacteriaceae*, *Beggiatoaceae*, and *Achromatiaceae* (*123*). This assemblage includes obligate aerobes, obligate anaerobes, facultative anaerobes, autotrophs, mixotrophs, and heterotrophs found in habitats ranging from arctic ice to hot springs and from alkaline lakes to acid-mine drainage. The existence of many of these bacteria in lakes is well known (e.g., *67, 68, 124*), but the majority of previous studies have focused on photosynthetic bacteria in the water column of meromictic lakes (e.g., *55, 125–129*). The presence and activity of photosynthetic sulfur-oxidizing bacteria in holomictic lakes also is documented (e.g., *130–133*), and species of *Beggiatoa* and *Thiobacillus* frequently are observed in lake sediments (*56, 58, 76, 104, 134–137*) and water columns (*55, 138–140*).

Measured rates of microbial oxidation of sulfide in lakes range from 0 to over 100,000 mmol/m^2 per year (Table IV). These rates, which are comparable to measured rates of sulfate reduction (Table I), suggest that microbial oxidation of sulfide is capable of supplying sulfate at rates needed to sustain sulfate reduction. The majority of measurements are for photosynthetic bacteria in the water column. Symbiotic sulfate reduction and sulfide oxidation are known to occur and lead to dynamic cycling of S within anaerobic water

Table IV. Rates of Sulfide Oxidation

System	Oxidation Rate, mmol/m^2 per year (% of Sulfate Reduction)	References
Phototrophic oxidation		
L. Faro	3430	55
Priest Pot	110–3140	47
Knaack L.	220–1020	128
Green L.	10510	124
Solar L.	2190–5475	147
L. Vechten	1130	230
L. Kaiike	5110–28470	139
Japanese lakes	290–3650	241
Big Soda L.	0–12780	138
L. Waldsea	0–24090	242
L. Ciso	0–131400	131
Aerobic microbial oxidation		
Beggiatoa mat	4380	144
Coastal marine sediment	2740	135
Coastal marine sediment	12400	145
Estuarine sediments	<365–16400	142
Thiobacilli in water of L. Faro	2280	55
S oxidation in water of Big Soda L.	1825–40150	138
Chemical oxidation by oxygen[a]		
Black Sea	3650	84
Calculated		
Hypolimnetic	6	
Sediment	11	
Oxidation by manganese oxides		
Laboratory system	(100%)	151
Marine sediments	73–1460 (20%)	146
Oxidation by iron oxides	20–1,000,000[b]	162, 163
Unidentified anaerobic oxidation		
Lake sediments	8400	58

NOTE: Blank spaces mean that no data are available.
[a]$R = k \cdot [HS^-]^a \cdot [O_2]^b$ where R is oxidation rate (mmol/m^2 per year); and k, a, and b are constants. The rate law is given in references 115–118.
[b]Extrapolation from laboratory studies is very sensitive to assumed surface areas of iron oxides.

columns (e.g., *47, 55, 124, 125, 127, 128, 141*). Rates appear to be comparable in sediments, but measurements are too few to assess how representative these rates are. Coupled reduction and oxidation of S within estuarine and marine sediments have been reported (e.g., *93, 135, 142–146*). Rates of sulfide oxidation nearly equal to rates of sulfate reduction have been reported by Jorgensen (*58*) for river and lake sediments. In Jorgensen's study, much of the sulfide was oxidized anaerobically to thiosulfate that in turn was subject to oxidation, reduction, and fermentative disproportionation. Similar results have been reported for marine sediments (*93*). These studies suggest that freshwater sediments may represent sulfuretums (ecosystems in which redox reactions with S are the dominant mechanism for energy flow) with rapid internal cycling of S. It remains to be determined how widespread this phenomenon is, what compounds serve as electron acceptors for sulfide oxidation, what end products are produced, what bacteria are involved, and what is the significance of the associated carbon production and respiration.

Possible electron acceptors for the oxidation of reduced sulfur include oxygen, iron oxides, and manganese oxides. Oxygen clearly is the electron acceptor in lake sediments populated with *Beggiatoa*. There has been no systematic study, but isolated reports show *Beggiatoa* to be quite commonly encountered. *Beggiatoa* mats have been observed in several hard-water lakes in Switzerland (*56, 104, 136, 137*). Strohl and Larkin (*134*) were able to isolate *Beggiatoa* from sediments of numerous lakes and bayous in Louisiana. A *Beggiatoa*-like bacterial mat was observed to develop even on sediments from soft-water Little Rock Lake when incubated under elevated sulfate concentrations (Urban, unpublished data). Jorgensen (*143*) and Strohl and Larkin (*134*) point out that *Beggiatoa* frequently occurs dispersed throughout the top few centimeters of sediments rather than as a visible mat over the surface. *Thiobacilli* are believed to be even more widespread than *Beggiatoa* (*76, 123, 143*). Mats of *Beggiatoa* can maintain high rates of sulfide oxidation (e.g., *144, 147*) that are limited only by diffusion over very short distances (*114, 147*). The relative importance of aerobic sulfide oxidation at sediment surfaces and anaerobic oxidation remains to be clarified.

Clearly oxygen is not the direct electron acceptor for S oxidation occurring below 1–2 cm in sediments, but even in such cases it may be the ultimate electron sink. Fe and Mn may serve as electron shuttles that carry electrons from anaerobic zones of sulfide oxidation to the sediment surface (*148*). The reduced Fe and Mn are oxidized at the sediment surface by oxygen. The high rates of sulfide oxidation measured by Jorgensen (*58*) and inferred to occur in other lakes imply that manganese and iron oxides are regenerated by such oxidation with oxygen in sediments. Phenomenologically, such regeneration is well known (e.g., *34, 149, 150*). Hence, oxygen in bottom waters may have a larger effect on sulfur cycling within sediments than is indicated only by the extent of direct oxidation of sulfide. Without oxygen in bottom waters, reduced Fe and Mn could not be oxidized at the sediment surface and sulfide oxidation might become limited by the supply of electron acceptors.

All measured profiles of sulfate reduction in sediments indicate that much sulfide production and, by inference, oxidation occurs in permanently anaerobic sediments (*78, 73, 90, 101*). The two most likely electron acceptors for anaerobic sulfide oxidation are manganese and iron oxides. Burdige and Nealson (*151*) demonstrated rapid chemical as well as microbially catalyzed oxidation of sulfide by crystalline manganese oxide (δ-MnO_2), although elemental S was the inferred end product. Aller and Rude (*146*) documented microbial oxidation of sulfide to sulfate accompanied by reductive dissolution of Mn oxides. Significantly, the sulfide source in these experiments was solid-phase iron sulfides, not dissolved sulfide. No sulfide oxidation was caused by addition of iron oxides. The authors (146) postulate that the reaction with manganese oxides is mediated by *Thiobacilli* during chemoautotrophic growth and, on the basis of observed rates of Mn dissolution and sulfate reduction in marine sediments, that this reaction could oxidize perhaps 20% of the sulfide produced by sulfate reduction.

Iron frequently has been postulated to be an important electron acceptor for oxidation of sulfide (58, 84, 119, 142, 152). Experimental and theoretical studies have demonstrated that Fe(III) will oxidize pyrite (153–157). Reductive dissolution of iron oxides by sulfide also is well documented. Progressive depletion of iron oxides often is coincident with increases in iron sulfides in marine sediments (94, 158, 159). Low concentrations of sulfide even in zones of rapid sulfide formation were attributed to reactions with iron oxides (94). Pyzik and Sommer (160) and Rickard (161) studied the kinetics of goethite reduction by sulfide; thiosulfate and elemental S were the oxidized S species identified. Recent investigations of reductive dissolution of hematite and lepidocrocite found polysulfides, thiosulfate, sulfite, and sulfate as end products (162, 163).

It is not clear if rates of sulfide oxidation by iron are rapid enough to account for the sulfate regeneration discussed. Numerous pore-water profiles show a gradual increase in Fe^{2+} with depth (e.g., 111, 164–166) resulting from reductive dissolution of iron oxides. Reduction of iron limits pyrite formation deep in saltmarsh sediments (167). Although iron typically is 10–100-fold more abundant than manganese in lake sediments (e.g., 23, 56, 168), fluxes of dissolved Mn^{2+} from sediments to the water columns of lakes frequently are greater than those of iron. Furthermore, most iron released to solution is attributed to microbial reduction, not reduction by sulfide (94, 97, 166, 169). When adequate weakly crystalline forms of iron oxides are present, iron-reducing bacteria can outcompete sulfate-reducing bacteria and methanogens (96, 170). Bacteria responsible for iron reduction are thought to obtain electrons from organic fermentation products or hydrogen, not from sulfide (96, 170). It remains to be clarified whether electron flow from sulfide to iron oxides is quantitatively important (cf. 94).

Oxidation of sulfide will affect rates of sulfate reduction only if sulfate is the end product of such oxidation. Many compounds with oxidation states intermediate between sulfide and sulfate may be formed instead. Although many details of the oxidation pathways remain to be clarified, evidence suggests that sulfate is formed. Oxidation of sulfide by phototrophic microorganisms results in production of elemental sulfur, sulfate, or polythionates (e.g., 171). Members of each of the three families of phototrophic sulfur-oxidizing bacteria are capable of carrying the oxidation all the way to sulfate; elemental sulfur and polythionates are intermediates that are stored until lower concentrations of sulfide are encountered (131, 171). Colorless sulfur bacteria can oxidize sulfide to elemental sulfur, thiosulfate, polythionates, sulfite, or sulfate. The predominant end product in sediments is not known, although thiosulfate has been implicated as such (*123*). Abiotic oxidation of sulfide by Fe and Mn oxides may result in a range of products including elemental sulfur, thiosulfate, sulfite, and sulfate (e.g., *151, 160–163*). Microbial oxidation of sulfide using iron or manganese oxides as the electron acceptor may result in a similar range of products (e.g., *123, 146, 172, 173*). Clearly, however, these intermediate oxidation states of sulfur do not accumulate within the sediments; if they are formed they must be transformed rapidly. Rapid microbial transformations (including oxidation, reduction, and disproportionation) of thiosulfate and elemental sulfur are well documented (e.g., *93, 173, 174*). The exact pathways that are followed and the rates of sulfate production under different environmental conditions remain to be clarified.

Measured rates of sulfate reduction can be sustained only if rapid reoxidation of reduced S to sulfate occurs. A variety of mechanisms for oxidation of reduced S under aerobic and anaerobic conditions are known. Existing measurements of sulfide oxidation under aerobic conditions suggest that each known pathway is rapid enough to resupply the sulfate required for sulfate reduction if sulfate is the major end product of the oxidation (Table IV). Clearly, different pathways will be important in different lakes, depending on the depth of the anoxic zone and the availability of light. All measurements of sulfate reduction in intact cores point to the importance of anaerobic reoxidation of sulfide. Little is known about anaerobic oxidation of sulfide in fresh waters. There are no measurements of rates of different pathways, and it is not yet clear whether iron or manganese oxides are the primary electron acceptors.

Other Early Diagenetic Reactions.   Within sediments S may undergo many diagenetic reactions. Reactants include inorganic species, a variety of mineral phases, a variety of organic compounds present in seston, and secondary organic S compounds formed in the sediments. Reactions may be abiotic (e.g., precipitation of iron sulfides), extracellular enzymatic reactions (e.g., hydrolysis of sulfate esters), or intracellular microbial processes (e.g., formation of polythionates in S-oxidizing bacteria). In addition to chemical reactions, physical transport of dissolved species may destroy the chronological accuracy of the stratigraphic record. An understanding of rates and pathways of diagenesis is necessary to interpret stratigraphic records of past environmental conditions.

Concentrations of H_2S typically are maintained low by a variety of reactions including oxidation, precipitation with iron, and interactions with organic matter. Sulfide may be oxidized biologically or abiotically. Elemental S reacts readily with H_2S to form polysulfides (*175*) and is oxidized and reduced by a variety of bacteria. In radiotracer experiments, $^{35}S^0$ was in isotopic equilibrium with H_2S and FeS so that the pathways by which it reacted were difficult to follow (176). Thiosulfate is probably an important intermediate of sulfide oxidation. Jorgensen (173) measured rapid rates of formation and disappearance of thiosulfate, although concentrations were too low to measure in sediments of a eutrophic lake (cf. 57). Thiosulfate is readily oxidized and reduced by a variety of bacteria (173), as well as disproportionated through fermentation by sulfate-reducing bacteria (173, 177–179). Polysulfides, formed by reaction of S^0 with H_2S, are important intermediates in the rapid formation of pyrite (161, 180) and may react with organic matter to form a variety of compounds (181–185).

Precipitation of iron sulfides is a major mechanism for maintaining low concentrations of H_2S in interstitial and hypolimnetic waters (94, 186). Kinetics of precipitation of amorphous iron sulfide are rapid (160). Subsequent aging produces more ordered and less soluble crystalline phases (mackinawite, pyrrhotite, and troilite) that react with elemental S to form greigite (Fe_3S_4) or pyrite (FeS_2). Various pathways and mechanisms have been proposed for interconversion of iron sulfide phases (e.g., 152, 187, 188), and interpretation of the occurrence of a particular phase in terms of environmental conditions is difficult. In a review of the literature, Davison (186) found all anoxic hypolimnia with confirmed precipitation of iron sulfides to be in equilibrium with amorphous FeS. Within pore-water profiles, ion activity products for FeS often increase with depth (e.g., 56, 111, 165). This increase typically is interpreted as evidence for diagenetic conversion to more crystalline sulfides. Davison (186) argued that this change may result from an increased rate of supply of iron or sulfide in surface sediments. At high rates of supply of H_2S and Fe^{2+} only formation of amorphous FeS is sufficiently rapid to control aqueous concentrations, whereas at lower rates of supply more stable phases may have time to form.

Pyrite is formed by two mechanisms in freshwater sediments. Framboidal pyrite results from reaction of iron monosulfides with S^0 (15), a slow reaction leading to gradual conversion of iron monosulfides to pyrite. In contrast, single crystals of pyrite are formed rapidly through reaction of Fe^{2+} and polysulfides (161). Framboidal pyrite has been reported in lake sediments (37, 189), where it appears to form in microenvironments of plant or animal skeletons (cf. 35, 36). Rapid formation of pyrite has been observed in short-term measurements of sulfate reduction with $^{35}SO_4^{2-}$. Up to 90% of reduced ^{35}S has been observed in pyrite after incubations of 1–24 h (72, 79, 98). A large fraction of inorganic S in the form of pyrite in surface sediments also has been interpreted to indicate rapid formation (112, 190). As discussed later, there is little evidence for extensive conversion of monosulfides to pyrite.

Stability of iron sulfides in lake sediments has not been thoroughly examined. Pyrite undergoes dynamic seasonal oxidation in salt marshes and coastal marine sediments (142, 184, 191–195). Pyrite oxidation is mediated by Fe(III), but may require oxygen (*153–157, 191*). High oxygen concentrations in bottom waters and sediments caused by low benthic respiration during winter, lake overturn, or bioturbation may promote sulfide oxidation in profundal lake sediments (*192–194*). In short-term experiments with ^{35}S, pyrite was in isotopic disequilibrium with respect to AVS, S^0, and H_2S (*176*). Similarly, pyrite is commonly the S pool most depleted in ^{34}S (*27, 196–198*). These observations argue for the stability of pyrite once formed. However, 55% of the iron sulfides (AVS and pyrite) formed in the sediments of Lake 302 disappeared over a 6-month period; pyrite was the dominant iron sulfide formed initially (*98*). In laboratory microcosms, Urban and Brezonik (*38*) observed formation and disappearance of $Fe^{35}S_2$ and $Fe^{35}S$ over periods of a few weeks. Introduction of manganese oxides led to oxidation of solid-phase iron sulfides in marine sediments (*146, 172*). Dissolution of Fe oxides and precipitation of sulfide in sediments of Big Moose Lake caused asynchronous increases in S concentrations among cores from different water depths (*34*). These observations suggest that iron sulfides are not stable, and hence the quantitative integrity of historical records of atmospheric deposition of sulfate may not be retained in lake sediments (cf. *29, 30, 50, 61, 199*).

Little is known about diagenetic transformations of organic sulfur compounds. Proteins are relatively labile and quickly degraded. King and Klug (*73*) found that 75% of protein S was converted to other forms or released from sediments, compared to a release of only 46% of total S. Hydrolysis of sulfate esters is known to occur, but rates are low and less than half are hydrolyzed within 10–100 years (*59, 72, 73, 83*). Formation of sulfate esters also may occur within sediments (*38, 79*), although the mechanisms are unknown. Little organic S in deep Everglades peat is present as sulfate esters (<10%) compared to recent peat and lake sediments (30–80%; *36, 49, 51, 59, 72*); this observation suggests that hydrolysis and reduction of ester sulfate may persist at slow rates for prolonged periods.

In addition to changes in existing organic S compounds, formation of new organic S compounds occurs. Evidence of this phenomenon has been gathered in at least five different ways. Incorporation of ^{35}S (either as $H_2{}^{35}S$ or as $^{35}S^0$) into natural organic matter has been widely reported (*38, 73, 79, 98, 200–202*). Mild heating of carbohydrates (*203*) and ethylbenzene (*204*) with sulfide produced numerous thiols, disulfides, thiophenes, dithiophenes, and thiolanes. Stable isotope ratios in sediment organic matter often are observed to be intermediate between that of pyrite and sulfate in the overlying water (*27, 196–198*). Because little isotope fractionation is caused by sulfate assimilation in seston, incorporation of microbially reduced S into organic matter is the probable cause. Ratios of C:S in bulk organic matter (*34, 61*) and humic acid and humin fractions (*56, 181, 205–207*) are observed to decrease with depth in sediments, presumably because of diagenetic enrichment. Numerous specific S-containing compounds have been identified that can only have arisen diagenetically. Cyclic disulfides and trisulfides have been attributed to formation and condensation of organic polysulfides (183). The existence of numerous isoprenoid thiophenes and thiolanes in marine and Black Sea sediments is attributed to incorporation of S into specific precursor molecules (32, 208, 209). Coexistence of polyenes and corresponding unsaturated thiophenes with two fewer double bonds is attributed to S addition across C=C double bonds (32). Partial oxidation products (sulfones) of these compounds also were identified, although it was not clear if oxidation occurred in the field or laboratory. Finally, numerous measurements of low-molecular-weight thiols in estuarine and marine pore waters suggest dynamic formation and degradation of these compounds in conjunction with seasonal oxidation of pyrite (120, 182, 191, 195, 210–215).

The exact formation pathways of much of the organic S remains to be clarified. The mechanism most often postulated is addition of sulfide or polysulfide across C=C double bonds of olefins (e.g., 182, 185, 216, 217). Subsequent condensation or disulfide formation also seems likely, as evidenced by release of thiols upon addition of tributylphosphine (182), release of DMS during treatment with Cr(II) (217), and the existence of cyclic and acyclic di- and trithienes (32, 183). Formation of large aggregates by weak disulfide linkages in coal and oil is known (32, 218). Profiles of S species in lake sediments led Losher and Kelts (190) to suggest that S-oxidizing bacteria incorporate oxidized forms of S into organic compounds. Peaks in toluene-extractable S at the oxic–anoxic interface were attributed by Losher and Kelts to formation of polythionates by *Beggiatoa* or other S-oxidizing bacteria. Subsurface increases in sulfate esters also were attributed to microbial formation.

The lability of organic S to oxidation and recycling is not understood. Short-chain thiols appear to be dynamically cycled in marine and saltmarsh sediments (120, 211). In contrast, sulfones may represent a stable oxidized product that could provide a useful signature of sediment oxygenation (121, 184). Sulfones have been identified in saltmarsh and ancient marine sediments (32, 207). Other (unidentified) forms of organic S appear to yield acidity and sulfate upon oxidation (38, 192). Much more research is needed to elucidate the mechanisms of organic S formation and the relationships between specific organic S compounds and lake conditions.

Retention of S in Sediments

An understanding of the retention of sulfur in lake sediments is important for paleolimnological reconstructions, for an understanding of the alkalinity balance of lakes, for determination of rates and pathways of diagenesis, and for an understanding of the dynamics of the microbial loop in sediments. Profiles of rates and forms of S accumulation in sediments may preserve records of past conditions of climate, lake chemistry, or atmospheric deposition. Sulfur retained in sediments represents a permanent gain in lake alkalinity. Whatever the mechanism, loss of SO_4^{2-} must be accompanied by an equivalent loss of cations or an increase in anions (*219*). Sulfur retained in sediments is also removed from cycling and participation in electron and energy flow. In the short term, S may participate in a dynamic cycle of reduction and oxidation, but ultimately it is buried to a depth at which rates are lower and it is permanently retained. Some of the very factors that promote retention of S in sediments also inhibit the reduction–oxidation cycle. Hence high rates of S retention may imply low rates of S cycling within sediments. Paradoxically, the highest rates of S cycling may occur in sediments with the lowest S concentrations. High rates of primary production and highly reducing conditions in hypolimnia and sediments may restrict reoxidation of S and increase the relative importance of methanogenesis. For S cycling to continue, either light or oxygen is required to provide an electron sink. (In the case of photosynthetic S oxidation, the sink is CO_2.)

Rates of sulfur retention have been measured by mass balance and by dating of sediment cores. Sulfur budgets have been calculated for entire basins or hypolimnia of more than 18 lakes (*62, 72, 100, 122, 193, 220–224*), and S accumulation rates in sediments have been published for more than 20 lakes (*29, 30, 59, 72, 199*). Lakewide rates of S accumulation range from 9 to 128 mmol/m^2 per year, and rates in hypolimnia can reach as high as 2920 mmol/m^2 per year (*221*). The range of rates reported in sediment cores is 6–159 mmol/m^2 per year. Rates of S accumulation in single cores may not represent lakewide rates. S concentrations and accumulation rates typically increase with increasing water depth (*25, 30, 61*), although Cook et al. (*100*) reported that alkalinity generation resulting from S retention in epilimnetic sediments was greater than that in hypolimnetic sediments. Accumulation rates are low relative to rates of sulfate reduction (Tables I and III), but because both processes have been measured in only three lakes, any relationship between rates of reduction and accumulation remains obscure. Rates of S accumulation correspond to rates of alkalinity generation of 12 to almost 500 mequiv/m^2 per year. These values are significant relative to rates of acid deposition (20–150 mequiv/m^2 per year).

Factors Affecting Retention Rates.   Factors affecting seston deposition, sulfate diffusion into sediments, and S recycling to the water column will affect S retention in sediments. The major factor controlling seston deposition is primary production. Lake depth, by controlling the extent of mineralization within the water column, also may affect the magnitude of seston S inputs. Factors influencing diffusive fluxes into sediments include sulfate concentrations in bottom waters, the extent of lake mixing, and the activity of sulfate-reducing bacteria. Recycling to the water column includes both diffusion and reoxidation. Oxidation may occur in either the water column or sediments, and hence diffusion of sulfide and sulfate may be important. Diffusion of sulfide from sediments will be enhanced by low concentrations in the overlying water and high concentrations in pore waters. Hence such factors as bottom-water oxygen content, mixing of bottom waters, quantity and lability of iron and manganese oxides, concentrations of Fe^{2+} in pore waters, quantity and reactivity of organic matter, and microbial consumption of sulfide within or above sediments will be important. Interactions of these environmental variables with each other and with S retention are complex.

A major factor governing diffusive fluxes of sulfate into sediments is lake sulfate concentration. A linear relationship exists between lake sulfate concentrations and diffusive fluxes calculated from pore-water profiles (Figure 5). The relationship extends over a range of 3 orders of magnitude in sulfate

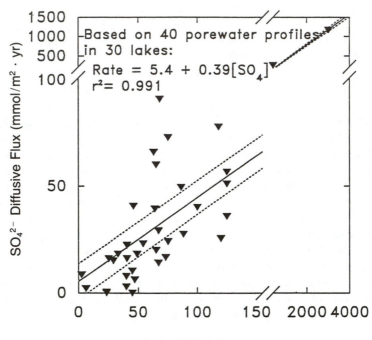

Figure 5. Data from the literature (56, 80, 99, 164, 195, 220, 222, 223, 243) indicate that diffusive fluxes of sulfate (calculated from 40 pore-water profiles measured with pore-water equilibrators) are linearly related to concentrations of sulfate in the overlying lake water. The correlation is significant ($p < 0.05$) both with ($r^2 = 0.991$) and without ($r^2 = 0.42$) the two lakes with high sulfate concentrations. The strong correlation suggests that variations in the depth interval within which sulfate is consumed and in the minimum sulfate concentration defining the gradient are relatively unimportant in determining the flux, compared to variations in sulfate concentrations defining the upper end of the gradient.

concentrations, although much scatter exists at low concentrations. Similar relationships were used to model the alkalinity generation associated with burial of S in sediments (220, 223). The common interpretation of this relationship is that sulfate reduction, a first-order reaction with respect to SO_4^{2-} concentration, determines the diffusive flux (99, 220, 223). The data discussed clearly demonstrate that this hypothesis is inaccurate; rates of sulfate reduction are much greater than rates of sulfate diffusion into sediments (Table III). Diffusive fluxes are controlled by diffusion gradients, temperature, and sediment porosity and tortuosity. The diffusion gradient has three components: the concentration in the water column, the depth of sediment over which sulfate is consumed, and the minimum sulfate concentration in the pore waters. Strong correlations between flux and lakewater sulfate concentration suggest that variations in other parameters are relatively minor, although they may contribute scatter to the data.

Despite the strong relationship between sulfate concentrations and diffusive fluxes, there is no universal relationship between lake sulfate concentrations and concentrations of S in sediments (Figure 1A; cf. 24, 26). Concentrations of S in sediments are the net result of inputs from seston, diffusive inputs, recycling to the water, and dilution by other materials. Mathematically this quantity may be expressed as

$$c = \frac{(s_{seston} + s_{diffusive} - s_{recycle})}{\omega}$$

where c is S concentration (mmol/g), s_{seston} is flux from seston (mmol/m^2 per year), $s_{diffusive}$ is diffusive flux (mmol/m^2 per year), $s_{recycle}$ is the rate of recycling back to the water (mmol/m^2 per year), and ω is the sedimentation rate (g/m^2 per year). Only the diffusive flux is a direct function of lake sulfate concentration. Hence only if diffusion is the predominant flux or if other terms are relatively constant will sediment S concentrations be proportional to sulfate in the water column. It has been argued that accumulation rates of S in sediments rather than concentrations are the appropriate parameter to examine for a relationship with lake-water sulfate concentrations (225). However, the sulfur accumulation rate is equal to the concentration (c) times the sediment accumulation rate (ω); hence a relationship between accumulation rates and sulfate concentrations is not to be expected unless diffusion is the predominant flux or unless other terms are relatively constant. Figure 1B suggests that seston deposition is the dominant term for approximately half of the 78 lakes shown.

Availability of reactive iron in sediments also has been postulated to control S retention. Reactive iron may limit fluxes of S recycled from sediments by rendering sulfide immobile and less amenable to oxidation by bacteria or chemical agents. Availability of iron strongly influences total S content and isotopic signature of marine sediments (198). Canfield (94) observed that as long as abundant reactive iron was available, sulfide concentrations were maintained at low concentrations in pore water because of reductive dissolution of iron and precipitation of iron sulfides. Carignan and Tessier (50) observed a strong correlation between burdens of iron and S in recent sediments from eight Canadian lakes. From 16 to 79% (mean = 55%) of total S was in the form of iron sulfides, but this fraction increased toward the sediment surface. Giblin et al. (30) also noted that most of the increased S content in New England lakes occurred as iron sulfides. In lake sediments with high organic matter content, 90% of acid-leachable iron was in the form of iron sulfides. Both studies observed that current rates of iron accumulation are less than rates of S accumulation in these lakes and suggested that the capacity of such lakes to retain S eventually may be exhausted.

Three other lines of evidence may support the hypothesis that availability of reactive iron limits S retention.

1. Few lakes exist with molar ratios of S to Fe greater than 1 (Figure 1C). Ratios of inorganic S to Fe are, of course, even lower.

2. The degree of pyritization of iron (DOP) or the fraction of reactive iron that is bound with sulfur appears to increase rapidly with increasing S:Fe ratios and may approach an asymptotic value of about 75% at S:Fe ratios of 3 (Figure 6). Both features suggest that iron reacts readily with increasing S inputs or that other mechanisms for binding S are relatively unimportant.

3. A relationship between the fraction of S in inorganic forms and Fe content exists at depths below 30 cm in sediment cores (Figure 7) but not in surface sediments. The $C:S_{total}$ (S_{total} is the sum of the organic S and inorganic S concentrations) ratios at these depths suggest that the sulfur is derived primarily from seston. The fraction of seston S that is mineralized and retained in these deeper sediments where S inputs are relatively low appears to be determined by the availability of iron.

Together with the studies of Carignan and Tessier (50) and Giblin et al. (30), these data indicate that increasing the Fe content of sediments will increase the sulfur-binding capacity and, at low rates of sulfur input, will increase the proportion of S bound to Fe (Figure 7). Increasing S inputs relative to Fe inputs will cause progressive sulfidation of Fe (pyritization in the sense of ref. 226; Figure 6). However, when the rates of sulfide production exceed the rates of Fe reduction, no further relationship will be observed between Fe content and the fraction of S bound by Fe.

Although the evidence cited indicates that iron influences retention of S in lake sediments, other factors may obscure this influence. First, organic

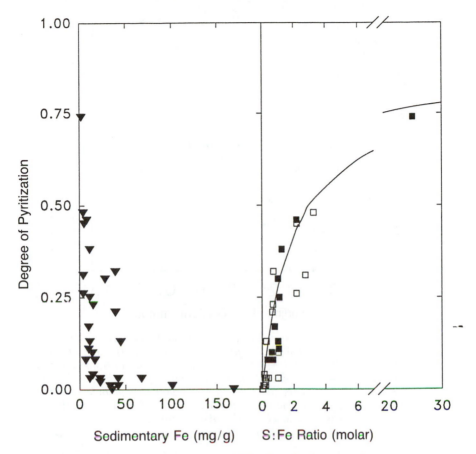

Figure 6. The degree of pyritization, defined as the fraction of reactive iron present as pyrite, is a measure of the extent to which available iron has reacted with sulfur (226). In lake sediments, iron monosulfides frequently are as abundant as pyrite and hence were included with pyrite in the values calculated for surface sediments from 13 lakes and presented here. Even this correction neglects Fe(II) that may have been reduced by sulfide but may be present as siderite. Availability of iron appears to be more important than bottom-water oxygenation in determining the degree of pyritization. In the right-hand graph, darkened squares represent sediments known to experience seasonal anoxia; only the uppermost point experiences permanent anoxia. (Data are from references 30, 34, 56, and 61.)

matter competes with Fe for binding of sulfide; the pool of secondary organic S (i.e., that not derived from seston) in surface sediments ranges from 0.01 to 30 times the size of the pool of Fe-bound sulfur (data from refs. 56 and 72). The factors determining the outcome of this competition are not yet clear, but appear to include the relative abundances (72, *198*) and relative reactivities toward sulfide of Fe and organic matter. Hence, as indicated by

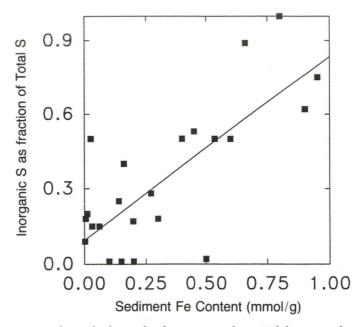

Figure 7. Only at the base of sediment cores from 27 lakes is a relationship
(r² = 0.755) observed between the iron content and the fraction of sulfur
present as iron sulfides (AVS + CRS). A similar relationship is not observed
in surface sediments. As discussed in the text, much of the sulfur at the base
of the cores appears to have originated from organic compounds in seston.
The relationship may indicate that retention of H₂S released during decom-
position of seston is determined by the availability of iron. References are
given in Figure 1.

Lago di Cadagno, where the pool of organic sulfur is 10-fold greater than
the pool of iron sulfide (56), sediments may have a large capacity to retain
sulfur even after all available Fe has reacted with sulfide. Second, the relative
susceptibility of organic S and Fe-bound S toward oxidation also will affect
the sulfur speciation in sediments. Several studies have demonstrated that
iron sulfides are more readily oxidized to soluble sulfur species than are
organic sulfur compounds (38, 98, 193). No generalities may yet be drawn
about the extent of iron sulfide oxidation in various lakes.

    Primary production (trophic state) affects S retention and speciation in
several ways. As primary production increases, inputs of organic S to sediments in seston increase. Hence, as the organic carbon content of sediments
increases, S content would be expected to increase proportionally. Such a
simple relationship is not observed among nearly 80 lakes for which sediment
S and C content are available (Figure 1B). However, a line defining the
minimum S content does increase linearly with increasing carbon content.
The slope of this line corresponds to the mean C:S ratio measured in seston

(27, 59, 72, 86). Approximately half of the lake sediments in Figure 1B have significant sources of S other than seston, but in all lakes seston S in sediments increases with increasing C content.

Primary production affects input and retention of nonseston S through the interplay between sulfate reduction and sulfate diffusion. In very oligotrophic lakes with low rates of carbon sedimentation and aerobic hypolimnia, little or no sulfate reduction occurs in sediments because the small amount of carbon is oxidized by oxygen (e.g., *113*). Sulfate is not attenuated in pore waters of such lakes (*1, 80, 113*), and hence diffusive influx to the sediments is minimal. Because seston is the major source of S to such sediments, most S will be organic (59) and C:S ratios will be close to values in seston. If sulfate reducers are C-limited, increasing C inputs will stimulate sulfate reduction in surface sediments, consume available sulfate in a shallower zone of sediments, and thereby sharpen the concentration gradient and enhance diffusion of sulfate into the sediment. Rates of sulfate reduction quickly become limited by rates of sulfate regeneration and do not increase in proportion to further increases in carbon inputs (Table I, Figure 3).

However, increasing organic carbon flux to sediments also increases oxygen demand in both the sediments and water column. This change reduces the depth of oxygen penetration into sediments and leads to consumption of other electron acceptors within thinner zones of sediment, thereby further sharpening concentration gradients and enhancing diffusive influxes of sulfate. Thus, in the English Lake District the lowest sediment C:S ratios occur in the most productive lakes (*24*), and in 11 Swiss lakes there is a strong correlation between accumulation rates of carbon, total S, and sulfide (Figure 8A). The low C:S ratios in these sediments indicate that the correlation is not driven by inputs of seston S. Rather, increased carbon inputs cause increased diffusion of sulfate into sediments and increased S retention and content in sediments.

Increased primary production has three other potential effects on retention and speciation of S in sediments. Increasing C sedimentation may decrease S recycling from sediments by reducing reoxidation of sulfide, by increasing dissolution of Fe and promoting iron sulfide formation, and by enhancing organic S formation. Decreased oxygen content of bottom waters will reduce chemical oxidation of sulfide and may curtail the activities of aerobic S-oxidizing bacteria. Increased productivity may lead to dissolution of manganese oxides in the water column or enhance microbial reduction of iron and manganese in sediments; in both cases, sulfide oxidation by these oxides may be reduced. Losher (56) attributed decreased S retention in surface sediments of Lake Geneva to the high concentrations of Mn oxides. Increased rates of Fe dissolution in sediments resulting from enhanced microbial activity may increase precipitation of iron sulfides (*186*). Among three English lakes AVS content of sediments was proportional to rates of primary production (*47*), although it is not clear whether production of H_2S, Fe^{2+},

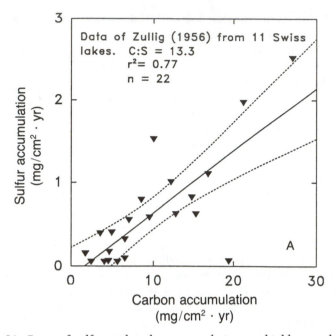

Figure 8A. Rates of sulfur and carbon accumulation are highly correlated in surface sediments of 11 Swiss lakes (23). The solid line is the regression line and the dotted lines represent 95% confidence intervals. Variations in the carbon accumulation rates represent differences in trophic status and lake depth.

or both was enhanced. Existing data do not support the hypothesis that increasing organic matter content of sediments leads to increased incorporation of S into organic compounds. Among 80 lakes in North America and Europe, there is no tendency toward increasing S enrichment with increasing carbon content (Figure 1B); the highest enrichments are noted at lower carbon contents.

Total S content cannot indicate whether increased carbon inputs to sediments cause increased diffusion of sulfate into sediments or restrict reoxidation and release of S from sediments, because the net effect is the same. In a survey of 14 lakes, Rudd et al. (80) did not observe a strong correlation between organic matter content per volume and net diffusive flux of sulfate. However, in English lakes the lowest C:S ratios occur in the most productive lakes (24); whether this represents enhanced influx or retarded release is not clear. Among 11 Swiss lakes, ratios of C to S sedimentation rates are relatively constant and substantially below C:S ratios in seston; net S fluxes

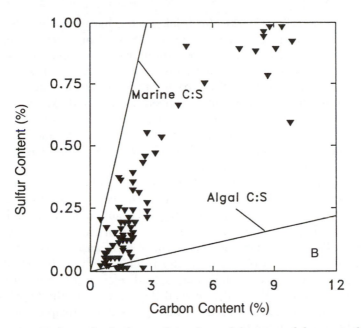

Figure 8B. Within individual cores from three of these same lakes, a similarly strong correlation is observed between S and C concentrations. Within each core, C concentrations increase toward the surface because of increasing eutrophication in recent years. The C : S ratios indicate that most of the sulfur is not derived from seston (ratio indicated by line labeled algal C : S), but from sulfate reduction. Increasing inputs of carbon cause increases in S from both seston and sulfate reduction. Even after eutrophication, C : S values remain below the ratio of 2.5 (marine line) typically observed in marine sediments (20).

into sediments increase as carbon content increases. Similarly, in lakes Constance and Zug, eutrophication has increased the carbon content of the sediments two- to fourfold but caused a proportionally greater (10- to 20-fold) increase in both total S and sulfide S (data from ref. 23). Existing data indicate that increasing carbon inputs to sediments will increase total S content, but the relative importance of various mechanisms remains unclear.

Sulfur-oxidizing bacteria may limit S retention in sediments by maintaining low sulfide concentrations in pore waters (thereby inhibiting precipitation of iron sulfides) as well as by converting sulfide to sulfate (a form not retained in sediments). It was hypothesized that such bacteria promote rapid internal cycling of S within sediments (or the water column; 47, 227); the effect on net retention is not so clear. Sulfur-oxidizing bacteria are commonly thought of in association with meromictic lakes with high sulfide concentrations (e.g., *55, 127, 227–229*). Such lakes often have very high S concentrations in sediments (e.g., *190*). However, S-oxidizing bacteria occur in a great variety of lakes. Some photosynthetic bacteria occur in oxic waters (e.g., *147, 229*), others in seasonally anoxic waters (*132, 230*). Photosynthetic and nonphotosynthetic S-oxidizing bacteria are present in soft-water Little Rock Lake despite very low sulfate concentrations (*133*). Many nonphotosynthetic S-oxidizing bacteria require oxygen (e.g., *Beggiatoa, Thiovulum, and Thiothrix*), and species of *Thiobacilli* exist that use oxygen, nitrate, and metal oxides as electron acceptors (e.g., *76, 146*). No study has yet shown whether S-oxidizing bacteria reduce S retention in sediments.

Lake morphometry, particularly depth, may influence S retention and speciation in several ways. Within a single lake, concentrations of S in sediments increase with increasing water depth because of focusing and greater oxidation in shallow sediments (*25, 30, 61*). Among different lakes, depth indirectly affects retention and speciation of S because of its influence on the quantity and quality of organic matter reaching sediments, extent of lake mixis, oxygen content of bottom waters, and relative penetration of light. The shallow depth of Priest Pot allowed the presence in the hypolimnia of photosynthetic S-oxidizing bacteria that prevented precipitation of FeS within the water column (*47*).

There is an interplay between lake mixis, lake productivity, oxygen content of bottom waters, the presence of S-oxidizing bacteria, lake depth, and S retention. Inadequate mixing of bottom waters can lead to depletion of sulfate and a decreased diffusive flux into sediments. Decreasing mixing also lowers the oxygen content of bottom waters and results in a range of conditions from permanently oxic to periodically to permanently anaerobic. Increasing anaerobiosis tends to increase carbon storage and, concomitantly, storage of S. Lack of oxygen can retard recycling by decreasing reoxidation of sulfide. In the absence of oxygen, only photosynthetic bacteria and those utilizing iron or manganese oxides can oxidize sulfide. Increased productivity can lead to decreased oxygenation of bottom waters and ultimately to decreased mixing. High concentrations of free H_2S under anoxic hypolimnia or in deep, poorly mixed lakes (*56, 104, 111, 168*) may enhance S retention in sediments. Such conditions preclude bioturbation, which can promote reoxidation of reduced S. Berner and Westrich (*194*) observed that the fraction of pyrite accumulated in sediments of Long Island Sound was much less than the gross rate of H_2S formation and that the fraction decreased with increasing bioturbation. Similar studies are lacking in fresh waters. Clearly, S retention is influenced by many factors, and simple relationships between single factors are unlikely to be observed.

Factors Controlling Speciation of S.   Speciation of S in sediments is influenced by many of the same factors that control S retention. However, S speciation may be a more sensitive indicator of many of these variables.

Sulfur content has been reported in the literature for approximately 80 lakes; speciation has been reported for only about half of them. These data point to some important factors controlling S speciation, but generally are inadequate to test hypotheses rigorously.

It has been hypothesized that as primary production increases, the relative importance of seston deposition as a source of S to sediments should increase (72). A corollary would seem to be that as primary production increases, the fraction of sedimentary S present as organic S should increase. This hypothesis and its corollary are not supported by existing data. The fraction of S present as organic S is similar in lakes Wintergreen (73) and South (59) despite a nearly 10-fold difference in carbon sedimentation rates. Within three Swiss lakes, C:S ratios have remained essentially constant and well below the value in seston despite a significant increase in carbon sedimentation and sediment carbon content (Figure 8B). Among world lakes there is no clear tendency for C:S ratios to approach the seston input value as sediment carbon content increases (Figure 1B). There is no relationship between the fraction of S as organic S and sediment C content (figure not shown).

Two factors negate the hypothesis that the importance of seston S should increase with increasing primary productivity. As discussed, the input or retention of microbially reduced S also appears to increase as C sedimentation rates increase. Second, interconversions between organic and inorganic S blur the distinction between seston S and microbially reduced S. Much microbially reduced S is incorporated into organic matter (72, 98). Consequently, S introduced into sediments via dissimilatory reduction can be present in both organic and inorganic forms. Similarly, H_2S released through putrefaction can precipitate with iron. In microcosms containing only lake sediments and water, up to 15% of ^{35}S present originally within algae was found present in iron sulfides after 15–200 days (38). At depths of 30–50 cm in lake sediments, ratios of $C:S_{total}$ equal the ratio in seston, but ratios of $C:S_{organic}$ are higher (Figure 9); a fraction of the organic sulfur from seston has been converted to iron sulfides. Stable isotope measurements (27, 196–198) have documented a similar interconversion of forms. Hence the relative importance of seston deposition and sulfate deposition cannot be determined merely from relative abundances of organic and inorganic S (cf. 49, 51, 59, 73).

If iron limits retention of S in sediments (cf. 50, 30) it would be expected that the fraction of S present as iron sulfides would increase with increasing Fe content of sediments. Although this relationship is observed in deep sediments (Figure 7), fractionation of S between organic and inorganic forms is not determined by iron content in surface sediments. Nor is there any relationship between Fe content and total S content in surface sediments for all lakes reported in the literature (Figure 1c). In deep sediments where C:S ratios indicate that seston was the major source of sedimentary sulfur

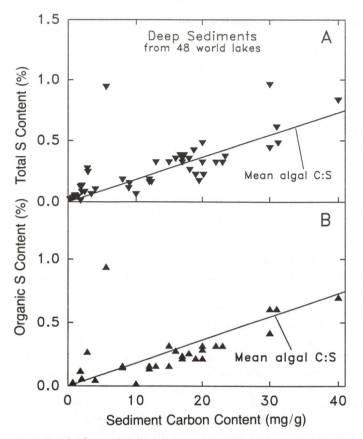

Figure 9. A, At the base of sediment cores (30–50 cm) from 48 lakes ratios of
$C:S_{total}$ *nearly equal the ratio found in seston (indicated by the line labeled*
mean algal C:S). A simplistic explanation is that most of the S is derived from
seston, and that C and S are mineralized and lost from sediments at similar
rates. B, Within the same cores for which data were available, ratios of $C:S_{org}$
tend to be lower than the ratio in seston (19 of 28 points lie below the line).
Together, the figures suggest that much of the mineralized S is retained within
the sediments. Figure 7 suggests that such retention is dependent on the avail-
ability of iron. References are given in Figure 1.

(Figure 9a), a relationship between inorganic S and Fe content may indicate
that transformation of seston S to inorganic forms depends on the availability
of iron (*see also* refs. 35–37). Alternatively, it may indicate that in oligotrophic
lakes rates of putrefaction are lower than rates of Fe^{2+} formation (*186*); as
eutrophication proceeds, rates of sulfide production exceed rates of Fe reduction and the relationship between inorganic S and Fe contents is lost.

The influence of S-oxidizing bacteria on speciation of S in sediments has
not yet been proven. Losher (*56*) postulated that layers of S-oxidizing bacteria
in lakes Cadagno, Zurich, and Geneva caused a layer rich in toluene-soluble

S (attributed to polythionates or elemental S and to subsurface production of ester sulfates). Davison and Finlay (47) noted that photosynthetic bacteria, by maintaining low sulfide concentrations, prevented precipitation of FeS in the water column, but they did not evaluate the effect of the bacteria on S speciation in the sediments. Whereas specific S bacteria can occur only in well-defined conditions, a link between S speciation and the presence of these bacteria could be a useful paleolimnological tool for reconstructing previous lake conditions.

It has been suggested that sulfones represent a stable form of oxidized organic S (*184*). Because of a lack of analytical techniques, such forms have been tentatively identified in only two studies (*32, 207*). Additional data are needed to determine if the abundance of these forms can serve as an indicator of bottom-water oxygenation.

Interpretation of Stratigraphic Records of S in Sediments

Profiles of rates and forms of S accumulation in sediments may preserve records of past conditions of climate, lake chemistry, or atmospheric deposition. Specifically, it has been suggested that S content, speciation, and isotopic ratio can serve as paleosalinity indicators (*18, 20, 21, 231, 232*); that S forms can indicate oxygenation of bottom waters (*18, 226*); that S content and isotopic signature can indicate historical rates of atmospheric deposition of sulfate (*26, 29, 30, 49-51, 61, 199*); and that S content records the eutrophication history of lakes (*24, 25*). These hypotheses are examined briefly in the following sections.

Paleosalinity.   Four indices have been proposed as measures of paleosalinity: $CRS:AVS$, $C:S_{total}$, $C:S_{pyr}$, and $S_{org}:S_{pyr}$, wherein S_{total} represents total S, S_{pyr} represents pyritic S, and S_{org} represents organic S. Berner et al. (*20*) noted that most S in marine sediments is present as pyrite and proposed that ratios of $CRS:AVS$ in the range of 0.2–100 (g of CRS S per g of AVS S) indicated marine environments. However, among 18 lakes reported in recent literature, ratios range from 0.01 to 69 with an average of 11 (Table V). Clearly this ratio cannot be used alone to infer paleosalinity.

Berner et al. (*20*) proposed that ratios of $C:S_{total}$ distinguish marine from freshwater sediments. In general, marine sediments have much higher total S concentrations than do freshwater sediments; this difference is attributed to the abundance of SO_4^{2-} in salt water (100–1000 times that of fresh waters). However, recent data indicate that $C:S_{total}$ ratios of freshwater sediments overlap with those of marine sediments. Nriagu and Soon (*27*) reported $C:S_{total}$ ratios of 2–3 (mass basis) in two Canadian lakes and Losher and Kelts (*190*) reported ratios of 2–8 among three Swiss lakes. The ratio typical of noneuxinic marine sediments is 2.8 (*18, 20*). Ratios of $C:S_{total}$ also appear to be inadequate indicators of salinity (cf. Figure 1b).

Table V. Relative Abundance of Pyrite and Iron Monosulfides in Lakes

Lake	Trophic State	CRS:AVS[a] Surface	Deep	References
Mendota	eutrophic	0.02	0.2	25
Blelham Tarn	eutrophic	0.08		112
Lago di Cadagno	meromictic	0.12	1.3	56
Cone		0.3	2	30
Sempach	eutrophic	0.43		unpublished
Horwer Bucht	eutrophic	0.71		unpublished
Zurich	mesotrophic	1.4	1.5	56
Geneva	mesotrophic		3	56
Kelly		1.8	7.4	27
Batchawanna	oligotrophic	2	2	27
Wintergreen	eutrophic	2.6		73
Ennerdale	oligotrophic	3		112
Turkey	oligotrophic	4	2	27
Windermere	eutrophic	5		112
McFarlane		7.2	1	27
Little Rock 5m	oligotrophic	12	6	7
Mirror	oligotrophic	15		30
South	oligotrophic	25		51
Miles		50	11	30
Little Rock 7m	oligotrophic	69		7

NOTE: Blank spaces mean that no data are available.
[a]Expressed as weight ratios of S (grams of CRS S:grams of AVS S).

In marine sediments most S is present as pyrite and only a fraction as organic S. Because of the comparative abundance of organic matter and the relative shortage of iron in many lake sediments, the situation is reversed. Berner and Raiswell (231) proposed that ratios of $C:S_{pyr}$ distinguish marine and freshwater sediments. However, even in marine sediments, pyrite is not always the dominant S species present. Therefore, use of this ratio is restricted to noneuxinic marine sediments with adequate iron content (22, 232, 233). Within these limitations, this ratio appears to be generally valid. Only one of 38 lakes reported in the literature has a value (3.9) within the range typical of marine sediments (0.5–5; 231, 234); the remaining lakes have values between 11.7 and 3500. Davison (9) proposed a slight variation of this index, ratios of organic S to pyritic S ($S_{org}:S_{pyr}$). This ratio appears to be similarly valid; the range for freshwater lakes (0.2–300; Table VI) is above that for marine sediments (<0.1; 9). Neither ratio can distinguish sediments of freshwater lakes and wetlands from estuaries and brackish wetlands (207, 235, 236).

Atmospheric S Deposition.   Increased S content of recent sediments is frequently attributed to increased atmospheric deposition of SO_4^{2-} (e.g., 7, 28–30, 49–51, 199). Increased supply of sulfate is thought to have in-

Table VI. Salinity Indices (Mass Ratios) Based on S Speciation in Lake Sediments

Lake	$C{:}S_{total}$	$C{:}S_{pyr}$	$S_{org}{:}S_{pyr}$	References
10636	14.0	25.0	0.80	50
78042	12.0	11.7		50
79269	31.0	62.5	1.0	50
79245	34.0	45.8	0.33	50
79398	23.0	469	19.0	50
96760	10.0	13.5	0.33	50
96791	38.0	375	9.0	50
77978	36.0	719	19.0	50
Blelham Tarn	18.0–20.3	556	16.8	112, 24
Windermere	15.0–40.3	562	36.8	112, 24
Ennerdale	35.0–64.8	1146	28.9	112, 24
Grass Pond	59.0			28
Wintergreen	25.0	238	8.14	73
South			6.56	59
Superior	236			26
Michigan	15.0			26
Huron	44.0			26
Georgian Bay	22.0			26
Erie	20.0			26
Ontario	23.0			26
Mendota		2344	24.0	25
MacFarlane	3.5–6.25	54.1	15.2	27, 48
Kelly	1.6–2.8	3.9	0.87	27, 48
Lohi	9.7			48
Ramsey	4.6			48
Opeongo	14.0			48
Big Moose	38–45	124–390	2.2–7.7	34
Turkey	88.0	1034	10.5	27
Batchawana	72–131	3000–3500	22–47	27
Geneva	95	800	6.57	56
Zurich	9.6–27	71.4	1.53	56, 23
Little Rock	30–54	145–352	2.3–5.4	61
Orajarvi	26.1	580		60
Munajarvi	48	521		60
Victoria	1.0–101			71
Zug	6.58			23
Wastwater	83.7			24
Thirlmere	36.8			24
Buttermere	71.6			24
Crummock	36.0			24
Hawes water	18.8			24
Coniston	44.4			24
Rydal water	66.5			24
Derwentwater	44.9			24
Loweswater	19.8			24
Loughrigg	9.14			24
Bassenthwaite	20.9			24
Esthwaite water	21.2			24
Miles	28.8	144	3.98	30
Spectacle	31.2	78.1	1.5	30
Mares	46.9	93.8	1.00	30
Mirror	45–88	78–312	1.5–4.6	30
Cone	37.5	750	15.5	30
Clouds	93.8	141	0.50	30
Cadagno	5	215	40.8	56

NOTE: Blank spaces mean that no data are available.

creased rates of sulfate reduction and hence S incorporation into lake sediments. Decreasing ratios of $^{34}S:^{32}S$ (27, 34, 196, 197) are cited in support of this postulated mechanism. Because of the mobility of sulfide in pore waters and diagenetic interactions with iron, depths of S increases do not necessarily correspond to sediments deposited since the increase in SO_4^{2-} deposition (28, 34).

Although abundant literature adequately demonstrates that S accumulation rates in sediments have increased in recent decades, this increase must be understood in light of the mechanisms already discussed. First, rates of sulfate reduction are not limited by concentrations of SO_4^{2-} in lake water, and hence it is doubtful whether rates of sulfate reduction have increased in recent decades as a result of increased atmospheric deposition of S. Diffusive fluxes and net retention of S in sediments do, however, appear to be limited by lake concentrations of SO_4^{2-} and may reasonably be expected to have increased. Second, iron sulfides comprise the increase in S content of recent sediments (30, 50, 61). This increase in S content in recent sediments may be attributable to increased inputs of reactive iron into lakes in recent decades (225). Evidence exists of such increases in iron inputs from acid deposition (237, 238). As discussed, increased iron availability can shift S speciation to favor iron sulfides. If Fe is limiting S retention, increased Fe inputs could cause increased S contents and changed isotopic signatures in sediments. Third, eutrophication also can lead to increased S retention and presumably to altered isotopic ratios. Widespread eutrophication has occurred in response to human settlement, agricultural activities, and increased atmospheric deposition of NO_3^-. Few studies implicating acid deposition as the cause of increased sediment S content have demonstrated that carbon accumulation rates have been constant over the past century (see refs. 108 and 225). Finally, not all lakes exhibit increased S retention in response to increased concentrations of SO_4^{2-} in lake water. In contrast to other nearby lakes, sediments of McNearney Lake do not show an increased S content (1); primary productivity is thought to be too low to support sulfate reduction. This anomaly suggests that changes in relative abundance of electron acceptors or primary production also could increase S accumulation rates in sediments. Thus increased atmospheric deposition of sulfate may have caused an increase in S accumulation in sediments of some, but not all, impacted lakes. Other factors may have augmented these increases or caused similar increases in sediments of other lakes.

Paleolimnological Conditions.   Because of the interplay between primary production, oxygen content of bottom waters, and the sulfur content and speciation of sediments, sediment profiles of S probably preserve records of paleolimnological conditions. Several studies (23–25, 205) point to increased S content of sediments as a result of eutrophication. Mechanisms involve both rates of S supply to sediments (seston deposition and diffusive gradients) and rates of S reduction and oxidation. The relative S enrichment of humic acids, fulvic acids, and humin in sediments of Greifensee changed as the lake eutrophied (205). Further work is needed to determine if any of these fractions could serve as indices of lake trophic state or oxygenation. It seems likely that other phenomena besides eutrophication could induce changes in S cycling and retention. Levels of bioturbation were shown to influence S retention in coastal marine sediments (194). Changes in lake-water temperature, lake depth, light penetration, or lake mixing also could lead to changes in processes of S deposition and recycling from sediments.

Degree of pyritization (DOP) of iron was proposed as an index of oxygenation of marine bottom waters (226). In oxic bottom waters only a small fraction (<42%) of reactive iron reacts with sulfide to form pyrite; in partially anoxic waters the fraction increases (46–80%), and in permanently anoxic, euxinic sediments the fraction is highest (55–93%; 226). Because iron monosulfides are often as abundant as pyrite in lake sediments, the index must be revised to incorporate these forms together with pyritic iron. This revised index does not appear to be a sensitive index of bottom-water oxygenation in lakes. Values are identical in surface sediments of permanently oxic and seasonally anoxic areas of Little Rock Lake (data from 61, 239), and these values are higher than values in eutrophic and seasonally anoxic lakes Zurich and Horwer Bucht (Urban, unpublished data). The highest value found in the literature (56) occurs in a meromictic lake; however, low inputs of iron may be as important as permanent anoxia in this case. For lakes reported in the literature, values of DOP appear to be determined by the relative abundance of S and Fe in lake sediments (Figure 6). Low availability of iron even under oxic conditions causes high values of DOP.

Ratios of pyrite to monosulfides also may indicate oxygenation of bottom waters or redox level of sediments. Formation of pyrite requires some oxidized S, either polysulfides or elemental S. Hence, formation of pyrite may be enhanced at oxic–anoxic interfaces or in sediments whose redox potential is poised relatively high by iron and manganese oxides (e.g., Ennerdale Water in ref. 112; lakes Zurich and Geneva in ref. 56). A corollary would seem to be that formation of AVS would be favored over pyrite in surface sediments under anoxic hypolimnia or permanently anoxic waters. Such a relationship is not universally observed, however. In lakes Mendota and 223, both of which experience summer anoxia, ratios of pyrite to AVS are very low (<0.1; 25, 111). However, in eutrophic Wintergreen Lake, which also experiences summer anoxia, the ratio is 2.6 (73) and the ratio is 0.12–1.3 in meromictic Lago di Cadagno (56). S-oxidizing bacteria, present in Lago di Cadagno and Wintergreen Lake (133, 137) may enhance formation of pyrite in these lakes by producing elemental S. Among the lakes in Table V, few show evidence of diagenetic conversion of AVS to pyrite. Only four have higher ratios of CRS:AVS in deep relative to surface sediments, and all four are eutrophic or meromictic lakes with relatively little pyrite in surface sediments; changes downcore may reflect historical changes in lake conditions rather than diagenesis.

Other factors that may influence ratios of pyrite to AVS are sediment pH, rates of benthic respiration, availability of Fe^{2+}, and forms of iron oxides present. Rapid formation of pyrite is enhanced at low (<7) pH (*161*), whereas precipitation of AVS is enhanced by high pH (*160*). Rates of AVS formation are governed by rates of H_2S production and availability of Fe^{2+} (*94, 112, 160*). Davison et al. (*112*) observed that AVS contents of sediments from three lakes were proportional to rates of carbon sedimentation. Pyrite content was independent of carbon supply in these lakes and was thought to be governed by availability of partially oxidized S. High carbon supply also might favor microbial reduction of iron oxides (*96, 97*) over the competing reduction by sulfide. Reduction of iron oxides by sulfide produces S^0 (*160*), which may enhance pyrite formation. Hence low carbon supply and an abundance of highly crystalline iron oxides (cf. *94, 96, 97, 170*) may enhance pyrite over AVS formation. The limited data set (Table V) suggests that the ratio is a reasonable indicator of trophic state or availability of oxidants in sediments.

Summary

A wealth of data assembled over the past 20 years allows an assessment of what factors control retention of S in lake sediments. A notable finding is that rates of S accumulation are generally much lower than S fluxes to and from sediments and much lower than rates of cycling within sediments. Retention of S represents a small difference between large opposing processes. Factors affecting these opposing processes of seston deposition, sulfate reduction, and S recycling determine the rate of S accumulation. Fluxes of S to sediments are controlled by rates of primary productivity and by the steepness of gradients of sulfate in pore-water profiles. Concentrations of sulfate in lakes appear to be the major factor determining the magnitude of the diffusive flux. Recycling of S from sediments is regulated by the presence of S-oxidizing bacteria and suitable electron acceptors (oxygen, iron, and manganese oxides). The availability of iron does appear to influence the speciation of S within sediments, but further evidence is needed to determine whether it limits retention of S. Sulfur speciation is a useful index of paleolimnological conditions. Ratios of CRS:AVS give a general indication of bottom-water oxygenation, and ratios of organic to pyritic S can distinguish lacustrine from marine sediments. Accumulation rates and speciation of S are also useful indicators of lake eutrophication.

Acknowledgments

This work was funded in part by the National Science Foundation (Grant No. INT–8909806) and by the Swiss Federal Institute for Water Research and Pollution Control. I am grateful to Stefan Peiffer and Bernhard Wehrli for helpful reviews of the manuscript.

References

1. Cook, R. B.; Kelly, C. A. In *Sulfur Cycling on the Continents;* Howarth, R. W.; Stewart, J. W.; Ivanov, M. V., Eds.; Wiley: New York, 1991; pp 145–188.
2. Beauchamp, R. S. A. *Nature (London)* **1953**, *171*, 769–771.
3. Ingvorsen, K.; Brock, T. D. *Limnol. Oceanogr.* **1982**, *27*, 559–564.
4. Lovley, D. R.; Klug, M. J. *Geochim. Cosmochim. Acta* **1986**, *50*, 11–18.
5. Capone, D. G.; Kiene, R. P. *Limnol. Oceanogr.* **1988**, *33*, 725–749.
6. Schindler, D. W.; Turner, M. A.; Stainton, M. P.; Linsey, G. A. *Science (Washington, D.C.)* **1986**, *232*, 844–847.
7. Brezonik, P. L.; Baker, L. A.; Perry, T. E. In *Sources and Fates of Aquatic Pollutants;* Hites, R. A.; Eisenreich, S. J., Eds.; Advances in Chemistry 216; American Chemical Society: Washington, DC, 1987; pp 229–262.
8. Caraco, N. F.; Cole, J. J.; Likens, G. E. *Nature (London)* **1989**, *341*, 316–318.
9. Davison, W. *Geol. Soc. Spec. Publ. London* **1988**, *40*, 131–137.
10. Sholkovitz, E. R. In *Chemical Processes in Lakes;* Stumm, W., Ed.; Wiley-Interscience: New York, 1985; pp 119–142.
11. Shoenheit, P.; Kristjansson, J. K.; Thauer, R. K. *Arch. Microbiol.* **1982**, *132*, 285-288.
12. Kristjansson, J. K.; Schoenheit, P.; Thauer, R. K. *Arch. Microbiol.* **1982**, *131*, 278–282.
13. Lovley, D. R.; Klug, M. J. *Appl. Environ. Microbiol.* **1983**, *45*, 187–192.
14. Ingvorsen, K.; Jorgensen, B. B. *Arch. Microbiol.* **1984**, *139*, 61–66.
15. Berner, R. A. *Am. J. Sci.* **1970**, *268*, 1–23.
16. Berner, R. A. *Geochim. Cosmochim. Acta* **1984**, *48*, 605–615.
17. Berner, R. A. *Am. J. Sci.* **1987**, *287*, 177–196.
18. Berner, R. A.; Raiswell, R. *Geochim. Cosmochim. Acta* **1983**, *47*, 855–862.
19. Garrels, R. M.; Lerman, A. *Am. J. Sci.* **1984**, *284*, 989–1007.
20. Berner, R. A.; Baldwin, T.; Holdren, G. R. *J. Sediment. Petrol.* **1979**, *49*, 1345–1350.
21. Whittaker, S. G.; Kyser, T. K. *Geochim. Cosmochim. Acta* **1990**, *54*, 2799–2810.
22. Zaback, D. A.; Pratt, L. M. *Geochim. Cosmochim. Acta* **1992**, *56*, 763–774.
23. Zullig, H. *Schweiz. Z. Hydrol.* **1956**, *18*, 5–143.
24. Gorham, E.; Lund, J. W.; Sanger, J. E.; Dean, W. E. *Limnol. Oceanogr.* **1974**, *19*, 601–617.
25. Nriagu, J. O. *Limnol. Oceanogr.* **1968**, *13*, 430–439.
26. Nriagu, J. O. *Sci. Total Environ.* **1984**, *38*, 7–13.
27. Nriagu, J. O.; Soon, Y. K. *Geochim. Cosmochim. Acta* **1985**, *49*, 823–834.
28. Holdren, G. R.; Brunelle, T. M.; Matisoff, G.; Wahlen, M. *Nature (London)* **1984**, *311*, 245–247.
29. Mitchell, M. J.; Schindler, S. C.; Owen, J. S.; Norton, S. A. *Hydrobiologia* **1988**, *157*, 219–229.
30. Giblin, A. E.; Likens, G. E.; White, D.; Howarth, R. W. *Limnol. Oceanogr.* **1990**, *35*, 852–869.
31. Damste, J. S. S.; Rijpstra, I. C.; Kock-van Dalen, A. C.; de Leeuw, J. W.; Schenck, P. A. *Geochim. Cosmochim. Acta* **1989**, *53*, 1343–1355.
32. Kohnen, M. E. L.; Damste, J. S. S.; Kock-van Dalen, A. C.; ten Haven, H. L.; Rullkotter, J.; de Leeuw, J. W. *Geochim. Cosmochim. Acta* **1990**, *54*, 3053–3063.
33. Kohnen, M. E. L.; Damste, J. S. S.; Kock-van Dalen, A. C.; de Leeuw, J. W. *Geochim. Cosmochim. Acta* **1990**, *55*, 1375–1394.
34. White, J. R.; Gubala, C. P.; Fry, B.; Owen, J.; Mitchell, M. J. *Geochim. Cosmochim. Acta* **1989**, *53*, 2547–2560.

35. Berner, R. A. *Am. J. Sci.* **1969**, *267*, 19–42.
36. Altschuler, Z. S.; Schnepfe, M. M.; Silber, C. C.; Simon, F. O. *Science (Washington, D.C.)* **1983**, *221*, 221–227.
37. Psenner, R. *Schweiz. Z. Hydrol.* **1983**, *45*, 219–232.
38. Urban, N. R.; Brezonik, P. L. *Can. J. Fish. Aquat. Sci.* **1993**, in press.
39. Brunskill, G. J.; Povoledo, D.; Graham, B. W.; Stainton, M. P. *J. Fish. Res. Board. Can.* **1971**, *28*, 277–294.
40. Zehnder, A. J. B.; Zinder, S. H. In *The Handbook of Environmental Chemistry*, Vol. 1., Part A; Hutzinger, O., Ed.; Springer-Verlag: New York, 1980; pp 105–145.
41. Fitzgerald, J. W.; Watwood, M. E.; Rose, F. A. *Soil Biol. Biochem.* **1985**, *17*, 885–887.
42. David, M. B.; Mitchell, M. J.; Nakas, J. P. *Soil Sci. Soc. Am. J.* **1982**, *46*, 847–852.
43. Freney, J. R.; Melville, G. E.; Williams, C. H. *Soil Biol. Biochem.* **1975**, *7*, 217–221.
44. Bettany, J. R.; Stewart, J. W.; Halstead, E. H. *Soil Sci. Soc. Am. Proc.* **1973**, *37*, 915–918.
45. Tabatabai, M. A.; Bremner, J. M. *Soil Sci.* **1972**, *114*, 380–386.
46. Losher, A. J.; Kelts, K. R. *Terra Nova* **1989**, *1*, 253–261.
47. Davison, W.; Finlay, B. J. *J. Ecol.* **1986**, *74*, 663–673.
48. Nriagu, J. O.; Coker, R. D. *Nature (London)* **1983**, *303*, 692–694.
49. Mitchell, M. J.; David, M. B.; Uutala, A. J. *Hydrobiologia* **1985**, *121*, 121–127.
50. Carignan, R.; Tessier, A. *Geochim. Cosmochim. Acta* **1988**, *52*, 1179–1188.
51. Mitchell, M. J.; Landers, D. H.; Brodowski, F.; Lawrence, G. B.; David, M. B. *Water Air Soil Pollut.* **1984**, *21*, 231–245.
52. Zhabina, N. N.; Volkov, I. I. In *Environmental Biogeochemistry and Geomicrobiology*; Krumbein, W. E., Ed.; Ann Arbor Science Publishers: Ann Arbor, MI, 1978; Vol. 3, pp 735–745.
53. Lowe, L. E.; Bustin, R. M. *Can. J. Soil Sci.* **1989**, *69*, 287–293.
54. Johnson, C. M.; Nishita, H. *Anal. Chem.* **1952**, *24*, 736–742.
55. Sorokin, J. I.; Donato, N. *Hydrobiologia* **1975**, *47*, 241–252.
56. Losher, A. Ph.D. Dissertation, Swiss Federal Institute of Technology, Zurich, 1989.
57. Nriagu, J. O.; Coker, R. D.; Kemp, A. L. W. *Limnol. Oceanogr.* **1979**, *24*, 383–389.
58. Jorgensen, B. B. *Limnol. Oceanogr.* **1990**, *35*, 1329–1342.
59. David, M. B.; Mitchell, M. J. *Limnol. Oceanogr.* **1985**, *30*, 1196–1207.
60. Kokkonen, P.; Tolonen, K. *Water Air Soil Pollut.* **1987**, *35*, 157–170.
61. Baker, L. A.; Engstrom, D. R.; Brezonik, P. L. *Limnol. Oceanogr.* **1992**, *37*, 689–702.
62. Stuiver, M. *Geochim. Cosmochim. Acta* **1967**, *31*, 2151–2167.
63. Wetzel, R. G. *Limnology*; W. B. Saunders: Toronto, Canada, 1975; 743 pp.
64. Zinder, S. H.; Deomel, W. N.; Brock, T. D. *Appl. Environ. Microbiol.* **1977**, *34*, 859–860.
65. Zinder, S. H.; Brock, T. D. *Appl. Environ. Microbiol.* **1978**, *35*, 344–352.
66. Nriagu, J. O.; Holdway, D. A. *Tellus* **1989**, *41B*, 161–169.
67. Winogradski, S. *Beitraege zur Morphologie und Physiologie der Bakterien*; Verlag A. Felix: Leipzig, Germany, 1888; Vol. 1, pp 120.
68. Kuznetsov, S. I. *Limnol. Oceanogr.* **1968**, *13*, 211–224.
69. Pfennig. N. *Ann. Rev. Microbiol.* **1977**, *31*, 275–290.
70. Goldman, J. C.; Porcella, D. B.; Middlebrooks, E. J.; Toerin, D. F. *Water Res.* **1972**, *6*, 637–179.

71. Hesse, P. R. *Hydrobiologia* **1958**, *11*, 29–39.
72. Baker, L. A.; Urban, N. R.; Sherman, L. A.; Brezonik, P. L. In *Biogenic Sulfur in the Environment;* Saltzman, E.; Cooper, W., Eds.; ACS Symposium Series 393; American Chemical Society: Washington, DC, 1989; pp 79–100.
73. King, G. M.; Klug, M. J. *Appl. Environ. Microbiol.* **1982**, *43*, 1406–1412.
74. O'Kelly, J. C. In *Algal Physiology and Biochemistry;* Stewart, W. D. P., Ed.; Blackwell Scientific: London, 1974; Chapter 22.
75. Schiff, J. A. In *Physiology and Biochemistry of Algae;* Lewin, R. A., Ed.; Academic: New York, 1962; Chapter 14.
76. Brock, T. D. *Biology of Microorganisms;* Prentice-Hall: Englewood Cliffs, NJ, 1979.
77. Monheimer, R. H. *Limnol. Oceanogr.* **1975**, *20*, 183–190.
78. Smith, R. L.; Klug, M. J. *Appl. Environ. Microbiol.* **1981**, *41*, 1230–1237.
79. Landers, D. H.; Mitchell, M. J. *Hydrobiologia* **1988**, *160*, 85–95.
80. Rudd, J. W.; Kelly, C. A.; St. Louis, V.; Hesslein, R. H.; Furutani, A.; Holoka, M. H. *Limnol. Oceanogr.* **1986**, *31*, 1267–1280.
81. Jones, J. G.; Simon, B. M.; Roscoe, J. V. *J. Gen. Microbiol.* **1982**, *128*, 2833–2839.
82. Dunnette, D. A. In *Biogenic Sulfur in the Environment;* Saltzman, E.; Cooper, W., Eds.; ACS Symposium Series 393; American Chemical Society: Washington, DC, 1989; pp 72–78.
83. King, G. M; Klug, M. J. *Appl. Environ. Microbiol.* **1980**, *39*, 950–956.
84. Jorgensen, B. B. *Philos. Trans. R. Soc. London* **1982**, *B298*, 543–561.
85. Ingvorsen, K.; Zeikus, J. G.; Brock, T. D. *Appl. Environ. Microbiol.* **1981**, *42*, 1029–1036.
86. Smith, R. L.; Klug, M. J. *Appl. Environ. Microbiol.* **1981**, *42*, 116–121.
87. Lovley, D. R.; Dwyer, D. F.; Klug, M. J. *Appl. Environ. Microbiol.* **1982**, *43*, 1373–1379.
88. Lovley, D. R.; Klug, M. J. *Appl. Environ. Microbiol.* **1982**, *43*, 552–560.
89. Pfennig, N.; Widdel, F. *Philos. Trans. R. Soc. London* **1982**, *B298*, 433–441.
90. Kuivila, K. M.; Murray, J. W.; Devol, A. H. *Geochim. Cosmochim. Acta* **1989**, *53*, 409–416.
91. Stumm, W.; Morgan, J. J. *Aquatic Chemistry;* Wiley-Interscience: New York, 1981; pp 458–461.
92. Canfield, D. E.; Des Marais, D. J. *Science (Washington, D.C.)* **1991**, *251*, 1471–1473.
93. Jorgensen, B. B.; Bak, F. *Appl. Environ. Microbiol.* **1991**, *57*, 847–856.
94. Canfield, D. E. *Geochim. Cosmochim. Acta* **1989**, *53*, 619–632.
95. Lovley, D. R.; Goodwin, S. *Geochim. Cosmochim. Acta* **1988**, *52*, 2993–3003.
96. Lovley, D. R.; Phillips, E. J. P. *Appl. Environ. Microbiol.* **1986**, *51*, 683–689.
97. Lovley, D. R. *Geomicrobiol. J.* **1987**, *5*, 375–399.
98. Rudd, J. W.; Kelly, C. A.; Furutani, A. *Limnol. Oceanogr.* **1986**, *31*, 1281–1291.
99. Cook, R. B.; Kelley, C. A.; Kingston, J. G.; Kreis, R. G. *Biogeochemistry* **1987**, *4*, 97–118.
100. Cook, R. B.; Kelly, C. A.; Schindler, D. W.; Turner, M. A. *Limnol. Oceanogr.* **1986**, *31*, 134–148.
101. Urban, N. R.; Baker, L. A.; Sherman, L. A.; Brezonik, P. L. *Limnol. Oceanogr.* **1993**, in press.
102. Fossing, H.; Jorgensen, B. B. *Biogeochemistry* **1989**, *8*, 205–222.
103. Skyring, G. W. *Geomicrobiol. J.* **1987**, *5*, 295–374.
104. Sorokin, J. I. *Hydrobiologia* **1975**, *47*, 231–240.
105. Nielsen, P. H. *Appl. Environ. Microbiol.* **1987**, *53*, 27–32.
106. Westrich, J. T.; Berner, R. A. *Limnol. Oceanogr.* **1984**, 29, 236–249.

107. Kelly, C. A.; Rudd, J. W. M. *Biogeochemistry* **1984**, *1*, 63–78.
108. Giblin, A. E.; Likens, G. E.; Howarth, R. W. *Limnol. Oceanogr.* **1991**, *36*, 1265–1270.
109. Stauffer, R. E. *Limnol. Oceanogr.* **1991**, *36*, 1263–1264.
110. Cook, R. B. Ph.D. Thesis, Columbia University, New York, 1981.
111. Cook, R. B. *Can. J. Fish. Aquat. Sci.* **1984**, *41*, 286–293.
112. Davison, W.; Lishman, J. P.; Hilton, J. *Geochim. Cosmochim. Acta* **1985**, *49*, 1615–1620.
113. Carlton, R. G.; Walker, G. S.; Klug, M. J.; Wetzel, R. G. *J. Great Lakes Res.* **1989**, *15*, 133–140.
114. Jorgensen, B. B.; Marais, D. J. *Limnol. Oceanogr.* **1990**, *35*, 1343–1355.
115. O'Brien, D. J.; Birkner, F. B. *Environ. Sci. Technol.* **1977**, *11*, 1114–1120.
116. Millero, F. J.; Hubinger, S.; Fernandez, M.; Garnett, S. *Environ. Sci. Technol.* **1987**, *21*, 439–443.
117. Jolley, R. A.; Forster, C. F. *Environ. Technol. Lett.* **1985**, *6*, 1–10.
118. Chen, K. Y.; Morris, J. C. *Environ. Sci. Technol.* **1972**, *6*, 529–537.
119. Jorgensen, B. B. In *The Nitrogen and Sulfur Cycles;* Society of General Microbiology Symposium 42; Society of General Microbiology: Cambridge, MA, 1988; pp 31–63.
120. Kiene, R. P.; Taylor, B. F. *Nature (London)* **1988**, *332*, 148–150.
121. Kuenen, J. G.; Bos, P. In *Autrophic Bacteria;* Schlegel, H. G.; Bowien, B., Eds.; Springer-Verlag: New York, 1989; pp 53–80.
122. Cook, R. B.; Schindler, D. W. *Ecol. Bull. (Stockholm)* **1983**, *35*, 115–127.
123. Kuenen, J. G. *Plant Soil* **1975**, *43*, 49–76.
124. Pfennig, N. L. *Plant Soil* **1975**, *43*, 1–16.
125. Van Gemerden, H. Ph.D. Thesis, University of Leiden, 1967.
126. Culver, D. A.; Brunskill, G. J. *Limnol. Oceanogr.* **1969**, *14*, 862–873.
127. Sorokin, J. I. *Arch. Hydrobiol.* **1970**, *66*, 391–446.
128. Parkin, T. B.; Brock, T. D. *Limnol. Oceanogr.* **1981**, *26*, 880–890.
129. Guerrero, R.; Montesinos, E.; Pedros-Alio, C.; Esteve, I.; Mas, J.; van Gemerden, H.; Hofman, P. *Limnol. Oceanogr.* **1985**, *30*, 919–931.
130. Steenbergen, C. L. M.; Korthals, H. J. *Arch. Hydrobiol. Beih. Ergeb. Limnol.* **1988**, *31*, 45–53.
131. Van Gemerden, H.; Montesinos, E.; Mas, J.; Guerrero, R. *Limnol. Oceanogr.* **1985**, *30*, 932–943.
132. Caldwell, D. E.; Tiedje, J. M. *Can. J. Microbiol.* **1975**, *21*, 377–385.
133. Hurley, J. P.; Watras, C. J. *Limnol. Oceanogr.* **1991**, *36*, 307–314.
134. Strohl, W. R.; Larkin, J. M. *Appl. Environ. Microbiol.* **1978**, *36*, 755–770.
135. Jorgensen, B. B. *Limnol. Oceanogr.* **1977**, *22*, 814–832.
136. Hanselmann, K. *Neue Zuercher Zeitung* **1985**, *121*, 8.
137. Gachter, R.; Meyer, J. S. In *Fates and Effects of In-Place Pollutants in Aquatic Ecosystems;* Bands, R.; Giesy, J., Eds; Lewis: Ann Arbor, MI, 1990; pp 131–162.
138. Cloern, J. E.; Cole, B. E.; Oremland, R. S. *Limnol. Oceanogr.* **1983**, *28*, 1049–1061.
139. Matsuyama, M.; Shirouzu, E. *Jpn. J. Limnol.* **1978**, *39*, 103–111.
140. Cohen, Y.; Krumbein, W. E.; Shilo, M. *Limnol. Oceanogr.* **1977**, *22*, 609–634.
141. Biebl, H.; Pfennig, H. *Arch. Mikrobiol.* **1978**, *117*, 9–16.
142. Howarth, R. W. *Biogeochemistry* **1984**, *1*, 5–28.
143. Jorgensen, B. B. *Mar. Biol. Berlin* **1977**, *41*, 19–28.
144. Jorgensen, B. B.; Revsbech, N. P. *Appl. Environ. Microbiol.* **1983**, *45*, 1261–1270.
145. Grant, J.; Bathmann, U. V. *Science (Washington, D.C.)* **1987**, *236*, 1472–1474.

146. Aller, R. C.; Rude, P. D. *Geochim. Cosmochim. Acta* **1988**, *52*, 751–765.
147. Jorgensen, B. B.; Kuenen, J. G.; Cohen, Y. *Limnol. Oceanogr.* **1979**, *24*, 799–822.
148. Jorgensen, B. B. In *Autrophic Bacteria;* Schlegel, H. G.; Bowien, B., Eds.; Springer-Verlag: New York, 1989; pp 117–146.
149. Berner, R. A. *Early Diagenesis: A Theoretical Approach;* Princeton University Press: Princeton, NJ, 1980.
150. Robbins, J. A.; Callender, E. *Am. J. Sci.* **1975**, *275*, 512–533.
151. Burdige, D. J.; Nealson, K. H. *Geomicrobiol. J.* **1986**, *4*, 361–387.
152. Berner, R. A. In *The Changing Chemistry of the Oceans;* Nobel Symposium 20; Dyrssen, D.; Jagner, D., Eds.; Almqvist and Wiksell: Stockholm, Sweden, 1972; pp 347–361.
153. Luther, G. W. *Geochim. Cosmochim. Acta* **1987**, *51*, 3193–3199.
154. Brown, A. D.; Jurinak, J. J. *J. Environ. Qual.* **1989**, *18*, 545–550.
155. Moses, C. O.; Nordstrom, D. K.; Herman, J. S.; Mills, A. L. *Geochim. Cosmochim. Acta* **1987**, *51*, 1561–1571.
156. Moses, C. O.; Herman, J. S. *Geochim. Cosmochim. Acta* **1991**, *55*, 471–482.
157. McKibben, M. A.; Barnes, H. L. *Geochim. Cosmochim. Acta* **1986**, *50*, 1509–1520.
158. Canfield, D.; Berner, R. A. *Geochim. Cosmochim. Acta* **1987**, *51*, 645–659.
159. Karlin, R.; Levi, S. *Nature (London)* **1983**, *303*, 327–330.
160. Pyzik, A. J.; Sommer, S. E. *Geochim. Cosmochim. Acta* **1981**, *45*, 687–698.
161. Rickard, D. T. *Am. J. Sci.* **1975**, *275*, 636–652.
162. Dos Santos Afonso, M.; Stumm, W. *Langmuir* **1992**, *8*, 1671–1675.
163. Peiffer, S.; Dos Santos Afonso, M.; Wehrli, B.; Gaechter, R. *Environ. Sci. Technol.* **1992**, *26*, 2408–2412.
164. Sherman, L. A. Masters Thesis, University of Minnesota, Minneapolis, 1988.
165. Emerson, S. *Geochim. Cosmochim. Acta* **1976**, *40*, 925–934.
166. Froehlich, P. N.; Klinkhammer, G. P.; Bender, M. L.; Luedtke, N. A.; Heath, G.; Cullen, D. *Geochim. Cosmochim. Acta* **1979**, *43*, 1075–1090.
167. Lord, C. J.; Church, T. M. *Geochim. Cosmochim. Acta* **1983**, *47*, 1381–1391.
168. Wersin, P.; Hoehener, P.; Giovanoli, R.; Stumm, W. *Chem. Geol.* **1991**, *90*, 233–252.
169. Lovely, D. R.; Phillips, E. J. P.; Longergan, D. J. *Environ. Sci. Technol.* **1991**, *25*, 1062–1067.
170. Lovley, D. R.; Phillips, E. J. P. *Appl. Environ. Microbiol.* **1987**, *53*, 2636–2641.
171. Pfennig, N. In *Autotrophic Bacteria;* Schlegel, H. G.; Bowien, B., Eds.; Springer-Verlag: New York, 1989; pp 97–115.
172. King, G. M. *FEMS Microbiol. Ecol.* **1990**, *73*, 131–138.
173. Jorgensen, B. B. *Science (Washington, D.C.)* **1990**, *249*, 152–154.
174. Troelsen, H.; Jorgensen, B. B. *Estuarine Coastal Shelf Sci.* **1982**, *15*, 255–260.
175. Hoffmann, M. R. *Environ. Sci. Technol.* **1977**, *11*, 61–66.
176. Fossing, H.; Jorgensen, B. B. *Biogeochemistry* **1990**, *9*, 223–246.
177. Bak, F.; Cypionka, H. *Nature (London)* **1987**, *326*, 891–892.
178. Bak, F.; Pfennig, N. *Arch. Mikrobiol.* **1987**, *147*, 184–189.
179. Kramer, M.; Cypionka, H. *Arch. Mikrobiol.* **1989**, *151*, 232–237.
180. Giblin, A. E. *Geomicrobiol. J.* **1988**, *6*, 77–97.
181. Francois, R. *Geochim. Cosmochim. Acta* **1987**, *51*, 17–28.
182. Vairavamurthy, A.; Mopper, K. In *Biogenic Sulfur in the Environment;* Saltzman, E. S.; Cooper, W., Eds.; ACS Symposium Series 393; American Chemical Society: Washington, DC, 1989; pp 231–242.
183. Kohnen, M. E. L.; Damste, J. S. S.; ten Haven, H. L.; de Leeuw, J. W. *Nature (London)* **1989**, *341*, 640–641.

184. Luther, G. W.; Church, T. M. In *Sulfur Cycling on the Continents;* Howarth, R. W.; Stewart, J. W. B.; Ivanov, M. V., Eds.; Wiley-Interscience: New York, 1992; pp 125–144.
185. Lalonde, R. T. In *Geochemistry of Sulfur in Fossil Fuels;* Orr, W. L.; White, C. M., Eds.; ACS Symposium Series 429; American Chemical Society: Washington, DC, 1990; pp 68–82.
186. Davison, W. *Schweiz. Z. Hydrol.* **1991**, *53*, 309–329.
187. Sweeney, R. E.; Kaplan, I. R. *Econ. Geol.* **1973**, *68*, 618–634.
188. Rickard, D. T. *Stockholm Contrib. Geol.* **1969**, *20*, 67–95.
189. Vallentyne, J. R. *Limnol. Oceanogr.* **1963**, *8*, 16–30.
190. Losher, A. J.; Kelts, K. R. In *Water-Rock Interaction;* Miles, D. L., Ed.; Balkema: Rotterdam, Netherlands, 1989; pp 449–452.
191. Luther, G. W. *Geochim. Cosmochim. Acta* **1991**, *55*, 2839–2850.
192. Kling, G. W.; Giblin, A. E.; Fry, B.; Peterson, B. J. *Limnol. Oceanogr.* **1991**, *36*, 106–122.
193. Rudd, J. W.; Kelly, C. A.; Schindler, D. W.; Turner, M. A. *Limnol. Oceanogr.* **1990**, *35*, 663–679.
194. Berner, R. A.; Westrich, J. T. *Am. J. Sci.* **1985**, *285*, 193–206.
195. Boulegue, J.; Lord, C. J.; Church, T. M. *Geochim. Cosmochim. Acta* **1982**, *46*, 453–464.
196. Fry, B. *Biogeochemistry* **1986**, *2*, 329–343.
197. Thode, H. G.; Dickman, M. D.; Rao, S. S. *Arch. Hydrobiol. Suppl. 74*, **1987**, *4*, 397–422.
198. Dinur, D.; Spiro, B.; Aizenshtat, Z. *Chem. Geol.* **1980**, *31*, 37–51.
199. Norton, S. A.; Mitchell, M. J.; Kahl, J. S.; Brewer, G. F. *Water Air Soil Pollut.* **1988**, *39*, 33–45.
200. Brown, K. A. *Soil Biol. Biochem.* **1986**, *18*, 131–140.
201. Casagrande, D. J.; Idowu, G.; Friedman, A.; Rickert, P.; Siefert, K.; Schlenz, D. *Nature (London)* **1979**, *282*, 599–600.
202. Casagrande, D. J.; Ng, L. *Nature (London)* **1979**, *282*, 598–599.
203. Mango, F. D. *Geochim. Cosmochim. Acta* **1983**, *47*, 1433–1441.
204. De Roo, J.; Hodgson, G. W. *Chem. Geol.* **1980**, *22*, 71–78.
205. Hollander, D. J. Doctoral Dissertation, Swiss Federal Technical Institute, Zurich, 1989.
206. Francois, R. *Limnol. Oceanogr.* **1987**, *32*, 964–972.
207. Ferdelman, T. G.; Church, T. M.; Luther, G. W. *Geochim. Cosmochim. Acta* **1991**, *55*, 979–988.
208. Brassell, S. C.; Lewis, C. A.; de Leeuw, J. W.; de Lange, F.; Damste, J. S. S. *Nature (London)* **1986**, *320*, 160–162.
209. Damste, J. S. S.; de Leeuw, J. W.; Kock-van Dalen, A. C.; de Zeeuw, M. A.; Lange, F. D.; Rijpstra, I. C. *Geochim. Cosmochim. Acta* **1987**, *51*, 2369–2391.
210. Luther, G. W.; Church, T. M.; Scudlark, J. R.; Cosman, C. *Science (Washington, D.C.)* **1986**, *232*, 746–749.
211. Luther, G. W.; Church, T. M. *Mar. Chem.* **1988**, *23*, 295–309.
212. Mopper, K.; Taylor, B. In *Organic Marine Geochemistry;* Sohn, M., Ed.; ACS Symposium Series 305; American Chemical Society: Washington, DC, 1986; pp 324–339.
213. Shea, D.; MacCrehan, W. A. *Sci. Total Environ.* **1988**, *73*, 135–141.
214. Vetter, R. D.; Matrai, P. A.; Javor, B.; O'Brien, J. In *Biogenic Sulfur in the Environment;* Saltzman, E. C.; Cooper, W., Eds.; ACS Symposium Series 393; American Chemical Society: Washington, DC, 1989; pp 243–261.
215. Vairavamurthy, A.; Mopper, K. *Nature (London)* **1987**, *329*, 623–625.

216. LaLonde, R. T.; Ferrara, L. M.; Hayes, M. P. *Org. Geochem.* **1987**, *11*, 563–571.
217. Guerin, W. F.; Braman, R. S. *Org. Geochem.* **1985**, *8*, 259–268.
218. Mojelski, T. W.; Ignasiak, T. M.; Frakman, Z.; McIntyre, D.; Lown, E.; Montgomery, D.; Strausz, O. *Abstr. Pap. Am. Chem. Soc. 200th* **1990**, GEOC 0035.
219. Urban, N. R.; Baker, L. A. *Limnol. Oceanogr.* **1989**, *34*, 1144–1146.
220. Kelly, C. A.; Rudd, J. W.; Hesslein, R. H.; Schindler, D. W.; Dillon, P. J.; Driscoll, C. T.; Gherini, S. A. *Biogeochemistry* **1987**, *3*, 129–140.
221. Brock, T. D. *A Eutrophic Lake;* Ecol. Studies 55; Springer-Verlag: New York, 1985; pp 181–187.
222. Baker, L. A.; Brezonik, P. L.; Edgerton, E. S.; Ogburn, W. O. *Water Air Soil Pollut.* **1985**, *25*, 215–230.
223. Baker, L. A.; Brezonik, P. L.; Pollman, C. D. *Water Air Soil Pollut.* **1986**, *31*, 89–94.
224. Lin, J. C.; Schnoor, J. L.; Glass, G. E. In *Sources and Fates of Aquatic Pollutants;* Hites, R. A.; Eisenreich, S. J., Eds.; Advances in Chemistry 216; American Chemical Society: Washington, DC, 1987; pp 209–228.
225. Norton, S. A.; Kahl, J. S.; Mitchell, M. J.; Owen, J. S. *Limnol. Oceanogr.* **1991**, *36*, 1271–1274.
226. Raiswell, R.; Buckley, F.; Berner, R.; Anderson, T. F. *J. Sediment. Petrol.* **1988**, *58*, 812–819.
227. Parkin, T. B.; Brock, T. D. *Arch. Hydrobiol.* **1981**, *91*, 366–382.
228. Truper, H. G.; Genovese, S. *Limnol. Oceanogr.* **1968**, *13*, 225–232.
229. Brown, S. R. *Limnol. Oceanogr.* **1968**, *13*, 233–241.
230. Steenbergen, C. L. M.; Korthals, H. J. *Limnol. Oceanogr.* **1982**, *27*, 883–895.
231. Berner, R. A.; Raiswell, R. *Geology* **1984**, *12*, 365–368.
232. Raiswell, R.; Berner, R. A. *Geochim. Cosmochim. Acta* **1986**, *50*, 1967–1976.
233. Bein, A.; Almogi-Labin, A.; Sass, E. *Am. J. Sci.* **1990**, *290*, 228–911.
234. Dean, W. E.; Arthur, M. A. *Am. J. Sci.* **1989**, *289*, 708–143.
235. Lowe, L. E.; Bustin, R. M. *Can. J. Soil Sci.* **1985**, *65*, 531–541.
236. Urban, N. R.; Eisenreich, S. J.; Grigal, D. F. *Biogeochemistry* **1989**, *7*, 81–109.
237. Urban, N. R.; Gorham, E.; Underwood, J. K.; Martin, F. B.; Ogden, J. G. *Limnol. Oceanogr.* **1990**, *35*, 1516–1534.
238. Dillon, P. J.; Evans, H. E.; Scholer, P. J. *Biogeochemistry* **1988**, *5*, 201–220.
239. Downing, G. M. M.S. Thesis, University of Minnesota, Minneapolis, 1986.
240. Jones, J. G.; Simon, B. M. *J. Ecol.* **1980**, *68*, 493–512.
241. Takahashi, M.; Ichimura, S. *Limnol. Oceanogr.* **1968**, *13*, 644–655.
242. Lawrence, J. R.; Haynes, R. C.; Hammer, U. T. *Verh. Int. Ver. Theor. Angew. Limnol.* **1978**, *20*, 201–207.
243. Greb, S. R.; Sherman, L. A.; Perry, T. E.; Baker, L. A., unpublished.

RECEIVED for review September 26, 1991. ACCEPTED revised manuscript June 9, 1992.

Reaction of H_2S with Ferric Oxides

Some Conceptual Ideas on Its Significance for Sediment–Water Interactions

Stefan Peiffer

Limnological Research Station, University of Bayreuth, P.O. Box 101251, D–8580 Bayreuth, Germany

Conceptual ideas based on laboratory studies explore the relevance of the interaction of H_2S with reactive iron in freshwater sediments. The experimental findings suggest a mechanism for sulfate recycling within sediments that involves anoxic oxidation of sulfide by reactive ferric oxides and that may help to explain high sulfate reduction rates. The transport limited supply of the redox equivalents into deeper sediment layers, however, requires very steep microscale horizontal gradients of substances along with the measurable macroscale vertical pore-water gradients. Such a microstructure can be adequately represented by a biofilm model. In addition to its role in sulfate recycling, oxidation of H_2S by reactive iron also may contribute to pyrite formation. In this process ferric oxides will replace elemental sulfur as the oxidant to form polysulfides, which further react with FeS to form pyrite.

SULFUR GENERALLY IS NOT A LIMITING NUTRIENT in either aquatic (1) or terrestrial ecosystems (2), unlike nitrogen or phosphorus. Considering the various redox states in which sulfur can be found in nature (3), its ecological role could be described as an electron mediator. This property can be observed particularly at interfaces with steep redox gradients, such as the borderline between sediment and water, where an intense cycling of sulfur compounds occurs (4, 5). Consumption and production of biomass are frequently accompanied by sulfate reduction and reoxidation of sulfide via several intermediates by chemoautotrophic organisms (6, 7).

0065–2393/94/0237–0371$06.00/0

Attention has been focused recently on sedimentary microbial reduction of sulfate. This process neutralizes atmospheric sulfuric acid deposited into soft-water lakes (8–10) through the production of two equivalents of alkalinity per mole of sulfate reduced (11).

$$SO_4^{2-} + 2<CH_2O> \longrightarrow H_2S + 2HCO_3^- \tag{1}$$

where $<CH_2O>$ denotes organic matter.

The strong coupling between sulfur and iron chemistry becomes obvious in this example. Conservation of alkalinity within the system is achieved only if the sulfide formed is prevented from reoxidation, a process that would restore the acidity. Prevention of reoxidation occurs through the ultimate storage of sulfide in sediments, either as organic sulfur or as iron sulfides (12, 13). The overall reaction of pyrite formation proceeds via formation of FeS:

$$Fe^{2+} + H_2S + 2H_2O \rightleftharpoons FeS + 2H_3O^+ \tag{2}$$

and a subsequent reaction with elemental sulfur

$$H_2S + \frac{1}{2}O_2 \longrightarrow S^0 + H_2O \tag{3}$$

$$S^0 + FeS \longrightarrow FeS_2 \tag{4}$$

which gives in total

$$2SO_4^{2-} + 4<CH_2O> + Fe^{2+} + \frac{1}{2}O_2 \longrightarrow$$
$$2HCO_3^- + FeS_2 + 2H_2CO_3 + H_2O \tag{5}$$

The ferrous iron supply stems from reductive dissolution of ferric oxides, another alkalinity-generating process:

$$FeOOH + e^- + 3H_3O^+ \longrightarrow Fe^{2+} + 5H_2O \tag{6}$$

For a detailed discussion of the various pathways and stoichiometries, see reference 14. However, the question remains open as to which redox process provides the electrons for reaction 6. When buried in sediments, ferric iron may be used by microorganisms as an electron acceptor (15–17). On the other hand, it also comes into contact with reductants like H_2S (18, 19). Although microbial reduction of ferric oxides using sulfide as the reductant has not yet been documented (17), various studies support a purely chemical interaction between these two compounds (20–22).

The dissolution rate of goethite by sulfide was found to increase with surface area and proton concentration. Pyzik and Sommer (*21*) suggested that HS^- is the reactive species that reduces surface ferric iron after exchanging versus OH^-. A subsequent protonation of surface ferrous hydroxide would lead to dissolution of a surface layer. Elemental sulfur was the prominent oxidation product; polysulfides and thiosulfate were found to a lower extent. The dissolution rate R (in moles per square meter per second) of hematite by sulfide was demonstrated to be proportional to the surface concentration of the surface complexes >FeHS and >FeS⁻ (*22*).

$$R = k\{>FeS^-\} + k'\{>FeSH\} \tag{7}$$

where {>FeS⁻} and {>FeSH} are surface concentrations (moles per square meter) and k and k' denote the corresponding rate constants (per second).

Recently we presented (*23*) the results of an experimental study on the kinetics and mechanisms of the reaction of lepidocrocite (γ-FeOOH) with H₂S. With respect to the interaction between iron and sulfur, lepidocrocite merits special attention. It forms by reoxidation of ferrous iron under circumneutral pH conditions (*24*), and it can therefore be classified as a reactive iron oxide (*19*). The concept of reactive iron was established by Canfield (*19*), who differentiated between a residual iron fraction and a reactive iron fraction (operationally defined as soluble in ammonium oxalate). The reactive iron fraction is rapidly reduced by sulfide or by microorganisms.

This chapter presents some implications for sediment–water interactions derived from the findings of our experimental study. Some hypotheses are formulated concerning the coupling of iron and sulfur in sedimentary environments.

Reaction of H₂S with Lepidocrocite

In this study we performed initial rate experiments, reacting H₂S with lepidocrocite (*23*). The consumption of H₂S was measured continuously by using a pH₂S electrode cell (*25*). To avoid interferences of pH buffer solutions with the iron oxide surface, the pH was stabilized by using a pH-stat that added appropriate amounts of HCl to the solution. The added volume, which was also continuously monitored, provided information about the amount of protons consumed during the reaction. Dissolved iron was measured only in some runs.

The reaction was pseudo-first-order with respect to dissolved sulfide. It was surface controlled (Figure 1), and the reaction rate showed a strong dependence on pH. Figure 2 depicts the influence of pH on the pseudo-first-order experimental rate constant k_{obs}. The initial amount of dissolved sulfide was 10^{-4} M in each experiment. The rate constant increased up to

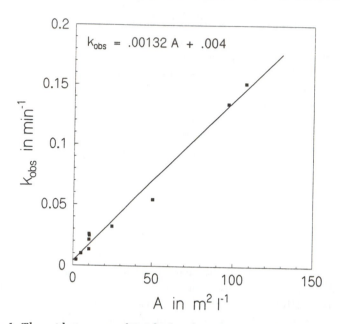

Figure 1. The oxidation rate of H_2S by lepidocrocite is pseudo-first-order with respect to H_2S. The experimental pseudo-first-order rate constant k_{obs} is plotted as a function of the surface area concentration of γ-FeOOH. The reaction rate depends on the surface area (A).

about pH 7 and decreased again at higher pH. An empirical rate law can be derived as follows:

For $5 < pH < 6$,

$$\frac{d[S(-II)]}{dt} = k_a[H^+]^{-2}[S(-II)]A \tag{8}$$

For $7 < pH < 8.6$,

$$\frac{d[S(-II)]}{dt} = k_b[H^+]^1[S(-II)]A \tag{9}$$

where $[S(-II)]$ is the concentration of total dissolved sulfide, A is the surface concentration in square meters per liter, t is time, and k_a and k_b are rate constants; $k_a = 1.5 \times 10^{-13}$ M^2 L m^{-2} min^{-1}, and $k_b = 2.1 \times 10^6$ M^{-1} m^{-2} min^{-1} for the acidic and alkaline ranges, respectively.

The observed rate constants in the pH range $6 < pH < 7$ may not be determined precisely because of the short time intervals. Probably, our values reflect only the lower limits of the fast reaction rates at neutral pH.

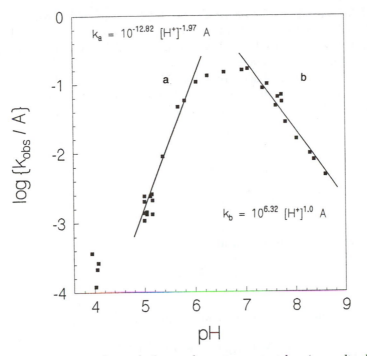

Figure 2. Experimental pseudo-first-order rate constant k_{obs} (normalized to the surface area concentration A) for the reaction of H_2S with lepidocrocite plotted as a function of pH. Straight lines a and b correspond to eqs 8 and 9, respectively. k_a and k_b are the empirical rate constants.

By following the reaction scheme proposed by dos Santos Afonso and Stumm (22) for the reductive dissolution of hematite surface sites (Scheme 1), we were able to explain perfectly the observed pH pattern of the oxidation rate of H_2S. The rate is proportional to the concentration of inner-sphere surface complexes of HS^- formed with either the neutral ($>FeOH$) or the protonated ($>FeOH_2^+$) ferric oxide surface sites.

$$-\frac{d\,[HS^-]}{dt} = k\{>FeS^-\} + k'\{>FeSH\} \qquad (10)$$

We were not able to measure an increase of dissolved iron at a pH higher than 5.7. Even at this pH, the recovery of dissolved iron accounted for only 28% of the dissolved sulfide consumed. This finding agrees with other studies of the reductive dissolution of ferric oxides in which dissolution rates frequently are not detectable at pH 7 (26, 27).

Although we did not measure the oxidized products of H_2S, these products can be deduced from the ratio of consumed protons per mole of total sulfide consumed. In Table I the reaction of H_2S with ferric oxide is formulated for various oxidation products of HS⁻. The wide span of ratios (~1–15) makes it possible to differentiate clearly among the products. Figure 3 shows that the measured ratios range mostly between 0.5 and 3.5, and it reveals a distinct pH dependence.

Sulfate does not appear to be a major product in these initial rate experiments. This conclusion is in contrast to the findings of dos Santos Afonso and Stumm (22), who used steady-state experiments to measure mainly

---

- reversible adsorption of HS⁻

$$>FeOH + HS^- \underset{k_{-1}}{\overset{k_1}{\rightleftharpoons}} >FeS^- + H_2O$$

- reversible electron transfer

$$>FeS^- \underset{k_{-et}}{\overset{k_{et}}{\rightleftharpoons}} >Fe^{II}S^{\cdot}$$

- reversible release of the oxidized product

$$>Fe^{II}S + H_2O \underset{k_{-2}}{\overset{k_2}{\rightleftharpoons}} >Fe^{II}OH_2^+ + S^{\cdot-}$$

- detachment of Fe^{2+}

$$>Fe^{II}OH_2^+ \xrightarrow[k_3]{H^+} \text{new surface site} + Fe^{2+}$$

Scheme I. Proposed mechanism for the reaction of H_2S with ferric hydroxide surface, according to dos Santos Afonso and Stumm (22).

Table I. $\Delta H^+ : \Delta H_2 S_{TOT}$ Ratio for Products of the Reaction of H_2S with γ-FeOOH

Reaction	$\Delta H^+ : \Delta H_2 S_{TOT}$
$6FeOOH + 4HS^- + 2H_2O \rightarrow S_4^{2-} + 6Fe^{2+} + 14OH^-$	3.5
$8FeOOH + 5HS^- + 3H_2O \rightarrow S_5^{2-} + 8Fe^{2+} + 19OH^-$	3.8
$2FeOOH + HS^- + H_2O \rightarrow S^0 + 2Fe^{2+} + 5OH^-$	5.0
$8FeOOH + 2HS^- + 3H_2O \rightarrow S_2O_3^{2-} + 8Fe^{2+} + 16OH^-$	8.0
$8FeOOH + HS^- + 3H_2O \rightarrow SO_4^{2-} + 8Fe^{2+} + 15OH^-$	15.0
Formation of solid-phase bound Fe^{2+}	
$6FeOOH + 10HS^- \rightarrow 6FeS + S_4^{2-} + 8OH^- + 4 H_2O$	0.8
$6FeOOH + 6>FeOH + 4HS^- \rightarrow$	
$\quad 6FeO-Fe^+ + S_4^{2-} + 8OH^- + 4H_2O$	2.0

SOURCE: Reproduced from reference 23. Copyright 1992 American Chemical Society.

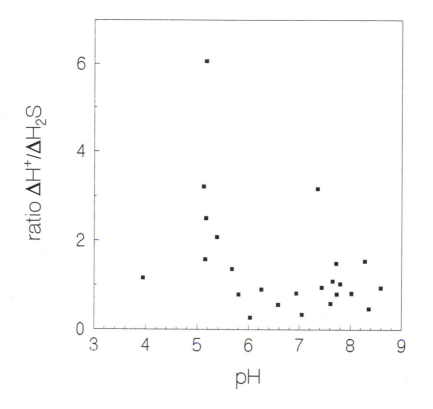

Figure 3. Measured ratio ΔH⁺ : ΔH₂S for the reaction of H₂S with γ-FeOOH plotted as a function of pH.

sulfate, thiosulfate, and traces of sulfite as oxidation products of H_2S. Our study reflects more the experimental conditions described in Pyzik and Sommer's work (*21*). Thus we may also assume elemental sulfur or poly-sulfides (S_4^{2-} and S_5^{2-}) to be the main products.

A black coloration that appeared during the experiments at pH values >6.5 indicated the formation of FeS. However, the black disappeared again toward the end of the experiments. Apparently most of the sulfide stored in FeS was also oxidized, and only a small portion of the sulfide may have remained as FeS. This development is not surprising because we worked with excess ferric oxide, in contrast to Pyzik and Sommer's study (*21*) in which FeS could accumulate during the experiments. We concluded that the ferrous iron released after redissolution of FeS adsorbs to the ferric oxide surface and forms a surface complex >FeO–Fe⁺ (reaction 7), to which polysulfides may bond.

A combination of the various processes (formation of polysulfides and elemental sulfur, precipitation of FeS, and adsorption of Fe^{2+}) leads to the low $\Delta H^+ : \Delta H_2 S$ ratios observed at pH > 6. At lower pH values FeS does not form and adsorption of Fe^{2+} decreases; these conditions provide an increased $\Delta H^+ : \Delta H_2 S$ ratio.

In summary, the reaction of $H_2 S$ with γ-FeOOH is a fast surface-controlled process. Equations 8 and 9 can be used to estimate an upper limit of sulfide oxidation rates in sediments with reactive iron (assuming reactive iron to be represented by lepidocrocite). The surface-area concentration A of reactive iron can be calculated according to

$$A = Fe_{reac}(1 - \phi)\rho S\Theta \tag{11}$$

where Fe_{reac} is the reactive iron content (between 0.01 and 1 mg/g; cf. ref. 19); ϕ is porosity (0.8); ρ is density (1.5 kg/L); S is specific surface area (10^5–3×10^5 m^2/kg); and Θ is surface coverage resulting from surface precipitates and adsorption of dissolved organic carbon (DOC) and other sorbing substances (0.1–0.01) so that A ranges between 0.3 and 360 m^2/L. Theoretical sulfide oxidation rates range between 9×10^{-7} and 1.8 M per day (assuming S(–II) concentrations ranging between 10^{-6} and 10^{-4} M in sediment pore waters and a pH of 7). The upper value is certainly an overestimate. However, aerobic microbial sulfide oxidation rates on the order of 10^{-3} M per day range within the limits calculated previously and indicate the environmental relevance of this process. The values are taken from reference 28, Table IV, and converted from flux densities into rates by assuming the oxidation takes place within the upper centimeter.

The extent to which $H_2 S$ contributes to the release of ferrous iron into pore-water solution through dissolution of reactive ferric oxides such as lepidocrocite or amorphous ferrihydrite remains unclear. According to Canfield (*19*), liberation of ferrous iron in sediments stems mainly from microbial dissolution of ferric oxides. The release rates of Fe^{2+} measured in his study range between 3×10^{-6} and 4×10^{-5} M per day, at the lower limit of the theoretical interval.

The observed reaction rate maximum around pH 7 corresponds to the pH usually found in anoxic sulfide-bearing sediment pore waters (*29*). In addition, the formation of FeS is favored under these conditions. Polysulfides may be expected to be at least an intermediate product of the reaction (*21*), which may be further oxidized to sulfate on a longer time scale (*22*).

Sediment and Pore-Water Data

The vertical profiles of solid and dissolved substances in sediments usually are interpreted according to a vertical sequence of organic matter decomposing processes (*11, 30*). The reason for such a succession is generally explained in terms of decreasing metabolic free energy gain for successive microbiological reactions (e.g., *31, 32*).

As long as oxygen can diffuse into a certain depth of the sediment, it will be the predominant terminal electron acceptor. Once the oxygen is

exhausted, other electron acceptors are used. As a consequence, characteristic reaction products such as ferrous iron, H_2S, CH_4, or other fermentation
products allow the identification of vertical zones with respect to the predominant redox process. The individual microbial processes are considered
exclusive at a certain sediment depth (33, 34). In other words, certain microorganisms are able to outcompete other microorganisms if their metabolism is linked to a higher gain in free energy.

On the basis of this model, Lovley et al. (17) argued that reductive
dissolution of ferric oxides must be a microbiological process because the
zone of sulfide generation is distinct from the zone of maximum ferric oxide
reduction. Highly eutrophic environments would be an exception. In these
systems the zone of decomposition with oxygen as terminal electron acceptor
directly overlies the zone of sulfate reduction.

Pore-water profiles are frequently interpreted according to this concept.
For example, White et al. (35) described a conceptual model of biogeochemical processes of sediments in an acidic lake (cf. Figure 4). They discussed the numbered points in Figure 4 as follows: Diffusion of dissolved
oxygen across the sediment–water interface leads to oxidation of ferrous iron
and to an enrichment of ferric oxide (point 1). Bacterial reductive dissolution
of the ferric oxides in the deeper zones releases ferrous iron (point 2). The
decrease in sulfate concentration stems from sulfate reduction, which produces H_2S to react with ferrous iron to form mostly pyrite in the zone below
the ferric oxide accumulation (point 3).

Interpretation of these data suggests a question: What is the oxidation
process leading to the formation of pyrite (analytically determined as
chromium-reducible sulfide, CRS) instead of simple precipitation of FeS
(analytically determined as acid-volatile sulfide, AVS)? AVS constituted less
than 10% in the study of White et al. (35).

As Berner (36) pointed out in his classic work, the formation of pyrite
is coupled to a process in which free sulfide is oxidized to form polysulfides,
which again react with FeS to form pyrite. In this study elemental sulfur
was the oxidant. However, elemental sulfur was always less than 1% of the
total sulfur content in the study of White et al. (35). The findings of the
experimental studies discussed on the interaction between H_2S and ferric
oxides (20–23), in combination with the field observations, suggest a mechanism in which ferric iron oxides are the oxidants to form polysulfides and
subsequently pyrite.

$$8FeOOH + 5H_2S \longrightarrow 8Fe^{2+} + S_5^{2-} + 2H_2O + 14OH^- \quad (12a)$$

$$4FeS + S_5^{2-} + 2H^+ \longrightarrow 4FeS_2 + H_2S \quad (12b)$$

In summary,

$$8FeOOH + 8H_2S \longrightarrow 4FeS_2 + 4Fe^{2+} + 8H_2O + 8OH^- \quad (13)$$

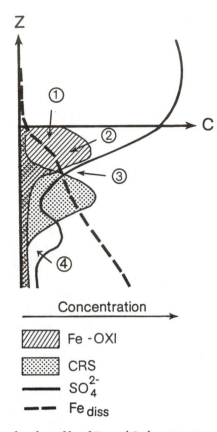

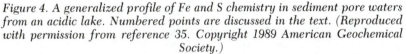

Figure 4. A generalized profile of Fe and S chemistry in sediment pore waters from an acidic lake. Numbered points are discussed in the text. (Reproduced with permission from reference 35. Copyright 1989 American Geochemical Society.)

The importance of polysulfides in the pyrite formation process was outlined by several studies (37, 38). Schoonen and Barnes (37) showed that no precipitation from homogeneous solution can be observed within a reasonable time scale, even in solutions highly supersaturated with respect to pyrite, unless pyrite seeds are already existing. Therefore future studies should address the role of ferric oxide surfaces in promoting the nucleation of pyrite.

Reactions 12a and 12b consume dissolved sulfide. This fact fits nicely with the data of White et al. (35), who could not detect free sulfide in their study. Dissolved sulfide is frequently absent in freshwater sediments (e.g., 39; *see* Urban, Chapter 10, for a discussion). This lack of sulfide is explained by an excess of reactive iron over the total sulfide concentration (19, 40).

Probably the H_2S produced at a certain depth diffuses both up and down in the sediment. As long as the requirements for FeS precipitation are fulfilled, reaction 12b and subsequently reaction 13 will proceed. However, reaction 12b will stop at sulfide activities lower than the solubility product of FeS, whereas reaction 12a may continue. Under such circumstances the polysulfides formed may further react with ferric oxides in a redox process and be slowly reoxidized to sulfate (22). It is therefore not surprising to find a sulfate peak below the pyrite peak (point 4 in Figure 4). Subsurface sulfate peaks also were reported in other lakes (10, 12).

In this context, the concept of sulfate recycling seems to be helpful (28, 41). Urban (Chapter 10) pointed to the fact that all measured sulfate reduction rates in freshwater sediments indicate a much higher turnover of sulfate than would be predicted by calculation of diffusive fluxes from the concentration gradients. More than 50% of sulfate reduction occurs below a depth of 2 cm, where diffusive gradients are negligible. Urban concluded that only sulfate regeneration resulting from reoxidation can explain sulfate reduction rates as high as those found under marine conditions, despite the low sulfate concentrations in freshwater systems.

Possible oxidants will be oxygen, ferric oxides, or manganese oxides. Oxygen usually shows very steep gradients at the sediment–water interface in both marine (42) and freshwater (43) systems. Therefore it will serve as an oxidant only in the upper few millimeters of a sediment (the redox process mediated, e.g., by *Beggiatoa*). Manganese was shown to oxidize sulfide (44). However, the pool of solid manganese in freshwater sediments is usually much smaller than that of solid iron (e.g., 35, 45–47). Because deeper layers are frequently impoverished with respect to manganese, it may well be that ferric oxides are the electron acceptors responsible for the postulated sulfate regeneration, particularly in deeper layers of sediment. Consequently, anoxic sulfide oxidation rates should counterbalance sulfate reduction rates. As an example, the maximum of the sulfate reduction rate measured in the pore waters of the FOAM site in the study of Canfield (reference 19, 3×10^{-4} M per day) corresponds nicely to theoretical oxidation rates calculated by using eqs 9 and 11 (2.4×10^{-5} to 6.7×10^{-4} M per day; $Fe_{reac} = 0.15$ mg/g, $\phi = 0.35$, pH was estimated to be 7, and dissolved sulfide concentration [S(–II)] = 10^{-6} M; for the range of the other parameters, cf. eq 11).

In summary, it seems meaningful to consider both the formation of pyrite from the reaction of H₂S with reactive ferric oxides and sulfate recycling as a result of this process in any discussion of the early diagenesis of sulfur and iron in sediments.

Formation of Pyrite in Sediments: A Kinetic Approach

The formation of pyrite in sediments depends on the availability of three parameters: iron, sulfate, and organic matter (48). Although organic matter content controls the formation rate under marine sulfate-rich conditions, sulfate concentration is usually regarded as the limiting factor under freshwater conditions. However, intensive recycling of sulfate has been found in both coastal (41) and limnetic (28) sediments. These data suggest that the recycling rate of sulfate controls the formation of pyrite more than its concentration. In contrast to sulfate, iron generally exists in sufficient amounts in sediments (with the exception of calcareous, iron-poor systems). The availability of ferrous iron for the formation of pyrite, however, will be controlled by the reactivity of ferric oxides (19).

As Schoonen and Barnes (49) pointed out, the existence of FeS is a necessary prerequisite for the formation of pyrite. However, high pyrite formation rates were observed, particularly in salt marshes or other systems exposed to temporary oxygen intrusion into sulfide-bearing sediments, although no FeS could be detected (50, 51). These phenomena can be explained by the growth of already-existing pyrite in favor of FeS (37). In other words, the rate of formation of Fe^{2+} and H_2S in such systems is high enough to cause precipitation of FeS. It is, however, lower than the consumption rate of both species by the two processes of reoxidation and precipitation as pyrite.

A parameter indicating the flux of Fe^{2+} and H_2S would be the measured ion activity product, IAP (52). A low pIAP value, corresponding to amorphous FeS, does not necessarily mean that other, more stable, solid FeS phases do not exist (the system would be supersaturated with respect to these phases), but it may indicate that the formation rate of both Fe^{2+} and H_2S is high. At low net fluxes, other solid phases have time to form. Consequently, inverse gradients can be observed in systems where the net fluxes of Fe^{2+} and H_2S are high (pIAP increases with depth) and in systems where the net fluxes of Fe^{2+} and H_2S are low at the sediment–water interface (pIAP decreases with depth) (cf. ref. 52).

Variation of pIAP depth profiles also occurs with time. Figure 5 presents pIAP values measured in Lake Kinneret sediments (53) after an algal bloom (May 30, 1988), during the stratification period (October 24, 1988), and after overturn (January 5, 1989). The organic matter decomposition after sedimentation of algae caused a buildup of dissolved sulfide (also in the hypolimnion) and therefore an increase in the rate of formation of amorphous FeS (pIAP < 3; 54) in the upper sediment layers. During the course of the stratification period, the formation rate decreased and the pIAP values increased to a more or less uniform value throughout the sediment, corresponding to a more crystalline FeS phase (mackinawite). After overturn, oxygen penetrates into the sediment; the sulfate reduction rate (and thus the sulfide supply) was decreased (Figure 5). The low sulfide concentration that diffusively accumulated in deeper layers was rapidly consumed through reaction with ferric oxides.

As this short discussion shows, the kinetics of formation of the single parameters (Fe^{2+} and H_2S) may control the extent and the pathway of pyrite formation. Oxidation of sulfide by elemental sulfur to form polysulfides (pathway 1) should predominate at the oxygen–sulfide interface of very productive

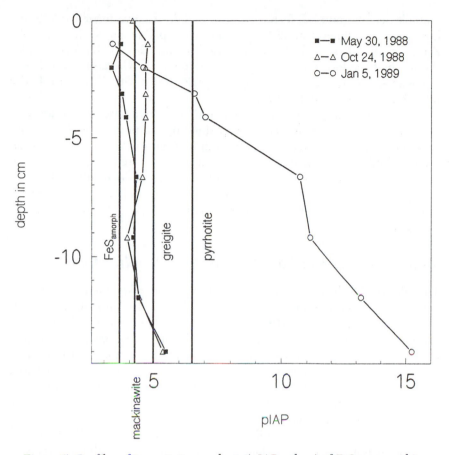

Figure 5. *Profiles of ion activity products (pIAP values) of FeS measured in Lake Kinneret sediments after the end of an algae bloom (May 30, 1988), during the stratification period (October 24, 1988), and after overturn (January 5, 1989). Straight lines correspond to solubility products of various FeS phases according to the reaction FeS + H⁺ ⇆ Fe²⁺ + HS⁻. (Based on data from ref. 53.)*

(eutrophic) environments of high organic matter supply (e.g., salt marshes; *51, 55*). Sulfur formation may be mediated by bacteria (e.g., *Beggiatoa, 56*). High rates of microbial reductive dissolution of ferric oxides together with high sulfate reduction rates cause a very sharp separation of reducing and oxidizing microenvironments. In contrast, oxidation of sulfide by ferric oxides to polysulfides (pathway 2) may occur in those environments where decomposition rates are not high enough for development of such a redox microstructure (e.g., sediment systems, as discussed in Figure 5).

FeS can still be found in such environments because of an insufficient supply of an oxidant to react with sulfide. This insufficiency may happen when either no sulfide-reactive iron exists or the rate of reoxidation of (microbiologically produced) ferrous iron to sulfide-reactive ferric oxide is too low (e.g., in sediments underlying an anoxic hypolimnion, such as in Lake Kinneret). In other words, the concentration of sulfide-reactive ferric oxides should limit the formation of pyrite in environments where pathway 2 predominates. The rate of the reaction of ferric oxides with H_2S controls the rate of pyrite formation from FeS.

The CRS:AVS ratio reflects the trophic state of a lake (cf. Urban (28), Table V). This observation may be explained by the preceding kinetic considerations. The high organic matter supply in eutrophic lakes leads to an intensive mineralization rate by both iron- and sulfate-reducing bacteria. However, reoxidation of Fe^{2+} to ferric oxide or of sulfide to sulfate does not take place because an anoxic hypolimnion prevents penetration of oxygen. Therefore FeS can build up, but the sediment becomes depleted with respect to reactive iron.

Sulfate Reduction in Lake Sediments: A Biofilm Model

The observed sulfate reduction rates in freshwater sediments cannot be explained by diffusion of sulfate from the lake water into the sediment, because much steeper sulfate concentration gradients should then be observed. Assuming diffusive supply alone, Urban (28) calculated that the change of sulfate concentration with depth should take place within 1 mm instead of several centimeters, which are usually measured. This assumption, however, also means that the sulfate recycling process and the sulfate reduction rate should not be limited by the vertical transport of sulfide to (frequently solid) oxidants or to the oxic boundary layer.

Instead of assuming that diffusive fluxes occur only vertically, following the macroscopic, measurable, large-scale concentration gradients, it seems reasonable to consider also lateral microscale concentration gradients within microniches (57–59).

A biofilm provides a very helpful approach to explaining the phenomenon of microniches. The biofilm concept, mostly applied in technical systems, describes the activity of microorganisms adhering to surfaces and thus separated from the bulk solution (60–63). Microorganisms in sediment pore waters benefit above all from the enhanced nutritional status at mineral surfaces (64). More generally, they exert better regulation or control of their microenvironment (65). Of particular interest is the formation of microbial consortia, the development of a syntrophic community of two or more bacterial species (66). Such consortia are often strikingly complex, showing temporal and structural heterogeneity, but exhibiting functional homogeneity based on intercellular fluxes of organic carbon compounds, inorganic electron acceptors, and reducing equivalents (65, 67).

Modeling of interactions between organisms attached to pore walls and the bulk solution of a porous medium such as a sediment has been done for groundwater systems (68, 69), but there is still a great lack of theoretical understanding of these interactions (W. Schäfer, University of Heidelberg, personal communication).

In theory, microbial activities within biofilms are regarded as diffusion-controlled (70). However, the diffusion length is shorter than the macroscale concentration gradient thickness, which is usually several centimeters. The separation of microenvironments within a biofilm, where organisms benefit from the metabolites of other organisms, leads to concentration gradients on a very small scale. The more intense the metabolic activity of a certain species is, the steeper the concentration gradients and the greater the fluxes of substances.

The intensity of decomposition depends on the availability of both organic matter and an adequate electron acceptor. The latter will be stimulated if the product of a certain organism group (e.g., sulfide) is rapidly reconverted to a reactant (e.g., sulfate) in close proximity to this organism group. Steep, but inverse, concentration gradients of both sulfide and sulfate will therefore enhance the decomposition rates of sulfate-reducing bacteria in close proximity to a sulfide-regenerating ferric oxide surface. A sufficient organic matter supply must be presumed in this model.

In lake sediments the bulk solution consists of lake water enclosed by the sedimented material. At the moment of burial the bulk solution has the same chemical composition as the lake water. It is, however, then exposed to early diagenetic processes, including growth of a biofilm on the pore walls. The right side of Figure 6 depicts a biofilm-covered, vertically directed sediment pore that is separated into five boxes, each representing a certain sediment depth. The change of concentration of the redox-dependent substances (DO, SO_4^{2-}, Fe^{2+}, and H_2S) with depth reflects the macroscale vertical concentration gradients within the bulk solution. Mn^{2+} was omitted for simplicity.

The pore wall consists of metal oxides. Its reactive part is consumed in proportion to depth because of metabolic activity within a biofilm. The biofilm separates the bulk solution from the pore wall. The horizontal biofilm concentration profiles on the left side of Figure 6 correspond to the center of each of the five boxes. (Concentration is on the vertical axis and biofilm thickness is on the horizontal axis.) Excess organic matter is assumed within the upper few centimeters. The arrows indicate net flux densities of various substances in and out of the biofilm.

Substances in the bulk solution diffuse into the biofilm, where they are consumed (such as oxygen, point 1 in Figure 6) or recycled (such as sulfate through stepwise reoxidation of H_2S from sulfate reduction, point 5). Within the biofilm, very steep gradients exist for oxygen or hydrogen sulfide and also for ferrous iron from reductive dissolution of ferric oxides. These gradients result from the coexistence of anaerobic and aerobic metabolisms such as aerobic respiration (point 1), reduction of ferric oxides (point 3), and sulfate

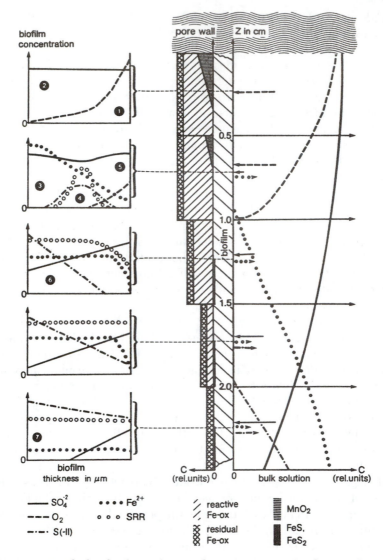

Figure 6. An idealized scheme for a sedimentary porous medium with pore walls covered by a biofilm. High sulfate reduction rates are maintained even in depths to which sulfate cannot diffuse because of recycling of sulfate within the biofilm. Numbered points (in black circles) denote the following processes: 1, Respiration consumes oxygen. 2, Microbial reduction of reactive metal oxides. Reduction of reactive ferric oxides is in equilibrium with reoxidation of ferrous iron by O_2. Thus, no net loss of reactive iron takes place in these layers. 3, Microbial reduction of ferric oxides. 4, Sulfate reduction rate (denoted as SRR). 5, Sulfide oxidation, either microbiologically or chemically. 6, Sulfide builds up within the biofilm, sulfate consumption increases, reactive iron pool decreases. 7, Formation of iron sulfides.

reduction (point 4). Within the biofilm, recycling of ferrous iron to ferric iron will take place as long as oxygen is available (point 2). In addition, H_2S is recycled to sulfate because of the reaction of H_2S with ferric oxides (point 5).

The bulk solution becomes impoverished with respect to oxygen. However, the sulfate concentration remains constant as long as the recycling rate of reduced sulfur to sulfate is higher than the sulfate reduction rate (denoted SRR in Figure 6). In the opposite case the sulfate concentration of the bulk solution also decreases, and H_2S is slowly enriched (point 6). The requirements for FeS precipitation and subsequent pyrite formation are then fulfilled (point 7).

In spite of its low concentration compared to ferric oxides, solid manganese also may play a role in recycling sulfide in the upper layers (44). The kinetics of homogeneous oxidation of Mn^{2+} to manganese oxide by dissolved oxygen are rather slow (32). Nevertheless, some recycling of Mn^{2+} because of heterogeneous catalysis of the oxidation process will occur at oxide surfaces (71). Field data imply a relatively fast depletion of solid manganese with depth (e.g., 47), so the recycling process should be restricted to the upper centimeter of a sediment. According to the model presented in Figure 6, reactive ferric oxides play a key role in the recycling of sulfate and therefore will control the gross sulfate reduction rate. If the reactive iron pool in boxes 2 and 3 (1.0 and 1.5 cm) did not exist, ferrous iron concentration would decrease in the bulk solution and sulfate would be consumed rapidly in the upper layers.

This model places special emphasis on the recovery of reactive metal oxides in the upper layers of the sediment by dissolved oxygen. In other words, the oxidation capacity of dissolved oxygen (DO) is transferred onto metal oxides, which are then buried by further sedimentation. The oxidation capacity is thus shuttled into deeper layers, where it will enhance the anaerobic turnover of organic carbon in sediment layers that could not be maintained by diffusive supply of sulfate alone. A shuttle of oxygen equivalents may also influence the pathways of organic matter decomposition.

On the other hand, a permanent supply of ferric oxides to the sediments is provided by sedimentation of allochthonous material. It is unknown to what extent these oxides are reactive with respect to sulfide or whether a predigestion of ferric oxides by bacteria is needed. Various studies indicate that ~50% of freshwater sediment iron exists as iron oxide and ~20% of the iron is reactive (72). Future studies should be directed to a better understanding of the existence of reactive iron.

A Personal View

Many of the hypotheses presented in this chapter are based on existing field and laboratory studies. As they are often not yet proven, they might stimulate controversial discussion. I hope they will provide a contribution to the development of future research objects in a field of great environmental relevance. The fate of substances of environmental concern may be closely linked to the processes discussed. Of particular relevance are substances that interact with the surface of ferric oxides, such as trace metals (73–76) or phosphate (77, 78).

In addition to a better understanding of the reaction of sulfide with ferric oxides and its role in pyrite formation, a more exact definition of the term "reactive iron" is critical. Does reactive iron mean a different iron oxide fraction for bacterial dissolution (e.g., weathering products such as goethite or hematite) than for reaction with sulfide (e.g., reoxidized lepidocrocite)? In other words, is there a predigestion of ferric oxides by bacteria that allows a subsequent rapid interaction of sulfide with ferric oxides?

The biofilm concept, applied to sediment–water interactions, breaks with classical strategies to model early diagenesis (i.e., the vertical redox zonation). Although far from completely developed, this concept may overcome modeling problems, such as an adequate description of recycling of substances.

Acknowledgments

This chapter benefited from stimulating discussions with Noel Urban and Bernhard Wehrli during a postdoctoral year at the Lake Research Institute of EAWAG, Kastanienbaum, Switzerland. I am grateful to Jeff White for the discussion of his data during the ACS meeting. I am also indebted to Wolfgang Durner for critically commenting on the biofilm concept and to two reviewers for their helpful comments. I further wish to thank Barbara Staudinger for making data available and Elisabeth Schill for preparing the drawings.

References

1. Wetzel, R. G. *Limnology;* W. B. Saunders: Philadelphia, PA, 1981.
2. Mengel, K.; Kirkby, E. A. *Principles of Plant Nutrition*, 4th ed.; International Potash Institute, Worblaufen-Bern, Switzerland, 1987.
3. Zinder, S. H.; Brock, T. D. In *Sulfur in the Environment. Part II. Ecological Impacts;* Nriagu, J. O., Ed; John Wiley and Sons: New York, 1978; Chapter 11, p 445.
4. Luther, G. W., III; Church, T. M. *Mar. Chem.* **1988**, *23*, 295.
5. Jørgensen, B. B. *Limnol. Oceanogr.* **1990**, *35*, 1329.
6. Kepkay, P. E.; Cooke, R. C.; Novitsky, J. A. *Science (Washington, D.C.)* **1979**, *204*, 68.
7. Kepkay, P. E.; Novitsky, J. A. *Mar. Biol. Berlin* **1980**, *55*, 261.
8. Schindler, D. W. In *Chemical Processes in Lakes;* Stumm, W., Ed.; John Wiley and Sons: New York, 1985; Chapter 11, pp 225–250.

9. Baker, L. A.; Urban, N. R.; Brezonik, P. L.; Sherman, L. A. In *Biogenic Sulfur in the Environment;* Saltzman, E. S.; Cooper, W. J., Eds.; ACS Symposium Series 393; American Chemical Society: Washington, DC, 1989; Chapter 7.

10. Baker, L. A.; Eilers, J. M.; Cook, R. B.; Kaufmann, P. R.; Herlihy, A. T. In *Acidic Deposition and Aquatic Ecosystems—Regional Case Studies;* Charles, D. F., Ed.; Springer: New York, 1991; Chapter 17, pp 567–613.

11. Froelich, P. N.; Klinkhammer, G. P.; Bender, M. L.; Luedtke, N. A.; Heath, G. R.; Cullen, D.; Dauphin, P. *Geochim. Cosmochim. Acta* **1979**, *43*, 1075.

12. Rudd, J. W. M.; Kelly, C. A.; Furutani, A. *Limnol. Oceanogr.* **1986**, *31*, 1281.

13. Giblin, A. E.; Likens, G. E.; White, D.; Howarth, R. W. *Limnol. Oceanogr.* **1990**, *35*, 852.

14. Anderson, R. F.; Schiff, S. L. *Can. J. Fish. Aquat. Sci.* **1987**, *44*(Suppl. 1), 188.

15. Nealson, K. H. In *Microbial Geochemistry;* Krumbein, W. E., Ed.; Blackwell: Oxford, England, 1983; Chapter 4, pp 159–221.

16. Lovley, D. R. *Geomicrobiol. J.* **1987**, *5*, 375.

17. Lovley, D. R.; Phillips, E. J. P.; Lonergan, D. J. *Environ. Sci. Technol.* **1991**, *25*, 1062.

18. De Vitre, R. R.; Buffle, J.; Perret, D.; Baudat, R. *Geochim. Cosmochim. Acta* **1988**, *52*, 1601.

19. Canfield, D. E. *Geochim. Cosmochim. Acta* **1989**, *53*, 619.

20. Rickard, T. *Am. J. Sci.* **1974**, *274*, 941.

21. Pyzik, A. J.; Sommer, S. E. *Geochim. Cosmochim. Acta* **1981**, *45*, 687.

22. Dos Santos Afonso, M.; Stumm, W. *Langmuir* **1992**, *8*, 1671.

23. Peiffer, S.; dos Santos Afonso, M.; Wehrli, B.; Gächter, R. *Environ. Sci. Technol.* **1992**, *26*, 2408.

24. Schwertmann, U.; Taylor, R. M. In *Minerals in the Soil Environment*, 2nd ed.; Dixon, J. B., Ed.; Soil Science Society of America: Madison, WI, 1989; Chapter 8.

25. Frevert, T.; Galster, H. *Schweiz. Z. Hydrol.* **1978**, *40*, 199.

26. LaKind, J. S.; Stone, A. T. *Geochim. Cosmochim. Acta* **1989**, *53*, 961.

27. Dos Santos Afonso, M.; Morando, P. J.; Blesa, M. A.; Banwart, S.; Stumm, W. *J. Colloid Interface Sci.* **1990**, *138*, 74.

28. Urban, N. R. In *Environmental Chemistry of Lakes and Reservoirs;* Baker, L. A., Ed.; Advances in Chemistry 237; American Chemical Society: Washington, DC, 1993; Chapter 10.

29. Ben-Yaakov, S. *Limnol. Oceanogr.* **1973**, *18*, 86.

30. Berner, R. A. *Early Diagenesis—A Theoretical Approach;* Princeton University Press: Princeton, NJ, 1980.

31. Claypool, G.; Kaplan, I. R. In *Natural Gases in Marine Sediments;* Kaplan, I. R., Ed.; Plenum: New York, 1974; pp 99–139.

32. Stumm, W.; Morgan, J. J. *Aquatic Chemistry*, 2nd ed.; John Wiley and Sons: New York, 1981.

33. Winfrey, M. R.; Zeikus, J. G. *Appl. Environ. Microbiol.* **1977**, *33*, 275.

34. Lovley, D. R.; Phillips, E. J. P. *Appl. Environ. Microbiol.* **1987**, *53*, 2636.

35. White, J. R.; Gubala, C. P.; Fry, B.; Owen, J.; Mitchell, M. J. *Geochim. Cosmochim. Acta* **1989**, *53*, 2547.

36. Berner, R. A. *Am. J. Sci.* **1970**, *268*, 1.

37. Schoonen, M. A. A.; Barnes, H. L. *Geochim. Cosmochim. Acta* **1991**, *55*, 1495.

38. Luther, G. W., III *Geochim. Cosmochim. Acta* **1991**, *55*, 2839.

39. Cook, R. B.; Kelley, C. A.; Kingston, J. C.; Kreis, R. G. *Biogeochem.* **1987**, *4*, 97.

40. Davies-Colley, R. J.; Nelson, P. O.; Williamson, K. J. *Mar. Chem.* **1985**, *16*, 173.

41. Fossing, H.; Jørgensen, B. B. *Geochim. Cosmochim. Acta* **1990**, *54*, 2731.
42. Jørgensen, B. B.; Revsbech, N. P. *Limnol. Oceanogr.* **1985**, *30*, 111.
43. Sweerts, J. R. A.; St. Louis, V.; Cappenberg, T. E. *Freshwater Biol.* **1989**, *21*, 401.
44. Aller, R. C.; Rude, P. D. *Geochim. Cosmochim. Acta* **1988**, *52*, 751.
45. Zullig, H. *Schweiz. Z. Hydrol.* **1956**, *18*, 5.
46. Losher, A. Ph.D. Thesis, ETH Zürich, Switzerland, 1989.
47. Wersin, P. Ph.D. Thesis, ETH Zürich, Switzerland, 1990.
48. Berner, R. A. *Geochim. Cosmochim. Acta* **1984**, *48*, 605.
49. Schoonen, M. A. A.; Barnes, H. L. *Geochim. Cosmochim. Acta* **1991**, *5*, 1505.
50. Howarth, R. W. *Science (Washington, D.C.)* **1979**, *203*, 49.
51. Howarth, R. W.; Jørgensen, B. B. *Geochim. Cosmochim. Acta* **1984**, *48*, 1807.
52. Davison, W. *Aquatic Sciences* **1991**, *53*, 309.
53. Staudinger, B. Master Thesis, University of Bayreuth, Bayreuth, Germany, 1989.
54. Davison, W. Geochim. Cosmochim. Acta **1980**, *44*, 803.
55. Oenema, O. *Biogeochem.* **1990**, *9*, 75.
56. Kuenen, J. G. *Plant Soil* **1975**, *43*, 49.
57. Emery, K. O.; Rittenberg, S. C. *Am. Assoc. Pet. Geol. Bull.* **1952**, *36*, 735.
58. Jørgensen, B. B. *Mar. Biol. Berlin* **1977**, *41*, 7.
59. Jahnke, R. *Limnol. Oceanogr.* **1985**, *30*, 956.
60. Bitton, G.; Marshall, K. C. *Adsorption of Microorganisms to Surfaces;* John Wiley and Sons: New York, 1980.
61. Marshall, K. C. *Microbial Adhesion and Aggregation;* Springer: Berlin, Germany, 1984; pp 317–330.
62. Harremoës, P. In *Water Pollution Microbiology;* Mitchell, R., Ed.; John Wiley and Sons: New York, 1978; Vol. 2, Chapter 4, pp 79–110.
63. Siegrist, H.; Gujer, W. *Water Res.* **1985**, *19*, 1369.
64. Marshall, K. C. In *Adsorption of Microorganisms to Surfaces;* Bitton, G.; Marshall, K. C., Eds.; John Wiley and Sons: New York, 1979; Chapter 9.
65. Breznak, J. A., et al. In *Microbial Adhesion and Aggregation;* Marshall, K. C., Ed.; Springer: Berlin, Germany, 1984; pp 203–222.
66. Dubourgier, H. C.; Archer, D. B.; Albagnac, G.; Prensier, G. In *Anaerobic Digestion 1988: Fifth International Symposium on Anaerobic Digestion;* Hall, E. R.; Hobson, P. N., Eds.; Pergamon: Oxford, England, 1988; pp 1–12.
67. McCarty, P. L.; Smith, D. P. *Environ. Sci. Technol.* **1986**, *20*, 1200.
68. Kindred, J. S.; Celia. M. A. *Water Resour. Res.* **1989**, *25*, 1149.
69. Kinzelbach, W.; Schäfer, W.; Herzer, J. *Water Resour. Res.* **1991**, *27*, 1123.
70. Atkinson, B. In *Microbial Adhesion and Aggregation;* Marshall, K. C., Ed.; Springer: Berlin, 1984; pp 351–372.
71. Davies, S. H. R.; Morgan, J. J. *J. Colloid Interface Sci.* **1989**, *129*, 63.
72. Davison, W.; de Vitre, R. R. In *Environmental Particles;* Buffle, J.; van Leeuwen, H. P., Eds.; Lewis: Chelsea, MI, 1992.
73. Sholkovitz, E. R.; Copland, D. *Geochim. Cosmochim. Acta* **1982**, *46*, 393.
74. Johnson, A. C. *Geochim. Cosmochim. Acta* **1986**, *50*, 2433.
75. Tessier, A.; Carignan, R.; Dubreuil, B.; Rapin, F. *Geochim. Cosmochim. Acta* **1989**, *53*, 1511.
76. Belzile, N.; Tessier, A. *Geochim. Cosmochim. Acta* **1990**, *54*, 103.
77. Boström, B.; Jansson, M.; Forsberg, C. *Ergeb Limnol.* **1982**, *18*, 5.
78. Frevert, T. *Arch Hydrobiol. Suppl.* **1979**, *55*, 278.

RECEIVED for review September 26, 1991. ACCEPTED revised manuscript July 24, 1992.

Factors Affecting the Distribution of H$_2$O$_2$ in Surface Waters

William J. Cooper[1], Chihwen Shao[1], David R. S. Lean[2], Andrew S. Gordon[3], and Frank E. Scully, Jr.[4]

[1]Drinking Water Research Center, Florida International University, Miami, FL 33199
[2]National Water Research Institute, Burlington, Ontario L7R 4A6, Canada
[3]Department of Biological Sciences and [4]Department of Chemical Sciences, Old Dominion University, Norfolk, VA 23529

This chapter presents a review of the factors that affect both the formation and the decay of hydrogen peroxide, H$_2$O$_2$, in fresh waters. Although biological and chemical (nonphotochemical) processes form H$_2$O$_2$, their contribution to surface waters is generally insignificant. The formation of hydrogen peroxide results principally from the UV portion of sunlight exciting humic substances in the water and thereby leads to the formation of superoxide ion, which reacts with itself to form H$_2$O$_2$. Because this production is limited to the depth of UV light penetration, its vertical distribution provides a sensitive tracer for mixing processes. The known chemical pathways for the decay of H$_2$O$_2$ appear to be insignificant in freshwater systems. Although some algae and zooplankton show catalase and peroxidase activity, the major organisms responsible for the decay of H$_2$O$_2$ are heterotrophic bacteria.

H YDROGEN PEROXIDE IN AQUEOUS SOLUTION can act as either an oxidizing or a reducing agent (1). The presence of hydrogen peroxide has been reported in fresh waters (2–18), marine waters (19–25), and estuarine environments (26–28). H$_2$O$_2$ affects redox chemistry in marine environments (15, 21, 24, 29–32) and may also be important in metal cycling in other natural waters (33–39) and in the fate of pollutants in the environment (7, 40–45). As a strong oxidant, it may affect both biological processes (46–48) and chemical

0065–2393/94/0237–0391$09.00/0

processes and, as a result, may shape ecosystem biogeochemistry. To understand its distribution in natural waters, it is necessary to understand the mechanisms of H_2O_2 formation and decay.

The major in situ process that results in the formation of H_2O_2 is undoubtedly photochemical (e.g., 12, 15, 49, 50). Photochemical formation of H_2O_2 in fresh and salt waters probably results from the disproportionation of the superoxide ion radical, $O_2^{\cdot -}$ (8, 9, 15, 51, 52). The kinetics of superoxide disproportionation are well established (53), and its steady-state concentration can be calculated. Because of the known effects of superoxide ion in cells (47), its presence in surface waters may be important in biologically mediated processes. However, other sources, such as biological formation (e.g., 45, 54), redox chemistry (21, 24, 29, 31, 32), wet (e.g., 55) and dry (50, 56, 57) deposition, and surfaces (e.g., 58) may also be important.

Several studies on the decomposition of H_2O_2 in natural waters strongly suggest that microbiological processes play an important role (13, 14, 16, 45). The relative importance of chemical processes in the decomposition of H_2O_2 has not been evaluated, but it appears to be small.

Although early studies in the marine environment showed little diel variability in the surface-water H_2O_2 concentration (22, 23), initial observations of rather high concentrations of H_2O_2 in surface waters of lakes led us to examine the factors affecting its variability and distribution. This chapter provides a critical review of the literature on H_2O_2 formation and decay, integrated with recent results of both field and laboratory studies.

The field studies will focus on the formation and distribution of H_2O_2 in a wide range of lakes compared with that obtained in an estuarine system, the Chesapeake Bay. Laboratory studies on the decomposition of H_2O_2 in natural waters, filtered natural waters (using various size filters), and waters with pure cultures of two bacteria [*Vibrio alginolyticus*, a common estuarine bacterium (59, 60), and *Enterobacter cloacae*, a common freshwater bacterium] will clarify the role of bacteria in the decay processes.

Materials and Methods

H_2O_2 Determination. The analytical method for determination of H_2O_2 measures the loss in fluorescence of scopoletin by the peroxidase-mediated decomposition of H_2O_2 (11, 61–64). Separate standard curves were determined for each water or culture studied because the slope of the curve is affected by dissolved organic carbon (65, 66).

Microbiological Studies. Bacterial numbers for the Chesapeake Bay samples and the pure cultures were determined by acridine orange direct counts (AODC) (67). Those for the lakes were determined by epifluorescence (68) using DAPI (4',6-diamidino-2-phenylindole) with a final stain concentration of 1.0 μg/mL, a process that is described in detail elsewhere (69).

Bacterial cells for the laboratory studies were prepared by growing *V. alginolyticus* in a M9 minimal salts bacterial growth medium with 8 mM glucose as the carbon and energy source (*70*). The standard M9 medium was modified by addition of 21 g/L of NaCl (SWM9). Water used for all studies was deionized and passed through a reverse osmosis membrane (Milli-Q).

Cells were transferred from an overnight liquid culture to fresh medium and grown overnight (~18 h) on a gyrorotary shaker at room temperature. The cells were harvested by centrifugation (11,000 g, 10 min) and resuspended in sterile assay buffer (0.01 M phosphate buffer, Na salt, 0.26 M NaCl, pH 6.78; the pH prior to sterilization was 7.20). The cell concentration was adjusted to approximately 5×10^8/mL as judged by optical density of the solution. The actual numbers in dilutions of the cell suspensions were determined by direct counts. *E. cloacae* was grown as described, except that the M9 contained 0.5 g/L of NaCl.

H₂O₂ Formation Studies.

Formation rates at specific depths or at the surface were determined by using quartz glass tubes with quartz glass stoppers. The tubes were filled with the selected waters and placed at the surface or at predetermined depths in the water column. Placing the glass tubes in the ambient water ensured that a constant temperature was maintained and minimized the tube wall effects. In all samples an initial H₂O₂ concentration was determined. The samples were then exposed to sunlight for 1 h, after which the H₂O₂ concentration was again determined. The formation rate (per hour) was calculated by subtracting the initial concentration from the final H₂O₂ concentration.

H₂O₂ Decay Studies.

Unfiltered lake waters were stored in washed 2-L polyethylene bottles in the dark, and the H₂O₂ concentrations were measured at specified intervals. Lake waters were filtered through 0.2-, 1.0-, 5.0-, and 12.0-μm microfilters (Nuclepore). Zooplankton were removed with 30-μm woven nylon (Nitex) screens. Generally 1 L of filtrate was used.

Samples for determining decay rates of H₂O₂ in Chesapeake Bay water were maintained as close to ambient water temperature as possible. The decay studies were conducted in 1-L glass bottles maintained in the dark without shaking.

The studies on pure cultures were conducted at 25 °C in the laboratory. Chesapeake Bay water samples were filtered through the following filters: 0.1-μm Tuffryn Membrane, HT100 (Gelman); 0.22-μm MPGL 06S H2 (Millipore); and 0.45-μm GN-6 (Gelman).

Sample Collection.

Water samples from the lakes and the Chesapeake Bay were obtained from either standard 6-L Niskin sample bottles or a pumping system employed to collect samples for rare earth elements (*71*).

Irradiation Measurements.   An Epply photometer with sensors for total global irradiance, photosynthetic active radiation (PAR), and ultraviolet (UV) radiation (295–385 nm) was used to measure irradiance continuously and averaged a measurement every 0.5 h.

Study Areas

The studies reported here were conducted in several different environments. The lakes studied included Lake Erie, Lake Ontario (Hamilton Harbor at the extreme westerly end of Lake Ontario), and Jacks Lake and Rice Lake in Ontario, Canada. The locations of the stations and study areas are shown in Figure 1; more details are described elsewhere (13, 14).

The study sites in the Chesapeake Bay are shown in Figure 2. Depth profiles were obtained at both the North and South Basin locations, where the water column was oxic throughout and had a fairly constant O_2 concentration of 5 mg/L. The surface-water temperature, approximately 10 °C, remained constant at the South Basin throughout the water column while decreasing in the North Basin to slightly less than 9 °C in the lower 18 m of the water column. At both locations the surface water (approximately the top 10 m) was relatively fresh, with a saltwater wedge at the bottom. The surface-water salinity increased from the North Basin (10.6‰) to the South Basin (12.3‰).

Figure 1. Lake study areas and sampling locations.

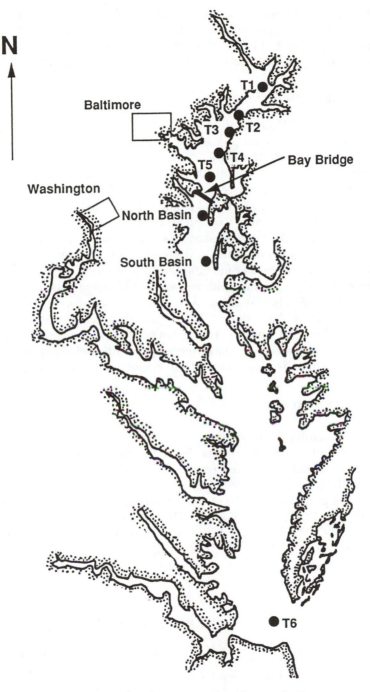

Figure 2. Sampling locations in the Chesapeake Bay.

The transect began in the upper reaches of the bay in water that had a surface salinity of 0.21‰ and proceeded in a southerly direction with sampling at five locations, T1–T5, where the salinity was 7.9‰. The day of the transect, April 13, 1988, was clear and sunny, with winds of 18–25 knots.

Results and Discussion

Detailed studies of H_2O_2 distribution in marine systems and factors affecting its formation have resulted primarily from the work of Zika and co-workers (9, 11, 12, 20–24, 30, 31). These studies indicated that the surface ocean, 5 m and less deep, was close to 100 nM in H_2O_2. Some diel variability was reported in subtropical surface waters, but the variation from night to day was approximately 10–20%. Initial studies of Jacks Lake, Ontario, Canada (13), showed surface water (1 m and less) diel variability of from <10 nM at night to >500 nM during the day. The contrasting diel pattern observed in the two systems led us to the next phase of our experiments.

We determined the net formation rates in discrete samples at several depths in the water column, with and without filtration, using waters from several lakes. We also investigated the decay processes, in the dark, for whole lake waters, lake waters filtered through varying mesh size filters, and in pure bacterial cultures. These results added to our understanding of the processes responsible for the observed distribution of H_2O_2. Our current research activities (18) are reviewed and synthesized in this chapter.

Formation of H_2O_2

Photochemical Formation.   The major pathway leading to the formation of H_2O_2 in surface waters (fresh and marine) results from photochemically (sunlight) mediated reactions of dissolved organic carbon (e.g., 12, 15). The portion of the sunlight most responsible for the formation of H_2O_2 is that portion below 400 nm. The intermediate in the process appears to be $O_2^{\cdot-}$ (8, 9, 12, 15, 51, 52, 72). Thus, those reactions that lead to the formation of $O_2^{\cdot-}$ in natural waters will increase the formation of H_2O_2, and those that lead to the loss of $O_2^{\cdot-}$ will result in lower concentrations of H_2O_2 in natural waters. The following generalized reaction mechanism has been proposed (12, 15):

$$_0^1DOC \xrightarrow{\text{light}} {}_1^1DOC^* \xrightarrow{\text{ISC}} {}_1^3DOC^* \tag{1}$$

$$_1^1DOC^* \text{ or } {}_1^3DOC^* \longrightarrow [DOC^{\cdot+} + e^-] \tag{2}$$

$$[DOC^{\cdot+} + e^-] \longrightarrow DOC^{\cdot+} + e_{aq}^- \tag{3}$$

$$O_2 + e_{aq}^- \longrightarrow O_2^{\cdot-} \tag{4}$$

$$\substack{3\\1}DOC^* + {}^3O_2 \longrightarrow DOC^{\cdot+} + O_2^{\cdot-} \tag{5}$$

$$\substack{3\\1}DOC^* + {}^3O_2 \longrightarrow \substack{1\\0}DOC + {}^1O_2 \tag{6}$$

$$\substack{1\\0}DOC + {}^1O_2 \longrightarrow DOC^{\cdot+} + O_2^{\cdot-} \tag{7}$$

$$\substack{3\\1}DOC^* + RNH_2 \longrightarrow DOC^{\cdot-} + RNH_2^{\cdot+} \tag{8}$$

$$DOC^{\cdot-} + O_2 \longrightarrow \substack{1\\0}DOC + O_2^{\cdot-} \tag{9}$$

$$\substack{1\\0}DOC + {}^1O_2 \longrightarrow endoperoxides \longrightarrow DOC + H_2O_2 \tag{10}$$

where DOC is the sunlight-absorbing dissolved organic carbon, $\substack{1\\0}DOC$ is the generalized electronic ground state, $\substack{1\\1}DOC^*$ is the generalized electronically excited singlet state, $\substack{3\\1}DOC^*$ is the generalized electronically excited triplet state, and ISC is intersystem crossing. The equations suggest that the rate of formation should be related to the concentration of dissolved organic carbon. This relationship has been demonstrated (*12*, *15*, *18*), although natural variability does exist. These results have been extended to several natural waters obtained from various temperate lakes and the Great Lakes. Data for net formation rates of H_2O_2 at the surface are summarized in Table I, but in each experiment incidental light was not controlled. Incidental light varies with latitude, season, and cloud cover.

In earlier studies it was presumed that filtration would not significantly affect formation rates because major decay processes were presumed to be chemically mediated (*11*, *12*). Therefore little attention was given to filtered versus unfiltered formation rates. However, we have shown that microbial processes are important (*16*). Therefore, the formation rate of H_2O_2 in both filtered (0.2-μm) and unfiltered (whole) lake water was determined in this study. In most of the samples the formation rate of the filtered water was higher than that of the whole (unfiltered) lake water, a result indicating that particle-mediated (biological) decay processes are important on 1-h time scales.

Quantum yields (reflecting the efficiency of converting sunlight energy to the formation of H_2O_2 at several wavelengths) in natural waters (Figure 3) suggest that in high-humic waters most of the H_2O_2 formation will occur in the near-surface water (*11*, *12*). That is, the quantum yields are highest at the low wavelengths that are absorbed in the waters near the surface. Figure 4 shows the H_2O_2 formation rate variation with depth (up to 1 m) in Sharpes Bay (DOC = 5.7 mg/L, with filtered water). The rate was determined by incubating the sample for 1 h at an integrated solar UV irradiation of 3.02 Langleys. The formation rate at depth is considerably slower than at the surface.

The effect of cloud cover on the formation of H_2O_2 was established during a diel study of H_2O_2 in surface waters (*13*). However, the effect of depth on the formation rate has not been reported. Figure 4 shows that on a hazy

Table I. H_2O_2 Formation Rate at the Surface in Natural Waters Under Natural Sunlight Conditions

Water	Weather	Sample Condition[a]	DOC (mg/L)	Formation Rate (nM/h)	Ref.
Jacks Lake					
Sharpes Bay		Un	6 ± 0.5	120–360	13
	sunny	F	5.7	126–221	this study
	cloudy	F		44	this study
Brooks Bay	sunny	F	7.2	310–350	this study
	sunny	Un		373	this study
	cloudy	F		102	this study
Hamilton Harbor		F	4.1	218	this study
		Un		202	this study
Lake Erie					
Station 23		F	4.2	51	this study
		Un		29	this study
Lake Ontario					
Station 41		F		70	this study
		Un		51	this study
Station 212		F	3.4	43	this study
		Un		31	this study
Rice Lake		F	7.9	526	this study
		Un		451	this study
Greifensee			4–4.5	160–300	15
Glatt			~4.5	180	15
Zurichsee			1.5	~30	15
Etang de la Gruere			13	1000	15

[a]Un is unfiltered; F is filtered (0.2-μm filter).

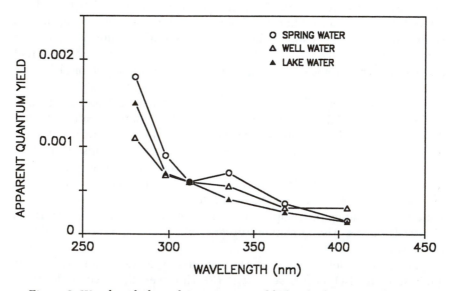

Figure 3. Wavelength-dependent quantum yields for the formation of H_2O_2 in three natural (fresh) waters. (Reproduced from reference 11. Copyright 1988 American Chemical Society.)

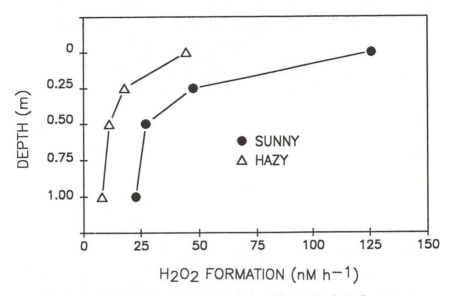

Figure 4. Rate of formation of H₂O₂ in Sharpes Bay of Jacks Lake, Ontario, Canada, at several depths, September 1990. (Reproduced with permission from reference 17. Copyright 1990 Academic Press.)

day (integrated solar irradiation for the day was 0.46 Langleys) the formation rate of the filtered lake water was reduced by approximately 33% between the surface and 1-m depth.

Several studies have concluded that $O_2^{\cdot-}$ leads to the formation of H_2O_2 in natural waters (8, 9, 15, 51, 52). This process has been demonstrated experimentally by using the enzyme superoxide dismutase (SOD); the addition of SOD (73–75) results in a more rapid formation rate and/or higher absolute concentrations of H_2O_2 in sunlight-irradiated natural waters. These observations were extended to Lake Greifensee in a comprehensive study of the factors affecting the formation of H_2O_2 in natural waters (15).

The relative importance of some of the reactions leading to the formation of $O_2^{\cdot-}$ was also studied (15). Radical scavenging experiments suggest that O_2 is not reduced directly by e_{aq}^-, but rather that direct electron transfer from the excited-state DOC to O_2 results in the formation of $O_2^{\cdot-}$. If the addition of a scavenger results in decreasing the concentration of H_2O_2 or the rate of its formation, then e_{aq}^- is involved directly in the formation of $O_2^{\cdot-}$. The scavengers, N_2O ($k = 6 \times 10^9$ M^{-1} s^{-1}) and 3 mM $CHCl_3$ ($k = 3 \times 10^{10}$ M^{-1} s^{-1}), had no effect on the formation or rate of formation of H_2O_2 (15). We confirmed this observation by using another e_{aq}^- scavenger, trichloroacetic acid (TCA), $k = 8.5 \times 10^9$ M^{-1} s^{-1} (76). When 1 mM TCA was added to filtered Lake Ontario water (Station 41) no change in the formation rate of H_2O_2, 68 nmoles per hour, was observed. Thus, we conclude on the basis of these studies that, under sunlight conditions in natural waters, O_2 is not reduced by e_{aq}^- (eq 4). On the contrary, direct electron transfer from the excited DOC (e.g., $_1^3$DOC) is the likely formation pathway (eq 5).

Equation 7 of the proposed mechanism suggests that $^1O_2$ could react with ground-state $_0^1$DOC and result in the formation of $O_2^{\cdot-}$. An effective $^1O_2$ quencher is N_3^- (77, 78; $k = 2.2 \times 10^8$ M^{-1} s^{-1}) and therefore the addition of N_3^- would result in a decreased formation rate and concentration of H_2O_2. Sturzenegger (15) showed an increase in H_2O_2 formation with time in the presence of 1 mM N_3^- in two waters, Greifensee and Etang de la Gruere.

In experiments similar to those reported by Sturzenegger (15), we added N_3^- to a groundwater exposed to sunlight. We observed an increase in H_2O_2 concentration, as did Sturzenegger. These results are the opposite of what would be expected if $^1O_2$ was involved in the formation of H_2O_2.

Another possibility is that $^1O_2$ might react with $_0^1$DOC to form an endoperoxide, in a process similar to the furfuryl alcohol reaction (78–81), and with decomposition it would result in the formation of H_2O_2 (eq 10). Sturzenegger (15), in studies comparing H_2O_2 formation from furfuryl alcohol and humic substances, concluded that this pathway was not significant.

Thus, we conclude from these studies that eqs 7 and 10 are both insignificant in the formation of H_2O_2 by sunlight-initiated reactions in natural waters. No published studies indicate that eqs 8 and 9 are of any significance in natural waters.

Biological Formation.   The formation of H_2O_2 in biological systems is well known (82–84). Studies of a phytoplankton (45) and of coastal water using $^{18}O_2$-labeling techniques (1) showed that H_2O_2 was formed through biologically mediated processes. The most comprehensive investigation of H_2O_2 formation via biologically mediated processes was reported by Palenik and co-workers (54, 85–89).

A mechanism leading to the extracellular formation of H_2O_2 in plankton involves L-amino acid oxidase, LAAO (87, 88):

$$\text{L-amino acid} + O_2 \xrightarrow{\text{LAAO}} \text{keto acid} + H_2O_2 + NH_4^+ \tag{11}$$

Data from three separate phytoplankton genera, two prymnesiophyte and one dinoflagellate, indicated that this mechanism may be more general than previously realized (88). The L-amino acid oxidase pathway is related to ammonia uptake and is not light dependent. Thus, this mechanism represents a possible light-independent source of H_2O_2. The results reported by Palenik and Morel (88) demonstrated that the addition of 1–5 μM amino acids to cultures results in the formation of H_2O_2. As yet these experiments have not been extended to natural waters, and the relative contribution of this biologically mediated source of H_2O_2 is not known. Our experiments on filtered and nonfiltered water samples show that the presence of algae and bacteria reduces rather than stimulates H_2O_2 formation.

Chemical Formation.   A potential chemically mediated source of H_2O_2 would be the presence of reduced metals in oxygenated waters (*15, 21, 32–34, 36–38*). This pathway has never been demonstrated in fresh waters, although Miles and Brezonik (*35*) showed that O_2 concentrations varied over 24 h in humic waters with iron present. No measurements of H_2O_2 were made, but most likely H_2O_2 was formed as the O_2 was consumed. The net impact of these processes on H_2O_2 concentration in fresh waters is not likely to be important in waters rich in humic substances. However, this assumption has not been verified experimentally.

Groundwater may be an important source of water in lakes, reservoirs, and other bodies of fresh water. If groundwater is a source of water in a lake and H_2O_2 has been reported in groundwater (*64*), it may be a direct source of H_2O_2. However, groundwater is unlikely to be a significant source of the total H_2O_2 found in a lake, for the concentrations reported in groundwater were very low. An alternative mechanism for the contribution of groundwater to H_2O_2 in aquatic systems involves the oxidation of reduced metals (e.g., Fe^{2+} and Cu^+) by O_2. Groundwater is often anaerobic; as such, the stable form of the dissolved or chelated metals would be the reduced state. When these react with the O_2 present in surface waters $O_2^{\bullet-}$ is formed:

$$M^{(n)+} + O_2 \longrightarrow M^{(n+1)+} + O_2^{\bullet-} \tag{12}$$

This process leads to the formation of H_2O_2. In summary, the chemical pathways leading to the formation of H_2O_2 in lakes and reservoirs are not thought to be significant in most waters.

Decay of H₂O₂

In the lakes we studied, the decay of H_2O_2 always fits a pseudo-first-order model (*13, 14, 16, 18*). That is, a plot of ln $[H_2O_2]$ versus time is linear and usually remains linear throughout the incubation (~24 h). We never observed an H_2O_2 concentration as low as our detection limit in natural waters or in pure cultures. The possibility that biological formation exists at low ambient H_2O_2 concentrations cannot be excluded. To better understand biologically mediated H_2O_2 decay in natural waters, we reviewed studies that examined the role of enzymes, the effect of filtration of lake waters through filters with various mesh sizes, and the loss of H_2O_2 in pure cultures of bacteria. Chemical decomposition, a possible factor in controlling H_2O_2 in natural waters, is also reviewed.

Biological Decay.   The two major enzyme systems that are used to control intracellular H_2O_2 concentrations in organisms are catalases and peroxidases (47). The different overall reactions for H_2O_2 decomposition mediated by these two enzyme systems can be illustrated as follows for catalase:

$$2H_2O_2 \longrightarrow 2H_2O + O_2 \qquad (13)$$

and for peroxidase:

$$peroxidase + H_2O_2 \longrightarrow perox\ I \qquad (14)$$

$$perox\ I + AH_2 \longrightarrow perox\ II + AH\cdot \qquad (15)$$

$$perox\ II + AH_2 \longrightarrow peroxidase + AH\cdot + H_2O \qquad (16)$$

where perox I and II are "activated" peroxidase states and AH_2 is a generalized hydrogen donor.

These enzyme systems are widespread in nature (90, 91). Several lines of investigation have led to the conclusion that these processes may account for most biologically mediated decay of H_2O_2 observed in natural waters.

Enzyme Decay.   Moffett and Zafiriou (1) differentiated catalase- and peroxidase-mediated decay in coastal (marine) waters by using ^{18}O-labeled H_2O_2 and O_2, and by determining the labeled end products. Equation 13 shows that the products of catalase decomposition are H_2O and O_2. In contrast, peroxidase decomposition results in the formation of H_2O without O_2. From the measurement of the relative amount of labeled products it is possible to determine the contribution of both enzymes in the decay of the H_2O_2. In the coastal water, 65–80% of the decomposition was attributed to catalase and the rest to peroxidase (1). These studies are the first to use this technique. The approach should be extended to freshwater ecosystems to see if the same pattern would be found.

Another approach uses the coupling reaction of p-anisidine. In the presence of H_2O_2 and peroxidase (16), an oxidation product that contains two aromatic rings, benzoquinone-4-methoxyaniline, is formed stoichiometrically (92). Equations 14–16 indicate that an electron donor or hydrogen donor is required for peroxidase-mediated decomposition of H_2O_2. In two natural waters and one soil suspension, peroxidatic activity was identified by the stoichiometric removal of p-anisidine by the addition of H_2O_2 (in the dark) (16). This procedure provides an independent corroboration of the results obtained by Moffett and Zafiriou (1). However, this method does not quantify the relative importance of peroxidases versus catalases in the decomposition of H_2O_2.

Filtration Studies.   We examined biologically mediated decay of H_2O_2 by measuring decay rates on lake waters filtered through filters with various mesh sizes. The filters used were 64 μm to remove zooplankton, 12 μm to remove large algae, 1.0 μm to remove small algae, and 0.2 μm to remove bacteria. Examples of the effect of filtration on the decay rate of H_2O_2 in natural waters are shown in Table II, and additional data are provided by Lean et al. (*18*). Apparently removal of the zooplankton and large algae causes little difference in H_2O_2 decay. However, when smaller particles (small algae and bacteria) are removed, H_2O_2 becomes quite stable in the water.

These observations are consistent with data obtained (*93*) when inhibitors of microbial activity were added to solutions of suspended sediments and natural waters. When formaldehyde or Hg(II) was added, the H_2O_2 was stable and no decay occurred. In other words, decay is related to particles in the water. These particles are likely to be small plankton and bacteria.

Pure Culture Studies.   Two bacteria were cultured to study the decomposition of H_2O_2, *Vibrio alginolyticus* and *Enterobacter cloacae*. *V. alginolyticus* was selected because it is an extremely common inhabitant of environments like the Chesapeake Bay. Vibrios may provide up to 50% of the culturable bacterial species in the bay (*60*). *E. cloacae* was studied because it is commonly found in freshwater environments (*94*).

Decay studies using *V. alginolyticus* were conducted at a constant H_2O_2 concentration as a function of cell concentration, and at a constant cell concentration as a function of H_2O_2 concentration. The five cell concentrations represent various environments, from highly productive areas to oligotrophic waters, 10^7 to 10^5 cells/mL, respectively. Triplicate experiments were conducted at each cell concentration, and the combined results are summarized in Table III. At each cell concentration of *V. alginolyticus* the rate of decomposition through the first one or two half-lives was obtained by least-

Table II. The Loss of H₂O₂ in Sharpes Bay (Jacks Lake) Water and Filtrates of Lake Water

	Lake Water			*Sharpes Bay Vertical Profile*	
Filter Size	*Organisms*	$t_{1/2}$ *(h)*	*Depth (m)*	*Rate Constant (h^{-1})*	$t_{1/2}$ *(h)*
Unfiltered	Natural assemblage	4.4	Surface	0.122	5.7
64 μm	Zooplankton removed	4.7	1	0.119	5.8
12 μm	Large algae removed	6.4	2	0.120	5.8
1.0 μm	Small algae removed	19.1	3	0.130	5.3
0.2 μm	Bacteria removed	58.7	4	0.110	6.3
			5	0.124	5.5
			6	0.115	6.0

NOTE: The half-life ($t_{1/2}$) was determined at 20 °C as it would be for a first-order kinetic process (i.e., the ln 2 divided by the decay rate constant).

**Table III. Hydrogen Peroxide Decomposition
by a Marine Bacterium *Vibrio alginolyticus* Culture**

$[H_2O_2]_o$ (nM)	Rate Constant (min^{-1})	$t_{1/2}$ (h)
2175	−0.015	0.79
1120	−0.016	0.72
1052	−0.015	0.78
530	−0.018	0.64
540	−0.017	0.67
114	−0.033	0.35
108	−0.028	0.42
31.9	−0.030	0.39
28.9	−0.017	0.69
Cells/mL ($\times$ 10^6)		
14	−0.024	0.49
7	−0.011	1.08
1.4	−0.0043	2.67
0.7	−0.00073	15.8
0.14	−0.00031	37.3

NOTE: In the experiments with varying $[H_2O_2]$, the *V. alginolyticus* concentration was 1.4×10^7 cells/mL. In the experiments with varying cell concentration, the $[H_2O_2]$ was 132 nM. The culture medium was 0.01 M phosphate, pH 6.78, containing 15 g/L of NaCl and kept at 25 °C.

squares regression analysis at time t of ln $([H_2O_2]_t)$. The decay rate constants are also summarized in Table III.

At a cell concentration of 1.4×10^7 cells/mL, the H_2O_2 decay rate was measured within the $[H_2O_2]_0$ range from 28.9 to 2175 nM. This range brackets surface water concentrations found in various environments, although the highest H_2O_2 concentrations studied have never been observed in natural waters. A decrease in decay rate was observed at higher H_2O_2 concentrations. This decrease is probably caused by some inhibition of the organisms at high H_2O_2 concentrations.

Several experiments, using solutions of H_2O_2 to which organisms had been added, were allowed to continue until no further decomposition was observed. In every case the H_2O_2 concentration was 2–5 nM. To determine whether this result was an artifact of the analytical method, excess catalase was added to several solutions and H_2O_2 was measured after several hours. In all cases in which excess catalase had been added, the H_2O_2 concentration was reduced to below the detection limit, 1 nM. This lower limit for H_2O_2 concentration may indicate biological formation of H_2O_2, as suggested by Palenik (87, 88). Further studies are required to determine the reason for the low steady-state concentrations of H_2O_2 in the cultures.

H_2O_2 decay in pure cultures of *V. alginolyticus* was shown to be second order overall and first order in $[H_2O_2]$ and cell concentration, as shown by eq 17:

$$\frac{-d[H_2O_2]}{dt} = k_{bact}[H_2O_2][cells] \tag{17}$$

where k_{bact} is the second-order rate constant for a specific bacterium, and [cells] is in organisms per milliliter.

For *V. alginolyticus* at 132 nM H_2O_2 and varying bacterial cell concentration, the second-order rate constant was:

$$k_{Va} = 2.3 \times 10^{-9} \text{ mL/cells·min} \tag{18}$$

Similar studies carried out using *E. cloacae* showed that decay was slower at the higher concentrations of H_2O_2 than at the lower concentrations (Table IV). However, the relatively high concentration of H_2O_2 may have inhibited microbial activity. At other concentrations the decay rate seemed to be first-order in H_2O_2 concentration and first-order in concentration of organisms, giving an overall rate constant at 25 °C:

$$k_{Ec} = 5.1 \times 10^{-9} \text{ mL/cells·min} \tag{19}$$

Thus, both pure cultures were efficient in removing H_2O_2 from solution and the overall rate constants were very similar.

Table IV. Hydrogen Peroxide Decomposition by a Freshwater Bacterium *Enterobacter cloacae* Culture

$[H_2O_2]_o$ (nM)	Rate Constant (min⁻¹)	$t_{1/2}$ (h)
1726	−0.00781	1.5
1176	−0.00962	1.2
558	−0.0110	1.1
121	−0.0218	0.53
17.5	−0.0196	0.59
Cells/mL ($\times$ 10^6)		
10	−0.0499	0.23
5	−0.0290	0.40
1	−0.00446	2.59
0.5	−0.00075	15.4
0.1	−0.00038	30.4

NOTE: In the experiments with varying $[H_2O_2]$, the *V. alginolyticus* concentration was 1.4×10^7 cells/mL. In the experiments with varying cell concentration, the $[H_2O_2]$ was 132 nM. The culture medium was 0.01 M phosphate, pH 6.8, containing 15 g/L of NaCl and kept at 25 °C.

Several experiments were conducted to characterize the H_2O_2 decay in the cultures and media. These experiments were all conducted at H_2O_2 concentrations likely to be found in natural waters. First, to test the particle-associated decomposition of H_2O_2, V. *alginolyticus* was cultured, centrifuged, and resuspended in fresh media. A portion of this resuspended culture was withdrawn, centrifuged, and filtered through a 0.1-μm filter. The cell-free filtrate (from a cell suspension that was 5.6×10^8 cells/mL) was diluted to give an equivalent of 1.4×10^7 cells/mL at 126 nM H_2O_2. The cell suspension that had not been filtered was also diluted in the same sterilized medium to give a cell concentration of 1.4×10^7 cells/mL. The H_2O_2 decomposition rate constant for the living cells was -0.0392 cell/min, with a half-life of 17.7 min (Table III). The filtrate showed no loss of H_2O_2 during a 50-h period. This result establishes that the processes for the decomposition of H_2O_2 in these pure cultures is associated with the living organism and is not associated with extracellular enzymes or metabolites.

Second, cultures of V. *alginolyticus* were heated in an incubator for 45 min at 62 °C. This relatively mild treatment would result in a minimum change to noncellular organic and inorganic compounds that may be responsible for H_2O_2 decomposition. An aliquot (1 mL) plated onto tryptic soy agar (Difco) showed no growth. The cell suspension was viscous after heating, a result that indicated that cell lysis probably had occurred. The heat-killed cell suspension was diluted in sterilized medium, 126 nM H_2O_2, to an equivalent of 10^7 cells/mL. No decomposition of H_2O_2 was observed over 50 h.

Third, azide ion was added to determine the effect of this inhibitor. The respiration rate of V. *alginolyticus* is slowed by the addition of 10 mM azide ion (92). Our studies suggest that the addition of azide ion inhibits most or all of the decomposition of H_2O_2 in some natural waters and/or soil suspensions (16). Therefore, solutions of H_2O_2 were prepared in sterilized medium. In one flask only NaN_3 (10 mM) was added. To the other two flasks, 1.4×10^7 cells/mL of V. *alginolyticus* were added. After the decomposition of H_2O_2 equivalent to one half-life, NaN_3 (10 mM) was added to one of these flasks. In the flask to which no bacteria were added, no decomposition of the H_2O_2 occurred over 4 h. The rate of H_2O_2 decomposition in the presence of 1.4×10^7 cells/mL was -0.0392 cell/min, and no change in the rate of decomposition was observed after the addition of azide ion. This result indicates that for these laboratory cultures N_3^- did not noticeably affect the mechanism responsible for the decay of H_2O_2.

Field Studies. Experiments conducted at Jacks Lake (13), Lake Erie and Lake Ontario (14, 18), and under laboratory conditions using suspended sediments and natural waters (16, 17) have all implicated biological processes in H_2O_2 decay. Unfiltered Jacks Lake water (Sharpes Bay) had a half-life ($t_{1/2}$) for H_2O_2 of 7.8 h. Filtrate from 64-, 12-, and 5-μm filters were all similar, with an average $t_{1/2}$ of 8.6 h. The filtrate of 1-μm filtration had a $t_{1/2}$ of 31 h, and the filtrate of a 0.45-μm filter showed no decay in 24 h (*13*).

To determine the H_2O_2 decay rate correlated with depth, two locations were studied: station 212 in Lake Ontario and Station 23 in Lake Erie. At both sampling locations the rate decreased with depth (i.e., the H_2O_2 half-life increased. For example, at station 212, the $t_{1/2}$ increased from 14.7 h in the surface-water sample to 21.6 h in the sample obtained at 10 m. At station 23 the surface $t_{1/2}$ was 9.6 h, and it increased to 20.2 h in the sample from a water depth of 16.4 m.

A summary of our most recent studies on biologically mediated decay is shown in Figure 5. In situ H_2O_2 decay was determined by using the natural logarithm of the nighttime areal concentration plotted versus time. The decay rate constant was taken as the slope of the line. Decay rate constants calculated using in situ values were similar to water samples incubated in bottles kept in the dark in the laboratory. The decay rate constants seemed to correlate with bacteria numbers.

The temperature effect on the decay rate of H_2O_2 in unfiltered Sharpes Bay water was determined by allowing samples to equilibrate at several temperatures (Figure 6). H_2O_2 was spiked in all samples, and the decay was followed for several half-lives. With lower water temperatures the decay rate constants were lower. As yet this has not been verified over a season. As water temperatures change the relationship may differ for cold-tolerant species. H_2O_2 decay studies have also been conducted in waters from the

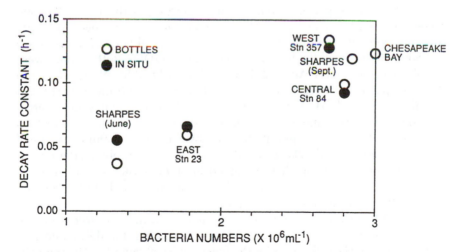

Figure 5. Hydrogen peroxide decay rate constants in surface-water samples from Sharpes Bay (in June and September), Jacks Lake, Ontario, Canada, from the East (Station 23), Central (Station 84), and West Basin (Station 357) of Lake Erie and the Chesapeake Bay, plotted as a function of bacteria numbers.

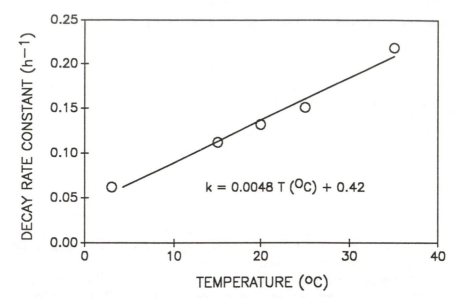

Figure 6. Decay rate constants for H_2O_2 in unfiltered Sharpes Bay lake water contained in glass bottles in the dark at temperatures from 3 to 35 °C. Initial concentrations were about 270 nM. (Reproduced with permission from reference 17. Copyright 1990 Academic Press.)

upper reaches of the Chesapeake Bay. Decay studies were performed on two depth profiles and for surface samples starting in the upper Bay and proceeding south (Figure 2). Table V summarizes the hydrogen peroxide concentration with depth, the decay rate constants obtained for the North and South Basins, and the bacterial cell counts (AODC). Both stations were sampled on cloudy days, April 12, 1988, and April 14, 1988, North and South Basin, respectively. Because of the low ambient concentrations of H_2O_2, it was necessary to add H_2O_2 prior to measuring the decrease in concentration with time. At both stations decay rates were slower in surface water than in deeper water, a trend opposite to that observed in the samples obtained from Lake Ontario and Lake Erie. All of the decay rates obtained are comparable to, or faster than, those obtained in other coastal (estuarine) environments (W. J. Cooper, unpublished data) and comparable to those found in some of the freshwater lakes sampled.

Table VI summarizes the sampling location and time, surface-water salinity, the measured surface-water H_2O_2 concentration, the decay rate constants, and bacterial cell counts for the five samples obtained on the transect. Although the ambient concentration of H_2O_2 remained similar throughout the transect, the decay rate decreased to give H_2O_2 half-lives of 2.5–12 h. The cell numbers were similar at all sampling locations, and the decay rate did not appear to correlate well. For some unexplained reason the increase

Table V. Hydrogen Peroxide Profile and Decay at Two
Stations in the Chesapeake Bay

Water Depth (m)	H$_2$O$_2$^{a} (nm)	Rate Constant (min^{-1})	t$_{1/2}$ (h)	Bacterial cells/mL ($\times$ 10^6)
North Basin, Lat 38°55'57" N; Lon 76°23'17" W				
Surface	38.0	−0.00183	6.3	2.7
2.8	32.0	−0.00152	7.6	3.1
5.7	6.3	−0.00176	6.6	3.1
8.7	4.3	−0.00211	5.5	3.4
12.0	2.8	−0.00258	4.5	3.3
15.3	3.3	−0.00257	4.5	3.2
South Basin, Lat 38°40'58" N; Lon 76°25'25" W				
Surface	24.0	−0.00242	4.8	2.8
5.0	12.2	−0.00273	4.2	3.9
10.0	6.3	−0.00289	4.0	4.1
12.1	10.3	−0.00255	4.5	4.0
15.1	NA^b	−0.00364	3.2	3.9
20.2	NA	−0.00298	3.9	4.5
25.0	NA	−0.00338	3.4	3.8
26.5	NA	−0.00533	2.2	5.4

NOTE: Decay was determined in the dark at approximately 10 °C.
^{a}Ambient concentration when water samples were brought on
board.
^{b}NA indicates not analyzed.

in salinity may have resulted in longer half-lives. Additional work is necessary
to better understand these results.

Several shipboard experiments were conducted in an attempt to determine the relative importance of biological and chemical processes in the
decay of H$_2$O$_2$. A water sample from 2.8 m (North Basin) was heated to
62–65 °C for 30 min, and the decay of H$_2$O$_2$ was measured. In a parallel
experiment, water from 15.3 m was boiled for 10 min and the H$_2$O$_2$ con-

Table VI. Hydrogen Peroxide Decay in Surface-Water Samples in the Upper
Chesapeake Bay

Sample (h)	Salinity	Decay Rate (min^{-1})	t$_{1/2}$ (h)	[H$_2$O$_2$]$_0$ (nM)	Bacterial cells/mL ($\times$ 10^6)
T1 1025	0.21	−0.00464	2.49	81.5	2.1
T1 1230	0.21	−0.00533	2.17	99.4	2.3
T2 1350	2.4	−0.00336	3.44	71.7	2.7
T3 1515	5.0	−0.00170	6.80	72.6	3.6
T4 1624	6.8	−0.00150	7.70	75.6	2.6
T5 1818	7.9	−0.00095	12.2	83.5	3.0

NOTE: Samples were taken within 1 m of the water surface.

centration was measured with time. In both cases the H_2O_2 concentration remained unchanged after the treatment, for the duration of the experiment, approximately 30 h.

Two water samples obtained at 5.7 m were filtered through membrane filters. Filtration through either 0.45- or 0.2-μm filters completely inhibited H_2O_2 decay over a 24-h period. At every sampling point along the transect an unfiltered and filtered (0.22-μm filter) sample was obtained. At all five locations the sample that had been passed through a 0.22-μm filter showed no loss of H_2O_2 over a 24-h period.

One additional surface-water sample was obtained from the lower reaches of the bay (location 6 in Figure 2) on April 23, 1988. The salinity was $19^o/_{oo}$ on an ebbing tide. The water, at 12 °C, contained 5.4×10^6 bacteria/mL. The sample was returned to the laboratory and allowed to come to room temperature (29 °C on the day of the study). Hydrogen peroxide was added at six concentrations (137, 216, 227, 541, 1076, and 2321 nM) and the decay rate was determined. The decay rate constant was -0.00417 ± 0.00016 min^{-1} ($t_{1/2} = 2.8$ h) for all samples, a result indicating little or no effect of H_2O_2 concentration on the decay rate over this range. This result confirms that the decay rate constant is a robust parameter for use in future models. Because it is constant, it also shows that the decay rate is not approaching the maximal rate, where a doubling of the concentration would halve the rate constant.

The H_2O_2 decay results obtained in the water samples of the Chesapeake Bay are consistent with biological processes in the short-term (<24 h) decay of H_2O_2. After filtration of the water with 0.1-, 0.22-, and 0.45-μm filters, no decomposition was observed within the 30-h duration of the experiment. Filtrates (0.1-μm filter) of pure cultures of V. alginolyticus gave similar results. These experiments strongly suggest that the processes leading to the decay of H_2O_2 are associated with the retained particles.

Samples heated to 62–65 °C and to boiling all resulted in stable (no loss of H_2O_2) solutions. This heating is considered a mild treatment that would minimize changes in the chemistry of the pure cultures and natural waters. The fact that the short-term decay of H_2O_2 was stopped strongly suggests that the processes are related to viable organisms.

Azide ion, which is known to reduce the respiration of V. alginolyticus by 50% at 10 mM (94), was added to pure cultures and to the natural waters. It did not detectably inhibit the decay of H_2O_2 in the pure cultures of V. alginolyticus. However, in the natural waters, H_2O_2 decay was nearly eliminated in one sample and the rate was decreased substantially in the second by the addition of 10 mM azide ion. The two samples were surface and 12.0-m water from the North Basin. The H_2O_2 concentration in the surface-water sample decreased from 91 to 81 nM and in the 12.0-m water sample it decreased from 73 to 45 nM, a drop of 11 and 38%, respectively. These two samples with no azide ion added had $t_{1/2}$ of 6.3 and 4.5 h. The reason for the differences in the N_3^- effect on pure cultures and natural waters is not understood. The N_3^- effect may be related to a natural assemblage versus cultured organisms.

In every case, heating or boiling unfiltered water samples resulted in the formation of H_2O_2. The processes that lead to this "abiotic" formation of H_2O_2 are not understood. Similar results have been reported for natural freshwater samples (*16*). Further work is necessary to determine whether this kind of result is an artifact of the heating procedure or is indeed a potential abiotic source of H_2O_2 in natural waters.

Chemical Decay.   In aquatic systems the presence of a strong oxidant, H_2O_2, may affect the redox chemistry of the environment. These effects may also result in the loss of H_2O_2 in the system.

H_2O_2 affects metal speciation in marine environments (*21, 24, 29–31*); however, little work has been reported for freshwater systems. A study by Sturzenegger (*15*) showed decomposition of H_2O_2 with both γ-MnOOH and β-MnO$_2$, in phosphate-buffered distilled water. Additional studies are necessary to determine the significance of these pathways of H_2O_2 in fresh waters.

The loss of H_2O_2 in natural waters by direct sunlight photolysis was considered of minor importance, but no experimental evidence existed until Moffett and Zafiriou (*1*) used $^{18}O_2$-labeled H_2O_2 in experiments conducted in coastal waters and no significant sunlight photolysis of H_2O_2 was observed.

Groundwater may be a significant source of water in a lake. This source of water may result in the decomposition of H_2O_2 in water through reactions of reduced metals with H_2O_2, such as Fenton's reactions (*95*):

$$M^{(n)+} + H_2O_2 \longrightarrow M^{(n+1)+} + OH\cdot + OH^- \qquad (20)$$

$$M^{(n)+} + OH\cdot \longrightarrow M^{(n+1)+} + OH^- \qquad (21)$$

If this mechanism is important in natural systems, it would also lead to the formation of the strongly oxidizing hydroxyl radical. This process could have a significant effect on the fate and transport of pollutants.

Distribution of H₂O₂

H_2O_2 has been reported in all surface waters studied. In lakes the distribution is limited to the epilimnion and is consistent with a sunlight (photochemical) pathway, a surface maximum, and decreasing concentration with increasing depth. We measured H_2O_2 in the metalimnion and observed it in the hypolimnion. This observation may be evidence of biologically mediated (dark) formation, or it may be an experimental artifact resulting from lowering the sampling equipment.

The sunlight-absorbing matter is substantially more concentrated in the small lakes we studied than in the marine systems (13). This concentration results in a distribution principally restricted to the epilimnion, which may be as shallow as several meters. That is, the total sunlight available for initiating reactions that lead to the formation of H_2O_2 is the same as in oceanic environments at similar latitudes, but it is absorbed nearer to the surface in lakes. As a result, higher concentrations are observed in the small high-humic (DOC) freshwater systems examined. The Great Lakes are closer to oceanic systems in terms of the dissolved organic carbon, but the epilimnion in Lake Erie and Lake Ontario is usually less than 20 m (14).

The decay rates we observed in the freshwater systems (13, 14, 18) are higher than those reported for oligotrophic marine systems but similar to near-shore (coastal) measurements (96). The increased decay rate with reduced light penetration leads to larger diel variability in H_2O_2 concentrations.

We measured H_2O_2 vertical profiles in Lake Erie (14, 18) and noted the similarity with oceanic profiles (23, 24). The major difference is the depth to which H_2O_2 is mixed in oceanic environments. To emphasize the effect of solar radiation and wind speed on the distribution of H_2O_2 in the epilimnion, we measured four vertical profiles of H_2O_2 concentration and temperature in Jacks Lake on 4 days, September 11–14, 1990, all at 4:00 p.m.

The 4 successive days were quite different in wind speed and solar radiation. Figure 7 shows the wind speed for the 4 successive days, and Table VII summarizes the solar radiation measured on the 4 days. The wind speed on the calm days was very similar, and, except for the more windy early morning on the cloudy and windy day, the windy days were quite similar. Figure 8 shows the vertical water temperature and H_2O_2 profiles obtained at 4:00 p.m. for the 4 days. The sunny and calm day resulted in some surface warming, but in general the profiles are those typical of a well-mixed epilimnion and provide no clue to the H_2O_2 profiles. Surface-water H_2O_2 concentration was elevated on the sunny and calm day (September 11, 1990) and rapidly decreased in concentration with depth, as would be predicted from the formation rate studies conducted in quartz tubes and presented in Figure 4. On the following day, when the wind speed was substantially higher and the insolation approximately the same, the H_2O_2 was mixed down through 7–8 m. Integration of the H_2O_2 concentration over the top 8 m of the water column for the 2 days (Table VII) shows that the total amount of H_2O_2 formed was very similar, 430 and 434 mg/m^2, and that the distribution was governed by physical mixing. The 2 days with clouds resulted in formation of a decreased integrated concentration of H_2O_2, 281 and 176 mg/m^2. The calm day (September 13, 1990) showed features similar to the sunny calm day except for the reduced surface H_2O_2 concentration. The integrated H_2O_2 concentration was higher for September 13, 1990, than for the cloudy windy day because of higher solar radiation.

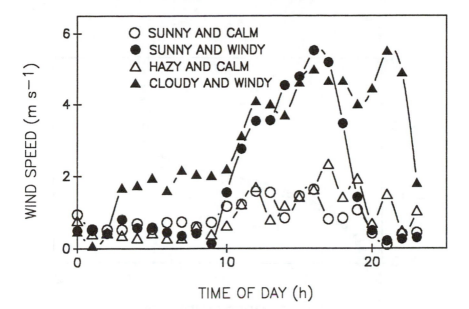

Figure 7. Wind speed in Sharpes Bay, Jacks Lake, September 11–14, 1990.

The integrated H_2O_2 concentration was directly related to the all of the radiation measurements, but correlated best with the ultraviolet portion ($r^2 = 0.985$). To use this limited data set, three assumptions need to satisfied:

1. that H_2O_2 decay is the same in the water column over the 4 days;

2. that H_2O_2 decay is the same with depth; and

3. that H_2O_2 decay is independent of H_2O_2 concentration over the concentration range observed.

Table VII. Daily Total Solar Radiation at the Surface of Jacks Lake and Integrated H₂O₂ Concentration

| | Energy (langleys) | | | $H_2O_2{}^{b}$ |
Date	Total Global	PAR[a]	UV	(mg m⁻²)
Sept. 11, 1990	505	254	22.1	430
Sept. 12, 1990	456	229	20.4	434
Sept. 13, 1990	241	129	11.8	281
Sept. 14, 1990	113	63	6.6	176

[a]PAR is photosynthetic active radiation.
[b]Calculated by integrating the H_2O_2 concentration through the water column to 8 m.

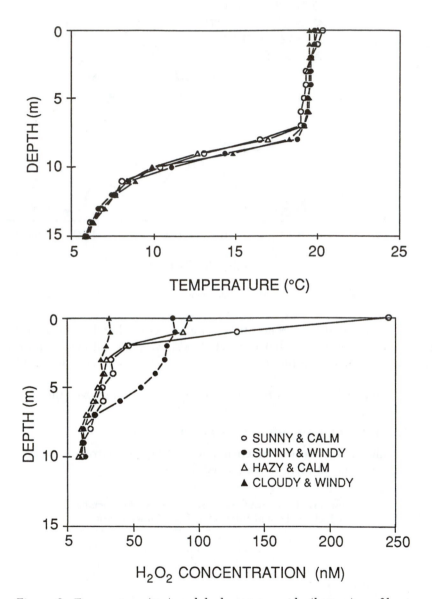

Figure 8. Temperature (top) and hydrogen peroxide (bottom) profiles in Sharpes Bay, Jacks Lake, Ontario, Canada, measured on September 11, 1990, when conditions were sunny with no wind and on subsequent days that were sunny and windy, hazy and fairly calm, and cloudy and windy.

We did not measure decay rates on successive days; however, other measurements give no reason to expect large variations over this short period. The H_2O_2 decay rate was determined at several depths on one of the days and was shown to be similar in all samples (Table II). We have shown on several occasions that, in the range of H_2O_2 concentrations observed, the decay rate was first order with respect to the observed H_2O_2 concentrations. Although this approach is greatly simplified, it appears that more detailed studies would help to quantify the relationship of solar radiation and H_2O_2 formation and cycling in natural waters.

The temperature data (Figure 8) suggest a well-mixed epilimnion. However, on a short time scale, the data from the H_2O_2 vertical profiles indicate that the epilimnion was not well mixed, even on the windy days. We can conclude from this data that low-resolution temperature-derived mixing rates are not applicable to H_2O_2 dynamics in these freshwater systems. In fact, many biological processes of interest occur on time scales far shorter than 24 h. The possibility of H_2O_2 as a short-term tracer is intriguing, because it is a sensitive tracer for vertical mixing processes.

Physical processes are important in determining the distribution of H_2O_2 in natural waters. In oceanic environments a model that includes photochemical formation and wind- and temperature-driven mixing has been developed (25). In freshwater systems, where the formation of H_2O_2 is restricted to the upper regions (<1 m in many cases), it is possible that H_2O_2 may serve as an in situ tracer for short-term mixing processes. The advantages are that it would behave similarly to the ambient water, its measurement is reasonably simple, and it is possible to adapt the analytical methods for continuous measurements (55) to be used in horizontal and vertical profiling. As yet no one has attempted such a study in fresh waters.

Significance of H₂O₂ in Surface Waters

Most likely, the precursor for the formation of H_2O_2 in natural waters is the superoxide ion ($O_2^{\cdot-}$), which may have an even greater potential than H_2O_2 to affect geochemical and biological processes in the ecosystem. Therefore estimates of its lifetime and steady-state concentration are important. The aqueous chemistry of $O_2^{\cdot-}$ was extensively reviewed by Bielski and coworkers (53). In aqueous solution $O_2^{\cdot-}$ is in equilibrium with the conjugate acid:

$$HO_2\cdot \rightleftharpoons O_2^{\cdot-} + H^+ \tag{22}$$

with a pK_a of 4.8. The autoredox dismutation, leading to the formation of H_2O_2, is complex at pH 6 and below:

$$HO_2 + HO_2 \longrightarrow H_2O_2 + O_2 \tag{23}$$

$$HO_2 + O_2^{\cdot-} + H_2O \longrightarrow H_2O_2 + O_2 + OH^- \tag{24}$$

However, at pH 6 and above, the rate can be described by the following equation:

$$\frac{d[O_2^{\cdot-}]}{dt} = 2k_2[O_2^{\cdot-}]^2 \tag{25}$$

where $2k_2 = 6 \times 10^{12}\,[H^+]$ (53). Millero (97) predicted that the rate would be slower in salt water as a result of ion-pair formation of MgO_2^+. If ion pairs formed, hard-water lakes might also result in slower autoredox dismutation. However, more recently, Zafiriou (98) showed experimentally that the autoredox dismutation rate in salt water is closely predicted by eq 25, with $2k_2 = (5 \pm 1) \times 10^{12}\,[H^+]$. The value of $2k_2$ would be 5–6 $\times 10^{12}$ $[H^+]$ in freshwater lakes, assuming no extremely fast reactions of unknown origin.

Thus, assuming $2k_2 = 6 \times 10^{12}\,[H^+]$ (53), the half-life, $t_{1/2}$, of $O_2^{\cdot-}$ can be estimated by the following equation (97, 99):

$$t_{1/2} = \frac{1}{2}\,k[HO_2]_T \tag{26}$$

where $[HO_2]_T$ is the total superoxide ion in solution. If $O_2^{\cdot-}$ is a comparatively long-lived intermediate, then the possibility exists that it may exert effects on the environment aside from those of H_2O_2. For instance, if we assume that the maximum concentration of $O_2^{\cdot-}$ is equal to the H_2O_2 concentration, then at a pH of 7.0 and a H_2O_2 concentration of 100 nM, the $t_{1/2}$ of $O_2^{\cdot-}$ is 0.08 s and for a H_2O_2 concentration of 500 nM the $t_{1/2}$ of $O_2^{\cdot-}$ is 0.2 s. These lifetimes are significant.

The biological implications of the presence of $O_2^{\cdot-}$ in surface waters is still speculative. The toxicity of $O_2^{\cdot-}$ is well-documented in other systems (e.g., 100–102) and this relatively reactive radical could have an impact on biological processes, considering its relatively long lifetime and concentration. The concentrations within the cell are even lower than in the media in which they live.

The presence of peroxidase has been reported in coastal oceanic environments (1) and in fresh waters (16). These reports present an intriguing possibility regarding the importance of H_2O_2 in natural waters. The reduction of H_2O_2 via peroxidase requires an electron or hydrogen donor. For example, with phenol the phenoxide radical is formed (eqs 14–16). If peroxidases are present in natural waters and H_2O_2 is also present, the reduction of H_2O_2 could result in the formation of free radicals from the naturally occurring organic compounds. Once free radicals are formed polymerization could result, which may lead to the in situ formation of higher weight organic compounds, such as humiclike substances and/or the surface microlayer. Another possibility is that pollutants could be transformed (16) or incorporated into existing humic substances that aggregate and settle to the bottom. These ideas are still speculative, but further study is warranted to assess the potential of peroxidatic activity in surface waters.

Summary

The major source of H_2O_2 in the epilimnion of lakes is a sunlight-initiated photochemical process. The mechanism for the photochemical formation of H_2O_2 can be simplified to the following equations:

$$_0^1DOC \xrightarrow{\text{light}} {}_1^1DOC^* \xrightarrow{\text{ISC}} {}_1^3DOC^* \tag{1}$$

$$_1^3DOC^* + {}^3O_2 \longrightarrow DOC^{\cdot+} + O_2^{\cdot-} \tag{5}$$

$$HO_2 + HO_2 \longrightarrow H_2O_2 + O_2 \tag{23}$$

$$HO_2 + O_2^{\cdot-} + H_2O \longrightarrow H_2O_2 + O_2 + OH^- \tag{24}$$

Neither the e_{aq}^- nor $^1O_2$ appears to be involved in the formation of H_2O_2 in sunlight-initiated reactions of humic substances in natural waters, and that finding simplifies the proposed mechanism. However, the exact nature of the initial reactions and electron transfer to oxygen are not as yet understood.

The biological and nonphotochemical pathways leading to the formation of H_2O_2 appear to be insignificant in the epilimnion of lakes and have not been demonstrated to occur in natural waters. Palenik and co-workers (87, 88) showed that L-amino acid oxidase, related to extracellular ammonia uptake, is a possible source of H_2O_2 that is independent of light. In the epilimnion this is probably not comparable to the photochemical formation of H_2O_2, but in low-light regions it may be significant.

The two major enzyme systems that lead to the decomposition of H_2O_2 are catalases and the peroxidases. By using $^{18}O_2$, Moffett and Zafiriou (1) showed that catalase is responsible for 65–80% of the decomposition of H_2O_2 and that peroxidase-mediated decay accounts for 20–35% of the loss. These experiments have not been extended to freshwater systems, and therefore the relative contribution of the two enzyme systems has not been established.

The data obtained from H_2O_2 decay rates in natural waters with different-size mesh filters has clearly established that H_2O_2 decay is particle-related. A series of biologically active inhibitors and heating or sterilization showed that the particles are live organisms, bacteria and/or plankton, that are retained by 1- and 0.2-μm filters.

The decay of H_2O_2 by two pure cultures of bacteria was studied, and rate equations were developed that describe the decay of H_2O_2. In both cases the decay was first-order in H_2O_2 and bacteria cell numbers. The rate equations for the two bacteria are for V. *alginolyticus*:

$$k_{Va} = 2.3 \times 10^{-9} \text{ mL/cells·min} \tag{18}$$

and for E. *cloacae*:

$$k_{Ec} = 5.1 \times 10^{-9} \text{ mL/cells·min} \tag{19}$$

The data are consistent with decay rates and bacterial numbers observed in natural waters. However, these data should be extended to include phytoplankton before the relative contribution of bacteria can be fully understood.

The chemical pathways for the decomposition of H_2O_2 appear to play a minor role in the overall decay processes. Direct-sunlight photolysis is not important in natural waters (1). The effect of H_2O_2 on metal speciation, and therefore on H_2O_2 decomposition, has been demonstrated in marine systems but not in lakes. Additional studies are required to better understand these processes in lake waters.

The distribution of H_2O_2 in lakes is limited to the epilimnion and is consistent with a sunlight (photochemical) pathway, with a surface maximum and decreasing concentration with increasing depth. During a 4-day study in which we measured H_2O_2 concentrations through the epilimnion, we showed that the distribution of H_2O_2 resulted from wind-driven mixing and that the integrated water column H_2O_2 concentration was directly related to the ultraviolet portion of the solar data results with a $r^2 = 0.985$.

During this 4-day study, the vertical H_2O_2 concentration profiles would not have been predicted from low-resolution vertical temperature profiles. H_2O_2 may be useful as an in situ tracer for mixing on short time scales (i.e., less than 24 h), and the development of on-line continuous analytical instrumentation for use in humic waters would be helpful.

The implications of H_2O_2 for lake biogeochemistry are mostly speculative at this time. Now that a general understanding of its formation, decay, and distribution is available, considerable work will be required to determine what effects it has on the ecosystem.

Acknowledgments

We acknowledge the helpful discussions of Ed Sholkovitz, Tim Shaw, Greg Cutter (also for the original of Figure 2), and Frances Pick during portions of this research. The captain and crew of the RV *Rigley Warfield* and the CSS *Limnos*, and the logistical support of the Johns Hopkins University,

Chesapeake Bay Institute, made the field portions of this study possible. The editorial assistance of L. Anita Holloway was invaluable in the final preparation of the manuscript. This work is a result of research sponsored in part by the Ohio Sea Grant College Program, Project R/PS–4, NA89AA–D–SG132 of the National Sea Grant College Program, National Oceanic and Atmospheric Administration in the U.S. Department of Commerce, and from the State of Ohio.

References

1. Moffett, J. W.; Zafiriou, O. C. *Limnol. Oceanogr.* **1990**, *35*, 1221–1228.
2. Sinel'nikov, V. E. *Gidrobiol. Zh.* **1971**, *7*, 115–119 (*Chem. Abstr.* **1971**, *75*, 25016a).
3. Sinel'nikov, V. E. *Tr. Inst. Biol. Vnutr. Vod Akad. Nauk SSSR* **1971**, *20*, 159–171 (*Chem. Abstr.* **1971**, *75*, 121176u).
4. Sinel'nikov, V. E.; Demina, V. I. *Gidrogeokhim. Mater.* **1974**, *60*, 30–40 (*Chem. Abstr.* **1976**, *83*, 151980j).
5. Sinel'nikov, V. E.; Liberman, A. *Sh. Tr. Inst. Biol. Vnutr. Vod Akad. Nauk SSSR* **1974**, *29*, 27–40 (*Chem. Abstr.* **1976**, *85*, 25170y).
6. Draper, W. M.; Crosby, D. G. *J. Agric. Food Chem.* **1981**, *29*, 699–702.
7. Draper, W. M.; Crosby, D. G. *Arch. Environ. Contam. Toxicol.* **1983**, *12*, 121–126.
8. Draper, W. M.; Crosby, D. G. *J. Agric. Food Chem.* **1983**, *31*, 734–737.
9. Cooper, W. J.; Zika, R. G. *Science (Washington, D.C.)* **1983**, *220*, 711–712.
10. Klockow, D.; Jacob, P. In *Chemistry of Multiphase Atmospheric Systems*; Jaeschke, W., Ed.; Springer-Verlag: New York, 1986; pp 117–130.
11. Cooper, W. J.; Zika, R. G.; Petasne, R. G.; Plane, J. M. C. *Environ. Sci. Technol.* **1988**, *22*, 1156–1160.
12. Cooper, W. J.; Zika, R. G.; Petasne, R. G.; Fischer, A. M. In *Aquatic Humic Substances: Influences on Fate and Treatment of Pollutants*; Suffet, I. H.; MacCarthy, P., Eds.; Advances in Chemistry 219; American Chemical Society: Washington, DC, 1989; pp 333–362.
13. Cooper, W. J.; Lean, D. R. S. *Environ. Sci. Technol.* **1989**, *23*, 1425–1428.
14. Cooper, W. J.; Lean D. R. S.; Carey, J. *Can. J. Fish. Aquat. Sci.* **1989**, *46*, 1227–1231.
15. Sturzenegger, V. T. Ph.D. Thesis ("Wasserstoffperoxid in Oberflaechengewaessern: Photochemische Producktion und Abbau"), Eidgenoessischen Technischen Hochschule, Zurich, Switzerland, 1989.
16. Cooper, W. J.; Zepp, R. G. *Can. J. Fish. Aquat. Sci.* **1990**, *47*, 888–893.
17. Cooper, W. J.; Lean, D. R. S. In *Encyclopedia of Earth System Science*; Nierenberg, W. A., Ed.; Academic: San Diego, CA, 1992; Vol. 2, pp 527–535.
18. Lean, D. R. S.; Cooper, W. J.; Pick, F. R., unpublished.
19. Van Baalen, C.; Marler, J. E. *Nature (London)* **1966**, *211*, 951.
20. Zika, R. G. In *Marine Organic Chemistry: Evolution, Composition, Interactions and Chemistry of Organic Matter in Seawater*; Duursma, E. K.; Dawson R., Eds.; Elsevier: Amsterdam, Netherlands, 1981; pp 299–325.
21. Moffett, J. W.; Zika, R. G. *Mar. Chem.* **1983**, *13*, 239–251.
22. Zika, R. G.; Moffett; J. W.; Petasne, R. G.; Cooper, W. J. *Geochim. Cosmochim. Acta* **1985**, *49*, 1173–1184.
23. Zika, R. G.; Saltzman, E. S.; Cooper, W. J. *Mar. Chem.* **1985**, *17*, 265–75.
24. Moffett, J. W.; Zika, R. G. *Environ. Sci. Technol.* **1987**, *21*, 804–810.

25. Plane, J. M. C.; Zika, R. G.; Zepp, R. G.; Burns, L. A. In *Photochemistry of Environmental Aquatic Systems;* Zika, R. G.; Cooper, W. J., Eds.; ACS Symposium Series 327; American Chemical Society: Washington, DC, 1987; pp 250–267.
26. Helz, G. R.; Kieber, R. J. In *Water Chlorination: Chemistry, Environmental Impact, and Health Effects;* Jolley, R. L.; Bull, R. J.; Davis, W. P.; Katz, S.; Roberts, M. H., Jr.; Jacobs, V. A., Eds.; Lewis: Chelsea, MI, 1985; pp 1033–1040.
27. Kieber, R. J.; Helz, G. R. *Anal. Chem.* **1986**, *58*, 2312–2315.
28. Szymczak, R.; Waite, T. D. *Aust. J. Mar. Freshwater Res.* **1988**, *39*, 289–299.
29. Moffett, J. W.; Zika, R. G. In *Photochemistry of Environmental Aquatic Systems;* Zika, R. G.; Cooper, W. J., Eds.; ACS Symposium Series 327; American Chemical Society: Washington, DC, 1987; pp 116–130.
30. Moffett, J. W.; Zika, R. G.; Petasne, R. G. *Anal. Chim. Acta* **1985**, *175*, 171–179.
31. Moffett, J. W.; Zika, R. G. *Geochim. Cosmochim. Acta* **1988**, *52*, 1849–1857.
32. Sunda, W. G.; Huntsman, S. A.; Harvey, G. R. *EOS.* **1983**, *64*, 1029.
33. McMahon, J. W. *Limnol. Oceanogr.* **1967**, *12*, 437–442.
34. McMahon, J. W. *Limnol. Oceanogr.* **1969**, *14*, 357–367.
35. Miles, C. J.; Brezonik, P. L. *Environ. Sci. Technol.* **1981**, *15*, 1089–1095.
36. Van der Weijden, C. H.; Reith, M. *Mar. Chem.* **1982**, *11*, 565–572.
37. Sunda, W. G.; Huntsman, S. A.; Harvey, G. R. *Nature (London)* **1983**, *301*, 234–236.
38. Collienne, R. H. *Limnol. Oceanogr.* **1983**, *28*, 83–100.
39. Baral, S.; Lume-Pereira, C.; Janata, E.; Henglein, A. *J. Phys. Chem.* **1985**, *89*, 5779–5783.
40. Draper, W. M.; Crosby, D. G. *J. Agric. Food Chem.* **1984**, *32*, 231–237.
41. Hoffman, M. R.; Edwards, J. O. *Inorg. Chem.* **1977**, *16*, 3333–3338.
42. Skurlatov, Y. I.; Zepp, R. G.; Baughman, G. L. *J. Agric. Food Chem.* **1983**, *31*, 1065–1071.
43. Zepp, R. G.; Baughman, G. L. In *Aquatic Pollutants: Transformation and Biological Effects;* Hutzinger, O.; van Lelyveld, I. H.; Zoeteman, B. C. J., Eds.; Pergamon: New York, 1978; pp 237–263.
44. Zepp, R. G.; Schlotzhauer, P. F.; Simmons, M. S.; Miller, G. C.; Baughman, G. L.; Wolfe, N. L. *Fresenius' Z. Anal. Chem.* **1984**, *319*, 119–25.
45. Zepp, R. G.; Skurlatov, Y. I.; Pierce, J. T. In *Photochemistry of Environmental Aquatic Systems;* Zika, R. G.; Cooper, W. J., Eds.; ACS Symposium Series 327, American Chemical Society: Washington, DC, 1987; pp 213–224.
46. Van Baalen, C. *J. Phycol.* **1965**, *1*, 19–22.
47. Fridovich, I. In *Free Radicals in Biology;* Pryor, W. A., Ed.; Academic: New York, 1976; pp 1, 239–277.
48. Morse, D. E.; Duncan, H.; Hooker, N.; Morse, A. *Science (Washington, D.C.)* **1977**, *196*, 297–300.
49. Mopper, K.; Zika, R. G. In *Photochemistry of Environmental Aquatic Systems;* Zika, R. G.; Cooper, W. J., Eds.; ACS Symposium Series 327; American Chemical Society: Washington, DC, 1987; pp 174–190.
50. Hoigne, J. In *Aquatic Chemical Kinetic;* Stumm, W., Ed.; John Wiley and Sons: New York, 1990; pp 43–69.
51. Baxter, R. M.; Carey, J. H. *Nature (London)* **1983**, *306*, 575–576.
52. Petasne, R. G.; Zika, R. G. *Nature (London)* **1987**, *325*, 516–518.
53. Bielski, B. H. J.; Cabelli, D. E.; Arudi, R. L.; Ross, A. B. *J. Phys. Chem. Ref. Data* **1985**, *14*, 1041–1100.

54. Palenik, B.; Zafiriou, O. C.; Morel, F. M. M. *Limnol. Oceanogr.* **1987**, *32*, 1365–1369.
55. Cooper, W. J.; Saltzman, E. S.; Zika, R. G. *J. Geophys. Res.* **1987**, *92*, 2970–2980.
56. Thompson, A. M.; Zafiriou, O. C. *J. Geophys. Res.* **1983**, *88*, 6696–6708.
57. Walcek, C. J. *Atmos. Environ.* **1987**, *21*, 2649–2659.
58. Kormann, C.; Bahnemann, D. W.; Hoffmann, M. R. *Environ. Sci. Technol.* **1988**, *22*, 798–806.
59. Baumann, P.; Baumann, L. In *The Prokaryotes*; Starr, M. P.; Stolp, H.; Truper, H. G.; Balows A.; Schlegel, H. G., Eds.; Springer-Verlag: New York, 1981; Vol. 2, pp 1302–1331.
60. Atlas, R. M.; Bartha, R. *Microbial Ecology*, 2nd ed.; Benjamin Cummings: Menlo Park, CA, 1987; p 288.
61. Andreae, W. A. *Nature (London)* **1955**, *175*, 859–860.
62. Perschke, H.; Broda, E. *Nature (London)* **1961**, *190*, 257–258.
63. Kieber, R. J.; Helz, G. R. *Anal. Chem.* **1986**, *58*, 2311–15.
64. Holm, T.; George, G. K.; Barcelona, M. J. *Anal. Chem.* **1987**, *59*, 582–86.
65. Zepp, R. G.; Skurlatov, Y. I.; Ritmiller, L. F. *Environ. Technol. Lett.* **1988**, *9*, 287–298.
66. Miller, W. L.; Kester, D. R. *Anal. Chem.* **1988**, *60*, 2711–2715.
67. Hobbie, J. E.; Daley, R. J.; Jasper, S. *Appl. Environ. Microbiol.* **1977**, *33*, 1225–1228.
68. Porter, K. G.; Feig, Y. S. *Limnol. Oceanogr.* **1980**, *25*, 943–948.
69. Pick, F. R.; Caron, D. A. Can. *J. Fish. Aquat. Sci.* **1987**, *44*, 2164–2172.
70. Schreiber, D. R.; Gordon, A. S.; Millero, F. J. Can. *J. Microbiol.* **1985**, *31*, 83–87.
71. Sholkovitz, E. R.; Elderfield, H. *Global Biogeochem. Cycles* **1988**, *2*, 157–176.
72. Zepp, R. G.; Braun, A. M.; Hoigne, J.; Leenheer, J. A. *Environ. Sci. Technol.* **1986**, *21*, 485–490.
73. Klug, D.; Rabani, J.; Fridovich, I. *J. Biol. Chem.* **1972**, *247*, 4839–4842.
74. Fridovich, I. *Acc. Chem. Res.* **1972**, *5*, 321–326.
75. Fridovich, I. *Annu. Rev. Biochem.* **1975**, *44*, 147–159.
76. Anbar, M; Hart, E. J. *J. Phys. Chem.* **1965**, *69*, 271–174.
77. Hasty, N.; Merkel, P. B.; Radlick, P.; Kearns, D. R. *Tetrahedron Lett.* **1972**, *1*, 49–52.
78. Haag, W. R.; Mill, T. *Photochem. Photobiol.* **1987**, *45*, 317–321.
79. Haag, W. R.; Hoigne, J.; Gassmann, E.; Braun, A. *Chemosphere* **1984**, *13*, 631–640.
80. Haag, W. R.; Hoigne, J.; Gassmann, E.; Braun, A. *Chemosphere* **1984**, *13*, 641–650.
81. Haag, W. R.; Hoigne, J. *Environ. Sci. Technol.* **1986**, *20*, 341–348.
82. Haag, W. R.; Hoigne, J. In *Water Chlorination: Chemistry, Environmental Impact, and Health Effects*; Jolley, R. L.; Bull, R. J.; Davis, W. P.; Katz, S.; Roberts, M. H., Jr.; Jacobs, V. A., Eds.; Lewis: Chelsea, MI, 1985; pp 5, 1011–1020.
83. Mehler, A. H. *Arch. Biochem. Biophys.* **1951**, *3*, 65–77.
84. Van Baalen, C. *J. Phycol.* **1965**, *1*, 19–22.
85. Stevens, S. E., Jr.; Patterson, C. O. P.; Myers, J. *J. Phycol.* **1973**, *9*, 427–430.
86. Palenik, B.; Morel, F. M. M. *Limnol. Oceanogr.* **1988**, *33*, 1606–1611.
87. Palenik, B.; Morel, F. M. M. *Limnol. Oceanogr.* **1990**, *35*, 260–269.
88. Palenik, B.; Kieber, D. J.; Morel, F. M. M. *Biol. Oceanogr.* **1989**, *6*, 347–354.
89. Palenik, B.; Morel, F. M. M. *Mar. Ecol. Prog. Ser.* **1990**, *59*, 195–201.

90. Roulier, M. A.; Palenik, B.; Morel, F. M. M. *Mar. Chem.* **1990**, *30*, 409–421.
91. Beyer, W. F., Jr.; Fridovich, I. In *Oxygen Radicals in Biology and Medicine;* Simic, M. G.; Taylor, K. A.; Ward, J. F.; von Sonntag, C., Eds.; Basic Life Sciences Series; Plenum: New York, 1988; Vol. 49, pp 651–661.
92. Saunders, B. C.; Holmes-Siedle, A. G.; Stark, B. P. *Peroxidase;* Butterworths: Washington, DC, 1964; 271 p.
93. Daniels, D. G. H.; Saunders, B. C. *J. Chem. Soc.* **1951**, 2112–2118.
94. Brock, T. D. In *Environmental Biogeochemistry and Geomicrobiology: Methods, Metals and Assessment;* Krumbein, W. E., Ed.; Ann Arbor Science Publishers: Ann Arbor, MI, 1978; Vol 3, pp 717–725.
95. Pelzar, M. J.; Charn, E. C. S.; Kreg, N. R. *Microbiology;* McGraw Hill: New York, 1986; 918 pp.
96. Walling, C. *Acc. Chem. Res.* **1975**, *8*, 125–131.
97. Petasne, R. G. M.S. Thesis ("Cycling of Hydrogen Peroxide in Seawater"), University of Miami, 1987.
98. Millero, F. J. *Geochim. Cosmochim. Acta* **1987**, *51*, 351–353.
99. Zafiriou, O. C. *Mar. Chem.* **1990**, *30*, 31–43.
100. Capallos, C.; Bielski, B. H. J. *Kinetic Systems;* Wiley Interscience: New York, 1972; 138 pp.
101. *Superoxide and Superoxide Dismutase in Chemistry, Biology, and Medicine;* Rotilio, G., Ed.; Elsevier: Amsterdam, Netherlands, 1986; 688 pp.
102. *Oxygen and Oxy-Radicals in Chemistry and Biology;* Rodgers, M. A. J.; Powers, E. L., Eds.; Academic: New York, 1981; 808 pp.
103. *Oxygen Radicals in Biology and Medicine;* Simic, M. G.; Taylor, K. A.; Ward, J. F.; von Sonntag, C., Eds.; Basic Life Sciences Series; Plenum: New York, 1988; Vol. 49, 1096 pp.

RECEIVED for review September 26, 1991. ACCEPTED revised manuscript May 7, 1992.

GEOCHEMISTRY OF TRACE METALS

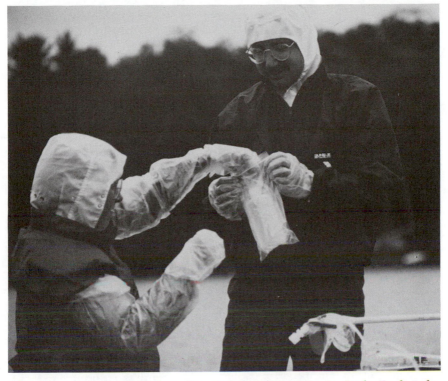

Ultra-clean sampling technique for analysis of mercury in Little Rock Lake. Photo courtesy of Bob Queen.

Trace metals are of pragmatic interest because many of them are toxic to humans and aquatic organisms. In many sites, metal contamination is the result of mining activity in an earlier era, although regional mercury contamination may be caused by current atmospheric mercury deposition. Additionally, manganese is a major problem in reservoirs because it causes taste and staining problems in water supplies. The following chapters provide a cross section of studies designed to improve our understanding of the biogeochemistry of metals. The complexity of trace metal behavior is revealed by the fact that trace metals are among the most difficult chemicals to represent in mathematical models of water quality.

Cycling of Mercury across the Sediment–Water Interface in Seepage Lakes

James P. Hurley[1,2], David P. Krabbenhoft[3], Christopher L. Babiarz[2], and Anders W. Andren[2]

[1]Bureau of Research, Wisconsin Department of Natural Resources, 1350 Femrite Drive, Monona, WI 53716
[2]Water Chemistry Program, 660 North Park Street, University of Wisconsin, Madison, WI 53706
[3]U.S. Geological Survey, 6417 Normandy Lane, Madison, WI 53719

The magnitude and direction of Hg fluxes across the sediment–water interface were estimated by groundwater, dry bulk sediment, sediment pore water, sediment trap, and water-column analyses in two northern Wisconsin seepage lakes. Little Rock Lake (Treatment Basin) received no groundwater discharge during the study period (1988–1990), and Pallette Lake received continuous groundwater discharge. In Little Rock Lake, settling of particulate matter accounted for the major Hg delivery mechanism to the sediment–water interface. Upward diffusion of Hg from sediment pore waters below 2–4-cm sediment depth was apparently a minor source during summer stratification. Time-series comparisons suggested that the observed buildup of Hg in the hypolimnion of Little Rock Lake was attributable to dissolution and diffusion of Hg from recently fallen particulate matter close to the sediment–water interface. Groundwater inflow represented an important source of new Hg, and groundwater outflow accounted for significant removal of Hg from Pallette Lake. Equilibrium speciation calculations revealed that association of Hg with organic matter may control solubility in well-oxygenated waters, whereas in anoxic environments sulfur (polysulfide and bisulfide) complexation governs dissolved total Hg levels.

SOLUTE EXCHANGE across the sediment–water interface serves as an important process in regulating water-column concentrations of metals in nat-

0065–2393/94/0237–0425$07.25/0

ural waters (1–3). Studies of solid-phase bulk sediments from lakes in Wisconsin and Minnesota (4–7) showed increases in Hg concentrations near the top of lake-sediment cores, and these increases were attributed to changes in atmospheric inputs following industrialization. However, estimating the amount of Hg remineralized after deposition at the sediment surface has been a difficult task.

The lack of reliable data on transport of Hg across the sediment–water interface arises from two factors. First, contamination may occur during sampling. Clean techniques for trace metals that are similar to techniques developed for sampling of lead in the mid-1970s (described in reference 8) must be followed during Hg sampling and analysis. The potential for contamination during sampling is high because of the small concentrations of Hg present in lake waters (typically 0.5–10 ng/L). Second, the relative insensitivity of previous analytical procedures made it difficult to adequately quantify concentrations of Hg in lake and sediment pore waters. Therefore, reasonable profiles of dissolved Hg were difficult to obtain, and calculation of flux rates across this important interface were similarly hampered.

Recent studies directed at assessing the fate and transport of Hg in natural waters used improved analytical methods and clean techniques for sampling and analysis (9, 10). Some lake studies (11, 12) were directed at assessing the effects of point-source Hg inputs, such as chloroalkali manufacturing plants and mining operations; other studies (13–16) were developed in response to concerns over recent observations of elevated Hg levels in fish from lakes remote from point sources.

Partly because of this concern, the Wisconsin Department of Natural Resources, in cooperation with the Electric Power Research Institute, initiated an extensive study of Hg cycling in seepage lakes of north-central Wisconsin (14). The mercury in temperate lakes (MTL) study used clean sampling and subnanogram analytical techniques for trace metals (10, 17) to quantify Hg in various lake "compartments" (gaseous phase, dissolved lake water, seston, sediment, and biota) and to estimate major Hg fluxes (atmospheric inputs, volatilization, incorporation into seston, sedimentation, and sediment release) in seven seepage lake systems.

A preliminary mass balance revealed the following interesting insights into Hg cycling in Little Rock Lake (18, 19).

1. Atmospheric deposition was the major external source of Hg to the lake.

2. Permanent accumulation of Hg in the bottom sediments was roughly balanced by atmospheric inputs on an annual basis.

3. Input from atmospheric deposition was sufficient to account for all of the Hg observed in fish, sediments, and water.

Although net accumulation of Hg in sediments roughly balanced atmospheric inputs, gross sedimentation, as measured by sedimentation traps, exceeded net accumulation by a factor of about 3. This observation, coupled with water-column profiles of Hg (*20, 21*), suggested strong recycling of Hg in the region of the sediment–water interface. Research efforts were therefore directed toward assessing factors that control Hg cycling near this important interface.

This chapter summarizes our results from two northern Wisconsin seepage lakes that were chosen to assess the importance of various processes controlling transport of Hg across the sediment–water interface. Total Hg (Hg_T) concentrations were determined as a function of depth in the solid and dissolved phases of the water column, and in littoral and profundal sediments. New sampling and analytical procedures allowed for the detection of low (picogram) levels of Hg. Measurements obtained in this phase of the study together with those obtained from previously published data on these lakes were used to make a preliminary examination of the relative importance of factors influencing Hg cycling at the sediment–water interface.

Site Description

Two lakes chosen for this study, Little Rock Lake Treatment Basin and Pallette Lake, are located in the Northern Highlands Lake District of Wisconsin. Little Rock Lake (45°50' N, 89°42' W) and Pallette Lake (46°04' N, 89°36' W) are soft-water seepage lakes, deriving water from only atmospheric and groundwater sources. Both lakes are remote from any point sources of Hg. Although experimental acidification of one of the two basins of Little Rock Lake offers a comparison of the effects of acidification (*22, 23*), we will limit our Hg discussion to the treatment portion of the lake. This basin, unlike the Reference Basin, is deep enough to exhibit strong hypolimnetic oxygen depletion and conditions more conducive to studying the release of redox-controlled constituents from profundal sediments. For the remainder of this chapter, we will refer to the Treatment Basin as Little Rock Lake.

A major aspect of this study was assessment of the role of groundwater transport in the overall Hg cycle. However, during the study period (1988–1990) Little Rock Lake was mounded (no groundwater inflow), but Pallette Lake had both groundwater inflow and outflow. Therefore, for the purposes of evaluating the importance of groundwater inflow and outflow on Hg transport, we extended our study to Pallette Lake.

Water Column.   Water-column profiles were taken at the deepest location (10 m) in Little Rock Lake. Details of the clean sampling techniques (*8*) that were used during sampling are given elsewhere (*18, 20, 21*). By following these stringent protocols, our group demonstrated (*18*) that typical epilimnetic Hg levels in seven northern Wisconsin lakes, including Pallette and Little Rock Lake (0.5–2 ng of Hg/L for unfiltered epilimnetic samples), were of magnitude similar to those levels observed in remote ocean sites

(24). Previous estimates of unfiltered total Hg in northern Wisconsin lakes were 2 orders of magnitude higher than our observed levels (18).

Our water-column sampling techniques include in-line filtration using an all-Teflon sampling device with quartz fiber filters (0.7-μm nominal size cutoff) to differentiate between dissolved and particulate phases (21). Particulate concentrations (nanograms per gram) and subsequent calculations of partitioning between particle and aqueous phases (log K_D) are based on this particle size division. This fractionation scheme precludes direct estimates of colloidal influences on Hg transport.

Sedimentation Traps. Sediment traps (25) were installed in the hypolimnion of Little Rock Lake to estimate the downward flux of Hg to the sediment surface. Traps were constructed of acrylic and Teflon following the design of Shafer (26). No metal components were used to avoid possible contamination artifacts. Traps were placed at 9 m at the 10-m-deep hole-sampling site. Traps were suspended from surface floats to prevent disturbing bottom sediments during deployment and retrieval.

Groundwater. Because sandy littoral sediments have greater hydraulic conductivity than silty profundal sediments (27), most of the exchange of water (and solutes) between groundwater systems and lakes occurs through the littoral zone (28). Therefore, efforts aimed at quantifying Hg transfer between lakes and their contiguous groundwater systems were focused in near-shore areas. Numerous groundwater sampling methods allowed for sampling of different features in the groundwater system near the study lake (Figure 1). The methods used (piezometers, wells dug to the water table,

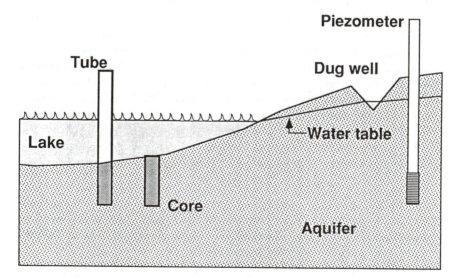

Figure 1. Schematic diagram of the sampling methods used to acquire groundwater samples near the aquifer–lake interface.

acrylic tubes inserted into littoral sediments, and pore-water extraction from sediment cores) allowed comparison of several aspects of Hg cycling as groundwater discharges to the lake.

The most common method for sampling chemical constituents in groundwater, piezometer sampling (29–31), was used in the initial stages. Precleaned acrylic piezometers with 0.3-m Teflon screens were installed by power auguring (hollow-stem auger). A threaded fitting used to join the screen and casing prevented potential contamination from solvents in piezometer construction. Wells were nested (several wells at one location with differing depths), with water-table wells at about 3 m and deeper wells at 5–15 m.

Dug wells were formed by trenching a small hole (about 0.3 m square and less than 0.5 m deep, 10 cm below the water table) with a precleaned plastic shovel. New wells were dug for each sampling period. The wells were located within 5 m of the shoreline, and samples from these wells were used to estimate the background Hg content of inflowing groundwater. After the wells were pumped for about 45 min (about two to three trench volumes), samples were taken with a peristaltic pump and Teflon line. Purging reduced the effects of possible contamination and particle suspension during well digging and provided a short hydraulic residence time in the well prior to sampling.

The tube method involved insertion of precleaned acrylic tubes (5-cm diameter) into the littoral zone sediments at a lake-water depth of about 1 m (about 15 cm into sediments). Lake water within the tube was removed by using a peristaltic pump and Teflon line. Once the tube was purged of lake water, groundwater was allowed to fill it to a depth of about 25 cm. The tube was then purged three times with groundwater before obtaining the sample.

Pore-water samples were obtained from littoral sediments by push-coring with precleaned acrylic core barrels (6.7-cm i.d., 7.6-cm o.d.). A device that eliminates air contact while sampling (32) was used for pore-water extraction. Teflon plungers at either end of the barrels were forced toward each other to pressurize the barrel. Interstitial water flowed out of sampling ports (2-cm intervals) in response to the external pressure. Samples were then filtered (0.4-μm filter; Nuclepore) and preserved with 6 N HCl before Hg analysis. Laboratory studies indicated no contribution of Hg from the filtering unit, tubing, or sample bottle. Because the filtering system removed interstitial water from the center of the core (>2 cm from the wall), low diffusion constants led to negligible contamination from the core barrel. Solid-phase materials (sands) in the littoral zones were not analyzed for particulate Hg.

Profundal Sediments.   Sediment cores were collected in precleaned acrylic tubes by scuba divers following similar clean sampling procedures for trace metals in the littoral-zone sediment sampling. Benefits of core sampling by scuba diving over other traditional methods (such as Jenkins coring and piston coring) included careful selection of the sampling site and the ability to observe whether mixing or disturbance of the core occurred during sampling. Cores were taken with minimal surface disruption and processed within 2–4 h of sampling. Profundal pore waters were sampled in a method similar to that used with littoral pore waters. Solid-phase samples were taken from separate cores, which were sectioned at 1-cm intervals to a depth of 30 cm. Care was taken to discard sediment in contact with the core barrel, in case smearing occurred during extrusion and slicing. Because organic-rich profundal sediments prevent groundwater inflow, other sampling methods used for groundwater sampling in littoral zones (such as tubes and piezometers) were not needed in profundal zones.

Various ancillary measurements from accompanying cores were needed to calculate accumulation rates and describe phase associations of Hg. ^{210}Pb and ^{137}Cs profiles in sediments were used to determine sedimentation rates from which historical interpretations could be made.

Laboratory Methods

Low-level (picogram) Hg analysis required preconcentration by two-stage gold amalgamation, followed by detection with a cold-vapor atomic fluorescence detection system (9, 33). Briefly, aqueous samples are treated with a strong oxidizing agent (BrCl) to destroy organo–Hg bonds and convert all Hg into the soluble Hg(II) state. Stannous chloride is added to reduce Hg(II) to the elemental (Hg^0) state. This volatile form is stripped from solution by nitrogen onto a gold-coated sand trap. The Hg is then thermally desorbed onto a second gold trap, and from this trap into the atomic fluorescence cell. Solid-phase sediment (about 1 g) required initial digestion in 5:2 HNO_3–H_2SO_4. In this study no distinction was made between total and methyl Hg.

Solid-phase sediment digestions were analyzed in triplicate, with one duplicate digestion and a spike recovery or analysis of reference material every 10 samples. Coefficients of variation (C.V.) for triplicates fell within 0.5–11.5% [$n = 60$, mean C.V. $= 3.6\% \pm 2.4\%$ (std)], and spike recoveries were within 90–103% [$n = 3$, mean $= 96\% \pm 7\%$]. Eight replicates of standard reference material [National Institute of Standards and Technology (NIST) Tennessee River sediment, Catalog No. 8406] were within 10% (0.053 ± 0.004 $\mu g/g$, C.V. $= 7.2\%$) of the recommended value of 0.06 $\mu g/g$. A standard reference for Hg in natural water was not available. Typical duplicates of small-volume pore waters (<30 mL) had an average C.V. of 30.8% $\pm$ 22% ($n = 30$).

Methods for dating sediment cores using ^{210}Pb and ^{137}Cs were similar to those described by Robbins and Edgington (34). ^{137}Cs content was determined by gamma spectroscopy and ^{210}Pb by alpha spectroscopy with surface-barrier detectors. Total carbon was determined by combustion techniques (Perkin-Elmer model 240C elemental analyzer).

Results

The distribution of Hg within seepage lakes is a net result of the processes that control Hg transport between the atmosphere, water column, seston, sediments, and groundwater. This discussion focuses on the processes that control the exchange of Hg between the sediments and lake water. We first present data on spatial and temporal concentrations in the water column, sediments, pore water, and groundwater. These data set the context for a subsequent discussion of the chemical and physical processes responsible for the transport of mercury across the sediment–water interface and are necessary for assessing transport rates.

Distribution of Hg.   *Hg in the Water Column.*   Water-column profiles help to illustrate the importance of Hg cycling in the region near the sediment–water interface. Data from 1989 in the Little Rock Treatment Basin (Figure 2) indicated strong seasonal variability in both dissolved and particulate Hg (21). Temperature and dissolved-oxygen profiles indicate strong thermal stratification, and low oxygen levels in the hypolimnion probably reflect oxidation and remineralization of organic matter. The lower oxygen levels in the deeper waters in April may have resulted from incomplete mixing or rapid depletion following the spring bloom, but temperature profiles do not suggest true stratification. Similarly, Hg profiles do not exhibit any near-bottom increases during April. However, as the hypolimnion of the lake becomes progressively anoxic throughout the stratification period, dissolved Hg concentrations increase dramatically. Epilimnetic Hg varies at about 1–2 ng/L throughout the year, whereas hypolimnetic levels increase to about 15 ng/L by late August, apparently through release from either sediments or sedimenting particles. Particulate Hg levels also increased in late summer in the hypolimnion.

Hg Concentrations in Sediments.   A typical Hg concentration profile in profundal sediment cores from Little Rock Lake Treatment Basin is shown in Figure 3. Mercury concentrations range from about 50–185 ng/g (dry weight). Similar concentrations were observed by Rada et al. (35) in Little Rock Reference Basin (6–205 ng/g for surface grabs across the lake, including sandy sediments) and by R. Rada (University of Wisconsin, LaCrosse, personal communication) for Little Rock Treatment Basin (3–220 ng/g for similarly retrieved surface grabs). The decrease in concentration toward the top

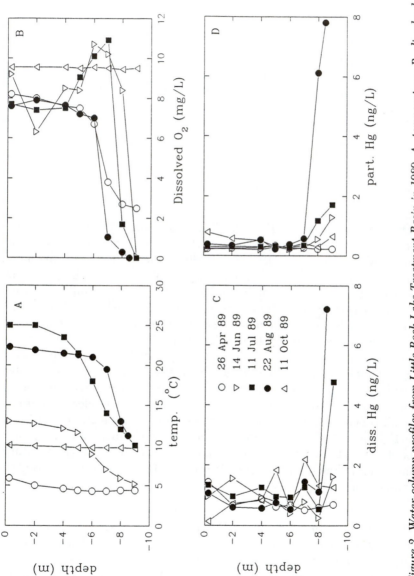

Figure 2. Water-column profiles from Little Rock Lake Treatment Basin in 1989. A, temperature; B, dissolved oxygen; C, dissolved Hg; and D, particulate Hg. (Adapted with permission from reference 21. Copyright 1991 D. Reidel Publishing Co.)

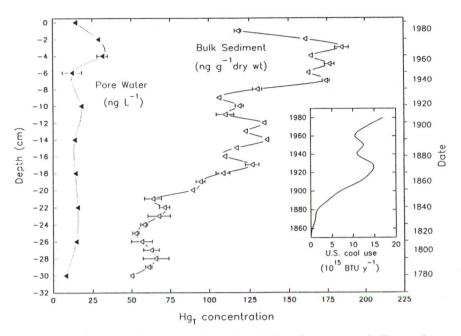

Figure 3. Sediment core from Little Rock Lake (3 m) depicting dry bulk particle (Δ) and pore-water (▲) Hg_T concentration. Sediments were dated by ^{137}Cs and ^{210}Pb. Inset: Coal use in the United States according to a 1986 report to the National Academy of Sciences.

few centimeters of the core is common to all cores taken in the profundal zone. A clear explanation for this observation is not immediately apparent. Several possibilities should be considered, such as lower Hg inputs in the last decade, postdepositional migration, and a nonequilibrium condition in the upper few centimeters.

^{210}Pb and ^{137}Cs dating of our sediment core indicate that preindustrial concentrations were about 50–70 ng/g, whereas contemporary levels range from 110 to 185 ng/g. Although some postdepositional migration probably occurs, the results of this profile suggest an approximate two- to fourfold increase in Hg concentrations since the beginning of industrialization. The factor of increase is similar to those observed in sediments of other lakes in the region by Rada et al. (4) and Engstrom et al. (7). Furthermore, trends in coal consumption over the past 150 years in the United States (36) somewhat parallel our observed increases (Figure 3 inset). Additional uses of mercury by other human activities (37) and a better understanding about the extent of postdepositional migration of Hg in sediments must be considered before it is possible to assign a definite causal relationship to the observed trend.

Hg Concentrations in Pore Waters. Profundal sediment pore-water concentrations varied from 10 to 30 ng/L throughout the profile (Figure 3).

A subsurface peak in Hg_T was evident at 2–4 cm in most cores. A mechanistic explanation for this observation is not clear. Dissolved organic carbon (DOC) concentrations ranged from about 2.5 to 4 mg/L and generally increased with increasing depth of the core. Thus, a clear correlation was not seen between Hg and DOC (38). Similarly, there was no apparent relationship between dissolved Hg and dissolved Fe or Mn. The distribution of Hg in aqueous and solid phases is the net result of many geochemical processes (e.g., redox, complexation, and solubility). Information available to our group thus far cannot explain the observed subsurface peak in the pore-water Hg profile.

Hg Concentrations in Groundwater.   Groundwater may represent both a delivery and a removal mechanism for Hg in lakes. Water flowing into a lake may transport Hg derived from atmospheric deposition or from dissolution of surrounding glacial deposits. Because no Hg-containing deposits are known to exist in this region (4), inflowing groundwater should represent Hg from atmospheric deposition that has passed through both the soil zone and the sandy aquifer. In areas of the lake bed where lake water seeps out, Hg levels might be assumed to be similar to those of lake water.

As with pore water, very few groundwater Hg concentrations have been reported in the literature; thus it is difficult to compare our values to what might be observed elsewhere. Piezometers were used initially to estimate the background Hg concentration in inflowing groundwater. The first Hg samples, taken in October 1988, were high (10–20 ng/L) relative to lake waters (1–2 ng/L) (18). The concentration was suspected to be a result of contamination from the well. A pumping test designed to evaluate contamination was conducted in July 1989. Hg levels at $t = 0$ (after pumping three well volumes), 24, and 48 h were 2.3, 2.6, and 2.5 ng/L, respectively. Although no significant trend toward pulse contamination was observed, continuous leaching of mercury from the piezometer could not be ruled out on the basis of this limited data set. If Hg leached from or sorbed to the acrylic walls of the piezometer, Hg samples may not be truly reflective of what one would consider background groundwater levels.

The dug-well technique was first tested at Pallette Lake in October 1989 because the hydraulically mounded condition of Little Rock Lake precluded testing of other groundwater sampling methods. Glacial-outwash sediments at Pallette Lake are similar to those at Little Rock Lake; therefore, the two sites can be compared. Mercury concentrations in samples from the dug wells were relatively consistent over an approximately 2-year sampling period (range 1.0–3.8, average 2.7 ng of Hg/L), and were similar to concentrations in samples from the piezometers. Therefore, two different sampling techniques, piezometers and dug wells, provided similar samples for estimating background groundwater Hg concentrations.

Analyses of littoral pore waters provided additional supporting evidence on levels of background Hg passing through the sediment–water interface (Figure 4). In addition, these profiles suggested that a near-surface mercury source existed in both inflow and outflow areas. In zones where groundwater flows upward toward the sediment–water interface (inflow areas) Hg concentrations at depth converge to about 3.5 ng/L, slightly higher than is found in dug wells and piezometers. On the basis of these data and two other independent sampling methods, the background Hg_T concentration in near-surface groundwater in this remote area of northern Wisconsin is assumed to be about 2–4 ng/L. These values are similar to those reported (39) in groundwater near Swedish lakes (2–8 ng/L), although Swedish in-

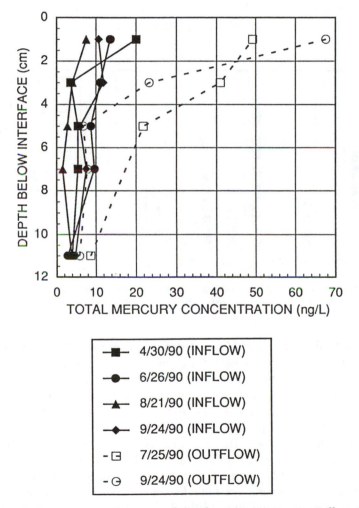

Figure 4. Mercury concentrations in littoral zone pore waters in Pallette Lake in 1990.

vestigators observed greater variability in groundwater Hg concentrations (A. Iverfeldt, Swedish Environmental Research Institute, personal communication).

Samples taken by the acrylic-tube sampling method provide further evidence for a near-surface Hg source (Figure 5). Mercury concentrations in tubes were consistently higher than those observed in dug wells, but less than those observed in near-surface pore-water profiles. This finding appears reasonable, because samples taken using the tube method represent a mixture of ambient groundwater and near-interface pore waters. Mercury levels in samples taken by the tube method are probably not representative of the actual concentrations of groundwater entering the lake. Flow rates induced after purging the tube before sampling result in refilling rates that are significantly greater than typical groundwater inflow rates.

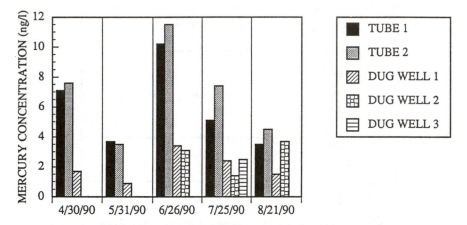

Figure 5. Comparison of groundwater mercury concentrations in samples taken from Pallette Lake using the dug-well and acrylic-tube sampling methods.

Chemical Controls of Hg in Bottom Sediments, Pore Water, and Groundwater.

Hg_T concentrations ranged by at least an order of magnitude among each of the various lake compartments studied (water column, pore waters, and groundwaters). Ambient groundwater ranged from 2 to 4 ng/L upgradient of Pallette Lake. In inflow regions near the sediment–water interface, levels of Hg_T peaked at 15–20 ng/L. In profundal pore waters and near-surface littoral zones, Hg_T concentrations were as high as 70 ng/L. Concentrations in the water column of Little Rock Lake ranged from about 0.5–15 ng/L. Each lake compartment clearly represents a distinct physical and chemical environment. Profundal pore waters were predominantly anoxic, as was the hypolimnion of Little Rock Lake in late summer. Groundwater and littoral pore waters were generally oxic, although subsurface oxygen depletion was observed in some littoral cores (D. Krabbenhoft, unpublished data).

The solubility of Hg(II) is controlled by chemical speciation in natural waters, and the availability of ligands for complexation shifts dramatically under varying redox conditions (40). Speciation of dissolved Hg(II) in anoxic environments, such as sediments or the hypolimnion, should be strongly influenced by reactions with reduced sulfur (40, 41), whereas organic complexation is potentially important under oxic conditions (42, 43).

Equilibrium speciation calculations were performed by using the software program Titrator (44). Dyrssen (45) and Dyrssen and Wedborg (46) presented the most important complexation reactions of Hg(II) with reduced sulfur and organic ligands in natural waters. For a model organic compound, we used Dyrssen and Wedborg's (46, 47) stability constants for thiols (SR). We calculated Hg species distribution at two different pH levels (pH 5.5 and 7.0) and at three different redox levels (pE = 2, 0, and −2) to illustrate the potential importance of these controlling geochemical parameters in aquatic systems. Concentrations of major anions were taken from the Project MTL database (14). Although we did not measure thiol concentrations directly, we assumed a concentration of 1.0×10^{-9} M. The following Hg species were included in our calculations: $HgHS^+$, $Hg(HS)_2$, HgS_2H^-, HgS_2^{2-}, $HgCl^+$, $HgCl_2$, $HgCl_3^-$, $HgCl_4^{2-}$, $HgOH^+$, $HgClHS$, $HgOHHS$ (or HgS; *see* ref. 46), $HgOHCl$, $Hg(aq)^0$, $HgSR^+$, $Hg(SR)_2$, and $HgS(s)$. Important Hg complexation reactions together with equilibrium constants are shown in Table I. The reaction of Hg(II) with S^{2-} to form $HgS(s)$ is also included.

Table I. Equilibrium Constants for Hg(II) and Reduced Sulfur Species

Equilibrium		log K
$Hg^{2+} + 2e^-$	$= Hg^{\circ}(aq)$	22.3
$Hg^{2+} + 2\,HS^-$	$= Hg(SH)_2$	37.72
$HgHS_2^- + H^+$	$= Hg(SH)_2$	6.19
$Hg^{2+} + 2\,S^{2-}$	$= HgS_2^{2-}$	51.53
$Hg^{2+} + RS^-$	$= HgRS^+$	22.1
$Hg^{2+} + 2\,RS^-$	$= Hg(RS)_2$	41.6
$HgS_2^{2-} + H^+$	$= HgHS_2^-$	8.30
$H^+ + HS^-$	$= H_2S$	6.88
$H^+ + RS^-$	$= RSH$	9.34
$H^+ + S^{2-}$	$= HS^-$	17.0
$HgS(s)$	$= Hg^{2+} + S^{2-}$	−55.9
$HgS(s)$	$= Hg^{2+} + HS^-$	−38.9
$HgS(s)$	$= HgS$	−10.0

NOTE: Constants were taken from references 45–52.

   Equilibrium speciation calculations for these conditions illustrate major differences in species dominance across pH and pE gradients (Table II). On the basis of these stability constants, the assumption that thiols are the organic complexation ligands, and our estimates of various ligand concentrations in our system, we conclude that organic matter complexation controls Hg(II) solubility at higher pE levels (i.e., well-oxygenated lake, ground, and pore waters). These calculations compliment studies that show a strong relationship between Hg and dissolved organic matter (42, 43), presumably humic substances. Mercury complexation by DOC in groundwater is also likely to be important. Samples taken from nine dug wells and analyzed for DOC ranged from 5.2 to 19.1 mg/L, and had a positive correlation coefficient with total mercury concentration of 0.71. Similar correlations were observed by Lindquist et al. (39) in Swedish groundwaters, although both Hg_T and DOC levels were higher in their systems.

   At lower pE levels, pH and sulfide levels determine the dominant dissolved species and complexation reactions. At pE 0 and pH 5.5, Hg(aq)^0 is the dominant species, although at pH 7 the pE level must approach -2 in order to observe a similar shift to Hg(aq)^0 dominance. At pH 5.5 and pE -2, bisulfide and polysulfide complexation dominates Hg speciation. These results are similar to calculations made by Gardner (41), who compared inorganic and organic complexation by trace metals. Gardner concluded that in sulfidic marine waters, complexation of Hg with bisulfides and polysulfides dominated over complexation with organic matter, including a variety of free amino acids and hydrocarboxylic acids. Although numerous assumptions have been made in these calculations (such as stability constant values and ambient concentrations of some ligands), these calculations serve as an ex-

**Table II. Percent Distribution of Hg Speciation
at Varying pH and pE Levels**

Species	pH 5.5	pH 7.0
	pE +2	
Hg^0(aq)	0.50	
Hg(RS)$_2$	0.01	
HgRS$^+$	99.5	100
	pE 0	
Hg^0(aq)	97.9	0.16
HgRS$^+$	2.09	99.8
	pE −2	
Hg^0(aq)	0.07	94.1
HgRS$^+$		5.87
HgS (HgOHSH)	0.01	
Hg(HS)$_2$	81.7	
HgS$_2$H$^-$	16.3	

cellent example of the nature of Hg speciation under conditions frequently observed in the lake environment.

Distribution coefficients (K_D) can also be used to gain a better understanding of factors controlling the partitioning of chemicals between solid and liquid phases (53). The affinity of a chemical constituent for particles is described by

$$K_D = \frac{C_S}{C_W} \tag{1}$$

where C_S and C_W are concentrations in the solid and aqueous phases, respectively. This parameter was used to investigate the differences in partitioning behavior of water-column and sediment particulate substances. With our filtration scheme, log K_D values ranged from about 3.4 to 4.1 in sediments, and log K_D in the water column ranged from 4.5 to about 5.7 (Figure 6). The differences observed between K_D values in the water column and in sediments are interpreted as dissimilar composition of suspended and bottom sediment. Additionally, K_D values have been shown to decrease with increasing concentrations of sorbate (54). This correlation could easily explain the sharp break between the water-column and bottom sediments. The observed K_D values in sediments may indicate that precipitation–dissolution reactions dominate in this region, whereas biological control and adsorption dominate partitioning in the water column. The narrow range of K_D values below a depth of about 10 cm in sediments could indicate a solid-phase solubility control (HgS(s)).

Transport Mechanisms.   The buildup in hypolimnetic Hg is the result of several transport mechanisms operating simultaneously. Numerous physical, chemical, and biological processes can affect downward and upward transport of metals across the sediment–water interface (55). Delivery of most heavy metals from the water column to the sediment–water interface is most likely particle-mediated (56), although direct adsorption of Hg to bottom sediments may occur to some degree. Thus, metals in the water column are incorporated into biogenic and nonbiogenic material such as detrital particulate matter, phytoplankton, zooplankton, bacteria, and fish. A portion of this particulate matter with associated Hg may settle to the sediment surface within time spans of days to weeks in lakes. However, some particulate matter may dissolve in the water column, releasing dissolved Hg back into the water column in the dissolved form. Sedimentation rates are also strong functions of season and are spatially nonuniform within the lake, with lowest rates of deposition in shallow sediments and highest rates in depressions.

Upward transport across the sediment–water interface may be physically, chemically, or biologically mediated. Periods of mixing in spring and

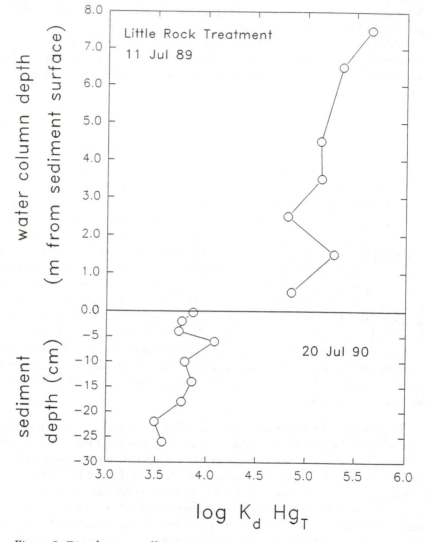

Figure 6. Distribution coefficients (K_D) for Hg in Little Rock Lake Treatment Basin water and sediments.

fall (a result of the breakdown of thermal stratification) can potentially cause sediment resuspension. In deep seepage lakes, this process is more common in littoral sediments, although intense mixing may cause resuspension of profundal sediments. Postdepositional particle migration may also focus material to the deepest point in the basin. Detritivores and other biological organisms living in the sediment can also serve as resuspension mechanisms. These biological and physical mixing processes may blur the sediment record

for paleolimnological interpretation. Advective flow may aid in upward transport in hydraulically conductive littoral sediments and act as a transport
mechanism for dissolved and colloidal species. Chemical and biochemical
processes such as redox shifts or methylation can cause Hg to become more
mobile. Diffusive Hg fluxes from sediments may influence Hg concentrations
in the water column near sediments. The importance of individual processes
in Hg transport are difficult to separate, but major processes were identified
in this investigation.

Sedimentation of Particles.   Gross deposition of particulate matter as
measured by sedimentation traps represents the settling of both allochthonous and autochthonous phases. Algal cells, zooplankton fecal material, inorganic phases, and other detrital material may all be present in trap material.
Conceptually, the flux of material collected in traps is considered gross
sedimentation, and the flux of material incorporated into the permanent
sediments is net sedimentation. We assume that most material collected in
sediment traps has fallen to the sediment–water interface. Traps were positioned 1 m above the sediment surface, so additional degradation may have
occurred in the water column below this point. However, calculated settling
velocities for nonturbulent conditions (53) suggest that the particles collected
in sediment traps will spend minimal additional time in the water column
below the trap. On the basis of these assumptions, the difference between
our calculated gross and net sedimentation is the amount recycled back into
the water column from recently deposited sediments.

Resuspension of bottom sediments presents a potential problem for flux
estimates. However, our results suggest minimal resuspension during stratification. As a part of a separate study, Hurley (unpublished data) measured
pigment fluxes to the sediment surface. Sediment trap material was dominated by chlorophyll *a* and pheophorbide *a* (a grazing indicator). Surface
sediments, however, were dominated by pheophytin *a*, a relatively stable
chlorophyll degradation product. The lack of any substantial amounts of
pheophytin in trap material suggested that if resuspension of particulates
from the surface sediment was important, it was probably minimal.

As with sediment trap studies on other lakes in the region (57), sedimentation trends in Little Rock Lake probably result from autochthonous
(in-lake) production settling from the water column. Mass deposition rates
ranged from about 0.25 to 2 g/m^2 per day and exhibited strong seasonal
variability (Figure 7). Organic matter deposition dominated; total C levels
were 310–460 mg/g, which corresponded to approximately 62–92% organic
matter. Pigment composition (J. P. Hurley, unpublished data) and microscopic identifications also supported our assumption of a predominance of
autochthonous carbon in sediment traps.

Gross deposition of Hg ranged from about 100 to 400 ng/m^2 per day
during the ice-free period. Fluxes were low in early spring, peaked in mid-

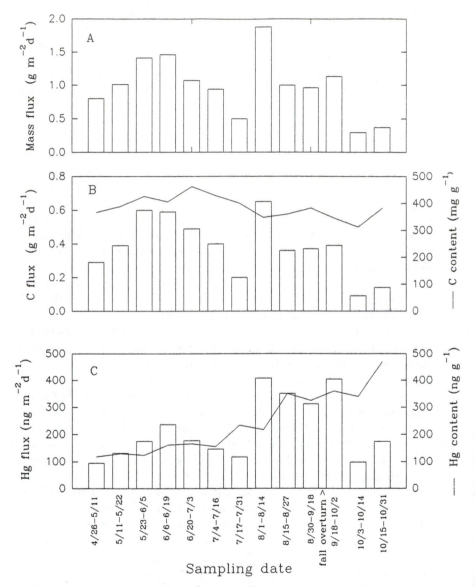

Figure 7. Deposition to the sediment surface of Little Rock Lake in 1989 as measured by sedimentation traps. A, mass flux; B, carbon flux: bars represent fluxes, lines are particle concentrations of carbon (percent); and C, Hg flux: bars represent flux, lines are particle concentrations of Hg in nanograms per gram. (Adapted with permission from reference 21. Copyright 1991 D. Reidel Publishing Co.)

June, decreased by late July, and then reached maximal levels during late stratification. These fluxes somewhat tracked mass and carbon deposition trends.

Particle-bound Hg concentrations of sediment trap material exhibited strong seasonal response and accounted for the differences between the Hg flux and mass and carbon fluxes late in the summer. Particle-bound Hg_T content in spring and early summer was below 200 ng/g, but during late summer stratification it reached levels between 200 and 400 ng/g. Levels were highest following breakdown of thermal stratification and remained high throughout the fall (>350 ng/g). The elevated Hg_T levels after overturn most likely represented a shift from dissolved to particle-bound Hg.

We assume that sediment resuspension was minimal after destratification of the water column. Our reasoning is as follows. First, temperature profiles in 1989 suggested that overturn occurred on or about September 15 (Mercury Cycling in Temperate Lakes and Long Term Ecological Research—Northern Temperate Lakes proprietary databases). The first post-overturn trap period (September 18–October 2) exhibited high Hg fluxes, yet was followed by two periods of low deposition. Second, particle Hg content (nanograms per gram) reached the highest levels during mixing. Because Hg levels in trap material are 2–4 times greater than sediment values (Figures 3 and 7), it can be assumed that sediment was not the source of the particles. Third, pheophytin was not important in September and October sediment traps. These independent observations strongly suggest that particles produced within the lake both during stratification and after thermal breakdown were settling to the sediment–water interface during late fall.

Net sedimentation is defined as the flux of material incorporated into the permanent sediment record. ^{210}Pb and ^{137}Cs geochronologies indicate a mass sedimentation rate of 103 g/m^2 per year for profundal sediments in Little Rock Lake. By using the mean Hg concentration (118 ng/g) in the top 1-cm slice of our bulk sediment profile, we estimated an annual net sedimentation of 12 μg of Hg_T/m^2 per year. This net accumulation rate is similar to the calculated atmospheric input rate of about 10 μg/m^2 per year (18, 19). Additionally, gross deposition rates (from sediment traps) exceeded these estimates by about a factor of 3; this rate suggests substantial internal recycling of material deposited at the sediment–water interface in this lake.

Advection and Diffusion of Hg from Sediments.   In lakes such as Pallette Lake, which receive continuous groundwater inflow and outflow, advection and diffusion of chemical constituents can be important for littoral sediments. To assess the importance of advective and diffusive Hg fluxes, information on background groundwater concentrations, levels of Hg at the sediment–water interface, and flow rates of water must be determined.

Although Hg_T concentrations in ambient groundwater were about 2–4 ng/L, the actual Hg_T levels in groundwaters that discharge to the lake averaged about 12 ng/L and at times were as high as 20 ng/L (Figures 3 and 4). This fact is important to consider for loading calculations. Advective groundwater Hg transport was estimated as the product of the average Hg concentration in near-surface pore waters and the estimated groundwater flow rate. Groundwater flow rates for Pallette Lake were estimated by using four methods: annual temperature profiles in groundwater inflow areas (58), infiltrometer tests (W. Rose, U.S. Geological Survey, personal communication), stable-isotope mass-balance calculations, and solute mass-balance calculations (59) yielding an average annual groundwater inflow rate of 5.5×10^4 m^3/yr. On the basis of the average near-surface Hg pore-water concentration (12 ng/L), this volume of groundwater discharge contributes about 0.7 g of Hg per year to Pallette Lake. About 17–33% (0.1–0.2 g) of this mercury load is "new" mercury from the discharge of ambient ground-water with an average concentration of 2–4 ng/L, whereas 67–83% (0.5–0.6 g) is recycled mercury from the near-surface pore waters. Thus, most of the mercury that discharges to the lake by groundwater flow is derived from near the interface and is effectively forced into the lake by advective discharge.

Groundwater Hg removal rates by seepage of water from the lake were calculated from the residual in the hydrologic budget and the pore-water Hg concentrations from the outflow zone. Because Pallette Lake experienced no change in net storage from ice-out to ice-on during 1990, groundwater outflow was estimated as the difference between precipitation (0.72 m) plus groundwater inflow (0.9 m/year) minus evaporation (about 0.5 m/year) or about 0.3 m/year (W. Rose, U.S. Geological Survey, personal communication). This area-averaged depth of water, when multiplied by the area of the lake, gave a volumetric flow rate of 2.25×10^5 m^3/year. Figure 4 indicates that the Hg concentration of groundwater outflow was about 7 ng/L, which gave a Hg removal rate of about 1.6 g/year. Therefore, the groundwater system around Pallette Lake was acting as a net Hg sink of about 0.9 g/year during the study period.

The relative importance of diffusion (F) can be examined with an advection–diffusion equation.

$$F = -D\frac{dC}{dZ} + VC \qquad (2)$$

where D (centimeters per second) is the diffusion coefficient; C is the solute concentration (micromoles per liter); dC is the change in solute concentration from the interface to the depth of the flux plane, dZ (centimeters); and V is the downward groundwater velocity (centimeters per second). In groundwater outflow areas where advection is downward and the diffusion gradient is upward toward the lake, this equation may be set to zero and solved for V to assess the minimum advective velocity necessary to overcome back diffusion. If a diffusion coefficient of 1×10^{-6} cm^2/s is assumed (corrected for tortuosity and temperature) (53, 55), if an average concentration gradient of 58 ng/L is used over a distance of 1 cm (the highest concentration in the uppermost pore-water sample in the outflow profiles; Figure 4), and if the mercury concentration in lake water is assumed to be 1 ng/L, this analysis shows that a minimum downward advection rate of about 18 m/year would be necessary to counteract the diffusive flux.

The average downward advection rate at Pallette Lake (intergranular groundwater velocity) is estimated by using the volume of groundwater discharged, the estimated area through which outflow occurs, and a porosity of 0.3. By using these data, we determined this rate to be 10 m/year. Considering the uncertainty in the parameters in eq 2 and the heterogeneous nature of groundwater flow, it is unclear whether a diffusive back-flux of mercury to the lake is occurring in groundwater-outflow areas. If, however, it is assumed that diffusion is occurring and the mitigating effects of downward advection are ignored, a maximum bound on this flux can be estimated. Under these assumed conditions, a diffusive flux would amount to about 5.5×10^{-8} ng/cm^2 per second or 1.2 g/year (assumed outflow area = 7.0×10^{4} m^2). Obviously, diffusive back-flux is a potentially important mercury-cycling mechanism.

Profundal Diffusion. In lakes that develop strong thermal stratification, our ability to estimate profundal diffusion rates for chemical constituents is enhanced. Such is the case for Hg in Little Rock Lake. Detailed time-incremented Hg measurements and pore-water profiles were needed to calculate the importance of this flux. We chose the time periods from June 14 to July 11 and from July 11 to August 22 to describe the importance of various fluxes and to compare those fluxes to observed trends in hypolimnetic enrichment (Figure 2). The total mass of Hg in the hypolimnion increased by 47 mg during the June–July period and by 148 mg during July–August. During those periods, gross deposition of particle-bound Hg (calculated from sediment traps) exceeded the observed buildup and could potentially account for the observed increases. However, this calculation assumes that all of the deposited particles are rapidly remineralized and released in dissolved forms at the interface, without accumulation in bottom sediments. Because total remineralization is unlikely (net sedimentation is evident in Figure 3), we calculated the potential flux of pore-water Hg from sediments in accounting for the observed hypolimnetic buildup.

The subsurface maximum in pore-water Hg$_T$ (Figure 3) suggested that diffusion from the profundal sediments to the overlying water column could be important. Fickian diffusive flux calculations (eq 2) were used to estimate Hg loading from pore waters. Diffusion coefficients for mercury in pore waters were not available. However, free-water diffusion coefficients for monovalent anions (*see* Table I) averaged about 5×10^{-6} cm^2/s (53, 55) and were probably applicable to these sediments, which were 95% water. In addition, the diffusivity constant is an average value and there are actually many species capable of diffusing independently (cf. Table I). The concentration gradient of 35 ng/L at 4 cm to 15 ng/L at the interface was used to derive a flux rate of 1.4×10^{-8} ng/cm^2 per second. This flux was then multiplied by the profundal area of the hypolimnion (1.86 ha). On the basis of these calculations, we estimated a profundal diffusion rate of 0.23 mg of Hg per day.

Comparison of profundal diffusion rates with observed increases in the hypolimnion (Table III) indicated that pore-water diffusion calculated from these profiles was probably not an important transport mechanism for Hg in this seepage lake. For the June–July period, pore-water diffusion accounted for only 13% of the hypolimnetic increase. For the July–August interval, pore-water diffusion could account for only 7% of the observed increase. Therefore, we can assume that the buildup in the hypolimnion is more likely a result of redissolution of recently fallen particulate matter at the sediment surface than of direct pore-water diffusion. Our present sampling scheme (2-cm intervals) precludes evaluation of dissolution in the uppermost sediments and would require much more detail (<1 cm) in the sediment–water interfacial zone.

**Table III. Hypolimnetic Hg, Depositional Fluxes,
and Calculated Diffusion from Bottom Sediments
in Little Rock Lake in 1989**

Parameter	June 14–July 11 (27 days)	July 11–August 22 (42 days)
Hypolimnetic Change (mg)	+47	+148
Gross Deposition from Epilimnion (mg)	89	206
Net Accumulated in Sediments (mg)	15	23
Calculated Profundal Diffusion (mg)	6	10

Other Mechanisms. We acknowledge that numerous other processes (such as detritivore activity and microbial transformations) may affect transport across the interface, but our techniques could evaluate only the processes previously discussed. An obvious area for future research is microbial degradation and release of methylmercury from sediments. The assessment of factors regulating this transformation and release is essential for predictive models on Hg transport and bioaccumulation.

Conclusions

Knowledge of cycling processes and transport mechanisms in the region of the sediment–water interface is essential for understanding the behavior of Hg in lakes. Although accumulation of Hg in bottom sediments roughly balances atmospheric inputs and sedimentary Hg profiles showed increasing concentrations at the tops of cores (correlated with industrialization and other anthropogenic inputs), significant recycling of Hg occurred prior to incorporation in the sedimentary record. Our flux calculations revealed that gross sedimentation of Hg exceeded net accumulation by about a factor of 3 in Little Rock Lake. This inequality suggested substantial remineralization and redeposition near the sediment–water interface.

Processes and mechanisms responsible for recycling at the sediment–water interface cannot be explained by a single process, but are most likely a combination of many biogeochemical processes. Although pore-water Hg_T concentrations were higher than in lake waters, direct release of pore waters below about 2 cm could not totally account for the observed buildup in the hypolimnion of Little Rock Lake. Remineralization of recently deposited biogenic particulate matter and release of particle-bound Hg from this source most likely accounted for the observed water-column buildup.

Groundwater inflow and outflow can also affect Hg cycling in seepage lakes. Our results from Pallette Lake indicated that advective flow across the sediment–water interface is important in Hg transport. Ambient groundwater (derived solely from precipitation) Hg_T concentrations ranged from about 2 to 4 ng/L; groundwater from outflow zones was about a factor of 2 higher. Flux estimates for Pallette Lake indicated that groundwater acts as a net sink for Hg_T, as more Hg is removed from the lake with outflow than is delivered to the lake via inflow.

Equilibrium calculations suggested that Hg complexation varies greatly among redox and pH levels typical of the regions of lakes sampled during this study. In an oxic lake, pore water, and groundwater, Hg complexation with organic matter most likely dominates. Under anoxic conditions in the hypolimnion and pore waters, Hg most likely forms soluble bisulfide and polysulfide complexes.

Acknowledgments

We thank S. Claas, S. Greb, T. Hoffman, and K. Morrison for their assistance in field sampling and sample processing. N. Bloom conducted the analyses of groundwater in the early phase of the study. We thank C. Watras, D. Engstrom, R. Striegl, T. McConnaughey, and two anonymous reviewers of this manuscript. This research was supported by the Electric Power Research Institute (D. Porcella, project manager), the Wisconsin Department of Natural Resources, and the U.S. Geological Survey.

References

1. Mortimer, C. H. *J. Ecol.* **1941**, *29*, 280–329.
2. Davison, W.; Woof, C. *Water Res.* **1984**, *18*, 727–734.
3. Davison, W. In *Chemical Processes in Lakes;* Stumm, W., Ed.; Wiley: New York, 1985; p 31–35.
4. Rada, R. G.; Wiener, J. G.; Winfrey, M. R; Powell, D. R. *Arch. Environ. Contam. Toxicol.* **1989**, *18*, 175–181.
5. Meger, S. A. *Water Air Soil Pollut.* **1986**, *50*, 411–419.
6. Swain, E. B.; Engstrom, D. E.; Brigham, M. E.; Henning, T. A.; Brezonik, P. L. *Science (Washington, D.C.)* **1992**, *257*, 785–787.
7. Engstrom, D. E.; Swain, E. B.; Brigham, M. E.; Henning, T. A.; Brezonik, P. L. In *Environmental Chemistry of Lakes and Reservoirs;* Baker, L. A., Ed.; Advances in Chemistry 237; American Chemical Society: Washington, DC, 1993; Chapter 13.
8. Patterson, C. C.; Settle, D. M. In *Accuracy in Trace Analysis: Sampling, Sample Handling, and Analysis;* LaFleur, P. D., Ed.; U.S. National Bureau of Standards Special Publication 422, 1976; pp 321–351.
9. Fitzgerald, W. F.; Gill, G. A. *Anal. Chem.* **1979**, *51*, 1714–1720.
10. Bloom, N. S.; Crecelius, E. A. *Mar. Chem.* **1983**, *14*, 49–59.
11. Bloom, N. S.; Effler, S. W. *Water Air Soil Pollut.* **1990**, *53*, 251–265.
12. Gill, G. A.; Bruland K. W. *Environ. Sci. Technol.* **1990**, *24*, 1392–1400.
13. Evans, R. D. *Arch. Contam. Toxicol.* **1986**, *15*, 502–512.
14. Watras, C. J.; Andren, A. W.; Bloom, N. S.; Fitzgerald, W. F.; Hurley, J. P.; Krabbenhoft, D. P.; Rada, R. G.; Wiener, J. G. *Verh. Int. Ver. Theor. Angew. Limnol.* **1991**, *24*, 2119.
15. Lathrop, R. C.; Rasmussen P. W.; Knauer, D. R. *Water Air Soil Pollut.* **1991**, *56*, 295–308.
16. *Assessment of Mercury Contamination in Selected Minnesota Lakes and Streams;* Minnesota Pollution Control Agency: St. Paul, MN, 1989.
17. Bloom, N. S. *Can. J. Fish. Aquat. Sci.* **1989**, *46*, 1131–1140.
18. Fitzgerald, W. F.; Watras C. J. *Sci Total Environ.* **1989**, *87/88*, 223–232.
19. Wiener, J. G.; Fitzgerald, W. F.; Watras C. J.; Rada, R. G. *Environ. Toxicol. Chem.* **1990**, *9*, 909–918.
20. Bloom, N. S.; Watras, C. J.; Hurley J. P. *Water Air Soil Pollut.* **1991**, *56*, 477–491.
21. Hurley, J. P.; Watras C. J.; Bloom, N. S. *Water Air Soil Pollut.* **1991**, *56*, 543–551.
22. Watras, C. J.; Frost, T. M. *Arch. Environ. Contam. Toxicol.* **1989**, *18*, 157–165.
23. Watras, C. J.; Bloom, N. S. *Limnol. Oceanogr.* **1992**, *37*, 1313–1318.
24. Gill, G. A.; Fitzgerald, W. F. *Deep-Sea Res.* **1985**, *32*, 287–297.
25. Gardner, W. D. *J. Mar. Res.* **1980**, *38*, 41–52.
26. Shafer, M. M. PhD. Thesis. University of Wisconsin—Madison, 1988.
27. Freeze, R. A.; Cherry J. A. *Groundwater;* Prentice-Hall: Englewood Cliffs, NJ, 1979.
28. McBride, M. F.; Pfannkuch, H. O. *J. Res. U.S. Geol. Surv.* **1975**, *3*(5), 505–512.
29. Wood, W. W. *U.S. Geological Survey Techniques of Water-Resources Investigations;* Book 1, 1976; Chapter D2, 24 pp.
30. Driscoll, F. G. *Groundwater and Wells;* Johnson Division: St. Paul, MN, 1986; 1089 p.
31. Dominico, P. A.; Schwartz, F. W. *Physical and Chemical Hydrogeology;* John Wiley and Sons: New York, 1990; 824 p.
32. Jahnke, R. *Limnol. Oceanogr.* **1988**, *33*, 483–487.
33. Bloom, N. S.; Fitzgerald, W. F. *Anal. Chim. Acta* **1988**, *208*, 151–161.
34. Robbins, J. A.; Edgington, D. N. *Geochim. Cosmochim. Acta*, **1975**, *39*, 285–304.

35. Rada, R. G.; Powell, D. E.; Wiener, J. G. *Can. J. Fish. Aquat. Sci.*, in press.
36. Husar, R. B. In *Acid Deposition: Long Term Trends*; National Academy of Sciences: Washington, DC, 1986; pp 48–92.
37. Andren, A. W.; Nriagu, J. In *Biogeochemistry of Mercury in the Environment*; Nriagu, J., Ed.; Elsevier: Amsterdam, Netherlands; 1979.
38. Babiarz, C. L.; Andren, A. W. *Mercury in Temperate Lakes*; Annual Report, Electric Power Research Institute: Palo Alto, CA, 1990.
39. Lindquist, O.; Johansson, K.; Aastrup, M.; Andersson, A.; Bringmark, L.; Housenius, G.; Håkanson, L.; Iverfeldt, Å.; Meili, M.; Timm, B. *Water Air Soil Pollut.* 1991, 55, 73–100
40. Benes, P.; Havlik, B. In *Biogeochemistry of Mercury in the Environment*; Nriagu, J., Ed.; Elsevier, 1979.
41. Gardner, L. R. *Geochim. Cosmochim. Acta* 1974, 38, 1297–1302.
42. Andren A. W.; Harriss, R. C. *Geochim. Cosmochim. Acta*, 1975, 39, 1253–1257
43. Mierle, G.; Ingram, R. *Water Air Soil Pollut.* 1991, 56, 349–357.
44. Cabaniss, S. E. *Environ. Sci. Technol.* 1987, 21, 209–210.
45. Dyrssen D. *Mar. Chem.* 1988, 24, 143–153.
46. Dyrssen, D.; Wedborg, M. *Water Air Soil Pollut.* 1991, 56, 507–519.
47. Dyrssen, D.; Wedborg, M. *Anal. Chim. Acta* 1986, 180, 473–478.
48. Sillen, L. G.; Martell, A. E. *Stability Constants of Metal–Ion Complexes*; Special Publ. No. 17; Chemical Society: London, England, 1964.
49. Stumm, W.; Morgan, J. *J. Aquatic Chemistry*, 2nd ed.; Wiley: New York, 1981.
50. Hershey, J. P.; Pease, T.; Millero, F. J. *Geochim. Cosmochim. Acta* 1988, 52, 2047–2051.
51. Licht, S.; Forouzan, F.; Longo, K. *Anal. Chem.* 1990, 62, 1356–1360.
52. Schoonen, M. A. A.; Barnes, H. L. *Geochim. Cosmochim. Acta* 1988, 52, 649–654.
53. Lerman, A. *Geochemical Processes Water and Sediment Environments*; Wiley: New York, 1979; 481 pp.
54. O'Connor, D. J.; Conolly, J. P. *Water Res.* 1980, 14, 1517–1523.
55. Berner, R. *Early Diagenesis: A Theoretical Approach*; Princeton University Press: Princeton, NJ, 1980; 241 pp.
56. Sigg, L. In *Chemical Processes in Lakes*; Stumm, W., Ed.; Wiley: New York, 1985; pp 283–310.
57. Hurley, J. P. M.S. Thesis, University of Wisconsin—Madison.
58. Lapham, W. W. U.S. Geological Survey Water Supply Paper No. 2337, 1989.

RECEIVED for review October 23, 1991. ACCEPTED revised manuscript September 22, 1992.

Contaminant Mobilization Resulting from Redox Pumping in a Metal-Contaminated River–Reservoir System

Johnnie N. Moore

Department of Geology, University of Montana, Missoula, MT 59812

Both large- and small-scale extraction of metals in the northern Rocky Mountains of Montana has left a legacy of contaminated soil and river and reservoir sediments. The important processes that affect the fate of metals and metalloids involve the often-complicated oxidation–reduction reactions of sulfides and oxygen. Combined with dissolution–solution reactions, such redox reactions result in transferring contaminants from contaminated floodplain sediments to rivers in particulate and solute phases. Reservoirs intercept these contaminants and store them as a major secondary source of contamination. The seasonal change in oxidation state in reservoirs releases some components while fixing others—here the process is termed redox pumping—and leads to tertiary contamination of groundwater adjacent to reservoirs. The effect of this complex interplay between dissolution and redox reactions has extended contamination over 500 km from the primary sources.

EXTRACTION OF METALS FOR 125 YEARS generated extensive hazardous waste in the area around the Montana Rocky Mountains. About 5–7% of all rivers in Montana—more than 2000 km of streams—are contaminated by mining wastes at a level that impairs beneficial use of water (Montana Department of Health and Environmental Sciences, personal communication). Central to this contaminant burden is the wastes generated by mining and smelting in the Butte–Anaconda area, where more than 1 billion metric tons (MT) of ore and waste rock were produced. The wastes were discarded into the headwaters of the Clark Fork River, the largest tributary of the Columbia River (1).

Extraction of metals from Butte, the major mining area, began in 1864. By 1896 more than 4500 MT of ore was processed every day, with the wastes discarded directly on the land surface or into streams draining the area. At the turn of the century one of the largest smelters in the world had been constructed in Anaconda, and within 15 years it was processing more than 11,500 MT of ore per day. Depressed copper prices forced closure of that smelter in 1980, and large-scale mining ended 3 years later. After a brief hiatus, mining has resumed on a reduced scale.

Called the "richest hill on earth", Butte produced more metals than either the Leadville district in Colorado or the Comstock Lode in Nevada (2) and left a legacy of equally grand contamination. Mining and smelting operations left behind extensive waste deposits. Possibly as much as 1500 km^2 of land was contaminated within the Clark Fork River basin (1), including 35 km^2 of tailings ponds; 300 km^2 of soil contaminated by air pollution; tailings deposits along hundreds of kilometers of riverine habitat; more than 50 km^2 of contaminated, once-productive agricultural land; alluvial and bedrock aquifers contaminated with metals, sulfate, and arsenic; and downstream reservoirs containing thousands of metric tons of metal-contaminated sediment. This basin encompasses the largest complex of Superfund sites in the country, where fluvial and geochemical processes transport contaminants hundreds of kilometers from the source and affect the entire Clark Fork River system (1).

Contamination Process

Early in the history of mining and smelting in the Clark Fork basin, reservoirs were the first recipients of contaminants. These reservoirs were built to retain milling wastes for secondary recovery of metals, to limit effluent moving downstream into the Clark Fork River, and to serve as hydroelectric impoundments. They now make up a vast array of tailings ponds in the headwaters of the Clark Fork River and large downstream lakes that act as sinks and sources for contaminants to surface and groundwater in the basin.

The first hydroelectric reservoir (Milltown Reservoir) was built in 1906–1907 at a site approximately 200 river km downstream from the major mining and smelting operations supplying metal-contaminated sediment to the river system (Figure 1). This most-upstream hydroelectric reservoir was the primary catch basin for wastes transported by the river before tailings ponds were built in the headwaters in the mid-1900s. Continuing downstream, three additional reservoirs were built at 452, 516, and 556 km in

Figure 1. Map of the Clark Fork River basin showing the location of the operable units in the main designated Superfund sites and other features mentioned in the text.

1915, 1959, and 1952, respectively. These impoundments also have trapped wastes from upstream, but to a lesser extent than Milltown Reservoir (*1, 3*).

Metal Sulfide Wastes.   The contamination processes associated with this river–reservoir system are controlled by the characteristics of metal sulfide wastes left within the basin. High-grade veins in Butte underground mines contained a variety of metal sulfides, including chalcocite (Cu_2S), bornite (Cu_5FeS_4), chalcopyrite ($CuFeS_2$), enargite (Cu_3AsS_4), tennantite–tetrahedrite ($Cu_{12}[As, Sb]_4S_{13}$), sphalerite (ZnS), pyrite (FeS_2), acanthite (Ag_2S), galena (PbS), arsenopyrite (FeAsS), and greenockite (CdS).

Lower grade ore and waste rock contained a significant amount of pyrite (FeS_2), as well as other metal sulfide minerals.

Wastes generated by mining and milling released these metal sulfides as particulate substances (1). Mixed with uncontaminated sediment in the river system, these particulate wastes moved hundreds of kilometers downstream from their original source (4–7). Channel, floodplains, and reservoir sediments throughout the river now contain much of this contamination.

Andrews (4) showed that fine-grained particulate contaminants (arsenic, cadmium, copper, lead, and zinc in sediment less than 0.016 mm in diameter) decreased downstream from the source and that the distribution could be explained solely by mixing of mill tailings with uncontaminated floodplain sediment. Work by Brook and Moore (5) and Moore et al. (6) showed that the sediments were enriched in arsenic, cadmium, copper, manganese, lead, and zinc. They also found that the contaminants were carried mostly in the reducible and oxidizable phases (operationally defined).

In size-fractionated samples, they found that concentrations of some metals generally increased with decreasing particle size. However, more upstream samples (nearer to the source) contained anomalously high concentrations in the coarsest fractions. Because Clark Fork River sediment is predominantly coarse-grained, coarse fractions significantly add to the bulk contaminant content of the system.

Distribution of Contaminants.   Most recently, Axtmann and Luoma (7) showed that metal contaminants in bed sediments decreased in an exponential trend away from the source and predicted that elevated metal concentrations should occur more than 550 km downstream from the contaminant source. They ascribed the downstream trend to dilution from uncontaminated sediment mixed with mill tailings. Johns and Moore (3, 8) found that contaminants had moved through tailings ponds and upstream reservoirs to accumulate in downstream reservoirs; copper and zinc were enriched over background tributaries in reservoirs more than 556 km from the source. Metal concentrations of surface sediment from these reservoirs lay on the downstream exponential trend of bed sediment in the river (Figure 2). These data showed that fine-grained surficial sediments in the river basin were highly contaminated with particulate wastes from the mining and milling operations upstream, and that reservoirs now actively bypass contaminated sediment downstream.

This distribution has significant effects on the storage and remobilization of contaminants from reservoirs in the drainage. The farthest upstream reservoir in the Clark Fork drainage (Milltown) is nearly 200 km from the sources of contamination (Figure 2), yet it contains sediment contaminants many times over background values. Sediments in the reservoir contain significant concentrations (depth-averaged values) of arsenic (32 times background values), manganese (7 times), copper (62 times), zinc (67 times), lead

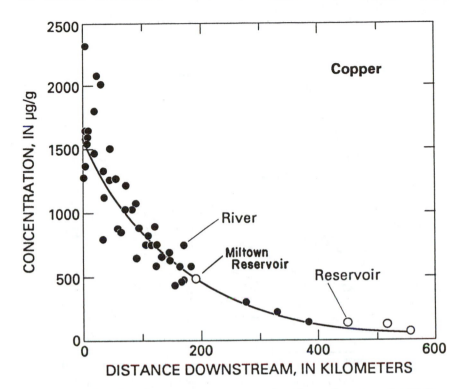

Figure 2. Plot of total copper concentration of surface sediments in the Clark Fork River versus distance downstream from the major sources of contaminants. Sediment came from the bed of the Clark Fork River channel and surface sediment from reservoirs. (Data are taken from references 5 and 7.)

(11 times), and cadmium (37 times). Tens of thousands of metric tons of these metals are stored in the reservoir sediment (9, 10).

The reservoir is efficient at trapping coarse-grained sediment (bed load) and is filled with a complex assemblage of sand and mud with abundant organic interlayers (9, 10). But fine-grained sediment (suspended load) now flows through the nearly filled reservoir during spring runoff when the river contaminant burden is greatest (9). This bypassing has led to the redistribution of fine-grained sediment to the downstream reservoirs; the farther from the source, the more the contaminant burden is transferred to the fine fraction of the sediment. This situation results in Milltown Reservoir containing the most complex distribution of contaminants in the reservoirs of the drainage basin.

Milltown Reservoir as a Model System

Milltown reservoir impounds about 180 acres of water at the confluence of the Blackfoot and Clark Fork rivers (Figure 3). The Clark Fork River at that

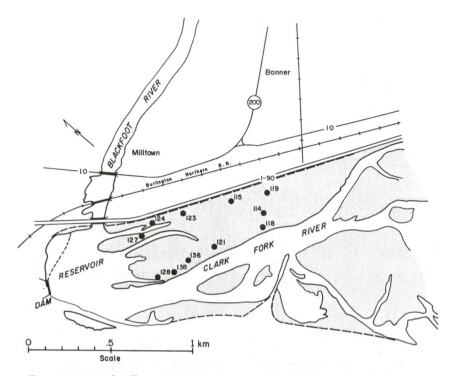

Figure 3. Map of Milltown Reservoir showing main tributaries and the location of some of the wells used for hydrogeologic and geochemical monitoring. (Data are taken from references 9 and 12.)

site drains about 15,000 km². When the dam was built, the sediment transport of the Blackfoot and Clark Fork rivers was disrupted. Both bedload and suspended load previously transported westward to the Columbia River were impounded in the reservoir, along with wastes from upstream. This sedimentation has nearly filled the reservoir, so that most of the present suspended load of the two rivers passes through the system.

Data collected by the Montana Power Company (the operators of the dam) show that the reservoir is extremely inefficient at trapping fine-grained sediment. During drawdowns of the reservoir stage, sediment is removed from the reservoir and transported downstream; during 1 day of drawdown in 1980 approximately 12,000 MT of sediment was removed. The reservoir contains about 1.9 million m³ (or approximately 3.8 million MT) of sediment, at a maximum thickness of about 8 m, and has accumulated contaminants for more than 80 years. The complex interplay of channels, floodplain, and thalweg environments of deposition as the reservoir filled has created a complex, interdigitizing mosaic of sediment types and grain sizes. Present main channels contain the coarsest surface sediment, and the swampy thalwegs and floodplains are rich in organic substances and mud. Stratigraphic cores of the reservoir sediment show a complex vertical intermixing of sandy, muddy, and organic-rich sediment throughout the reservoir (9).

Sediment Contamination.   Contaminated sediment in the reservoir was first identified by Bailey (*11*), when she found that the reservoir contained high concentrations of Cu and Zn. Five years later, in November 1981, the Montana Department of Health and Environmental Sciences determined that four community wells adjacent to the reservoir contained arsenic at levels above the drinking water standards recommended by the Environmental Protection Agency (EPA). Work funded by the EPA Superfund (9) subsequently identified the reservoir sediments impounded above the original alluvial valley aquifer (Figure 4) as the source of this contamination. Strong vertical hydraulic gradients in the reservoir drive groundwater in the contaminated sediments into the underlying, coarse-grained, alluvial aquifer. This system has resulted in the contamination of the domestic water supply in adjacent Milltown, Montana. Because the reservoir has a small storage capacity, significant stage fluctuations are common. This situation results in sediments being inundated for much of the year and exposed for a few weeks at most each year, depending on flow conditions and maintenance needs.

Because the reservoir is almost completely filled with sediment, at low stage only channels are filled with water. At high stage the broad floodplain flat is partially covered by water. During the full stage the groundwater system connects the reservoir sediment to the river through complex flow paths (Figure 5A). At the low stage groundwater flow changes, with some input back into the river channel and continued flow through the sediments into the adjacent alluvial aquifer (Figure 5B). The processes controlling the mobilization of arsenic in this system result from a complex interaction between this groundwater flow system and geochemistry of the contaminated sediments (*10*, *12*). Within this complexity a model can be developed that is generally applicable to reservoirs contaminated by the mining and smelting of base-metal, sulfide-rich ores.

Preliminary work (*10*) on the transition from oxidized surface sediment to reduced subsurface sediment in Milltown Reservoir showed that the redox transition occurs in the upper few tens of centimeters. Strong chemical gradients occur across this boundary. Ferrous iron in sediment pore water (groundwater and vadose water) is commonly below detection in the oxidizing surface zone and increases with depth. Arsenic is also low in pore water of the oxidized zone, but increases across the redox boundary, with As(III) as the dominant oxidation state in the reduced zone. Copper and zinc show the opposite trend, with relatively high concentrations in pore water of the oxidized surface sediment decreasing across the redox boundary.

Moore et al. (*10*) concluded that the formation of diagenetic sulfides provided an important control on metals and arsenic mobility in the sedi-

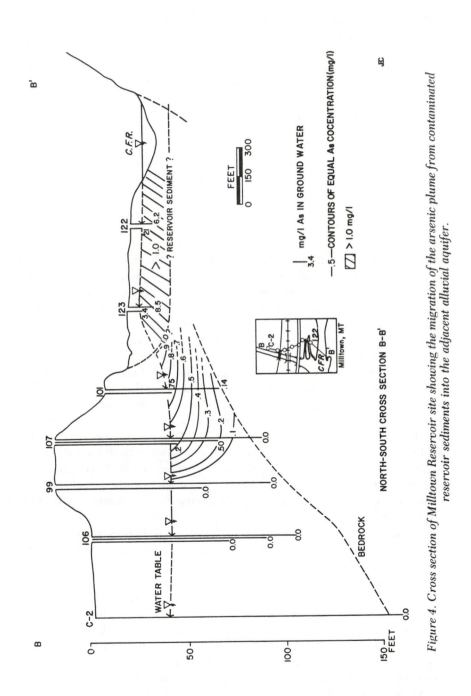

Figure 4. Cross section of Milltown Reservoir site showing the migration of the arsenic plume from contaminated reservoir sediments into the adjacent alluvial aquifer.

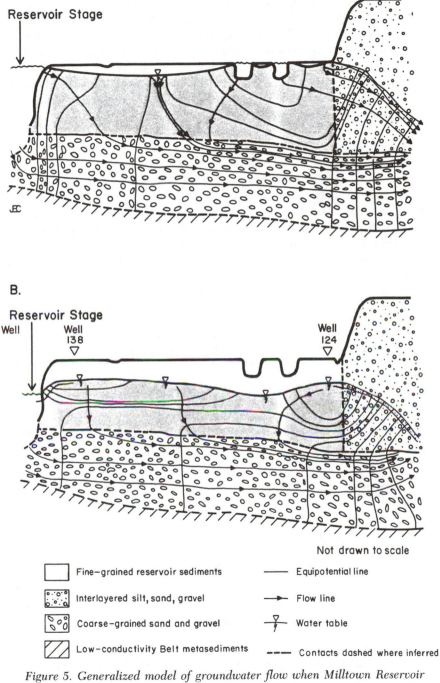

Figure 5. *Generalized model of groundwater flow when Milltown Reservoir is at full stage (A) and at low stage (B). (Modified from reference 12.)*

ment. Arsenic and metals are carried into the reservoir as primary sulfides (likely with oxidized crusts) and as coprecipitates and coatings on other grains (5). When buried and reduced, these oxyhydroxides of iron and manganese dissolve, then arsenic and metal sulfides precipitate. Arsenic is released to the groundwater system dominantly as As(III). The system is controlled by metal sulfide precipitation, which in turn is controlled by sulfate availability. Sulfate is supplied by oxidizing sulfides in the surface sediments. Although this early work examined the vertical changes across the redox boundary, no temporal understanding was gained.

Stage Changes.   An opportunity to study the temporal changes in detail arose in 1986 when the Milltown Dam was severely damaged by high, ice-laden flows from an early February thaw. Repairs to the dam required that the stage be drawn down approximately 2.5 m and held low for several months. In May 1986 the reservoir stage was lowered continuously, and after approximately 100 days it remained at a constant low stage for approximately 230 days (Figure 6). In the summer, the reservoir stage was rapidly elevated onto the newly repaired dam, resubmerging sediments. The stage was held constant at this higher elevation through the winter, until lowered in prelude to the spring runoff.

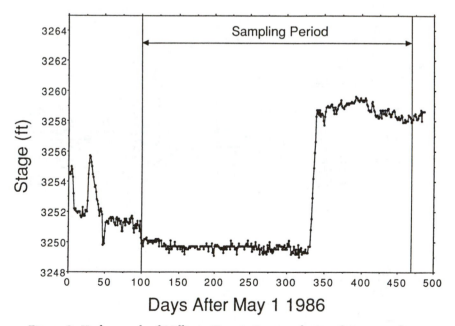

Figure 6. Hydrograph of Milltown Reservoir stage during the temporal geochemistry and hydrogeology study (12). (Data are from Montana Power Company records.)

This rapid change in stage had strong effects on the redox boundary and hence on contaminant mobility. The chemistry and hydrogeology of this system were examined tetraweekly, over a 370-day period from August 1986 to August 1987, to determine the role of redox fluctuations on the mobilization of arsenic into the underlying alluvial aquifer (*12*). Wells were installed in the contaminated sediment at different elevations to monitor the vertical and lateral chemical changes before and during refilling.

The Redox Pump

Mobilization of contaminants in Milltown Reservoir can be explained by a model in which groundwater composition is controlled by successive diagenetic reactions during the transition to and from oxic and anoxic environments as the reservoir stage changes (*12*). Several important reactions govern the mobility of contaminants in this system:

1. those involving metal sulfides;
2. those utilizing organic matter in the sediments, and
3. those producing or dissolving oxide and hydroxide coatings on grains.

All of these reactions are partially or strongly controlled by bacterial interaction. These reactions develop a general vertical zonation of oxic, anoxic sulfidic, and anoxic methanic environments (*13*) within the reservoir sediment (Figure 7) that migrates with the rise and fall of the reservoir stage. This fluctuation develops a "redox pump" that mobilizes contaminants with each successive stage cycle.

At full stage the sediments are saturated and reduced, with a thin oxic zone (variably between 0.1 and 0.5 m) corresponding to the vadose zone and uppermost groundwater (Figure 7A). When the reservoir stage drops and sediments are drained, oxygen-rich vadose water gains access to the underlying reduced sediments. The redox zonation moves downward along with the falling water table (Figure 7B). This system is driven by the mineralogy–chemistry of the contaminated sediment filling the reservoir. Residual sulfides transported downstream and authigenic sulfides formed in place (*10*) undergo oxidation in the vadose zone, releasing metals, sulfate, and hydrogen ions, according to the following general reaction (*14–17*).

$$2FeS_2 + 7O_2 + 2H_2O \longrightarrow 2Fe^{2+} + 4SO_4^{2-} + 4H^+ \qquad (1)$$

In Milltown Reservoir sediments and other metal sulfide contaminated systems (*18*), FeS_2 (pyrite) in this reaction can be replaced by any number of other metal sulfides (for example, arsenopyrite, chalcocite, galena, and

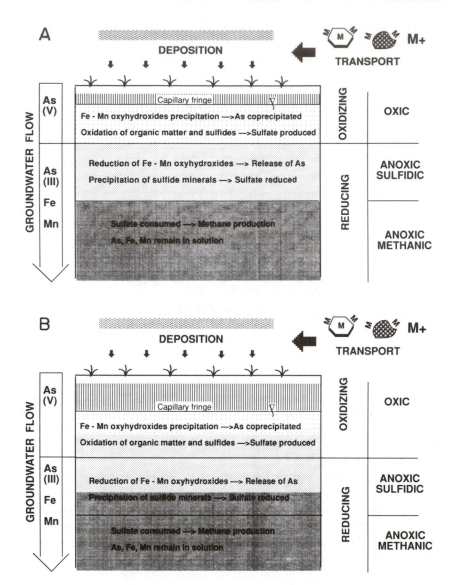

Figure 7. Geochemical model showing development of redox environments as the reservoir stage falls: A, reservoir at full stage; B, low reservoir stage. The scale of this system is highly variable. The oxic zone, depicted by the lightest stippled pattern, is generally 0.1–0.5 m thick at high stage and 0.5–1.5 m thick at low stage. The reducing zone, depicted by the medium stipples, extends to depths of approximately 5–8 m. The presence of the methanic zone (heavy stipples) is probably extremely variable.

sphalerite) so that other metals—As, Cd, Cu, Mn, Pb, and Zn—are released to the groundwater along with Fe^{2+} (*17*, *18*). This general reaction is catalyzed by bacteria (e.g., *Thiobacillus ferrooxidans*) (*16*) and continues as long as the pH remains near neutral (*19*, *20*). This reaction can be seen in the highest elevation wells at Milltown, the water-table wells. While the water table is below these wells, sulfides are oxidized in the vadose zone. As groundwater moves into this zone, sulfate concentration increases dramatically at the water table but remains below detection in deeper wells (Figure 8A). Arsenic, iron, and manganese concentrations also increase rapidly as the water table rises into wells (Figures 8B, 8C, and 8D), while remaining relatively high and constant in deeper zones that remain saturated.

The oxidation of organic matter via reactions similar to eq 2 is associated with metal sulfide oxidation and the release of metals (*21*).

$$(CH_2O)_{106}(NH_3)_{16}(H_3PO_4) + 138O_2 + 122CaCO_3 \longrightarrow$$

$$228HCO_3^- + 16NO_3^- + 122Ca^{2+} + 16H_2O + H_3PO_4 \quad (2)$$

Metals bound to organic substances are released as the organic material is oxidized. The carbonate–bicarbonate component of the system buffers the acid produced from sulfide oxidation reactions. These reactions can be seen readily in Milltown sediments as soon as the water-table wells receive groundwater during the rising stage. Bicarbonate and calcium concentrations rise immediately while pH remains fairly constant (Figures 9A, 9B, and 9C); bicarbonate then increases as calcium decreases. In the continuously saturated zones at depth, bicarbonate and calcium remain relatively unresponsive to water-table fluctuations. Sulfide and organic oxidation processes seem to be intimately joined by bacterially catalyzed reactions. A strong correlation between calcium and sulfate (r^2 = 0.903) suggests that the dissolution–precipitation of calcium sulfates is a cross product of these two redox-controlled reactions.

Oxides and hydroxides (oxyhydroxides) of iron and manganese also form during oxidation of sediment at the low reservoir stage. The upper oxic zone contains mottled, "rusty" sediment after only a few days of exposure. Ferrous iron and manganese released during the oxidation of pyrite and other sulfides (Figures 8C and 8D) will react with oxygen in the vadose zone to form iron and manganese oxyhydroxides via reactions similar to eqs 3 and 4 (*15*, *22*).

$$4Fe^{2+} + 10H_2O + O_2 \longrightarrow 4Fe(OH)_3 + 8H^+ \quad (3)$$

$$2Mn^{2+} + O_2 + 4OH^- \longrightarrow 2MnO_2 + 2H_2O \quad (4)$$

These oxyhydroxide compounds coprecipitate–adsorb metals and arsenic released from the sulfide and organic reactions, fixing some metals in the oxidized zone (*10*, *23*). These reactions can be seen in changes in iron

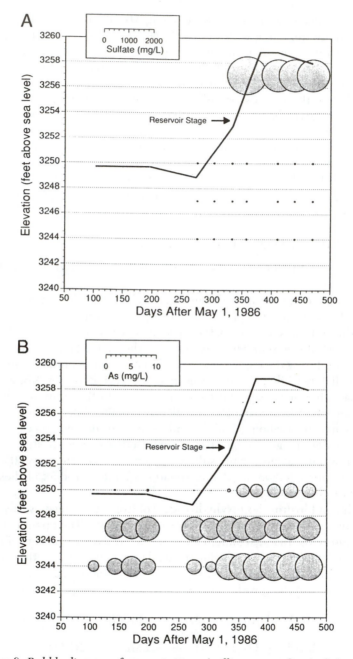

Figure 8. Bubble diagram of concentrations (milligrams per liter) of chemical components in well 138 (Figure 3) showing changes with time in wells of different depths; the upper scale bar (bubble diameter) is the concentration of the species as depicted by bubble, the vertical axis is the elevation of the screened interval of the well, and the horizontal axis is the days since drawdown: A, sulfate; and B, arsenic.

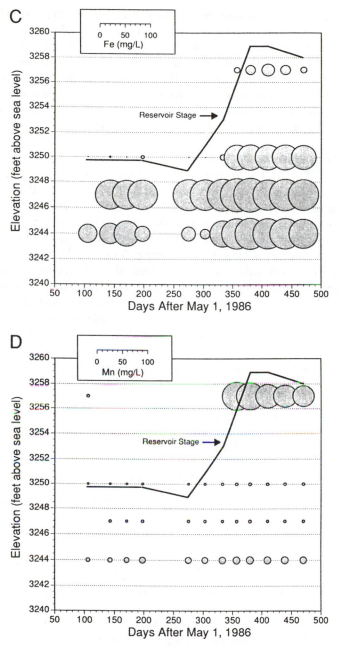

Figure 8. Continued. C, iron; and D, manganese.

and manganese concentration before stage rise (Figures 8C and 8D). Some manganese oxides can also control the oxidation state of arsenic, oxidizing it from III to V (24–26). In deep wells in the reservoir the dominant arsenic species is usually As(III), and total arsenic is strongly controlled by As(III) ($r^2 = 0.910$). However, shortly after the reservoir was filled, wells at several sites showed temporary increases in As(V):As(III) ratios. This response probably results from the oxidation of arsenate in the oxic zone and transport of As(V) into the groundwater system as anoxic groundwater rose into the oxic zone.

All these reactions are representative of the oxic environment described by Berner (13), with one main difference. Large amounts of organic matter remain to power anoxic reactions when the water table rises. Because of the fluctuating reservoir stage, organic material does not oxidize completely. Thus, it remains to power anoxic reactions when the stage rises once again. As the stage rises, the oxic reactions described are replaced by anoxic reactions as oxygen is consumed. In the upper levels of the reducing groundwater system, where sulfate is available from the oxic reactions, anoxic sulfidic reactions dominate (12, 13). At greater depths, where sulfate is used up in reduction reactions, a transition occurs through postoxic to sulfidic methanic environments (12, 13). In one deep well methane was produced in high enough quantities to blow off the slip-on poly(vinyl chloride) well cap.

As the water table rises with increasing reservoir stage, these processes control the upward shift of the redox zonation; anoxic reactions reestablish themselves in the once-oxic zone. Several main processes are important for mobilizing contaminants during this transition.

Manganese oxyhydroxides formed in the vadose zone undergo reduction by organic reduction similar to the general eq 5 (21).

$$(CH_2O)_{106}(NH_3)_{16}(H_3PO_4) + 236MnO_2 + 366Ca^{2+} + 260HCO_3^- \longrightarrow$$
$$366CaCO_3 + 236Mn^{2+} + 8N_2 + H_3PO_4 + 260H_2O \quad (5)$$

Similar reactions occur with iron oxyhydroxides (21).

$$(CH_2O)_{106}(NH_3)_{16}(H_3PO_4) + 424FeOOH + 758Ca^{2+} + 652HCO_3^- \longrightarrow$$
$$758CaCO_3 + 424Fe^{2+} + 16NH_4^- + H_3PO_4 + 636H_2O \quad (6)$$

These processes are catalyzed by bacteria and probably involve both inorganic and organic iron and manganese species (22). They may also be strongly controlled by microbial competition between Fe(III) and sulfate-reducing bacteria (27). Associated with these reduction reactions is the reduction of residual sulfate (produced in the oxic zone by bacterially catalyzed reactions) similar to eq 7 (21).

$$(CH_2O)_{106}(NH_3)_{16}(H_3PO_4) + 53SO_4^{2-} \longrightarrow$$

$$106CO_2 + 16NH_3 + 53S^{2-} + H_3PO_5 + 106H_2O \quad (7)$$

These reduction reactions result in free metal ions and sulfide ions that form diagenetic metal sulfide phases (*10*) via bacterially mediated reactions similar to eq 8 (*16*).

$$Fe^{2+} + H_2S \longrightarrow FeS + 2H^+ \quad (8)$$

Ferrous iron in this reaction can be replaced by any of the metal ions freed in the oxic reactions to form a number of metal sulfides. This system is highly dependent on the availability of sulfate. When sulfate is exhausted by precipitating metal sulfides, processes move into the anoxic methanic state (*13, 27*). Methane production is strongly bacterially mediated following a general reaction similar to eq 9 (*21, 28*).

$$(CH_2O)_{106}(NH_3)_{16}(H_3PO_4) \longrightarrow 53CO_2 + 53CH_4 + 16NH_3 + H_3PO_4 \quad (9)$$

The presence of these reactions in Milltown sediments is indicated by the vertical trends in groundwater chemistry. Even during times of recent stage rise, when the sulfate concentrations are highest in water-table wells, sulfate concentrations decrease rapidly with depth (Figure 8A). Some deeper wells maintain relatively constant chemical composition for iron and arsenic, but show fluctuations in response to stage rise. Other wells show that even at the deepest levels sulfate is transported into the groundwater as the stage rises. The concentration of sodium (Figure 9D), presumably a conservative tracer, shows little response to stage rise or depth. Apparently groundwater flow does not affect concentration, but changes in nonconservative species are the result of rapid diagenetic reactions.

Once the high stage has stabilized, the anoxic environment is rapidly reestablished at the higher elevations in the sedimentary column. When the reservoir is lowered again, the process is repeated. This fluctuation occurred over approximately 2–3 m with the stage changes seen in 1986. During every transition, metals bound to sulfides and organic substances are released in the oxic vadose zone. The strong downward groundwater flow gradient transports mobilized metals and arsenic into the underlying anoxic environments, where they are reprecipitated as sulfides until sulfate is consumed. When the sulfate is gone, methanic reduction takes over. Elements that were not efficiently removed in the sulfidic zone (i.e., did not react quickly enough) are free to move into the underlying alluvial aquifer. Chemistry of wells adjacent to Milltown Reservoir show that these elements are arsenic, iron, and manganese. Although this system is nowhere near equilibrium, control of what escapes the sulfidic zone seems related to the solubility

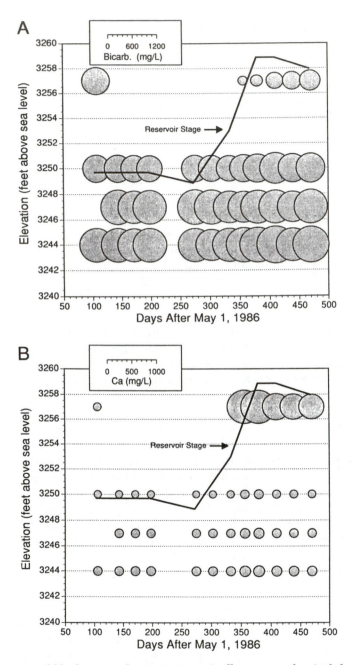

Figure 9. Bubble diagram of concentrations (milligrams per liter) of chemical components and pH in well 138 (Figure 3) showing changes with time in wells of different depths; the upper scale bar (bubble diameter) is the concentration of the species as depicted by bubble, the vertical axis is the elevation of the screened interval of the well, and the horizontal axis is the days since drawdown: A, bicarbonate; and B, calcium.

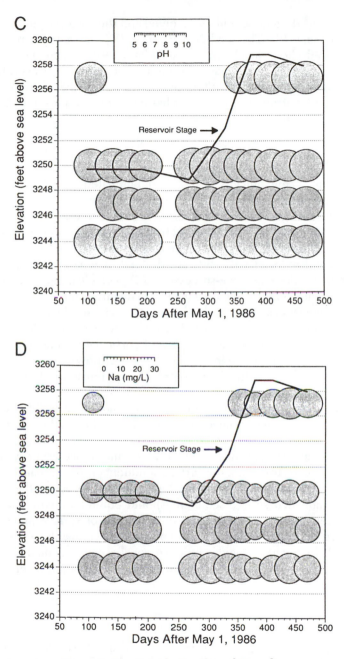

Figure 9. Continued. C, pH; and D, sodium.

products of the diagenetic sulfides involved and the bacterial mediation rates of the reactions (10). The solubility products (K_{sp}) for natural, poorly crystalline, mixed chemical phases are not characterized, and the kinetics of complex, competitive microbial reactions are not known. However, the thermodynamics and kinetics of sulfide precipitates and oxyhydroxide coatings probably have a strong control over what elements are most mobile during redox pumping (10).

Conclusions

Mobilization of contaminants in metal-contaminated reservoirs can be explained by a model in which pore water (i.e., groundwater) composition is controlled by diagenetic reactions involving the successive transition from oxic to anoxic environments as reservoir stage changes seasonally. Although the distribution of processes is not perfectly uniform, when the reservoir stage falls and reduced sediments are drained, a zonation typical of depth succession (13) becomes time-successive (12). The changing reservoir stage acts like a redox pump to displace contaminants from the surface sediments into underlying and adjacent sediments and alluvial aquifers. The fate of this high burden of contaminants to the groundwater system at Milltown Reservoir is not known. However, solute and particulate contaminants are remobilized from the reservoir sediment during draw-down events at the reservoir and added to the overall contaminant burden of the Clark Fork River system and possibly to downstream reservoirs.

Contaminants trapped in reservoirs are not necessarily fixed, even though sediment is not removed from storage. A fluctuating stage may be all that is needed to mobilize contaminants like arsenic into the adjacent groundwater and eventually into the downstream surface-water system. The redox pump tends to cleanse the surface sediment of contaminants as a result of a strong downward flow gradient that is needed to power the pump. In time, contaminated surface sediment should equilibrate with the surface-water system moving contaminants out of reach of surface processes and aquatic organisms, assuming that the source of contaminants to the reservoir is stopped. In any case, the dynamic, seasonal aspects of contaminated reservoirs must be taken into account in any long-term monitoring or successful remediation program.

References

1. Moore, J. N.; Luoma, S. N. *Environ. Sci. Technol.* **1990**, *24*(9), 1278–1285.
2. Lang, W. L. In *The Last Best Place*; Kitteredge, W.; Smith A., Eds.; Mont. Hist. Soc. Press, 1988; p 130.
3. Johns, C.; Moore, J. N. *Trace Metals in Reservoir Sediments of the Lower Clark Fork River, Montana*; Montana Water Resources Research Center: Bozeman, MT, 1986.

4. Andrews, E. D. In *Chemical Quality of Water and the Hydrologic Cycle*; Averett, R. C.; McKnight, D. M., Eds; Lewis: Chelsea, MI, 1987.
5. Brook, E. J.; Moore, J. N. *Sci. Total Environ.* **1988**, *76*, 247–266.
6. Moore, J. N.; Brook, E. J.; Johns, C. *Environ. Geol. Water Sci.* **1989**, *14*(1), 107–115.
7. Axtmann, E.; Luoma S. N. *Appl. Geochem.* **1990**, *6*, 75–88.
8. Johns, C.; Moore, J. N. In *Proceedings, Clark Fork River Symposium*; Carlson, C. E.; Bahls, L. L., Eds; Montana Academy of Science: Butte, MT, 1985; p 74.
9. Woessner, W.; Moore, J.; Johns, C.; Popoff, M.; Sartor, L.; Sullivan, M. *Arsenic Source and Water Supply Remedial Action Study, Milltown, Montana*; Solid Waste Bureau, Montana Department of Health and Environmental Science: Helena, MT, 1984.
10. Moore, J. N.; Ficklin, W. H.; Johns, C. *Environ. Sci. Technol.* **1988**, *22*(4), 432–437.
11. Bailey, A. K. *Proc. Mont. Acad. Sci.* **1976**, *36*, 165–170.
12. Udaloy, A. G. M.S. Thesis, University of Montana, Missoula, MT, 1988.
13. Berner, R. A. *J. Sediment. Petrol.* **1981**, *51*(2), 359–365.
14. Singer, P. C.; Stumm, W. *Science* **1970**, *167*, 1121–1123.
15. Nordstrom, D. K. *Acid Sulfate Weathering* (Special Publication No. 10); Soil Science Society of America: Madison, WI, 1982; p 37.
16. Jørgensen, B. B. In *Microbial Geochemistry*; Krumbein, W. E., Ed.; Blackwell Scientific: Boston, MA, 1983; pp 91–124.
17. Moses, C. O.; Nordstrom, D. K.; Herman, J. S.; Mills, A. L. *Geochem. Cosmochim. Acta* **1987**, *51*, 1561–1571.
18. Nimick, D. A.; Moore, J. N. *Appl. Geochem.* **1991**, *6*, 635–646.
19. Ludgren, D. G.; Stark, M. *Ann. Rev. Microbiol.* **1980**, *34*, 263–283.
20. Arkesteyn, G. J. *Antonie van Leeuwenhoek* **1979**, *45*, 423–435.
21. Froelich, P. N.; Klinkhamer, G. P.; Bender, M. L.; Luedtke, N. A.; Heath, G. R.; Cullen, D.; Dauphin, P.; Hammond, D.; Hartman, B.; Maynard, V. *Geochim. Cosmochim. Acta* **1979**, *43*, 1075–1090.
22. Nealson, K. H. In *Microbial Geochemistry*; Krumbein, W. E., Ed.; Blackwell Scientific: Boston, MA, 1983; pp 191–221.
23. Peterson, M. L.; Carpenter, R. *Mar. Chem.* **1983**, *12*, 295–321.
24. Oscarson, D.; Huang, P.; Liaw, W. *J. Environ. Qual.* **1980**, *9*(4), 700–703.
25. Oscarson, D.; Huang, P.; Liaw, W. *Clays Clay Miner.* **1981**, *29*(3), 219–225
26. Moore, J. N.; Walker, J. R.; Hayse, T. H. *Clays Clay Miner.* **1990**, *38*(5), 549–555.
27. Chapelle, F. H.; Lovely, D. R. *Ground Water* **1992**, *30*(1), 29–36.
28. Krumbein, W. E.; Swart, P. K. In *Microbial Geochemistry*; Krumbein, W. E., Ed.; Blackwell Scientific: Boston, MA, 1983; pp 5–62.

RECEIVED for review September 26, 1991. ACCEPTED revised manuscript April 7, 1992.

Cycles of Trace Elements in a Lake with a Seasonally Anoxic Hypolimnion

Annette Kuhn, C. Annette Johnson, and Laura Sigg*

Institute for Water Resources and Water Pollution Control (EAWAG), Swiss Federal Institute of Technology Zurich (ETHZ), CH–8600 Dübendorf, Switzerland

The cycles of trace elements (Mn, Fe, As, Cr, and Zn) in a eutrophic lake are influenced by biological productivity and the development of anoxic conditions in the hypolimnion. The occurrence of redox species [Mn(II)/Mn(III,IV), Fe(II)/Fe(III), As(III)/As(V), and Cr(III)/Cr(VI)] in the anoxic hypolimnion of Lake Greifen is compared to the thermodynamic redox sequence. The reduction and oxidation processes are discussed. As(III) appears in the anoxic hypolimnion together with Fe(II) and sulfide, whereas only indirect evidence for the reduction of Cr(VI) is found. The concentration of Zn is influenced by binding to algal material and to manganese oxides. The precipitation of manganese oxides during lake overturn affects trace elements by adsorption and oxidation reactions on the oxide surfaces. The cycles of As, Cr, and Zn are strongly coupled to the cycles of Mn and Fe.

STRONG SEASONAL VARIATIONS in the chemical and biological conditions within the water column are typical of eutrophic lakes. Eutrophication phenomena, a major perturbation of lake ecosystems, have been observed in various regions. A number of Swiss lakes in highly populated areas have been subject to eutrophication caused by an elevated supply of nutrients from sewage and agricultural activities (1, 2). The phosphate inputs into Swiss lakes are decreasing as a consequence of water pollution control measures, so that long-term changes are occurring in the eutrophication state

*Corresponding author

0065–2393/94/0237–0473$07.25/0

of the lakes (2, 3). The high productivity of algal biomass in eutrophic lakes causes varied biological and chemical responses. To predict the response to changes in nutrient load and the evolution of such lake systems, the chemical interactions involved need to be well understood.

Lakes in temperate regions are often characterized by seasonal temperature-induced mixing and stagnation processes. Summer stagnation, because of higher temperature and thus lower densities in upper water layers, hinders the resupply of oxygen to the deeper water layers. The high productivity of biomass and the subsequent consumption of oxygen during mineralization in the deeper water layers cause the development of anoxic conditions, which may occur seasonally in the hypolimnion and in the surficial sediments.

Although general phenomena occurring in eutrophic lakes are well known, many questions regarding the detailed chemical and biological mechanisms involved are still open. Under anoxic conditions, reduced chemical species [Mn(II), Fe(II), and S(–II)] are produced, following in principle the thermodynamic redox sequence (4, 5); nitrate and sulfate are effective electron acceptors in the absence of oxygen. However, most of the redox reactions involved take place at significant rates only if microbially mediated.

The occurrence of anoxic conditions causes cycling of iron and manganese at the oxic–anoxic interface (6–10). In lakes with a significant seasonal cycle, iron and manganese oxides are reduced during anoxia, and Fe(II) and Mn(II) are released into solution. The Fe(II) and Mn(II) species are reoxidized, and Fe(III) and Mn(III,IV) precipitate as oxides during lake overturn, when the reduced species come into contact with oxygen.

Few examples of studies on cycling of trace elements (other than iron and manganese) at oxic–anoxic interfaces are found in the literature (11–17). Trace element cycling in the water column of a eutrophic lake (Figure 1) is affected by a number of processes related to the redox conditions.

- Redox-sensitive trace elements may undergo changes in their redox states, according to the predominant redox intensities. The presence of suitable reductants (or oxidants) is essential. These effects are examined here for As(III)/As(V) and Cr(III)/Cr(VI). The redox states of trace elements affect their solubility (e.g., Cr(III)/Cr(VI)) and their affinity for binding to solid phases (e.g., arsenite and arsenate). Biological effects (e.g., uptake and toxicity) are also dependent on the redox state.

- Iron and manganese oxides are characterized by high specific surface areas and high affinity of their surface hydroxyl groups for adsorption of a variety of trace elements. In addition to adsorption processes, oxidation reactions are catalyzed by these surfaces (18–20). The in situ precipitation and dissolution of these oxides are thus significant for the fate of various trace elements. Other elements may be released from sediments together with Fe(II) and Mn(II).

- The occurrence of sulfide strongly affects the speciation and solubility of numerous trace elements. Solid sulfide phases may be precipitated, and dissolved sulfide and polysulfide complexes may be formed (*21–23*). The influence of sulfide on trace

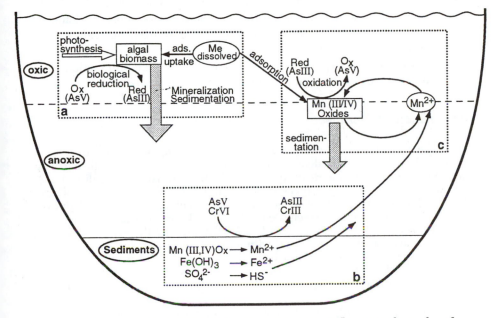

Figure 1. Schematic representation of the processes influencing the cycles of trace elements in Lake Greifen. During stagnation, the lake is divided into an oxic epilimnion and an anoxic hypolimnion. Three different stages can be distinguished in time and space:

a. *Large amounts of algal biomass are produced during spring and summer; trace elements are bound to this algal material by adsorption or by uptake. Biological reduction processes such as reduction of As(V) occur. The sedimentation of this algal material removes trace elements from the epilimnion.*

b. *Near the sediment–water interface, anoxic conditions are established in early summer as a consequence of the mineralization of the settling biomass; Mn(II) is released into the water column through reductive dissolution of manganese oxides. At a later stage of the stagnation period (October–November), Fe(II) and S(–II) are also released from the sediments. Reduction of As(V) and Cr(VI) is observed.*

c. *At the interface between oxic and anoxic water, the oxidation of Mn(II) to manganese oxides takes place. Large amounts of manganese oxides are precipitated during mixing of the lake in November–December. Adsorption of various elements and oxidation processes occurs at the manganese oxide surfaces.*

elements in strongly reducing environments has been demonstrated in the Black Sea and the Framvaren Fjord (*12, 14*).

This chapter discusses the chemical mechanisms influencing the fate of trace elements (arsenic, chromium, and zinc) in a small eutrophic lake with a seasonally anoxic hypolimnion (Lake Greifen). Arsenic and chromium are redox-sensitive trace elements that may be directly involved in redox cycles, whereas zinc is indirectly influenced by the redox conditions. We will illustrate how the seasonal cycles and the variations between oxic and anoxic conditions affect the concentrations and speciation of iron, manganese, arsenic, chromium, and zinc in the water column. The redox processes occurring in the anoxic hypolimnion are discussed in detail. Interactions between major redox species and trace elements are demonstrated.

Experimental Details

Lake Description. Lake Greifen is a small eutrophic lake on the Swiss Plateau. It is located in a densely populated area and receives high nutrient inputs from sewage and agriculture. A very high phosphate load is discharged into the lake, in spite of phosphate elimination in the sewage-treatment plants. The phosphate concentration in the mixed water column was 3.5 μM during spring 1990. Chlorophyll *a* reaches 0.01–0.04 mg/L during algae blooms (*24*).

The surface area of Lake Greifen is 8.5×10^6 m^2, and the volume is 150×10^6 m^3; its average depth is 17.7 m, with a maximum of 32.2 m. The residence time of water is 1.1 years. Thermal stratification lasts about from May to December, and lake overturn usually takes place in December–January. An anoxic hypolimnion develops during summer stagnation from about June to December.

Sampling. Monthly samplings were performed at the deepest point of the water column (depth = 31 m) over a period of 2 years. Because a number of labile reduced species occur in anoxic water samples, special attention has to be given to the sampling and sample pretreatment methods to obtain reliable results. Several approaches have been used to circumvent sampling and storage problems. Rapid field analysis of labile species has been recommended (e.g., ref. 9). This rapid analysis is particularly important in measuring the very labile species of Fe(II) and S(−II).

In our study, contact of the anoxic samples with oxygen was avoided. The samples were transferred from the samplers (Go-Flo, General Oceanics, 5 L) into bottles equipped with three-way taps under N$_2$ pressure. The bottles had previously been flushed with N$_2$, and they were completely filled. Filtration in the laboratory with acid-cleaned 0.45-μm cellulose nitrate filters (Sartorius) and a polycarbonate filtration unit (Sartorius) was also carried out under nitrogen gas. Most analytical determinations were performed in the laboratory on the sampling day.

Reliable results on trace elements require precautions to avoid contamination. Concentrations of trace elements determined in lake water are often in ranges similar to those measured in the oceans; the requirements for contamination-free sampling are thus similar (*25, 26*). Teflon-coated Go-Flo samplers were used. They were carefully cleaned with acid (0.01 M HNO$_3$) and high-purity water (Nanopure) before use, then rinsed with lake water at the beginning of each sampling. All bottles and devices coming into contact with the samples were made of high-quality polyethylene or glass. They were cleaned with acid and rinsed with high-purity water. Samples were protected with plastic bags during transport. Field blanks were obtained by filling ultra-pure water into bottles under field conditions and treating these samples like lake-water samples. Average field blanks obtained in this manner were the following: filtered Mn, <0.02 μM; particulate Mn, <0.02 μM; filtered Fe, 0.05 μM; particulate Fe, 0.03 μM; filtered Zn, 4 nM; Cr(VI), 0.2 nM; total filtered Cr, 1 nM; filtered As, 0.1 nM.

Sediment Traps.   Sediment traps with a height:diameter ratio of 10:1 were exposed in the deepest part of the lake close to the water-sampling site at 15- and 28-m depth for 15 months. The particulate material in the traps was collected approximately every 3 weeks and was subsequently freeze-dried until analysis.

Analytical Methods.   Temperature, pH, and oxygen were measured in situ by using a combined sensor (Züllig). Ammonium was determined by flow injection analysis (27), and nitrate and silicate by spectrophotometric methods (Auto-Analyzer) (28). Sulfide was determined by using a H_2S-specific electrode (29).

Mn(II) and Fe(II) were determined by differential pulse voltammetry (9, 30). The determinations were carried out within a few hours after sampling. Total dissolved and particulate Fe and Mn (after digestion of the particulate matter in a microwave digestion unit with HNO_3–HCl) were measured by flame or graphite furnace atomic absorption spectrometry.

Inorganic As(III) and As(V) were determined by atomic absorption spectrometry using the hydride technique. Total inorganic arsenic, As(III) + As(V), was measured after a prereduction reaction of As(V) to As(III) in acidic solution containing potassium iodide and ascorbic acid. For the selective hydride formation of As(III), samples were maintained at pH 5.0 during the hydride reaction (with 3% $NaBH_4$, 1% NaOH) with a citrate–sodium hydroxide buffer solution (31). As(V) was determined by difference between total As and As(III). The detection limit of As(III) and As(V) was 0.1 nM. The selectivity of this method was checked by additions of As(III) and As(V) to lake water; 95–100% recovery of As(III) and As(V) was found (32).

Dissolved Cr(VI) and Cr(III) were preconcentrated and separated by using ion exchange (33), and total dissolved Cr (filtered <0.45 μm) was preconcentrated by volume reduction. The samples were passed through an anion exchanger (Bio-Rad 1-X4) for collection of Cr(VI) and eluted with 5 M HNO_3. For collection of Cr(III) the samples were passed through a cation exchanger (Baker) and eluted with 1.0 M NH_4NO_3–0.1 M HNO_3. After elution the preconcentrated samples were analyzed by graphite furnace atomic absorption spectrometry. Selectivity for Cr(VI) and Cr(III) and recovery were checked by adding Cr(VI) and Cr(III) to lake-water samples. Recovery of Cr(VI) was 98%, and of Cr(III) it was 102%.

Zinc was determined in filtered samples (0.45 μm pore size) by flame atomic absorption spectrometry after preconcentration. For the preconcentration, 8-hydroxyquinoline was added as a chelating agent to the acidified samples and subsequently buffered to pH 8. The hydroxyquinoline complexes were collected on C-18 columns (Baker) and eluted with 0.6 M HCl (modified after ref. 34). The preconcentration factor was 40. Blank values of this method were ≤2–3 nM Zn; reproducibility was ±1.5 nM. Recovery of Zn spikes added to lake water was 90–100%.

The particulate material from the sediment traps was digested in aqua regia in a microwave digestion unit. Fe, Mn, Zn, Ca, Cr, and Cu were determined by inductively coupled plasma atomic emission spectrometry (ICP–AES); P was determined by the molybdate spectrophotometric method (28). A sediment standard (NBS No. 1645) was used regularly to check the accuracy of the sediment digestion procedure.

Thermodynamics and Kinetics of Redox Processes

Although the various redox species in a lake are unlikely to be in equilibrium with each other, the thermodynamic redox sequence gives a framework that helps to predict the sequence of redox reactions and the thermodynamically stable species under given conditions (4). The predicted sequence for lowering the redox intensity as a consequence of the mineralization of the settling biomass is shown in Figure 2 in the form of an electron activity (pϵ) versus pH diagram for the pH range typically encountered in eutrophic lakes. In the pH range of interest, nitrate is reduced to N_2 before Mn(II) is formed; the reduction of nitrate to ammonium takes place subsequently. The reduction to Fe(II) and S(–II) occurs in the lower redox potential range. The reduction of As(V) to As(III) is expected to occur at about the same redox potential as the reduction of Fe(III) to Fe(II). The reduction of Cr(VI) to Cr(III) should take place at a redox potential similar to that of the reduction of Mn(IV) to Mn(II).

Redox equilibrium is not achieved in natural waters, and no single pϵ can usually be derived from an analytical data set including several redox couples. The direct measurement of pϵ thus is usually not meaningful because only certain electrochemically reversible redox couples can establish the potential at an electrode (4, 35). However, pϵ is a useful concept that indicates the direction of redox reactions and defines the predominant redox conditions. Defining pϵ on the basis of the more abundant redox species like Mn(II) and Fe(II) gives the possibility of predicting the equilibrium redox state of other trace elements. The presence of suitable reductants (or oxidants) that enable an expedient electron transfer is, however, essential in establishing redox equilibria between trace elements and major redox couples. Slow reaction rates will in many cases lead to nonequilibrium situations with respect to the redox state of trace elements.

We use the occurrence of oxygen, Mn(II), ammonium, Fe(II), and sulfide to indicate the redox conditions in the lake. The conditions observed in Lake Greifen for the major redox couples correspond qualitatively to the thermodynamic redox sequence. The seasonal variations in the deep lake-water column are illustrated by Figure 3. In the mixed lake (winter–spring), oxygen is present in the whole water column at concentrations close to saturation, and very low concentrations of Mn and Fe (both dissolved and particulate forms) are present. During summer stagnation oxygen is consumed first in the deepest layers and later in the whole hypolimnion below about 10-m depth.

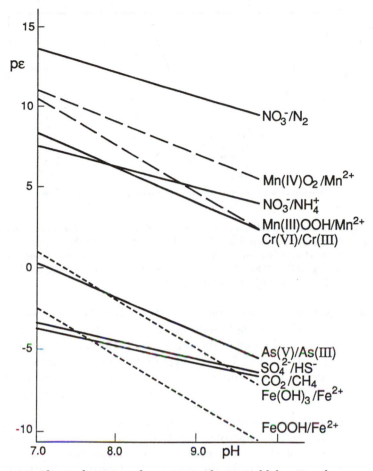

Figure 2. Thermodynamic redox sequence for typical lake pH and concentration conditions. For Mn, the pε versus pH relationships are shown assuming $[Mn^{2+}] = 1 \times 10^{-5}$ M and the solid phases $Mn(IV)O_2$ and $Mn(III)OOH$, and for Fe assuming $[Fe^{2+}] = 1 \times 10^{-6}$ M and amorphous $Fe(OH)_3(s)$ and $FeOOH(s)$ (goethite).

The different reduced species appear successively as a function of time and depth in the anoxic hypolimnion (Figure 3). Increased Mn(II) and $NH_4{}^+$ concentrations appear shortly after disappearance of oxygen from the deeper layers of the lake-water column and further increase in the hypolimnion during anoxia. Fe(II) and S(−II) are detectable at the end of the stagnation period only in the deepest water layers (28–31 m). Upon mixing of the lake and contact with oxygen during overturn, the reduced species are reoxidized. Mn(II) and $NH_4{}^+$ are only slowly oxidized and persist in the presence of oxygen (December 1989). The various species measured in the anoxic hypolimnion indicate the predominant pε range, although they are not in

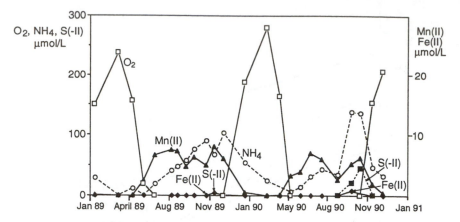

Figure 3. Seasonal variations of the major redox species in the water column of Lake Greifen at the depth 30–31 m: O_2, NH_4^+, $Mn(II)$, $Fe(II)$, and $S(-II)$ are represented as a function of time. $Fe(II)$ and $S(-II)$ are found only at the end of the stagnation time (October–November 1989 and 1990). $S(-II)$ was quantified only in the fall 1990 samples.

equilibrium with each other. The pε is estimated to be in the 6–10 range for the period in which the only reduced species are Mn(II) and NH_4^+, and in the -3–0 range at the end of the stagnation period in presence of Fe(II) and S(-II). The pH is 7.5–7.7 in the hypolimnion and increases to 8.5–8.7 in the epilimnion during summer stagnation. The major ion composition of the lake is dominated by calcium carbonate dissolution and precipitation.

Nitrate and Ammonium. The transformations of nitrogen species may occur under suitable microbial catalysis (5, 36). Nitrate reduction may result in formation of either elemental nitrogen or ammonium. Mass balances over a whole lake have indicated the importance of the denitrification process for the elimination of nitrogen from lakes (37). The conditions for the dissimilative ammonification of nitrate are poorly known (36). Ammonium is also released by the mineralization of biomass.

Because elemental nitrogen is not measured and no overall nitrogen mass balance was calculated, only indirect evidence for the denitrification process in Lake Greifen may be given. The mass balance of inorganic nitrogen (including ammonium and nitrate) in the anoxic hypolimnion of Lake Greifen over stagnation time indicates nitrogen losses that are attributed to denitrification (38). The accumulation of ammonium, observed in the anoxic hypolimnion (38), results in the overall replacement of a large fraction of nitrate by ammonium. Ammonium and nitrate are, from a kinetic point of view, not reactive for direct redox reactions with other elements.

Manganese. The reduction and oxidation processes of iron and manganese, which may be important under the conditions of the anoxic hypolimnion, are briefly discussed and compared with the observations in the lake. Processes occurring under photochemical conditions and under the influence of photosynthesis, as they may occur in the upper photic water layers of a lake, are not discussed here.

The release of Mn(II) from sediments and in hypolimnetic waters is well known and has often been described (6, 8). Several chemical and biological mechanisms may be involved in the reduction of manganese oxides at the sediment–water interface. Manganese oxides are reduced by various organic compounds (39, 40); phenols and quinones, which are degradation products of organic matter, are especially efficient reductants. The reduction reactions involving these organic ligands occur on a time scale of minutes to hours. The reduction of manganese oxides by sulfide is kinetically a very efficient reaction (41). Reduction by Fe(II) also occurs readily (42). The biological reduction of manganese oxides may be performed by a number of different organisms, although the relative importance of direct and indirect (by metabolic products) reduction pathways is unclear (42–45).

The oxidation of Mn(II) by oxygen in homogeneous solution is extremely slow (on a time scale of years) (46), but it is catalyzed by the adsorption of Mn(II) on oxide surfaces (47, 48). The main oxidation pathway under natural conditions, however, is assumed to be the microbially mediated oxidation that occurs on the time scale of hours to days. The importance of microbial oxidation of Mn(II) in natural environments has been demonstrated in a number of studies (49–54).

Mn(II) and particulate manganese (Mn_{part}) are depleted from the mixed lake-water column of Lake Greifen in spring (Mn(II) $<0.03 \times M$ and Mn_{part} ~0.2 $\times$ M). Typical depth profiles of Mn(II) and particulate manganese are observed during stagnation with increasing concentrations of Mn(II) toward the sediment–water interface and maxima of particulate manganese at the oxic–anoxic boundary (Figure 4). Mn(II) concentrations up to 10 $\times$ M are built up in the anoxic hypolimnion.

The reduction of Mn(IV) or Mn(III,IV) oxides probably takes place at the sediment–water interface. Mn(II) diffuses from there into the water column and is oxidized at the oxic–anoxic boundary. The reduction of manganese oxides in the water column itself is possible, but is probably not of quantitative importance. The profiles of Mn(II) and particulate manganese in Lake Greifen were modeled by taking into account the Mn(II) flux from sediments, the oxidation of Mn(II), and the settling of particulate manganese (55). The oxidation of Mn(II), which results in the manganese peaks at the oxic–anoxic boundary, can be explained only by biological oxidation; an oxidation rate constant of 0.2 per day (pseudo first-order) is estimated from the model calculations.

During mixing of the lake in November–December, the accumulated Mn(II) is reoxidized, and the manganese oxides formed are eliminated from the water column by sedimentation. This process results in the precipitation

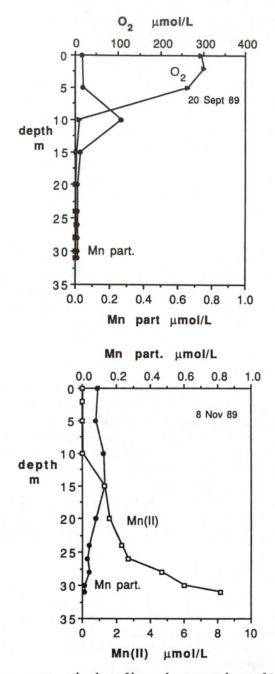

Figure 4. Concentration–depth profiles in the water column of Lake Greifen: particulate manganese (Mn$_{part}$) and O$_2$ during summer stagnation (September 20, 1989); Mn(II) and particulate manganese (Mn$_{part}$) at the end of stagnation (November 8, 1989).

of large amounts of manganese oxide particles, with maximum sedimentation rates on the order of 20–50 mg of Mn/m^2 per day. In contrast, the sedimentation rates during summer stagnation are only about 1 mg of Mn/m^2 per day. This precipitation of manganese oxides is of special interest with regard to the adsorption and oxidation of trace elements.

Iron. The reduction of Fe(III) oxides and release of dissolved Fe(II) is a well-known phenomenon in sedimentary environments. However, the detailed mechanisms of the reduction and dissolution processes are not fully elucidated. The dissolution rates of iron oxides in presence of organic ligands depend on surface speciation (56, 57). Chemical reduction by organic ligands has rarely been observed to occur at sufficiently fast rates in the neutral pH range (58, 59) to explain Fe(II) production in natural waters. Fe(III) oxides have been shown to be catalytically dissolved in the presence of a ligand and of Fe(II) (60). Sulfide readily reduces Fe(III) oxides (61). The reduction of Fe(III) oxides by microorganisms is known to be carried out by different genera of bacteria (44, 62). Direct reduction by enzymatic pathways and reduction by end products of metabolism are both possible. A recent study points out the importance of the enzymatic reduction of Fe(III) for sedimentary conditions (63).

On the other hand, the oxidation of Fe(II) by O_2 is very rapid in the neutral pH range (4, 48). Oxidation of Fe(II) on MnO_2 has also been observed to be a fast process and may be important in sedimentary environments (42, 64).

The cycling of Fe in the water column of Lake Greifen is much less marked than the cycling of Mn. Fe(II) appears only at the end of stagnation in the deepest layers of the lake (28–31 m) in concentrations up to 0.5–1 μM. Particulate iron is increasing over time in the anoxic hypolimnion. Particulate Fe and Fe in filtered samples increase up to about 1 μM (Figure 5). The increasing concentrations of Fe [but not of Fe(II)] in filtered samples may also result from the presence of small Fe hydroxide particles. The sedimentation rate of Fe (6–20 mg/m^2 per day) exhibits much smaller seasonal variations than the sedimentation rate of Mn. This difference results from larger allochthonous inputs of iron oxide particles than of manganese oxides and less recycling of Fe from sediments.

Fe(II) formed in the sediments may diffuse into the water column and be reoxidized by oxygen traces. Reoxidation on manganese oxide particles is a possible mechanism, but very low concentrations of particulate manganese are measured in the anoxic hypolimnion. Small iron hydroxide particles will settle only slowly and thus be accumulated in the hypolimnion. Iron oxide particles formed by reoxidation in the water column are potentially important for the scavenging of trace elements because small particles with large surface areas are expected to be formed. The occurrence of Fe(II) is a good indicator of the redox conditions, as its reoxidation both by oxygen and by manganese oxide would be fast.

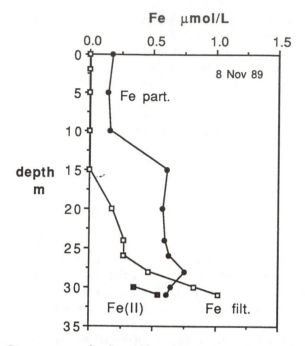

Figure 5. Concentration–depth profiles of filtered Fe (<0.45 μm filter pore size), particulate Fe, and Fe(II) at the end of stagnation (November 8, 1989). (Reproduced with permission from reference 38. Copyright 1991 Elsevier Science Publishers, Amsterdam.)

Sulfide.   Sulfide appears in the water column of Lake Greifen only at the end of stagnation time and in the deepest water layers. This distribution indicated the biological reduction of sulfate in sediments. The reactions involved in the sulfur cycle are described in ref. 65. The occurrence of sulfide indicates a very low pε (pε < 0); sulfide is an efficient reductant for many elements, including Fe(III), Mn(IV), As(V), and Cr(VI). The occurrence of sulfide also implies the possible precipitation of solid sulfide phases of various elements and the formation of dissolved complexes (21–23).

Trace Elements

Arsenic.   The inorganic species arsenate [As(V)] and arsenite [As(III)] were measured in the depth profile of the lake over the seasonal cycle (Figure 6) (32). The relevant reduction and oxidation processes will be briefly considered. The equilibrium constants for the various reactions are calculated on the basis of the thermodynamic data given in refs. 66 and 67. According to the thermodynamic sequence, the reduction of As(V) to As(III) occurs in a pε range similar to that of the reduction of $Fe(OH)_3(s)$ to Fe(II) (Figure 2).

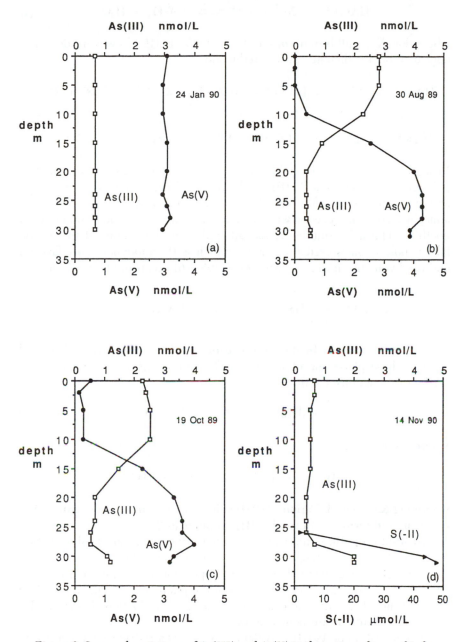

Figure 6. Seasonal variations of As(III) and As(V) in the water column of Lake Greifen: mixed lake (January 24, 1990); summer stagnation (August 30, 1989); end of stagnation (October 19, 1989, and November 14, 1990). On November 14, 1990, S(−II) is shown for comparison with As(III).

$$HAsO_4^{2-} + 2e^- + 4H^+ \rightleftharpoons H_3AsO_3 + H_2O \tag{1}$$

where $\log K$ (equilibrium constant) = 28.45. Fe(II), however, does not appear to be a sufficiently strong reductant.

$$HAsO_4^{2-} + 2Fe^{2+} + 5H_2O \rightleftharpoons H_3AsO_3 + 2Fe(III)(OH)_3(s) + 2H^+ \tag{2a}$$

$$HAsO_4^{2-} + 2Fe^{2+} + 3H_2O \rightleftharpoons H_3AsO_3 + 2FeOOH(s) + 2H^+ \tag{2b}$$

Reactions 2a and 2b only yield a negative free energy (ΔG) in the neutral pH range if a crystalline Fe(III) oxohydroxide (such as goethite) is formed as the reaction product. Inversely, the oxidation of As(III) is thermodynamically favored by reaction with an amorphous iron hydroxide, but not with goethite. The most suitable reductant under the conditions of the anoxic lake hypolimnion is certainly sulfide, which is thermodynamically favored and has been shown to react with As(V) (*68, 16*) according to reaction 3.

$$HAsO_4^{2-} + \frac{1}{4} HS^- + \frac{7}{4} H^+ \rightleftharpoons H_3AsO_3 + \frac{1}{4} SO_4^{2-} \tag{3}$$

where $\log K$ = 20.6. In the presence of sulfide the formation of a solid As(III) sulfide phase and of soluble As(III) sulfide species are possible according to reactions 4 and 5.

$$2H_3AsO_3 + 3HS^- + 3H^+ \rightleftharpoons As_2S_3(s) + 6H_2O \tag{4}$$

where $\log K$ = 64.3.

$$As_2S_3(s) + H_2S \rightleftharpoons 2AsS_2^- + 2H^+ \tag{5}$$

where $\log K$ = −24.6. Various As(III) sulfide species have been assumed to occur in equilibrium with solid As(III) sulfides (*69*).

In addition to these inorganic reduction processes, biological reduction processes by phytoplankton yielding either inorganic As(III) or various organic As(III) species were shown to be important in marine environments (*70, 71*). These processes are important in the photosynthetically productive layers of the lake. Microbial processes causing the reduction of As(V) under anaerobic conditions are poorly known.

The oxidation of As(III) to As(V) by oxygen is a rather slow process (*68, 72*). The oxidation of As(III) to As(V) was, however, shown to occur readily by reactions with manganese oxides (*19, 20, 73*). For the conditions encountered in Lake Greifen, the oxidation of As(III) by manganese oxides is likely to be an important oxidation mechanism. The role of iron oxides in the oxidation of As(III) is unclear because the thermodynamically predicted direction of the reaction depends on the type of the solid iron hydroxide phase involved (74). Both arsenate and arsenite may be adsorbed on iron oxides (75).

The seasonal cycle of dissolved As(III) and As(V) in Lake Greifen is illustrated by Figure 6. Total concentrations of dissolved inorganic As are in the range of 3–4 nM. In the mixed lake and in the presence of oxygen in the water column (January 24, 1990), As(V) is the predominant species, whereas As(III) is found at very low concentrations. During summer stratification major changes in redox speciation occur in the epilimnion, although As(V) remains the predominant redox species in the hypolimnion until late in the stagnation time. As(V) disappears from the productive epilimnion and As(III) becomes predominant (August 30, 1989). As(III) appears in the hypolimnion in the deepest layers at the end of stagnation time, simultaneously with Fe(II) and S(−II) (October 19, 1989, and November 14, 1990), but As(V) is still present in these samples. As(III) is then eliminated from the lake-water column during overturn. This pattern was observed consistently throughout both years of observation.

The processes occurring in the hypolimnion and in the epilimnion have to be considered separately. The appearance of As(III) in the anoxic hypolimnion is in qualitative agreement with the thermodynamic redox sequence, because it appears together with Fe(II) and S(−II). The pϵ calculated from the As(III)/As(V) couple at 30–31 m on October 19, 1989, is in agreement with pϵ calculated from Fe(II)/Fe(III) (p$\epsilon \sim 0$), but is higher than indicated by the presence of S(−II).

Reduction by sulfide seems a likely reaction for the formation of As(III). The ion product (Q) for reaction 3 was calculated from the measured concentrations of As(III), As(V), HS$^-$, and SO$_4$$^{2-}$ at 30–31 m depth in October–November 1990 (in the presence of sulfide); log Q = 13.1–13.4 was obtained. The reduction of As(V) is incomplete, probably because of slow reduction processes (16). According to the reactions listed, complexes of As(III) with sulfide (such as AsS$_2$$^-$ or H$_2$As$_3$S$_6$$^-$) may control the solubility of As in presence of sulfide. The kinetics of the reduction of As(V) by sulfide and of the formation of sulfide species are poorly known.

In the epilimnion, the observed speciation obviously deviates from the thermodynamic predictions. The predominance of As(III) during the summer stagnation time has to be explained by biological processes. During photosynthetic production, As(V) is taken up by algae, resulting in nutrientlike profiles of As(V). This uptake may be explained by the chemical similarity between arsenate and phosphate (71). As(V) is probably reduced to As(III) within the algae and released. As(III) appears in the water column and is only slowly reoxidized. The dependence of these biological processes on various factors is discussed in detail in ref. 32.

The reoxidation of As(III) formed both in the epilimnion and in the hypolimnion is completed within a few weeks during overturn of the lake. The most likely oxidation mechanism is the oxidation on manganese oxides precipitated during overturn. The epilimnion contents of As(V), As(III), and particulate manganese from November to December were considered in evaluating this oxidation process. The increase in As(V) is larger than calculated from mixing processes during sinking of the thermocline and correlates with the occurrence of particulate manganese (32). However, the As sedimentation rates do not appear to follow the sedimentation rates of manganese oxides. This independence indicates that the oxidation of As(III) occurs predominantly in presence of manganese oxides, but that binding of As(III) or As(V) to manganese oxides is not a significant sedimentation mechanism.

The observations in Lake Greifen are in line with other studies of As(III) and As(V) at oxic–anoxic boundaries (16, 17, 76). In Saanich Inlet (16) and in Lake Pavin (17), increasing As(III) concentrations were found below the O_2–H_2S boundary, although the reduction of As(V) was not complete. A similar situation is encountered in Lake Greifen, in which sulfide is only an intermittent species. The formation of reduced and methylated As species in the upper layers of various lakes has also been described in ref. 76.

Chromium.   The chemical properties of the two possible oxidation states Cr(VI) and Cr(III) are very different. Cr(VI) occurs as an anion, whereas Cr(III) is a strongly hydrolyzing cation with a strong tendency to bind to the surfaces of oxides and other particles (77). According to the thermodynamic sequence, the reduction of Cr(VI) to Cr(III) occurs in a pϵ range similar to that for the reduction of Mn(III,IV) to Mn(II) (Figure 2).

$$CrO_4^{2-} + 3e^- + 6H^+ \rightleftharpoons Cr(OH)_2^+ + 2H_2O \tag{6}$$

where $\log K = 66.8$. Possible reductants under conditions of the anoxic hypolimnion may thus be Fe(II), sulfide, and organic compounds (78–80). Little is known about the reactivity or availability of organic compounds for the reduction of chromate under natural water conditions.

The oxidation of Cr(III) by oxygen is a very slow reaction with a pseudo-first-order rate constant of approximately 0.4/year (0.2 atm of O_2, ref. 78). Other investigators could not detect oxidation under their experimental conditions (18, 81). Manganese oxides appear to be more efficient oxidants, though the oxidation rate depends on the type of oxide, the surface area, and the solution chemistry. With pyrolusite (18), approximately 10% of Cr(III) was oxidized after 100 h ([MnO_2] = 35.6 m^2/L; [Cr(III)] = 96 μM). Over 90% of Cr(III) was oxidized within 3 min by using manganite ([MnO_2] = 0.92 m^2/L; [Cr(III)] = 0.5 μM) (82). The strong binding of Cr(III) to particles and to organic ligands decreases its availability for oxidation reactions.

The seasonal variations of Cr(VI) in the water column of Lake Greifen (83) are illustrated by Figure 7. In the mixed lake, the concentration of Cr(VI) is uniform throughout the water column (approximately 2.5 nM). During stagnation, the concentration of Cr(VI) increases in the epilimnion and decreases in the hypolimnion. This separation can be explained by a slow removal process in the bottom waters, while at the same time Cr(VI) enters only the epilimnion and is cut off from the hypolimnetic waters. In the late stagnation time (November 1989), the concentration of Cr(VI) close to the sediment–water interface decreases. Cr(III), however, cannot be detected in solution.

These results indicate that Cr(VI) is reduced only in the presence of Fe(II) or sulfide as reductants. Reduction by organic compounds probably does not occur fast enough under the conditions of the lake. However, calculations indicate that a slow removal process, with a half-time of 100–230 days, removes Cr(VI) from the water column in both oxic and suboxic waters (83). It is not known whether this process is reduction, possibly by organic material, or a weak sorptive reaction.

The evidence for the reduction of Cr(VI) to Cr(III) is only indirect, because Cr(III) is not detected in solution. Cr(III) has a strong tendency to adsorb to particle surfaces and to precipitate as insoluble (hydr)oxide. Thus, Cr(III) produced within the water column by reduction is expected to bind to particles and to be found in the particulate phase. No evidence for release of Cr from sediments was found. Cr(III) is expected to be retained very strongly in sediments, so the release of Cr(III) under anoxic conditions is unlikely. Under oxic conditions the oxidation of Cr(III) by Mn oxides, for example, and release of Cr(VI) from the sediments is plausible; such a mechanism in sediment pore waters is indicated in ref. 84.

These results are in broad agreement with the findings in ref. 11, which indicate the reduction of Cr(VI) in the presence of H_2S. In Saanich Inlet H_2S is always present in the deeper water, whereas in Lake Greifen an intermediate situation is observed with the predominance of Mn(II) in the hypolimnion. The method used for the determination of Cr(III) in ref. 11 would probably include colloidal Cr(III); the present study attempted to determine dissolved Cr(III).

Zinc.   Although Zn is not a redox-active element, the zinc concentration in Lake Greifen is expected to be influenced by the processes represented in Figure 1. Zinc may be bound to algal material in the epilimnion by both uptake and adsorption, and it may be transported to the sediments together with biological material. The importance of binding of Zn to biological material has been shown for settling particles in Lake Zurich (85). The precipitation of manganese oxides within the water column at the end of stagnation provides additional sedimenting material with large surfaces to which zinc may be bound and sedimented. The role of the dissolution of

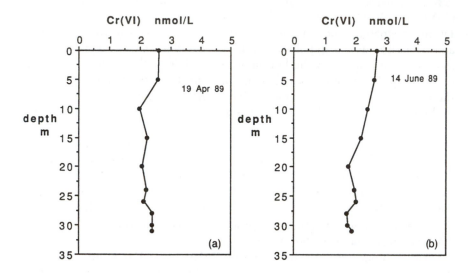

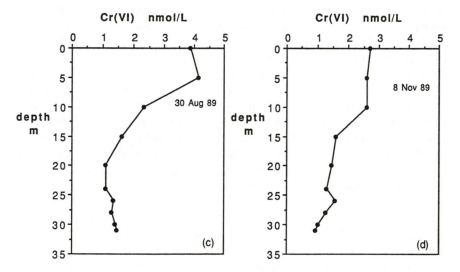

Figure 7. Seasonal variations of Cr(VI) in the water column of Lake Greifen: mixed lake (April 19, 1989); summer stagnation (June 14, 1989, and August 30, 1989); and end of stagnation (November 8, 1989).

manganese oxide has to be considered in this context. Finally, the presence of sulfide in the late stagnation stage affects the solubility of zinc and its speciation (*23*).

Typical concentration–depth profiles of dissolved zinc in the mixed lake (January 12, 1989) during summer stagnation (July 26 and September 20, 1989) and at the end of stagnation (November 14, 1990) are shown in Figure 8. The average concentration in the mixed water column is about 30 nM. Comparison of Zn with a typical nutrient element (Si) shows that Zn is depleted from the epilimnion in summer. Toward the end of the stagnation time, dissolved Zn is depleted from the whole water column; its concentration decreases especially close to the sediment–water interface. No evidence was found for increased concentrations of Zn close to the sediment–water interface during summer stagnation.

The depletion of Zn from the epilimnion in summer, which was observed consistently in the summers of 1989 and 1990, indicates the important role of phytoplankton in binding Zn. Although the major inputs into the epilimnion occur during stagnation, the removal processes are so efficient that Zn in the epilimnion decreases from 30 nM during mixing to an average of 15–20 nM during summer stagnation. If the removal of Zn occurs mostly through uptake by phytoplankton, the ratios of $\Delta Zn:\Delta Si$ and $\Delta Zn:\Delta P$ in the water column and the ratios of $Zn:Si$ and $Zn:P$, respectively, in the settling particles are expected to represent the ratios of Zn to these elements in phytoplankton. From the water column profiles in summer, we derive the ratios $\Delta Zn:\Delta Si = \sim 5 \times 10^{-4}$ and $\Delta Zn:\Delta P = \sim 0.01$ (mol/mol). In the settling particles the $Zn:P$ ratio varies between 0.01 and 0.03 in the summer samples, which are dominated by organic material and calcium carbonate (*86*); Si was not determined in the settling particles. The ratios in the particles during summer are thus similar to those in water.

The concentration of Zn in the hypolimnion remains rather constant until the end of stagnation time. At the end of stagnation, decreasing dissolved Zn concentrations are observed in the hypolimnion (November 1990); the low concentrations in the deepest three samples correspond to the samples in which sulfide was measurable. This observation is comparable to the decrease of Zn concentration that has been observed in other cases below an O_2–H_2S interface (*12*). According to ref. 23, the solubility of Zn in the presence of H_2S is given by the estimated constants in eqs 7 and 8.

$$ZnS(s) + H^+ \rightleftharpoons Zn^{2+} + HS^- \tag{7}$$

where $\log K = -10.93$ or -8.95, depending on the solid phase formed.

$$ZnS(s) + HS^- \rightleftharpoons ZnS_2H^- \tag{8}$$

where $\log K = -0.4$. The occurrence and stability of the zinc sulfide complexes is poorly known. According to reaction 8, the concentrations measured

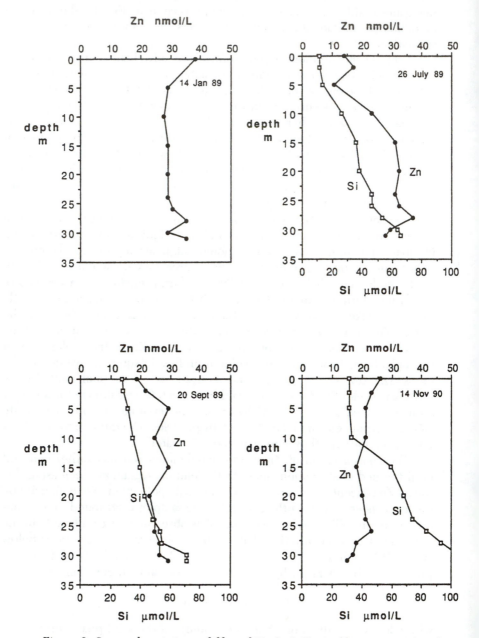

Figure 8. Seasonal variations of filtered Zn (<0.45 μm filter pore size) and of dissolved Si in the water column of Lake Greifen: mixed lake (January 14, 1989); summer stagnation (July 26 and September 20, 1989); end of stagnation (November 14, 1990).

in Lake Greifen are far below the solubility limit. Other processes, such as binding to other sulfide-containing particles, may thus be responsible for the decrease of Zn concentrations in the presence of H_2S.

For the removal of Zn from the water column by sedimentation, both algal material and manganese oxides are likely to be important carrier phases. The Zn sedimentation rates show maxima from June to August, at the time of the maximum sedimentation of P (indicating the sedimentation of algal material), and in December, in line with the sedimentation of manganese oxides (86).

There is no indication of a release of Zn from the sediments during the development of anoxia, unlike the release of phosphate and dissolved silicate. Zn bound to algal material may be dissolved upon mineralization of this material and Zn bound to manganese oxides upon dissolution of manganese oxides. It appears, however, that Zn is efficiently retained in the sediments, probably through bonding to other less soluble particles, such as iron oxides and silica parts of diatoms. In the presence of sulfide, Zn is probably retained in association with sulfide-containing particles.

Conclusions

High biological productivity and subsequent anoxic conditions in a eutrophic lake have significant effects on the speciation and overall cycling of trace elements such as arsenic, chromium, and zinc.

The photosynthetic production in the epilimnion drives the redox cycles and thus affects the cycling of trace elements. The settling phytoplankton acts as a scavenger of trace elements [e.g., Zn and As(V)] and carries them into the deeper water layers and into the sediments. Reduction processes, such as reduction of As(V) to As(III), occur in connection with photosynthetic production. The settling biomass in the deeper water layers is the primary reductant; by microbial mediation it consumes oxygen and reduces nitrate, Mn(III,IV) hydroxides, Fe(III) hydroxides, and sulfate. The products of these reduction processes, especially Fe(II) and sulfide, may then act as reductants, which can interact with As(V) and Cr(VI) to produce As(III) and Cr(III). The reoxidation of the reduced species by oxygen is in many cases a slow process [e.g., oxidation of Mn(II), Cr(III), and As(III)]. Mn(III,IV) hydroxides may act as intermediates for oxidation reactions; for example, As(III), Cr(III), Fe(II), and sulfide undergo fast oxidation reactions on manganese oxides.

The oxidation state of redox-sensitive trace elements such as As(III)/As(V) and Cr(III)/Cr(VI) is thus affected by the redox conditions, as indicated by the occurrence of major reduced species. Kinetic control of the redox reactions plays an important role. As(III) appears in the anoxic hypolimnion in agreement with the thermodynamic redox sequence together with Fe(II) and sulfide, although the reduction of As(V) is incomplete under these conditions. Whereas the reduced As(III) species can clearly be observed in the anoxic hypolimnion, only indirect evidence can be given for the reduction of chromate to Cr(III). The reduction of Cr(VI) is incomplete under these anoxic conditions; the decrease of Cr(VI) indicates the formation of Cr(III), which is bound to particles.

The cycling of manganese is a very significant process in the water column of Lake Greifen, whereas the cycling of iron is much less marked in this system and has only a minor impact on the trace elements investigated. The precipitation of large amounts of manganese oxides during overturn affects the speciation and the sedimentation rates of different elements. Significant amounts of zinc are sedimented in connection with the sedimentation of manganese oxides during overturn. During summer stagnation the sedimentation with algal material is important. The oxidation of As(III) to As(V) on the surfaces of manganese oxides is a likely mechanism. The sedimentation rates of arsenic, however, are relatively low and do not appear to be related to the sedimentation of manganese oxides.

The retention of zinc in the lake sediments appears to be efficient under both oxic and anoxic conditions; no indication of a release of zinc from the sediments into the water column was found. In a similar way, the retention of chromium in the sediments appears to be efficient. Under anoxic conditions Cr(III) is formed, which is strongly bound to particles and is thus retained in the sediments. The occurrence of anoxic conditions favors the retention in the sediments of chromium and of zinc, in contrast to the release of manganese and of iron.

The cycles of trace elements in eutrophic lakes are thus strongly connected, directly and indirectly, to the biological processes and to the cycles of major redox-sensitive elements.

Acknowledgment

We thank D. Kistler, C. Mäder, U. Lindauer, U. Müller, and Th. Rüttimann for sampling and analytical work. This work was supported by Schweizerischer Nationalfonds, Project No. 20–5607.88.

References

1. Ambühl, H. Z. *Wasser Abwasser Forschung* **1982**, *15*, 113–120.
2. Ambühl, H. *Gas Wasser Abwasser* **1987**, 433–439.
3. Gächter, R.; Imboden, D. In *Chemical Processes in Lakes*; Stumm, W., Ed.; Wiley: New York, 1985; pp 365–388.
4. Stumm, W.; Morgan, J. J. *Aquatic Chemistry*; Wiley: New York, 1981.
5. Zehnder, A. J. B.; Stumm, W. In *Biology of Anaerobic Microorganisms*; Zehnder, A. J. B., Ed.; Wiley: New York, 1988; Chapter 1.
6. Davison, W. In *Chemical Processes in Lakes*; Stumm, W., Ed.; Wiley: New York, 1985; pp 31–53.
7. Davison, W.; Woof, C.; Rigg, E. *Limnol. Oceanogr.* **1982**, *27*, 987–1003.
8. Davison, W.; Woof, C. *Water Res.* **1984**, *18*, 727–734.

9. De Vitre, R. R.; Buffle, J.; Perret, D.; Baudat, R. *Geochim. Cosmochim. Acta* **1988**, *52*, 1601–1613.
10. Egeberg, P. K.; Schaaning, M.; Naes, K. *Mar. Chem.* **1988**, *23*, 383–391.
11. Emerson, S.; Cranston, R. E.; Liss, P. S. *Deep-Sea Res.* **1979**, *26A*, 859–878.
12. Jacobs, L.; Emerson, E.; Skei, J. *Geochim. Cosmochim. Acta* **1985**, *49*, 1433–1444.
13. Westerlund, S. F. G.; Anderson, L. G.; Hall, P. O. J.; Iverfeldt, A.; Rutgers van der Loeff, M. M.; Sundby, B. *Geochim. Cosmochim. Acta* **1986**, *50*, 1289–1296.
14. Haraldsson, C.; Westerlund, S. *Mar. Chem.* **1988**, *23*, 417–424.
15. Morfett, K.; Davison, W.; Hamilton-Taylor, J. *Environ. Geol. Water Sci.* **1988**, *11*, 107–114.
16. Peterson, M. L.; Carpenter, R. *Mar. Chem.* **1983**, *12*, 295–321.
17. Seyler, P.; Martin, J.-M. *Environ. Sci. Technol.* **1989**, *23*, 1258–1263.
18. Eary, L. E.; Rai, D. *Environ. Sci. Technol.* **1987**, *21*, 1187–1193.
19. Oscarson, D. W.; Huang, P. M.; Defosse, C; Herbillon, A. *Nature (London)* **1981**, *291*, 50–51.
20. Oscarson, D. W.; Huang, P. M.; Liaw, W. K.; Hammer, V. T. *Soil Sci. Soc. Am. J.* **1983**, *47*, 644–648.
21. Dyrssen, D. *Mar. Chem.* **1985**, *15*, 285–293.
22. Dyrssen, D. *Mar. Chem.* **1988**, *24*, 143–153.
23. Dyrssen, D.; Kremling, K. *Mar. Chem.* **1990**, *30*, 193–204.
24. Ambühl, H. EAWAG, Dübendorf, Switzerland, personal communication.
25. Bruland, K. W.; Franks, R. P.; Knauer, G. A.; Martin, J. H. *Anal. Chim. Acta* **1979**, *105*, 233–245.
26. Mart, L.; *Fresenius' Z. Anal. Chem.* **1979**, *299*, 97–102.
27. Zürcher, F.; Gisler, B. In *Proc. Eur. Symp. Physico-Chemical Behaviour of Atmospheric Pollutants, 4th.*; Angeletti, G.; Restelli, G., Eds.; D. Reidel: Dordrecht, Netherlands, 1987.
28. *Standard Methods for the Examination of Water and Wastewater*, 17th ed.; American Public Health Association: Washington, DC, 1989.
29. Peiffer, S.; Frevert, T. *Analyst (London)* **1987**, *112*, 951–954.
30. Davison, W. *Limnol. Oceanogr.* **1977**, *22*, 746–753.
31. Yamamoto, M.; Urata, K.; Murashige, K., Yamamoto, Y. *Spectrochim. Acta* **1981**, *36B*, 671–677.
32. Kuhn, A. Ph.D. Thesis No. 9783, Swiss Federal Institute of Technology, Zurich, Switzerland, 1992. Kuhn, A.; Sigg, L. *Limnol. Oceanogr.* **1993**, in press.
33. Johnson, C. A. *Anal. Chim. Acta* **1990**, *238*, 273–278.
34. Watanabe, H.; Goto, K.; Taguchi, S.; McLaren, J. W.; Berman, S. S.; Russell, D. S. *Anal. Chem.* **1981**, *53*, 738–739.
35. Lindberg, R. D.; Runnells, D. D. *Science (Washington, D.C.)* **1984**, *225*, 925–927.
36. Tiedje, J. M. In *Biology of Anaerobic Microorganisms*; Zehnder, A. J. B., Ed.; Wiley: New York, 1988; Chapter 4.
37. Höhener, P. Ph.D. Thesis, Swiss Federal Institute of Technology, Zurich, Switzerland, 1990.
38. Sigg, L.; Johnson, C. A.; Kuhn, A. *Mar. Chem.* **1991**, *36*, 9–26.
39. Stone, A. T. *Environ. Sci. Technol.* **1987**, *21*, 979–988.
40. Stone, A. T.; Morgan, J. J. *Environ. Sci. Technol.* **1984**, *18*, 450–456.
41. Burdige, D. J.; Nealson, K. H. *Geomicrobiol. J.* **1986**, *4*, 361–387.
42. Myers, C. R.; Nealson, K. H. *Geochim. Cosmochim. Acta* **1988**, *52*, 2727–2732.
43. Ehrlich, H. L. *Geomicrobiol. J.* **1987**, *5*, 423–431.

44. Ghiorse, W. G. In *Biology of Anaerobic Microorganisms;* Zehnder, A. J. B., Ed.; Wiley: New York, 1988; Chapter 6.
45. Myers, C. R.; Nealson, K. H. *Science (Washington, D.C.)* **1988**, *240*, 1319–1321.
46. Diem, D.; Stumm, W. *Geochim. Cosmochim. Acta* **1984**, *48*, 1571–1573.
47. Davies, S. H. R.; Morgan, J. J. *J. Colloid Interface Sci.* **1989**, *129*, 63–77.
48. Wehrli, B. In *Aquatic Chemical Kinetics;* Stumm, W., Ed.; Wiley: New York, 1990; pp 311–336.
49. Emerson, S.; Kalhorn, S.; Jacobs, L.; Tebo, B. M.; Nealson, K. H.; Rosson, R. A. *Geochim. Cosmochim. Acta* **1982**, *46*, 1073–1079.
50. Tebo, B. M.; Nealson, K. H.; Emerson, S.; Jacobs, L. *Limnol. Oceanogr.* **1984**, *29*, 1247–1258.
51. Chapnick, S. D.; Moore, W. S.; Nealson, K. H. *Limnol. Oceanogr.* **1982**, *27*, 1004–1014.
52. Sunda, W. G.; Huntsman, S. A. *Limnol. Oceanogr.* **1987**, *32*, 552–564.
53. Hastings, D.; Emerson, S. *Geochim. Cosmochim. Acta* **1986**, *50*, 1891–1824.
54. Tipping, E.; Jones, J. G.; Woof, C. *Arch. Hydrobiol.* **1985**, *105*, 161–175.
55. Johnson, C. A.; Ulrich, M.; Sigg, L; Imboden, D. M. *Limnol. Oceanogr.* **1991**, *36*, 1415–1426.
56. Stumm, W.; Wieland, E. In *Aquatic Chemical Kinetics;* Stumm, W., Ed.; Wiley: New York, 1990; pp 367–400.
57. Suter, D.; Banwart, S.; Stumm, W. *Langmuir* **1991**, *7*, 809–813.
58. Hering, J. G.; Stumm, W. In *Mineral-Water Interface Geochemistry;* Hochella, M. F.; White, A. F., Eds.; *Rev. Mineral. (Washington)* **1990**, *23*.
59. La Kind, J. S.; Stone, A. T. *Geochim. Cosmochim. Acta* **1989**, *53*, 961–971.
60. Sulzberger, B.; Suter, D.; Siffert, C.; Banwart, S.; Stumm, W. *Mar. Chem.* **1989**, *28*, 127–144.
61. Peiffer, S. In *Environmental Chemistry of Lakes and Reservoirs;* Baker, L. A., Ed.; Advances in Chemistry 237; American Chemical Society: Washington, DC, 1993; Chapter 11.
62. Lovley, D. R. *Geomicrobiol. J.* **1987**, *5*, 375–399.
63. Lovley, D. R.; Philipps, E. J. P.; Lonergan, D. J. *Environ. Sci. Technol.* **1991**, *25*, 1062–1067.
64. Potsma, D. *Geochim. Cosmochim. Acta* **1985**, *49*, 1023–1033.
65. Urban, N. In *Environmental Chemistry of Lakes and Reservoirs;* Baker, L. A., Ed.; Advances in Chemistry 237; American Chemical Society: Washington, DC, 1993; Chapter 10.
66. Bard, A. J.; Parsons, R.; Jordan, J. *Standard Potentials in Aqueous Solution. IUPAC;* Dekker: New York, 1985.
67. Sillen, L. G.; Martell, A. E. *Stability Constants;* Special Pub. No. 17; Chemical Society: London, 1964.
68. Cherry, J. A.; Shaikh, A. U.; Tallman, D. E.; Nicholson, R. V. *J. Hydrol.* **1979**, *43*, 373–392.
69. Spycher, N. F.; Reed, M. H. *Geochim. Cosmochim. Acta* **1989**, *53*, 2185–2194.
70. Andreae, M. O. *Limnol. Oceanogr.* **1979**, *24*, 440–452.
71. Andreae, M. O. In *Arsenic: Industrial, Biomedical, Environmental Perspectives;* Lederer, E. H.; Fensterheim, R. J., Eds.; Van Nostrand Reinhold: New York, 1983.
72. Johnson, D. L.; Pilson, M. E. Q. *Environ. Lett.* **1975**, *8*, 157–171.
73. Scott, M. Ph.D. Thesis, California Institute of Technology, Pasadena, CA, 1991.
74. Belzile, N.; Tessier, A. *Geochim. Cosmochim. Acta* **1990**, *54*, 103–109.
75. Pierce, M. L.; Moore, C. B. *Water Res.* **1982**, *16*, 1247–1253.
76. Anderson, L. C. D.; Bruland, K. W. *Env. Sci. Technol.* **1991**, *25*, 420–427.
77. Wehrli, B.; Ibric, S.; Stumm, W. *Colloids Surf.* **1990**, *51*, 77–88.

78. Schroeder, D. C; Lee, G. F. *Water Air Soil Pollut.* **1975**, *4*, 355–365.
79. Eary, L. E.; Rai, D. *Am. J. Sci.* **1989**, *289*, 180–213.
80. Smillie, R. H.; Hunter, K.; Loutit, M. *Water Res.* **1981**, *15*, 1351–1354.
81. Van der Weijden, C. H.; Reith, M. *Mar. Chem.* **1982**, *11*, 565–572.
82. Johnson, C. A.; Xyla, A. G. *Geochim. Cosmochim. Acta* **1991**, *55*, 2861–2866.
83. Johnson, C. A.; Sigg, L.; Lindauer, U. *Limnol. Oceanogr.* **1992**, *37*, 315–321.
84. Shaw, T. J.; Gieskes, J. M.; Jahnke, R. A. *Geochim. Cosmochim. Acta* **1990**, *54*, 1233–1246.
85. Sigg, L; Sturm, M.; Kistler, D. *Limnol. Oceanogr.* **1987**, *32*, 112–130.
86. Sigg, L.; Kuhn, A.; Xue, H.; Kiefer, E.; Kistler, D. In Huang, C. P.; O'Melia, C., Eds; Advances in Chemistry 244; American Chemical Society: Washington, DC, in press.

RECEIVED for review September 26, 1991. ACCEPTED revised manuscript May 21, 1992.

Manganese Dynamics in Lake Richard B. Russell

Tung-Ming Hsiung and Thomas Tisue

Department of Chemistry, Clemson University, Clemson SC 29634–1905

Lake Richard B. Russell is an impoundment of the Savannah River. The lake's waters are soft and mildly acidic. Organic matter is abundant. Seasonal oxygen depletion occurs during stratification, and reduced species, including Mn^{2+}, accumulate in the hypolimnion. Electron paramagnetic resonance studies indicate that Mn in the water column is present almost entirely as soluble and colloidal Mn^{2+} species, except near the surface where particulate forms sometimes predominate. Field and laboratory studies were used to estimate the rate of oxidation of Mn^{2+} in the water column, and to characterize the flux of reduced Mn across the sediment–water interface. Surprisingly, incubating bottom deposits with oxygenated bathylimnetic water released more Mn than did maintaining anoxia.

MANGANESE EXHIBITS COMPLEX BEHAVIOR in natural water systems, cycling readily among various oxidation states in response to changing environmental conditions (*1, 2*). The behavior of manganese in seasonally anoxic hypolimnetic waters generally follows the model developed by Delfino and Lee (*3*), who traced the migration of the boundary between oxidized and reduced forms from below the sediment–water interface up into the water column as anoxia developed during stratification. Manganese thus resembles iron in its response to changing redox conditions, and the biogeochemistries of the two elements are closely linked (*4*).

As a rule, however, the oxidation of reduced Mn is slower than that of reduced Fe. The reverse is true for reduction; Mn is released first from sediments as anoxia develops (*5, 6*). The net result is that the proportion of particulate Fe to total Fe is generally larger than the proportion of particulate

0065–2393/94/0237–0499$07.50/0

Mn to total Mn. These differences in kinetics and speciation can cause fractionation of the two elements when redox potential changes rapidly, or when particle and solute transport are disjunct.

Manganese is essential for the oxidation of water to oxygen in the photosynthetic process, occurring at the active site of photosystem II where its redox changes are linked to the four-electron oxidation of water. It is the only metal that has been found to be associated with the water-splitting apparatus in all the oxygen-evolving organisms studied to date. Although it is an essential element in plants and animals, elevated concentrations of Mn are toxic to a variety of aquatic organisms (7). In addition, reduced Mn makes water unpalatable and causes fouling and corrosion in water systems and cooling towers.

Oxidation

Oxidation of Mn^{2+} in aqueous solution appears to occur in a stepwise fashion to MnO_2, with MnOOH (8) and Mn_3O_4 (9) as possible intermediates. The reaction exhibits the induction period and kinetics characteristic of an autocatalytic process (10). Mn(II) is strongly sorbed to the surfaces of the newly formed, insoluble oxides, where its oxidation is greatly facilitated.

Morgan (11) derived a rate law that adequately describes the observed kinetics and was able to extract rate constants for both homogeneous and particle-catalyzed reactions. In laboratory experiments with sterile, filtered synthetic solutions, Mn^{2+} oxidation proceeds much more slowly than in natural waters. It may not occur at all at neutral or acidic pH, especially in the absence of catalytically active surfaces such as preformed oxidation products (12, 13).

The conventional interpretation is that rapid oxidation observed under natural circumstances indicates mediation by microorganisms (14), although an important caveat was offered by Tipping et al. (12, 15). These authors pointed out that it is difficult to prove conclusively whether oxidation is strictly biologically mediated or is catalyzed by abiotic particulate matter as well. Filtration removes abiotic catalysts along with microorganisms, and poisons used to halt biological activity also change chemical properties such as pH, redox potential, and speciation.

Reduction

MnO_x can be reduced to Mn(II) in natural waters by several means, including direct reaction with Fe(II) species (16, 17). Stone and Morgan (18–21) and others (22) showed that reduction of MnO_2 is rapid in the presence of readily oxidizable organic compounds such as catechols, hydroquinones, and related compounds. Oxalic acid is an effective reductant under acidic conditions (23), converting hausmanite (Mn_3O_4) to manganite (MnOOH). Many MnO_2-

reactive functionalities are present in natural waters in appreciable concentrations as microorganismal metabolites or the degradation products of nonliving organic matter. Reduction by this means thus is a distinct possibility, although its occurrence by strictly abiotic means appears not to have been conclusively demonstrated. Certainly, however, the presence of large amounts of exogenous organic matter, such as tannery effluent (*24*), will potentiate Mn reduction (*25*).

Reduced forms of sulfur, such as sulfide and thiols, also react rapidly with MnO_2 (*26–28*) as well as with FeO_x. However, sulfur in fresh water is often present in substoichiometric amounts with respect to iron. Thus little or no free reduced S is present even under strongly anoxic conditions because of the formation of very insoluble FeS_x species. Our equilibrium calculations (*29*) indicate that complexation with reduced sulfur species is not a quantitatively important aspect of Mn speciation in Lake Richard B. Russell (RBR). However, this result does not rule out the occurrence of such species as transient intermediates.

A third mechanism for Mn reduction has been demonstrated more unequivocally. Reduction of manganese oxides by naturally occurring organic matter is promoted by sunlight. Sunda et al. (*30*) suggested that the surprising predominance of Mn(II) in ocean surface waters is attributable to the reaction of MnO_x with hydrogen peroxide, produced photochemically by sunlight in the presence of organic matter (fulvic acid, FA), perhaps according to eqs 1 and 2.

$$FA + O_2 \xrightarrow{\text{light}} O_2^{\cdot -} + FA^+ \qquad (1)$$

$$2O_2^{\cdot -} + 2H^+ \longrightarrow H_2O_2 + O_2 \qquad (2)$$

Hydrogen peroxide can function as a reducing agent with respect to Mn(III,IV), but the chemistry doesn't necessarily stop with reduction. MnO_2 can bring about the disproportionation of H_2O_2, as shown in eqs 3 and 4.

$$MnO_2 + H_2O_2 + 2H^+ = Mn^{2+} + O_2 + 2H_2O \qquad (3)$$

$$Mn^{2+} + H_2O_2 = MnO_2 + 2H^+ \qquad (4)$$

Both of these reactions are thermodynamically feasible in neutral solution and could play a role in Mn dynamics throughout the water column. However, we were unable to detect H_2O_2 in Lake RBR even with the luminol chemiluminescent method (*31*). Manganese in seawater acts as an effective scavenger of superoxide, whose disproportionation it also catalyzes.

$$Mn^{2+} + O_2^{\cdot -} + 2H^+ = Mn^{3+} + H_2O_2 \qquad (5)$$

$$Mn^{3+} + \frac{1}{2}H_2O_2 = Mn^{2+} + H^+ + \frac{1}{2}O_2 \qquad (6)$$

Adsorption

Adsorption phenomena are an important aspect of Mn dynamics in several ways. Because of their high surface activity, freshly formed hydrous manganese oxides can strongly influence the cycling of nutrients, heavy metals, and organic substances (32, 33). MnO_x has a greater affinity for H^+ and multivalent cations than for alkali metal cations. Murray (34) found the zero point of charge for a fresh MnO_2 suspension at pH 2.25. His electron microscope studies showed highly aggregated particles (0.2–1.0-μm diameter) that became less reactive upon aging, perhaps because of condensation and dehydration.

Manganese oxides formed as precipitates by raising the pH of Mn-containing solutions are highly hydrated and of variable and nonstoichiometric composition. Like the corresponding iron phases, the hydrous manganese oxides are highly sorptive of Mn(II). Because adsorption also enhances the rate of Mn(II) oxidation, it is difficult to distinguish the removal of Mn(II) from solution by each of the two mechanisms.

Many surfaces (including those of the oxides of Ti(IV), Si(IV), Sn(IV); calcite; clay minerals; and feldspar) may accelerate the oxidation of Mn(II) (35). Morris and Bale (36) and Coughlin and Matsui (37) noted that the removal of Mn from solution is associated with the presence of suspended particles. Wilson (38) presented evidence that this effect is attributable at least in part to catalysis of Mn(II) oxidation.

The effects of pH on Mn adsorption always must be taken into account because they can mask or mimic other effects, as Hoffmann and Eisenreich (39) demonstrated. For example, lowering pH releases soluble Mn^{2+} from adsorption sites on particles even when no reduction is involved.

The metal oxide surface apparently is not necessarily a passive participant in photoassisted reduction reactions involving humic and fulvic substances. Redox processes can be induced or enhanced by interaction of light with chromophores on the oxide surface itself. These processes lead to accelerated particle dissolution (40). Another factor to consider is that light absorption by organic matter may result in the production of cationic species that are bound more strongly than their neutral counterparts to the usually negatively charged surfaces of metal oxides (41). Suwanee River fulvic acid and MnO_2 underwent a redox reaction when illuminated at a rate significantly greater than in the dark; O_2 was not required (42). In soft acidic fresh waters, especially those high in suspended mineral matter or humic substances, photoassisted dissolution of manganese oxides may occur by a different mechanism than in ocean surface waters.

Microbial Mediation

Many of the reactions discussed so far are subject to microbial mediation or influence (43, 44). In fact, a common view is that Mn biogeochemistry in fresh water is dominated by the element's involvement in the life processes of various types of organisms. Originally, this hypothesis was based largely on evidence from field studies, such as the occurrence of temperature maxima for Mn oxidation rates (45), seasonal redox cycling in synchrony with microbial population fluctuations, and inhibition by well-known poisons such as azide (14).

The importance of microbiological involvement also has been strongly supported by more recent discoveries including, for example, bacterial strains that can use MnO_x as their sole terminal electron acceptor (46). Many different types of microorganisms oxidize manganese, including bacteria, yeast, and fungi (47). Richardson and Nealson (48) divided microbially related Mn redox reactions into conceptual categories, including direct oxidation involving specific proteins that are often extracellular; direct reduction, as when oxidized Mn is used as the terminal electron acceptor for anaerobic respiration; indirect chemical oxidation, associated with increases in the environmental pH or redox potential as a result of microbial activity; and indirect chemical reduction, carried out by reductants released from microbial cells, such as sulfide (27) or oxalate (23). These phenomena have been observed in both water and sediments. Alexander (49) pointed out that microbiological involvement is likely to be least prominent below pH 5.5, where exchangeable Mn(II) predominates, and above pH 8.0, where oxidation by oxygen becomes rapid.

Extreme anoxia is not required for and may actually inhibit the release of reduced forms of Mn near the sediment–water interface in lakes. Studies in Lake Constance, for example, revealed that Mn can be mobilized out of the sediment at oxygen concentrations as high as 4 mg/L (50). In Esthwaite Water, reduced Mn in the surficial sediment reached a maximum under well-mixed conditions and Mn(II) accumulated in the hypolimnion under oxic conditions (51). Two sediment-trap experiments showed that the flux of reduced Mn into the water column was actually higher during overturn than during the seasonal anoxic period (52, 53). These observations suggest that the release of Mn accompanying mineralization of organic matter may be a more important source at times than reductive dissolution of MnO_x. We will discuss this hypothesis in greater detail because it is consistent with the behavior of Mn in Lake RBR.

Lake Richard B. Russell

The structure impounding Lake RBR was completed in December 1983, and full pool elevation (145 m above sea level) was reached in November 1984. The impoundment inundates part of the Savannah River watershed between Lake Hartwell Dam and the headwaters of Strom Thurmond Lake, into which the tailrace discharges (*see* Figure 1). Maximum depth is about 40 m, and the surface area is approximately 105 km^2 (26,000 acres).

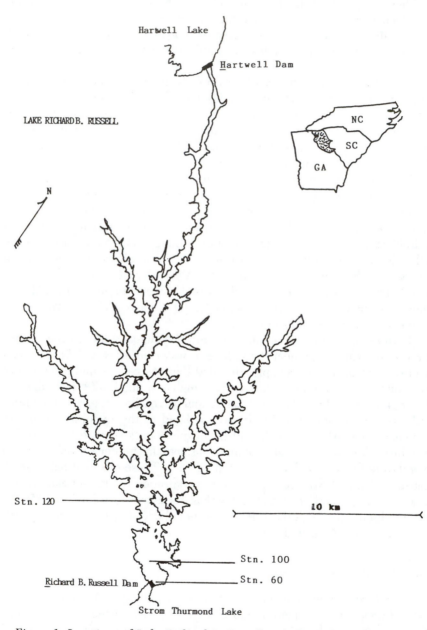

Figure 1. Locations of Lake Richard B. Russell and of sampling stations 60, 100, and 120.

Harvestable timber was removed from the lake bed during site preparation, but the remaining forestation was left untouched except for some shoreline clearing. The lake's waters are soft (conductivity around 50 μS/cm), slightly acidic (pH 6–7), and contain abundant organic matter; dissolved organic carbon (DOC) is 2–5 mg/L.

A thermocline develops at depths around 10 m beginning in March, and the lake remains stratified from then until November. Oxygen depletion occurs in the hypolimnion during stratification, but becomes pronounced (dissolved oxygen <5 mg/L) only in the bottom few meters and only during the warmest months.

An unusual feature of the lake is the oxygen injection system installed by the U.S. Army Corps of Engineers, which is constrained by the states of South Carolina and Georgia to maintain at least 6 mg/L of dissolved oxygen in the discharge waters. This criterion often is met by pulse injection of oxygen just prior to discharge, which minimizes the impact of oxygenation on the power pool as a whole.

To study the effectiveness of the oxygen injection system, as well as to investigate other water-quality issues, the Waterways Experiment Station (WES), U.S. Army Corps of Engineers, conducted extensive studies in both Lake RBR and Lake Thurmond. These ongoing studies, which began prior to creation of the impoundment, provide a detailed picture of the course of events since impoundment. Reports containing these data (54) form a valuable basis for other studies.

An interesting observation that emerged early in the WES studies was accumulation of reduced soluble Mn in the hypolimnion at concentrations up to several milligrams per liter. Our attention was drawn also to a preliminary report by Turner (55) indicating rapid rates of reoxidation of Mn^{2+} despite the lake's low pH.

Experimental Methods

Hydrological data including temperature, dissolved oxygen, pH, and conductivity were collected by using a water-quality instrument package (Hydrolab Surveyor II, Hydrolab Corp., Austin, Texas) on 11 occasions in 1988. Three sites were sampled: Stations 60, 100, and 120 are shown on the map in Figure 1. These stations correspond, respectively, to the power pool just behind the dam, the upstream end of the power pool, and a location further upstream at the head of the principal basin.

Water samples for chemical analyses were pumped through Tygon tubing by a submersible pump, then stored in linear polyethylene containers that had been cleaned by soaking in 10% nitric acid, followed by extensive rinsing with deionized (Nanopure system) water. Samples were kept on ice in the dark until they reached the lakeside field laboratory. Filtration through 0.45-μm cellulose acetate membranes and acidification took place within a few hours following collection. Subsamples for electron spin resonance (ESR) spectrometric determinations were usually frozen promptly in hematocrit tubes, which were thawed just prior to analysis and inserted directly into the spectrometer. Other subsamples for atomic absorption (AA) spectrophotometric determinations were preserved at pH 2 or below and stored at room temperature in precleaned linear polyethylene bottles. Standards treated in the same way as water samples showed no significant changes in Mn concentration or speciation.

A gravity dredge was used to recover samples of the soft, dark, organic-rich muds that have accumulated thinly over the inundated clay soils of the watershed. In some locations, soft deposits such as alluvial silts could be recovered with a 2.5-inch-diameter gravity corer.

An atomic absorption spectrophotometer (Perkin-Elmer 4000) was used with standard operating conditions to determine dissolved and total Mn, irrespective of oxidation state. Dissolved organic carbon was determined with an organic carbon analyzer (Beckman model 915B TOCmaster).

The basis of ESR spectrometry is the measurement of the energy required to reverse the spin of unpaired electrons in an external magnetic field. At a given external magnetic field strength H_0 (in gauss), the energy ΔE (in ergs) required for the transition between two quantized electron spin states (parallel and antiparallel to the field) is given as $\Delta E = g\mu_B H_0$, where g is the so-called spectroscopic splitting factor (2.0023 for the free electron) and μ_B, the Bohr magneton, equals 0.92371×10^{-20} erg/G. In the octahedral aquo complex, $Mn(H_2O)_6^{2+}$, manganese has a $3d^5$ electron configuration with a g-value close to 2.0.

Microwaves are conveniently generated with klystrons in the X-band region around 9 GHz. For g values around 2, these frequencies require magnetic fields of about 3 kG to produce spin transitions. Hyperfine structure in the spectrum results from interactions of the electron spin (S) with the nuclear spin (I), leading to $2I + 1$ transitions. In mobile fluids, anisotropic spin coupling averages to zero, leaving isotropic coupling as the only interaction observed. Because $I = 5/2$ for ^{55}Mn (100% abundance), aquo Mn^{2+} gives a six-line absorption spectrum.

Spectral line width varies inversely with the excited-state lifetime according to Heisenberg's principle, $\Delta T \times \Delta H = h/2\pi$, where ΔT is the lifetime of the excited spin state, h is Planck's constant, and ΔH is the effective width of the absorption signal. Excited-state lifetimes are subject to environmental (including chemical) influences. The resulting line-shape changes yield information about the chemical environment of the Mn atoms. Both spin–lattice and spin–spin relaxation mechanisms can contribute to the overall lifetime.

Spin–lattice relaxation (time T_1) results from dissipation of the excited-state energy among vibrational modes of the matrix. In mobile liquids, vibrational fluctuations are spread over a very wide frequency range. This configuration decreases the probability of spin–lattice coupling. As a result, T_1 is long and thus makes a negligible contribution to the line width.

Spin–spin relaxation (time T_2) can result from both intermolecular homonuclear exchange coupling and dipole–dipole interactions, but only the latter is observable at Mn^{2+} concentrations $<10^{-2}$ M. Mn^{2+} forms both inner- and outer-sphere complexes. In symmetrical inner-sphere complexes like $Mn(H_2O)_6^{2+}$, the spin–spin coupling is strongly forbidden, T_2 is long, and lines remain narrow. When nonsymmetric inner-sphere complexes form, the resulting anisotropy of the electric field leads to allowed spin–spin transitions that produce very small values of T_2 and very broad, perhaps even unobservable, lines (56).

Fortunately, Mn^{2+} does not form strong (inner-sphere) complexes at the ligand concentrations normally present in natural waters. For example, nitrate, chloride, bicarbonate, and sulfate do not form observable complexes in fresh water, and pH has no effect in the range from 2 to 7. Weakly interacting species that form only outer-sphere complexes, such as naturally occurring organic matter, could have some influence on line width, but solvent–ligand exchange in the absence of a strong ligand field produces little perturbation of the electron-spin system. Ligand exchange rates for outer-sphere complexes would be expected to be very close to solvent fluctuation rates. This similarity would make their line-broadening effect, if any, difficult to observe (57).

ESR offers compelling advantages in studies of Mn biogeochemistry (57, 58). Chiswell and Mokhtar (59) compared various means for studying Mn speciation. They concluded that ESR offers optimal specificity for Mn^{2+}, sufficient sensitivity, and minimal alteration of the sample. With its $3d^5$ electron configuration on Mn, $Mn(H_2O)_6^{2+}$ gives a strong, highly characteristic spectrum at a field strength well separated from other commonly occurring paramagnetic species. Several groups have reported (60–62) that the only ESR signal detectable in nonchemically isolated humic material was that of Mn^{2+}. Thus, association of Mn^{2+} with even relatively large colloids does not interfere with its determination by ESR spectroscopy.

An ESR spectrometer (Varian model E-3) was used to observe and quantify Mn^{2+} species at a field strength of 3155 ± 50 G and a frequency of 9.5 GHz. A flat fused silica "ribbon" cell (Wilmad Glass No. WG-812) was used at very low concentrations to optimize the signal-to-noise ratio by minimizing dielectric losses. Microwave power was set routinely to 4 mW, but was occasionally raised to optimize sensitivity at very low concentrations. Quantitation was based on the height of the lowest-field peak in the first derivative of the absorption spectrum. As reported by others (63), this technique is characterized by precision and accuracy of about 1% relative standard deviation over a linear range from $<10^{-6}$ to 10^{-4} M ($<0.05–5$ mg/L).

Two types of incubations were carried out to simulate in situ conditions. In the first, the kinetics of the oxidation of Mn^{2+} were observed by using an apparatus in which liter-sized lake-water samples were equilibrated with various gases as shown in Figure 2. The major features of this device were a 1-L perfluoroalkoxy Teflon reservoir that fed the sample to a borosilicate glass counterflow gas–liquid equilibration chamber by means of a recirculation pump operating at 120 mL/min. The liquid fell as a continuous film down the undulating inner surface of an Allihn condenser, where it met the upward-flowing (50 mL/min) gas stream. After passing through the equilibrator, the liquid stream returned to the reservoir via a pear-shaped stilling well. The entire apparatus was wrapped in Al foil to minimize photoreduction. The feed gases (oxygen, nitrogen, and air) were humidified to prevent evaporation, and CO_2 was scrubbed with 1 M NaOH to avoid large pH changes. Bubble formation was strictly avoided because particle nuclei may be formed by adsorption of dissolved organic compounds on bubble surfaces. Such adsorption would lead to irreproducible effects in successive experiments. With purified nitrogen as the feed gas, Mn^{2+} solutions underwent no concentration changes in this apparatus for periods of up to 1 week.

In the second incubation study we observed the rate of mobilization of Mn from bottom deposits as oxygen partial pressure was varied in the overlying water. Freshly collected muddy sediment was centrifuged briefly under nitrogen, then transferred into a 4-L high-density polyethylene jar to form a layer about 4 cm thick. This layer was then quickly frozen. About 3 L of freshly collected bathylimnetic water, chilled to 4 °C, was carefully added without disturbing the frozen mud surface. A lid was affixed with ports for a gas inlet and sampling and gas outlets. The contents were then allowed to come to room temperature while a slow stream of purified nitrogen or oxygen was admitted through a plastic tube 5 cm above the sediment–water interface to keep the aqueous phase homogeneous and gas-saturated. Samples were withdrawn at designated time intervals by closing the gas outlet to force liquid out through the sampling port. To assess the effect of biological activity, phenylmercuric acetate (3–5 mg/L) was added in some experiments.

Discussion of Results

The relationship between Mn^{2+} distribution and the usual hydrological variables is depicted in Figure 3 for station 120 during 1988. The three-

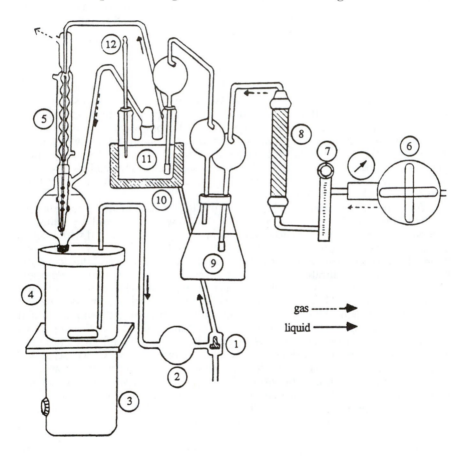

gas - - - - - ➤
liquid ———➤

1. 3-way valve
2. Masterflex pump
3. Magnetic stirrer
4. One liter teflon PFA vessel
5. Counterflow gas-liquid equilibrator
6. Gas pressure regulator and gauge

7. Rotameter
8. Activated carbon bed
9. 1.0 M NaOH
10. Heating mantle
11. Deionized water
12. Thermometer

Figure 2. Sketch of the falling-film gas–liquid equilibrator.

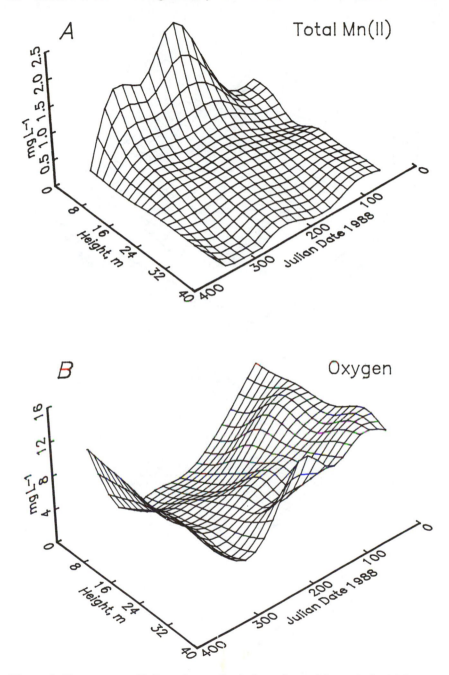

Figure 3. Variations in Mn^{2+} and some hydrological variables in Lake RBR during 1988. Height, distance above the sediment–water interface, is used in place of depth because the level of the reservoir fluctuates by ~3 m during the year. Continued on next page.

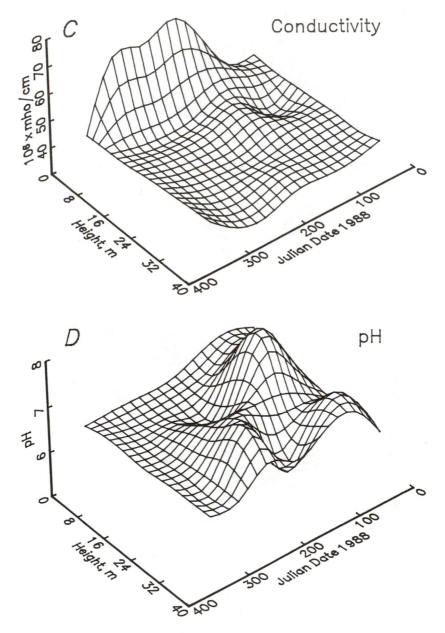

Figure 3. Continued.

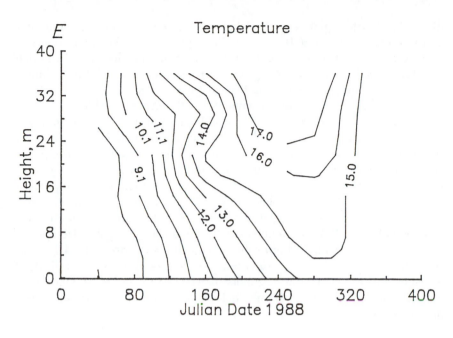

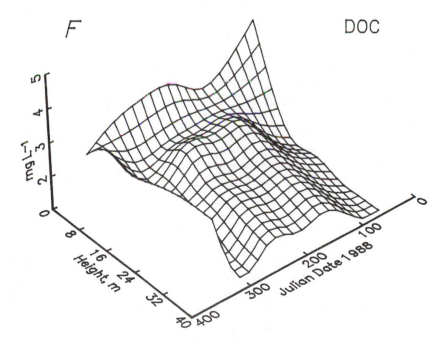

Figure 3. Continued.

dimensional surface plots and a two-dimensional contour plot were generated by proprietary software (64), using moderate tension in the smoothing routines to assist visualization of general trends. Some idealization of the data inevitably results from generating contours in either two or three dimensions. Temperature was plotted in two dimensions because the optimal perspective in that case was not the same as for the other variables.

Data were also collected at stations 60 and 100, but they are not presented here because of their similarity to the results at station 120. Because station 120 was well upstream from the site of oxygen injection in the power pool near stations 60 and 100, we conclude that operation of the oxygen injection system in 1988 had only minor effects on the chemical and physical variables investigated. A detailed study of Mn speciation was carried out at station 120 during the stratified period in 1990, and the results are summarized in Figure 4. A sediment core collected during the well-mixed period had subsurface maxima in both soluble–colloidal and adsorbed Mn^{2+} concentrations (Figure 5). The results obtained in laboratory studies of Mn^{2+} oxidation are displayed in the rate plots shown in Figures 6 and 7. Manganese behavior during laboratory simulations of forced benthic reoxygenation is presented in Figure 8.

Speciation.   Most of the Mn found in the water column of Lake RBR occurred in the Mn(II) oxidation state, as both soluble–colloidal and particulate species. This conclusion rests on the following evidence. At station 120, for all of 1988 at all depths ($N = 52$), there was no statistically significant difference between total Mn determined by AA spectrophotometry on unfiltered samples after acidification (0.62 mg/L), dissolved Mn determined by AA spectrophotometry on filtered samples after acidification (0.60 mg/L), and Mn^{2+} determined by ESR spectroscopy on unfiltered samples (0.63 mg/L).

A more detailed investigation was carried out at station 120 during stratification. Manganese concentration exhibited a nearly logarithmic profile spanning more than 2 orders of magnitude, decreasing with height above the bottom (Figure 4). (The slight bulge in the profile near the surface may reflect the influence of photoassisted reduction.) This study showed that most of the Mn in the water column was in the bathylimnion, in the form of soluble–colloidal species. Acidification of 0.45-μm filtrates from all depths led to no change in the ESR signal. Thus Mn passing through the filter must have been present either as $Mn(H_2O)_6^{2+}$ or as outer-sphere complexes with weakly interacting organic ligands. Nonsymmetric inner-sphere complexes give undetectably broad ESR signals, and outer-sphere complexes with strongly interacting ligands show decreased peak height. Both are converted to $Mn(H_2O)_6^{2+}$ in 1 M HNO_3.

Previous ESR determinations of stability constants for complexes of Mn^{2+} with commercial fulvic acid preparations (Aldrich) gave log K values

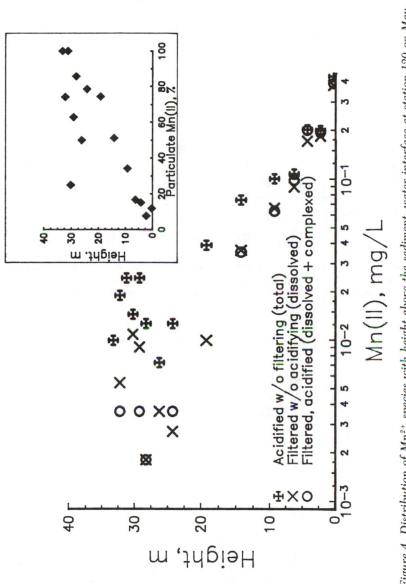

Figure 4. *Distribution of Mn²⁺ species with height above the sediment–water interface at station 120 on May 23, 1990. Inset: percent of Mn²⁺ retained on a 0.45-µm filter.*

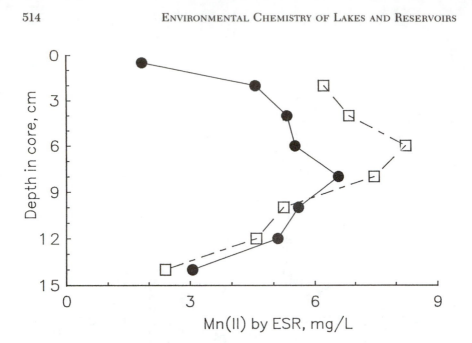

Figure 5. Mn²⁺ concentration in silt sediment core from Shuckpin Eddy. Key: ●, total Mn²⁺ in acidified pore fluids in milligrams per liter; and □, Cu²⁺-exchangeable Mn²⁺ in milligrams per liter in wet sediment.

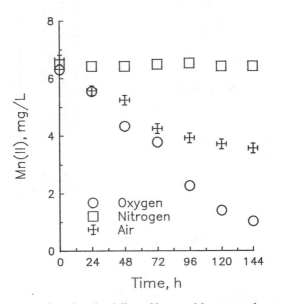

Figure 6. Kinetic data for the falling-film equilibrator with various gases.

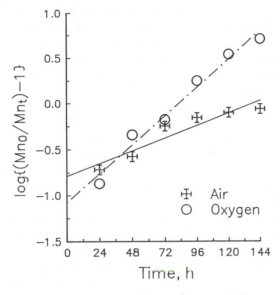

Figure 7. Linearized rate plots at 293 K.

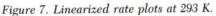

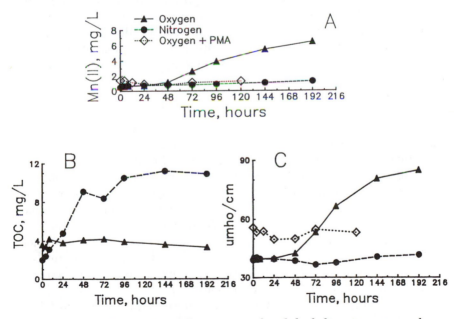

Figure 8. Incubation of initially anoxic mud with bathylimnetic water under various conditions.

of 4.2–4.8 in the pH range 5.5–6.5 (29), if one accepts the average molecular weight of 1000 Daltons determined by field-flow fractionation (65). These values agree well with stability constants reported in the literature (66) for complexes of Mn^{2+} with peat, lake, and marine fulvic acids.

ESR examination of nonchemically isolated fulvic acids showed that Mn^{2+} was the primary paramagnetic species observable (60, 61). Most likely, the soluble–colloidal fraction we identified in the speciation studies consisted primarily of such complexes. Because the ESR spectral characteristics of Mn in fulvic acid complexes are quite similar to $Mn(H_2O)_6^{2+}$, Alberts et al. (62) suggested that the metal–fulvate interaction was weak. Stronger interaction would be expected to lead to changes in peak shape. This view leaves unexplained the ability of the complexes to survive the isolation procedure's long ultrafiltration steps, because weak interactions are usually associated with reversible complexation.

Particulate forms of Mn^{2+} predominated (>60% of total Mn) above the thermocline, although the total concentration was low (Figure 4). The productivity, which was high in this zone, suggested incorporation of Mn into living biomass rather than adsorption onto abiotic particles. Stauber and Florence (67) noted that Mn associated with cells is in the II or III oxidation states, rather than as the fully oxidized MnO_2. Thus, desorption of Mn^{2+} from particulate matter occurred as detritus settled out of the epilimnion with little or no change in oxidation state, presumably in association with the onset of cell lysis and mineralization of organic matter. One is led to hypothesize that Mn transport at this time of year was dominated by association of reduced forms with biomass, rather than by classical cycling of oxidized particulate species and reduced soluble species between zones of high and low redox potential. We will show that a different situation may exist at other times.

Annual Cycling. The behavior of Mn in Lake RBR over an entire year seems to follow the classical cycle, as seen in Figure 3. Concentrations are relatively low in the epilimnion at all times, but they increase during the stratified period in the hypolimnion and especially the bathylimnion. This increase is correlated with falling dissolved oxygen concentration ($r = -0.81$), rising conductivity ($r = 0.93$), and rising dissolved and total organic carbon ($r = 0.84$ and 0.93, respectively). There is also some association with decreasing pH, but the variation in pH is small and the correlation is not significant for either pH or $[H_3O^+]$ ($r = -0.54$ and $+0.23$, respectively). In these soft waters with low alkalinity, pH changes alone cannot explain the observed variation in Mn concentrations (39).

After a short lag following the onset of destratification, Mn that had accumulated in the water column decreased sharply as oxygen levels increased and organic carbon declined. There was very little retention of Mn in the water column during the following well-mixed period. The efficient removal of Mn from the water column during the well-mixed period probably was associated with both an increase in the rate of oxidation to insoluble MnO_x and a decrease in the rate of reduction, as oxidizable organic matter became less abundant. This mechanism was not operating during stratification, however, because essentially all the Mn in the water column then was present as Mn^{2+} (Figure 4). Settling of particulate organic matter and detritus may remove Mn to the bathylimnion at this time. Oxidized forms of Mn that are formed transiently in the epilimnion could participate actively in this process by catalyzing the condensation of the precursors to humic substances (68, 69), while undergoing reduction in the process.

Another source of the Mn that accumulated in the bathylimnion during stratification is the Fe- and Mn-rich clay soils that characterize the region, plus the thin layer of organic–mineral ooze that has accumulated since impoundment. Little suspended sediment reaches the lake from the watershed during the warmer, drier months. In the inundated soils, Mn is abundant in the pore fluids and as exchangeable forms adsorbed to particles (Figure 5), even during the well-mixed period when the sediment–water interface is well oxygenated. The presence of a subsurface maximum in Mn concentration in the sediment interstitial fluids has been noted in other lacustrine and marine systems (70, 71). Our measurements indicate that in Lake RBR the oxic–anoxic boundary rises to near the sediment–water interface, or just above it, as stratification develops.

Oxygen concentration is not the only factor involved in releasing Mn. It exerted an indirect influence, perhaps by controlling the availability of oxidizable organic matter. Even the particulate forms present in the well-oxygenated epilimnion were primarily Mn(II) species. Any oxidized species that formed must have been reduced rapidly because their steady-state concentration remained very low. Accumulation of reduced Mn under suboxic conditions has been noted (51, 53). Ostendorp and Frevert (50) found that the reduction of Mn(IV) in lake sediment began at 2–3 mg/L of dissolved oxygen. In Rostherne Mere, United Kingdom, Mn accumulated in a hypolimnion that was not completely anoxic (72). Bacteria have been reported (73, 74) to reduce MnO_2 under both aerobic and anaerobic conditions.

One may hypothesize that the mobilization and accumulation of Mn we observed during stratification was linked to the degradation of organic matter. The degradation rate reaches a maximum in the bathylimnion during summer, as indicated by the sharp increase in conductivity. This process facilitates the release of Mn under suboxic, rather than anoxic, conditions, and is consistent with the incubation experiments reported here.

Oxidation Kinetics.   The results of kinetic experiments carried out in the laboratory in the gas–liquid equilibrator are illustrated in Figure 6. For these experiments, unfiltered epilimnetic water from Lake RBR was adjusted to pH 6.5 and amended with $Mn(ClO_4)_2$ to bring the concentration of Mn^{2+} to 6–7 mg/L. Samples were withdrawn at intervals from the apparatus, filtered (0.45 μm), and analyzed for Mn^{2+} by ESR spectroscopy. Stumm and Morgan (75) postulated a three-step oxidation process to account for the reaction's autocatalytic nature.

Step 1. Homogeneous oxidation

$$Mn^{2+}(aq) + O_2(aq) \xrightarrow{\text{slow}} MnO_2(s) \tag{7}$$

Step 2. Adsorption

$$Mn^{2+}(aq) + MnO_2(s) \xrightarrow{\text{fast}} Mn(II) \cdot MnO_2(s) \tag{8}$$

Step 3. Heterogeneous oxidation

$$Mn(II) \cdot MnO_2(s) + O_2(aq) \xrightarrow{\text{slow}} 2MnO_2(s) \tag{9}$$

The rate expression for this process at constant pH and O_2 partial pressure is

$$\frac{-d[Mn^{2+}]}{dt} = k_0[Mn^{2+}]_t + k[Mn^{2+}]_t[MnO_x] \tag{10}$$

where $[Mn^{2+}]_t$ is the concentration of Mn in the II+ oxidation state at any time t; $[MnO_x]$ is the concentration of Mn with oxidation number $>$II+ but $<$IV+; k_0 is the homogeneous first-order rate constant; and k is the heterogeneous second-order rate constant. After separating the variables, and assuming no MnO_x is present initially, integration gives

$$\log\left(\frac{[Mn^{2+}]_0}{k_0}\right) + \log\left(k_0 + \frac{k([Mn^{2+}]_0 - [Mn^{2+}]_t)}{[Mn^{2+}]_t}\right)$$

$$= \frac{[Mn^{2+}]_0 k + k_0}{2.3}t \tag{11}$$

If one also assumes that

$$k([Mn^{2+}]_0 - [Mn^{2+}]_t) >> k_0 \tag{12}$$

the integrated rate expression becomes

$$\log\left(\frac{k[Mn^{2+}]_0}{k_0}\right) + \log\left(\frac{[Mn^{2+}]_0}{[Mn^{2+}]_t} - 1\right) = \frac{k[Mn^{2+}]_0}{2.3}t \tag{13}$$

The condition required for determining the experimental rate constants from this expression is that a straight line be obtained when

$$\log \left(\frac{[Mn^{2+}]_0}{[Mn^{2+}]_t} - 1 \right)$$

is plotted against t. For this line the slope is $-k[Mn^{2+}]_0/2.3$ and the intercept is $-(\log[Mn^{2+}]_0 k/k_0)$.

We applied this approach to extract rate constants for the homogeneous and heterogeneous reactions from the kinetic data shown in Figure 6. The resulting linearized rate plots are shown in Figure 7, and the rate constants are given in Table I. The kinetics demonstrate the dependence on O_2 partial pressure that theory predicts. The value of k for O_2 partial pressure of 1 atm is approximately fivefold greater than the value for air.

Table I. Experimental Rate Constants

Equilibration Gas	k $(10^{-3}\ h^{-1}\ mg^{-1}\ L^{-1})$	k_0 $(10^{-3}\ h^{-1})$
N_2	0.0	0.0
Air	-0.9	-2.3
O_2	-4.8	-2.5

The oxidation reactions shown in Figure 6 were completely inhibited by even a few milligrams per liter of phenyl mercuric acetate, a broad-spectrum poison. Also, no oxidation occurred at pH values <5.5. We began with fresh, unfiltered epilimnetic water that contained the natural assemblage of particulate matter and microorganisms. Therefore, it seems likely that we were measuring the rates of reactions mediated by biological processes. This conclusion is consistent with the results of control experiments involving sterile filtered epilimnetic water. The oxidation rate was reduced until a barely significant decrease in Mn^{2+} was detectable only after 72 h of equilibration with air.

However, if one accepts the hypothesis that only microbially promoted processes are occurring, it becomes necessary to explain the approximately linear dependence of the oxidation rate on O_2 partial pressure as a conveniently exact effect on metabolic reaction rates. Ross and Bartlett (76) suggested that oxidation of Mn^{2+} is initiated by bacteria but that subsequent reaction is dominated by abiotic autocatalytic processes, once catalytically active colloids and particles are formed in sufficient numbers.

Simulation of Sediment–Water Interactions.   To investigate factors affecting the flux of Mn across the sediment–water interface, we incubated mud (freshly collected during stratification) with bathylimnetic water under both nitrogen and oxygen saturation. The mud was centrifuged briefly under N_2 to remove most of the interstitial fluid. Conductivity, pH, total organic carbon (TOC), and the concentrations of oxygen and Mn^{2+} were monitored in the overlying water.

The results of these experiments are depicted in Figure 8. Under nitrogen, little change took place in any of the variables after the first 12 h. An exception was TOC, which increased in the water, presumably with release of fermentation products from the solids. The mud retained its dark color and foul odor, and Mn^{2+} remained around 1 mg/L in the green-brown supernatant water. Saturation with CO_2-free oxygen caused the increase in conductivity and decrease in pH expected to accompany oxidative degradation of organic matter. TOC showed little change. The surface of the solids changed from black to brown, while the supernatant liquid lightened visibly and lost much of its stench.

Mn^{2+} increased severalfold in concert with these changes. As in the water column, there was no significant difference in Mn^{2+} determined by ESR spectroscopy and total Mn determined by AA spectrophotometry after acidification. Similarly, there was no difference in Mn concentration between filtered and whole-water samples. All the Mn released into the water column was present as soluble–colloidal Mn^{2+} species. Although surprising when it occurred, the mobilization of Mn^{2+} under oxygen actually is in accord with the reports of other groups, who have observed accumulation of Mn^{2+} under oxic conditions.

We surmise that under anoxic conditions, Mn^{2+} was associated primarily with particulate organic matter and organic–mineral aggregates in the solid phase. Aerobic degradation of this organic matter released Mn^{2+} into the supernate, presumably in the form of complexes with lower molecular weight humic substances. Phenylmercuric acetate at a few milligrams per liter strongly inhibited the release of soluble Mn (see Figure 8). Thus, it is likely that microorganisms are responsible for degradation of the organic matter. This release mechanism was probably abetted by the accompanying decrease in pH. Observation of this phenomenon was probably made easier by the very low rate of oxidation of Mn^{2+} that would be expected under these conditions of low pH and alkalinity. In other natural water systems where the oxidation rate is higher, this effect might manifest itself as an increase in the Mn cycling rate during the mixing period, as has been observed (52, 53).

Net fluxes of Mn^{2+} from the mud to the overlying water in the incubation experiments averaged 0.20 ± 0.01 g/m^2 per day under oxygen and 0.018 ± 0.006 g/m^2 per day under nitrogen ($N = 2$), over the period of the experiment. These values can be used to calculate apparent diffusion coefficients in the sediment. The mean calculated value, $68 \pm 31 \times 10^{-6}$ m^2/day, is close to Li and Gregory's reported value of 50×10^{-6} m^2/day at 25 °C near the water–sediment interface in marine sediments (77). This result indicates that physical resuspension and bioturbation effects were probably absent in our experiment.

Are the fluxes observed in the incubation experiments representative of conditions in situ? A linear regression of Mn^{2+} concentration on the logarithm of distance above the interface gives a good fit ($r > 0.99$) to the data collected for the speciation study shown in Figure 4. The regression equation yields a value for the gradient in Mn^{2+} concentration, $dC/dz = 0.1$ mg/L per centimeter at a height $z = 1$ cm above the interface. This gradient must be created by a flux, J, of Mn^{2+} out of the sediment, which is given by

$$J = -K \left(\frac{dC}{dz} \right) \qquad (14)$$

where K is the eddy diffusion coefficient. Thus, if the flux in the lake falls in the range observed in the incubation experiments (~ 0.02–0.20 g/m^2 per day), the observed gradient implies values of the eddy diffusion coefficient in the range $10^{-3} < K < 10^{-2}$ m^2/day. Unfortunately, there have been very few direct determinations of eddy diffusivity in lakes and reservoirs. Under winter ice cover, Coleman and Armstrong's ^{222}Rn profiles yielded values of $10^{-3} < K < 10$ (78). The values implied by our data for Lake RBR fall toward the lower end of this range, as perhaps they should, given the generally quiescent conditions associated with the stratified period. Additional work clearly is needed in this area. The ^{222}Rn profile method described in detail by Craig (79) could yield K values of improved accuracy and greatly facilitate calculation of sediment–water exchange fluxes.

Acknowledgments

We are grateful to Ray Turner, Physics Department, Clemson University, for his assistance with the ESR measurements. J. Hains, J. Carroll, W. Jabour, and M. Potter, of the U.S. Army Waterways Experiment Station–Calhoun Falls, provided sampling gear and instrumentation, laboratory space, many useful discussions, and much help in the field. Thoughtful reviews by the editor and three reviewers and a careful reading of the manuscript by S. Libes contributed greatly to accuracy and completeness. We acknowledge support of the initial parts of this work by a grant from the U.S. Geological Survey through Clemson University's Water Resources Research Institute. T. Tisue thanks the Center for Great Lakes Studies, University of Wisconsin–Milwaukee, for a Shaw Visiting Professorship.

References

1. Crerar, D. A.; Cormick, R. K.; Barnes, H. L. In *Geology and Geochemistry of Manganese* Vol. I: General Problems; Varentsov, I.; Grasselly, G., Eds.; Springer Verlag: Stuttgart, Germany, 1980.

2. Stumm, W. *Chemical Processes in Lakes*; Wiley: New York, 1985.
3. Delfino, J. J.; Lee, G. F. *Environ. Sci. Technol.* **1968**, *2*, 1094.
4. Balzer, W. *Geochim. Cosmochim. Acta* **1982**, *46*, 1153.
5. Thibodeaux, L. J. *Chemodynamics, Environmental Movement of Chemicals in Air, Water, and Soil*; Wiley: New York, 1979.
6. Koyama, T.; Tomino, T. *Geochem. J.* **1967**, *1*, 109.
7. *Nutritional Bioavailability of Manganese*; Kies, C., Ed.; ACS Symposium Series 354; American Chemical Society: Washington, DC, 1987.
8. Kessick, M. A.; Morgan, J. J. *Environ. Sci. Technol.* **1975**, 9, 157.
9. Klinkhammer, G. P.; Bender, M. L. *Earth Planet. Sci. Lett.* **1980**, *46*, 361.
10. Hsiung, T. M. "Manganese Dynamics in the Richard B. Russell Impoundment." Thesis, Clemson University, Clemson, SC, 1987.
11. Morgan, J. J. In *Principles and Applications of Water Chemistry*; Faust, S. D.; Hunter, J. V., Eds.; Wiley: New York, 1967; pp 561–624.
12. Tipping, E.; Thompson, D.; Davidson, W. *Chem. Geol.* **1984**, *44*, 359.
13. Diem, D.; Stumm, W. *Geochim. Cosmochim. Acta* **1984**, *48*, 1571.
14. Chapnic, S. D.; Moore, W. S.; Nealson, K. H. *Limnol. Oceanogr.* **1982**, *27*, 1004.
15. Tipping, E. *Geochim. Cosmochim. Acta* **1984**, *48*, 1353.
16. Stone, A. T.; Morgan, J. J. In *Aquatic Surface Chemistry*; Stumm, W., Ed.; Wiley: New York, 1987.
17. Postma, D. *Geochim. Cosmochim. Acta* **1985**, *49*, 1023.
18. Stone, A. T.; Morgan, J. J. *Environ. Sci. Technol.* **1984**, *18*, 450.
19. Stone, A. T.; Morgan, J. J. *Environ. Sci. Technol.* **1984**, *18*, 617.
20. Stone, A. T. *Environ. Sci. Technol.* **1987**, *21*, 979.
21. Stone, A. T.; Morgan, J. J. In *Aquatic Surface Chemistry*; Stumm, W., Ed.; Wiley: New York, 1987.
22. Norwood, D. L.; Johnson, J. D.; Christman, R. F.; Hass, J. R.; Bobenrieth, M. J. *Environ. Sci. Technol.* **1980**, *14*, 187.
23. Lind, C. J. *Environ. Sci. Technol.* **1988**, *22*, 62.
24. Yoshimura, T.; Ozaki, T.; Okuno, T. *Anal. Chim. Acta* **1986**, *186*, 115.
25. Stone, A. T. In *Aquatic Surface Chemistry*; Stumm, W., Ed.; Wiley: New York, 1987.
26. Emerson, S.; Jacobs, L.; Tebo, B. In *NATO Conference Series IV:9, Trace Metals in Seawater*; Wong, C. S.; Boyle, E.; Bruland, K. W.; Burton, J. D.; Goldberg, E. D., Eds.; Plenum: New York, 1981.
27. Burdige, D. J.; Nealson, K. H. *Geomicrobiol. J.* **1986**, *4*, 361.
28. Kremling, K. *Mar. Chem.* **1983**, *13*, 87.
29. Hsiung, T.-M. "The Manganese Chemistry at Lake Richard B. Russell." Dissertation, Clemson University, Clemson, SC, 1990.
30. Sunda, W. G.; Huntsman, S. A.; Harvey, G. R. *Nature (London)* **1983**, *301*, 234.
31. Shaw, F. *Analyst* **1980**, *105*, 950.
32. Jenne, E. A. In *Trace Inorganics in Water*; Gould, R. F., Ed.; Advances in Chemistry 73; American Chemical Society: Washington, DC, 1968.
33. Tessenow, U. *Arch. Hydrobiol. Suppl.* **1974**, *47*, 1.
34. Murray, J. W. *J. Colloid Interface Sci.* **1974**, *46*, 357.
35. Davis, S. H. R. In *Geochemical Processes at Mineral Surfaces*; Davis, J. A.; Hayes, K. F., Eds.; ACS Symposium Series 323; American Chemical Society; Washington, DC, 1986.
36. Morris, A. W.; Bale, A. J. *Nature (London)* **1979**, *279*, 318.
37. Coughlin, R. W.; Matsui, I. *J. Catal.* **1976**, *41*, 108.
38. Wilson, D. E. *Geochim. Cosmochim. Acta* **1980**, *44*, 1311.
39. Hoffmann, M. R.; Eisenreich, S. J. *Environ. Sci. Technol.* **1981**, *15*, 339.

40. Waite, T. D. In *Geochemical Processes at Mineral Surfaces;* Davis, J. A.; Hayes, K. F., Eds.; ACS Symposium Series 323; American Chemical Society: Washington, DC, 1986.
41. Morgan, J. J.; Stumm, W. *J. Colloid Sci.* **1964**, *19*, 347.
42. Waite, T. D.; Wrigley, I. C.; Szymczak, R. *Environ. Sci. Technol.* **1988**, *22*, 778.
43. Nealson, K. H.; Ford, J. *Geomicrobiol. J.* **1980**, *2*, 21.
44. Balikungeri, A.; Robin, D.; Haerdi, W. *Toxicol. Environ. Chem.* **1985**, *9*, 309.
45. Tipping, E. *Geochim. Cosmochim. Acta* **1984**, *48*, 1353.
46. Myers, C. R.; Nealson, K. H. *Science (Washington, D.C.)* **1988**, *240*, 1319.
47. Nealson, K. H.; Tebo, B. M.; Rosson, R. A. *Adv. Appl. Microbiol.* **1988**, *33*, 279.
48. Richardson, L. L.; Nealson, K. H. *J. Great Lakes Res.* **1989**, *15*, 123.
49. Alexander, M. *Introduction to Soil Microbiology;* Wiley: New York, 1977.
50. Ostendorp, V. W.; Frevert, T. *Arch. Hydrobiol. Suppl.* **1979**, *55*, 255.
51. Hamilton-Taylor, J.; Morris, E. B. *Arch. Hydrobiol.* **1985**, *72*, 135.
52. Yagi, A.; Shimodaira, I. *Jpn. J. Limnol.* **1986**, *47*, 279, 291.
53. Davison, W.; Woof, C.; Rigg, E. *Limnol. Oceanogr.* **1982**, *27*, 987.
54. James, W. F.; Kennedy, R. H.; Schreiner, S. P.; Ashby, S.; Carroll, J. H. "Water Quality Studies: Richard B. Russell and Clarks Hill Lakes; First Annual Interim Report"; Misc. Paper EL-85-9, U.S. Army Engineers Waterways Experiment Station, Vicksburg, MS, 1985, and subsequent papers in the same series.
55. Turner, R. C. "Supplemental Limnological Studies at Richard B. Russell and Clarks Hill Lakes, 1983–1985"; Kennedy, R. H., Ed.; Misc. Paper EL-87-2, U.S. Army Engineers Waterways Experiment Station, Vicksburg, MS, 1987; pp 128–139.
56. Hayes, R. G.; Myers, R. J. *J. Chem. Phys.* **1964**, *40*, 877.
57. Carpenter, R. *Geochim. Cosmochim. Acta* **1983**, *47*, 875.
58. Gamble, D. S.; Schnitzer, M.; Skinner, D. S. *Can. J. Soil Sci.* **1977**, *57*, 1239.
59. Chiswell, B.; Mokhtar, M. B. *Talanta* **1986**, *33*, 669.
60. Abdult-Halim, A. L.; Evans, J. C.; Rowlands, C. C.; Thomas, J. H. *Geochim. Cosmochim. Acta* **1981**, *45*, 481.
61. Cheshire, M. V.; Berrow, M. L.; Goodman, B. A.; Munde, C. M. *Geochim. Cosmochim. Acta* **1977**, *41*, 1131.
62. Alberts, J. J.; Schindler, J. E.; Nutter, D. E., Jr.; Davis, E. *Geochim. Cosmochim. Acta* **1976**, *40*, 369.
63. Guilbault, G. G.; Lubrano, G. J. *Anal. Lett.* **1968**, *1*, 725.
64. *Axum Technical Graphics and Data Analysis;* TriMetrix, Inc., Seattle, WA, 1989.
65. Beckett, R.; Bigelow, J. C.; Jue, Z.; Giddings, C. In *Aquatic Humic Substances;* Suffet, I. H.; McCarthy, P., Eds.; Advances in Chemistry 219; American Chemical Society: Washington, DC, 1989.
66. Mantoura, R. F. C.; Dikson, A.; Riley, J. P. *Estuarine Coastal Mar. Sci.* **1978**, *6*, 387.
67. Stauber, J. L.; Florence, T. M. *Aquat. Toxicol.* **1985**, *7*, 241.
68. Schindo, H.; Huang, P. M. *Soil Sci. Soc. Am. J.* **1984**, *48*, 927.
69. Schindo, H.; Huang, P. M. *Nature (London)* **1982**, *298*, 363.
70. Robbins, J. A.; Callender, E. *Am. J. Sci.* **1975**, *275*, 512.
71. Burdige, D. J.; Gieskes, J. M. *Am. J. Sci.* **1983**, *283*, 29.
72. Davison, W.; Woof, C. *Water Res.* **1984**, *18*, 727.
73. Bromfield, S. M.; David, D. J. *Soil Biol. Biochem.* **1976**, *8*, 37.
74. Ehrlich, H. L. In *Biogeochemistry of Ancient and Modern Environments;* Trudinger, P. A.; Walter, W. R.; Ralph, B. J., Eds.; Australian Academy of Science, Canberra, and Springer-Verlag: New York, 1980.

75. Stumm, W.; Morgan, J. J. *Aquatic Chemistry*, 2nd ed.; Wiley: New York, 1981; pp 465–468.
76. Ross, D. S.; Bartlett, R. J. *Soil Sci.* **1981**, *132*, 153.
77. Li, Y. H.; Gregory, S. *Geochim. Cosmochim. Acta* **1974**, *38*, 703.
78. Coleman, J. A.; Armstrong, D. E. *Limnol. Oceanogr.* **1987**, *32*, 577.
79. Craig, H. *J. Geophys. Res.* **1969**, *74*, 5491.

RECEIVED for review September 26, 1991. ACCEPTED revised manuscript August 14, 1992.

BEHAVIOR OF ORGANIC POLLUTANTS

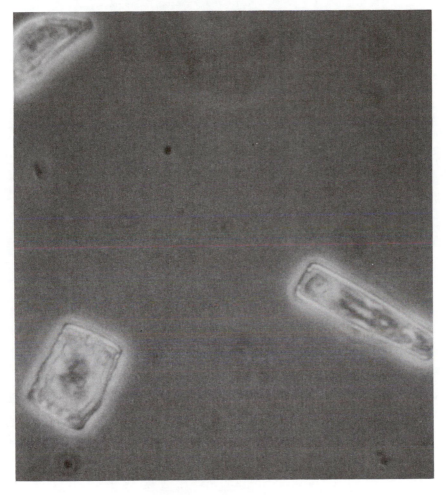

Algae under the microscope.

Hundreds of potentially toxic organic compounds enter the environment from anthropogenic activities. Accordingly, studies of toxic organic substances has been a major focus of environmental chemistry in recent years. The three chapters in this section examine the biogeochemical processes by which organic pollutants are removed from aquatic systems.

Environmental Behavior and Fate of Anionic Surfactants

Nicholas J. Fendinger, Donald J. Versteeg, Els Weeg, Scott Dyer, and Robert A. Rapaport

Procter and Gamble Company, Ivorydale Technical Center, Cincinnati, OH 45217

Linear alkylbenzenesulfonates, alcohol sulfates, and alcohol eth-oxysulfates are used in a variety of household cleaning and personal care products that are generally disposed of "down the drain". Even though these surfactants are characterized as highly biodegradable and are not expected to persist in aquatic environments, the current use of anionic surfactants in consumer products requires increased understanding of anionic surfactant degradation, toxicity, and environmental behavior. This chapter reviews anionic surfactant biodegradation, removal during sewage treatment, environmental concentrations, and aquatic-effects data that illustrate the environmental safety of these materials.

ANIONIC SURFACTANTS USED IN SHAMPOOS, cosmetics, toothpaste, and laundry products include linear alkylbenzenesulfonates (LAS), alcohol sulfates (AS), alcohol ethoxysulfates (AES), alcohol glycerol ether sulfonates, and alpha-olefin sulfates. Household end use of anionic surfactants in the United States was 7.3×10^5 metric tons in 1987; LAS, AS, and AES accounted for 98% of the total (1).

LAS, the major anionic surfactant used in the world today, accounts for approximately 28% of all synthetic surfactant use. In the United States it is widely included in detergent formulations (about 270,000 metric tons per year) (1). It is used in almost every household cleaning application except automatic dishwasher detergents (2). LAS represents a mixture of homologs with alkyl chain lengths ranging from 10 to 15 carbon units and with isomers of varying phenyl position (structure 1). Dodecylbenzenesulfonate ($C_{12}LAS$)

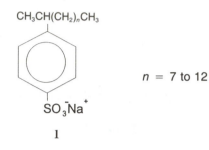

$n = 7$ to 12

1

is the homolog most widely used in detergent applications. Therefore the following discussion of aquatic hazard will focus on this homolog.

Alcohol sulfates (AS) are currently used in shampoos, bath preparations, cosmetics, medicines, toothpaste, rug shampoos, hard-surface cleaners, and light- and heavy-duty laundry applications (2). Total AS use in the United States is about 136,000 metric tons per year. The 91,000 metric tons per year used in household products (3) ranges from C_{12} to C_{18} in alkyl chain lengths (structure 2) and may contain some ethyl or methyl branching. The $C_{14}AS$ homolog is widely used in detergent formulations and approximates the average commercial AS chain length. Therefore, our aquatic hazard discussion will focus on the C_{14} homolog.

R—alkyl chain length from 10 to 18 carbons

2

Alcohol ethoxysulfates (AES) are widely used in household and personal care applications (dishwasher detergents, laundry detergents, and shampoos; 204,000 metric tons per year in the United States) (3) (structure 3). AES are described by the dominant alkyl chain length (C_x) and number of ethoxylate units (E_y) in the technical-grade material. Because the volume of AES used in household cleaning applications is divided almost equally between $C_{12-13}E_3S$ and $C_{14-15}E_3S$, the hazard assessment will focus on the approximate average chain length of $C_{13}E_3S$.

$$RO\text{-}(CH_2CH_2O)_n\text{-}SO_3^-$$

R—alkyl chain length from 10 to 18 carbons
$n = 1$ to 5

3

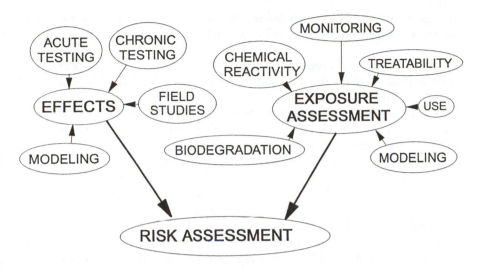

Figure 1. Factors considered in determination of environmental hazard assessment for anionic surfactants.

Given the current use of LAS, AS, and AES in consumer products, understanding of anionic surfactant degradation, toxicity, and environmental behavior is needed. Even though these surfactants are all readily biodegradable (easily converted to carbon dioxide and cellular material) and are not expected to persist in the environment, integration of material usage, environmental fate, exposure, and aquatic toxicity data is necessary to obtain a comprehensive environmental hazard assessment (*4*) (Figure 1). This chapter presents information on the use, biodegradation or removal during sewage treatment, environmental concentration, and aquatic effects of LAS, AS, and AES. These data are then used to calculate protection factors.

Fate Studies

Biodegradation, hydrolysis, and sorption influence the environmental fate of LAS, AS, and AES. Primary degradation of surfactants is important because this process usually results in loss of surfactancy and reduced toxicity (*5, 6*). Complete mineralization ensures that persistent intermediates will not be formed and that biodegradation will be an effective mass-removal mechanism in the environment. Sorption and association of surfactants with particles or dissolved organic substances are processes that decrease bioavailability and can be correlated with decreased surfactant toxicity (*7*).

Photolysis also may degrade surfactants in surface waters. It is not likely to be a direct degradation mechanism for AS and AES because these molecules lack a chromophoric group, but it could affect the fate of LAS. Because surfactants are introduced into the environment mainly through waste-treatment systems, biodegradation is expected to be the major environmental removal mechanism. Thus, it will be discussed in the most detail.

LAS Biodegradation.   The initial enzymatic attack of LAS occurs by omega oxidation of the terminal carbon of the alkyl side chain. The enzymes involved in this reaction, although not yet isolated, are probably associated with cell membranes. This enzymatic attack results in a carboxylated alkyl chain or sulfophenylcarboxylate. The alkyl chain biodegrades further through beta oxidation, with two carbon units converted into acetic acid at a time (8). Once carboxylated, the molecule loses its surfactant properties because it no longer has a hydrophobic side chain. Following complete mineralization of the alkyl chain, the benzene ring is desulfonated and cleaved (9, 10).

Mineralization of the LAS alkyl chain can be accomplished by pure bacterial cultures (11). However, complete LAS mineralization in natural systems is most likely to be accomplished by a consortium of microorganisms (12). Jimenez et al. (13) identified a consortium of four bacterial species that are capable of complete LAS mineralization from activated sludge. Other types of consortia that are capable of LAS mineralization probably also exist.

Biodegradation Rate.   The biodegradation rate in surface water depends on many variables such as biomass and substrate concentration, presence of suspended material, and availability of nutrients. Larson and Payne (14) and Larson (15) investigated factors that controlled LAS mineralization in river water by following $^{14}CO_2$ evolution and ^{14}C incorporation into biomass from samples amended with ^{14}C ring-labeled LAS homologs and phenyl isomers (2-phenyl C_{10-14} and 5-phenyl C_{10-12}LAS). The river-water LAS half-life was approximately 24 h, did not vary significantly as a function of LAS homolog or isomer, and was not affected by high concentrations of competing homologs. Intermediates from LAS biodegradation did not appear to accumulate. In addition, the rate and extent of LAS mineralization were not significantly different between river water with a low suspended-solids content and river water with up to 1000 mg/L of suspended sediment, despite the fact that significant amounts of LAS homologs with the longer alkyl chain length were sorbed on suspended sediment. Larson (15) theorized that sorption did not affect the extent or rate of biodegradation as long as LAS sorption to solids was reversible and the desorption kinetics were more rapid than the kinetics of biodegradation.

Addition of sediment solids to overlying surface water increases biomass and may increase the rate of LAS degradation if the system biomass is limited. For example, Larson and Payne (14) correlated increases in the LAS bio-degradation rate constants in river water with increased bacterial populations resulting from addition of sediment solids. Similarly, Yediler et al. (16) found that primary LAS degradation in lake water was affected by the size of the microbial population present.

Laundromat Pond. Federle and Pastwa (*17*) compared LAS biodegradation in sediments from a pond that received discharge from a local laundromat to biodegradation that occurred in a pristine control pond. Sediment cores collected from each of the ponds were used to determine sediment depth profiles for surfactant concentrations, bacterial number and activity, and ability to mineralize ^{14}C ring-labeled C_{13}LAS. Biomass and microbial activity were found to decrease with depth in the sediments for both ponds, with higher biomass and activity measured in the laundromat pond. LAS was mineralized without a lag period by the laundromat pond sediments at all depths with half-lives that ranged between 3.2 and 16.5 days. In the pristine control pond, LAS was mineralized only after a lag period of 3.2–40 days. The mineralization progressed more slowly than that observed in the laundromat pond (calculated half-lives were from 5.2 to 1540 days).

Clearly, LAS undergoes rapid (within days) and complete biodegradation in natural water systems. Although many factors influence LAS degradation in natural waters, the governing factor is probably acclimation that results from previous LAS exposure. However, because LAS has been in use for approximately 25 years, both sewage-treatment operations and the rivers that receive treated sewage are already acclimated and contain LAS-mineralizing consortia that are capable of rapid LAS degradation. For example, Moreno et al. (*18*) found that LAS removal by biodegradation in aerated sewers can be as high as 50%.

Anaerobic Environments. Because the initial attack of the LAS molecule is oxidative, LAS does not biodegrade under anaerobic conditions (*19*). Therefore concerns are sometimes expressed that LAS may accumulate in deep anaerobic sediment layers, where it will not biodegrade further. However, given the high rate of LAS removal during sewage treatment combined with in-stream degradation, it is unlikely that LAS sediment accumulation will occur unless there is rapid deposition into an anaerobic environment.

AS Biodegradation. The primary step in the aerobic degradation of AS is the sulfatase-catalyzed hydrolysis of the sulfate ester from the hydrophobic group to form inorganic sulfate and an alcohol (*20*). Degradation proceeds through oxidation of the alcohol catalyzed by dehydrogenases (*12*), to give first the corresponding aldehyde and then the corresponding fatty acid. The final degradation of the fatty acid is by beta oxidation. Thomas and White (*21*) indicated that elongation and desaturation of the fatty acid chain may also occur with fatty acid residues that are rapidly incorporated into lipid fractions. They also demonstrated (*21*) that hydrophobic metabolites of the AS alkyl chain can be incorporated into cellular components (lipid membranes) without prior degradation by beta oxidation.

Measurement of Biodegradation.   Numerous studies have documented the aerobic biodegradability of various AS compounds (*see* ref. 12). Most of these studies used methylene blue active substance (MBAS) and other colorimetric determinations, change in surface tension, foaming capacity, and sulfate formation as an indication of primary AS degradation.

Ultimate biodegradation of AS has been measured by biological oxygen demand (BOD), chemical oxygen demand (COD), carbon loss, CO_2 production, and ^{14}C-labeled AS. These tests also indicate total degradation of primary straight- and branched-chain AS in laboratory experiments with Mediterranean Sea water (*22*), Black Sea water (*23*), activated sludge (*24–26*), forest soil (*27*), and river water (*12*). However, highly branched AS (secondary and tertiary branched) are reported (*28–30*) to be more resistant to biodegradation than linear primary AS. AS manufactured from natural oils (animal and vegetable) do not contain any branched components, and AS manufactured from petroleum-derived oils will generally contain less than 20% branched material. Virtually all of the branched material that is present in petroleum-derived AS consists of primary methyl or ethyl branched material (Shell Development Corporation, unpublished data). $C_{11-15}AS$ with this type of branching underwent rapid and complete degradation (>80–111% by CO_2 production; Procter and Gamble, unpublished data).

Studies of AS degradation during simulated sewage treatment also show rapid and complete biodegradation. For example, McGauhey and Klein (*31*) and Klein and McGauhey (*32*) found complete primary degradation of AS based on formation of $^{35}SO_4^{2-}$ in model septic tank systems. Steber et al. (*26*) used uniformly labeled $C_{16}AS$ to demonstrate ultimate degradation in a model activated-sludge treatment system. An average of 60% of the AS fed into the system was measured as carbon dioxide. The effluent contained 0.3% of the original feed as intact AS and less than 10% of the total ^{14}C spike. This analysis indicates >90% carbon elimination during wastewater treatment.

Bacterial enzymes capable of initiating AS hydrolysis are widespread in the environment. Sulfatase has been isolated from soil, activated sludge, river water, and raw sewage bacteria (*33–38*). Although White et al. (*37*) found more AS-resistant strains and a greater proportion of alkylsulfatase-producing bacteria at polluted sites than at clean sites in the River Ely, 29% of all bacterial isolates tested were alkylsulfatase producers. Anderson et al. (*39*) isolated epilithic bacteria from riverbed stones collected from polluted and clean sites from the South Wales River and assessed their ability to produce alkylsulfatase. The number of alkylsulfatase-producing bacteria was greater in polluted sites, but they were present at both locations and varied as a function of season and water quality. Greater numbers of alkylsulfatase bacteria were present at sampling sites during late summer than during winter; the count was lower at sewage discharge sites than at downstream locations. In addition, because of enzyme composition (far more constitutive than inducible enzymes), the epilithon would not lose its ability to degrade AS even if levels fluctuated.

Microbial Degradation.   Identification of bacteria responsible for AS degradation was attempted in part to understand what conditions favor microbial degradation, either in nature or under controlled conditions. Pure cultures of *Escherichia coli, Serratia marcescens, Proteus vulgaris,* and *Pseudomonas fluorescens* degraded $C_{12}AS$ (8). Other researchers (36, 37) showed that *Pseudomonas* strains are also capable of degrading $C_{12}AS$. Extensive degradation of tallow $C_{16-18}AS$ occurred with 35 out of 47 strains of bacteria from the following genera: *Acetobacter, Chromobacterium, Bacillus, Corynebacterium, Escherichia, Micrococcus, Mycobacterium, Pseudomonas, Staphylococcus,* and *Vibrio.*

Anaerobic degradation was also demonstrated for AS (12, 26). Oba et al. (40) suggested that the anaerobic degradation of AS may consist of sulfate ester hydrolysis, with formation of a fairly inert alcohol. Wagner and Schink (41) measured accumulation of sulfide and fatty acids during the anaerobic degradation of $C_{12}AS$. Degradation of the AS was initiated by the hydrolytic cleavage of the sulfate. Because there was no other source of sulfur in the system, Wagner and Schink speculated that the sulfide resulted from reduction of the liberated sulfate. High concentrations of acetate and valerate indicated that the AS residues were at least partially degraded. However, Steber et al. (26) found 90% of the ^{14}C label from stearyl sulfate and dodecyl sulfate in the final degradation products (CO_2 and CH_4) when they tested for them in a simulated anaerobic treatment system.

AES Biodegradation.   Primary degradation of $A_{12}E_3S$ was accomplished enzymatically by sulfatase or etherase, or by omega and beta oxidation. Sulfatase hydrolyzed the AES into its corresponding alcohol ethoxylate (AE) and inorganic sulfate. Liberation of the sulfate from primary AS was shown to be accomplished by primary alkyl sufatases (12). Although linear alcohol ethoxysulfates are structurally similar to primary AS, the sulfate may not be hydrolyzed by the same enzyme. The rate of AES degradation by sulfatase varies as a function of hydrophobe structure from rapid and complete to no observed degradation, based on the "fit" of the enzyme (12). Four strains of bacteria (three *Pseudomonas* and one unidentified strain) were shown to utilize this pathway. Their extent of degradation varies from 6% for TES5 to as high as 39% for C12B (42, 43).

The rate of primary AES degradation by etherase is also strain-specific. For example, Hales et al. (42, 43) reported that etherases of three strains of *Pseudomonas* and an unidentified strain isolated from sewage effluent could cleave at each of the three available ether linkages. Extent of degradation (1–72%) varied as a function of the bacterial strain and ether linkage affected. Glycol sulfates that result from etherase activity are degraded further via oxidation of the terminal alcohol to a carboxylic acid. Terminal C–C units are sequentially removed via hydrolysis to yield the next shorter polyglycol sulfate (*12*).

Degradation by omega, beta oxidation was identified as a mechanism that shortens the hydrophobe chain by two carbon increments. This route is responsible for 11% of the parent AES degradation (*43*).

Exposure Assessments

Exposure of an organism to a surfactant in surface water will depend on the amount of material used, disposal practice, removal rate during sewage treatment, dilution in the receiving stream, and sorption on particles or aquatic dissolved organic carbon. The exposure component of an environmental hazard assessment utilizes information from the fate studies described in the previous section, mathematical modeling to predict environmental concentrations, and environmental monitoring to verify model predictions.

Predicted Wastewater Concentrations of LAS, AS, and AES. Wastewater concentrations of LAS, AS, and AES in the United States were calculated by using the following equation.

$$C = \frac{X \text{ (mg)}}{540 \text{ L/person-day} \cdot 365 \text{ days} \cdot 240 \times 10^6 \text{ persons}} \quad (1)$$

where X is the amount of material used in down-the-drain applications (milligrams) and C is the wastewater concentration in milligrams per liter. The per capita water use of 540 L/person-day, obtained from the Environmental Protection Agency (EPA) needs survey, was calculated by dividing treatment-plant flow rates (obtained from survey information) by population serviced by the same plants (*44*). This calculation yields a higher flow rate than the individual home per capita flow rate of 200 L/person-day because of contributions from nondomestic wastewater sources. Thus it is a more realistic flow determination for treatment-plant-modeling purposes than the individual per capita home flow (*45*). Annual United States LAS, AS, and AES use in down-the-drain applications were stated previously. On the basis of these values, predicted wastewater concentrations of LAS, AS, and AES are 4.5, 1.8, and 4.3 ppm, respectively (*45*).

LAS Treatability and Environmental Concentrations. The removal of LAS during sewage treatment was confirmed by monitoring studies in both the United States and Europe. Numerous studies reported anionic surfactant concentrations in surface waters measured by nonspecific analytical techniques such as methylene blue active substance (MBAS). However, the correlation between MBAS and LAS concentrations determined by specific analytical techniques are variable and, in general, not useful. Therefore, only those studies that used specific analytical techniques for LAS are reviewed here.

In the United States the most extensive LAS monitoring studies were reported by Rapaport and Eckhoff (46) and McAvoy et al. (47). Rapaport and Eckhoff (46) described the results of 13 years (1973–1986) of LAS monitoring at 36 sewage-treatment plants, 35 river water–sediment sites, and 5 sludge-amended soil sites. A microdesulfonation–GC analysis technique with a detection limit of 0.01 mg/L was used to measure LAS in the collected samples. Influent sewage LAS (average chain length C_{12}) concentration averaged 3.5 ± 1.2 mg/L. Effluent LAS levels varied as a function of treatment type. For example, LAS concentrations in activated-sludge effluents averaged 0.06 mg/L and ranged from 0.01 to 0.13 mg/L, whereas primary treatment effluents averaged 2.1 mg/L and ranged from 1.7 to 2.5 mg/L.

Rapaport and Eckhoff (46) also reported results from an extensive LAS study in Rapid Creek, South Dakota. LAS river-water concentrations were rapidly attenuated as a function of distance downstream from a trickling-filter treatment plant. LAS concentration in sediments downstream was probably diminished by movement of the surficial sediments, biodegradation, and dilution by bank solids. The LAS concentration of 190 ± 95 mg/kg in sediments immediately below the plant outfall was at the low range of the calculated steady-state concentrations, which ranged from 190 to 740 mg/kg.

McAvoy et al. (47) monitored LAS levels from several drainage basins at a variety of treatment plants in 11 states. The plant types included activated sludge (15 sites), trickling filter (12 sites), oxidation ditch (6 sites), lagoon (8 sites), and rotating biological contactor (9 sites). Influent wastewater, effluent, upstream and downstream river water, and sediment samples were collected at each site. Sampling was conducted during periods of low river flow, when effluent dilution would be lowest. Influent and effluent samples were collected as 24-h composites and then combined into 3-day flow-averaged composites. River-water samples were collected as grab samples across a horizontal transect of the river to assess mixing of the effluent plume.

Solid-phase extraction and cleanup were used, followed by high-pressure liquid chromatography (HPLC)–fluorescence detection for quantitation of LAS levels (48). Detection limits in water samples and sediments were 0.010 mg/L and 1 mg/kg, respectively. The average influent concentration was 5 mg/L. Effluent LAS concentrations varied as a function of treatment type (Table I). Average effluent concentrations ranged from 0.04 mg/L for activated-sludge plants to 1 mg/L for trickling-filter plants. In terms of removal, this concentration corresponds to >99% for activated-sludge plants and an average of 77% for trickling-filter plants. The average removal for other types of treatment plants ranged from 96 to 98%.

**Table I. Sewage Treatment Plant Influent and Effluent
Concentrations of LAS for Different Types
of Sewage Treatment**

Treatment Type	Average Conc. (mg/L)	Removal Rate (%)
Influent LAS concentrations	4.8 ± 2.0	
Effluent LAS concentrations:		
Primary treatment		27 ± 19
Activated sludge	0.04 ± 0.3	99.3 ± 0.61
Trickling filter	1.04 ± 0.98	77.4 ± 15.46
Lagoons	0.06 ± 0.1	98.5 ± 1.81
Oxidation ditch	0.12 ± 0.27	98.0 ± 4.24
Rotating biological contactor	0.19 ± 0.38	96.2 ± 6.10

SOURCE: Data are taken from refs. 46 and 47.

Figure 2 shows a histogram recording the frequency of observed in-stream concentration of LAS measured below sewage-treatment outfalls. Concentrations of LAS measured in samples from the left, middle, and right stream channels from sampling locations were consistent, which indicated complete mixing. Stream concentrations of LAS were generally less than

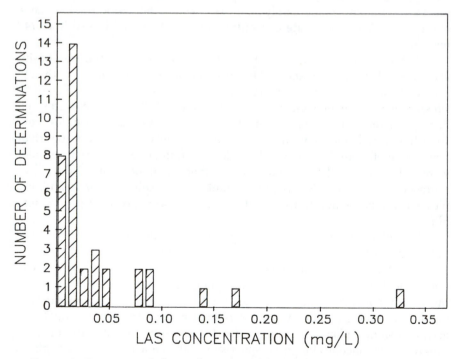

*Figure 2. Frequency of observed in-stream LAS concentrations below sewage
outfalls. (Data are from ref. 47.)*

0.05 mg/L. However, a concentration of 0.320 mg/L was measured at one location in an irrigation ditch that served as a receiving stream for a trickling-filter plant discharge. The stream-water concentrations represent worse-case conditions because the samples were collected under low-flow conditions at locations with minimal effluent dilution.

DeHenau and Matthijs (*49*) reported LAS sediment monitoring results from Germany (14 sites). Sediment concentrations of LAS were measured by HPLC–UV spectroscopy and were found to be dependent on the distance between sampling location and sewage-treatment-plant (STP) outfall. Measured concentrations ranged from a few milligrams per kilogram of dried sediment at 48 km downstream to a maximum of 275 mg/kg at only 1 km downstream from the STP effluent.

AS Treatability and Environmental Concentrations. Studies of AS degradation in the environment or during actual sewage treatment have been limited because specific analytical methods to measure AS were not available until recently. We developed a method that isolated AS from water samples on a strong anion-exchange column. The AS were then hydrolyzed to a fatty acid and analyzed by gas chromatography with flame-ionization detection (GC–FID). The method has a detection limit of 5 μg/L per component (*50*).

Levels of AS in wastewater flowing into sewage-treatment plants that discharge to surface water were predicted to be near 1.7 ppm (*see* eq 1). However, AS levels in wastewater samples collected as 1-h composite samples from two STPs near Cincinnati, Ohio, averaged 0.27 ± 0.18 mg/L and ranged from 0.04 to 0.53 mg/L. Rapid loss of AS from wastewater by either microbial or enzymatic hydrolysis may account for the measured AS concentrations being lower than expected.

Analyses of AS in AS-amended river water and filtered and unfiltered wastewater as a function of time confirm this hypothesis. For example, AS levels in unfiltered wastewater decreased to less than 40% of the original spike after only 6 h (Figure 3). After 24 h the AS levels remained at less than 40% of the original spike. AS concentrations in river water (Figure 4) amended with AS also decreased to levels less than 40% of the original spike after 24 h. Reduced AS degradation after addition of formaldehyde to the samples indicated that the AS loss was biologically mediated.

AS levels in effluent from an activated-sludge plant near Cincinnati, Ohio, averaged 0.0178 ± 0.0091 mg/L (*n* = 5) with a removal rate of 93% (*50*). Mann and Reid (*51*) reported near complete primary degradation of AS during trickling-filter treatment.

Given the high rate of AS degradation in natural waters and the high removal during sewage treatment, surface-water concentrations are expected to be very low. Even direct discharge of AS-containing wastewater to a river or stream that feeds a lake or reservoir is not likely to transport measurable

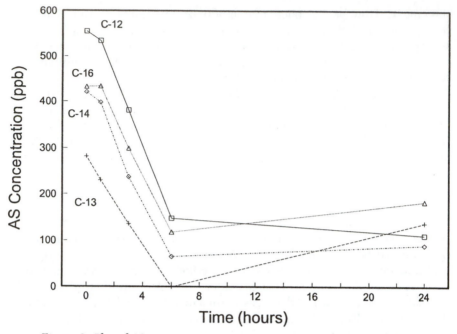

Figure 3. Plot of AS concentrations in wastewater as a function of time.

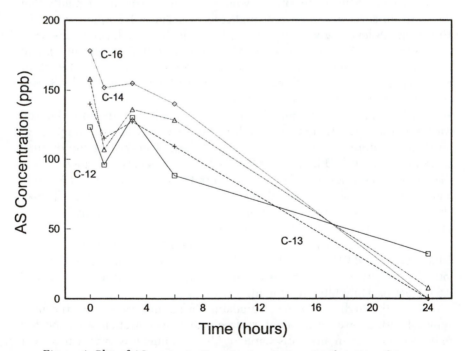

Figure 4. Plot of AS concentrations in river water as a function of time.

amounts of AS because of the abundance of sulfatase-producing organisms in the environment.

AES Treatability and Environmental Concentrations.   Removal of $C_{12-15}E_3S$ during laboratory-scale semicontinuous activated-sludge testing was greater than 95%, with greater than 98% removal by activated-sludge treatment (52). Rapid loss was also observed in river water and estuarine water. For example, Kikuchi (53) found 80–100% of AES degraded in river water within 3 to 5 days, as measured by MBAS. In addition, Vashon and Schwab (54) measured 80–100% AES mineralization by $^{14}CO_2$ evolution within 15 days in estuarine water.

AES are typically measured in environmental matrices by nonspecific colorimetric analyses (MBAS) that collectively measure LAS, AS, and naturally occurring anionic surfactants. Alternatively, a specific gas chromatographic method for AES, developed by Neubecker (55), was employed to measure AES concentrations in influent and effluent from STPs and river water. Total AES measured in influent wastewater to a STP was 1.88 mg/L. AES removal of 94–100% was measured during actual sewage treatment by activated sludge; the resulting effluent concentration was 0.06 mg/L. Total AES levels in river water were less than 0.01 mg/L. AES accounted for 6–13% of MBAS measured in natural water.

Predicted Anionic Surfactant Concentrations in Surface Water. Modeling of surfactant concentrations in surface water provides exposure estimates when large-scale monitoring data are not available. The models we used to estimate surfactant concentrations in surface waters are USTEST (56) and PG-ROUT (57). For both models only wastewater levels from actual monitoring are used, along with removal rates from actual sewage treatment (Table I). AS and AES removal rates for treatment types other than activated sludge are not available. For these surfactants, LAS removal rates were used to estimate removal during primary and various forms of secondary sewage treatment. This assumption provides a conservative estimate of AS and AES removal because these surfactants are expected to be removed from wastewater at least as efficiently as LAS.

USTEST, developed by Rapaport (56) is a river concentration model applicable throughout the United States. The model predicts concentrations below the mixing zones of sewage-treatment plants. It is built on three large databases that link river flow, treatment type, and sewage-discharge volume. Thus dilution factors are predicted for each of the 11,500 publicly owned treatment works (POTW) in the United States. The model links the treatment type to the dilution factor so that different removal rates for the chemical being modeled can be assigned for each treatment. The result is a national frequency distribution of river concentrations just below the mixing zones of treatment-plant outfalls. These predictions can be done under mean or critical-low river flow rates. The latter closely approximates 7-consecutive-day, 10-year low-flow (7Q10) conditions (56). Concentrations predicted by USTEST are typically the highest that are expected to occur in the environment because the model does not take into account in-stream removal processes such as biodegradation and sorption onto particles with settling out of the water column. Model input includes wastewater concentration and removal by primary, trickling-filter, and activated-sludge treatment. Trickling-filter removal, used in USTEST predictions, was calculated as the flow-weighted (United States) average removals for trickling-filter, rotating biological contactor, lagoon, and oxidation-ditch treatment systems.

Figure 5 shows the frequency distribution for LAS below the 11,500 POTWs in the United States under mean-flow and low-flow conditions, plus ranked distribution of the actual river-monitoring data from Rapaport and Eckhoff (46) and McAvoy et al. (47). The USTEST model predicts that concentrations will be less than 0.148 and 0.038 mg/L for critical low-flow and mean-flow conditions, respectively, at 90% of the locations. Concentrations at mean-flow conditions are lower because of greater in-stream dilution. The monitoring results correspond closely to the predicted low-flow con-

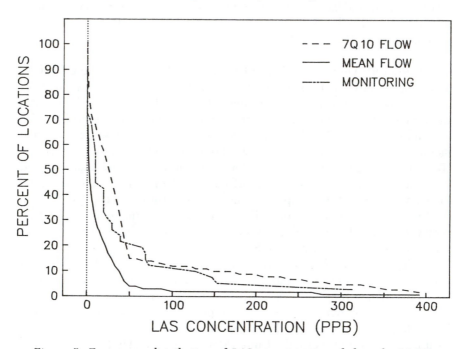

Figure 5. Frequency distribution of LAS concentrations below the 11,500 publicly owned treatment works in the United States under mean-flow and low-flow conditions plus ranked distribution of actual river-monitoring data. (Data are from ref. 47.)

centrations. These results indicate that the model is a good predictor of actual LAS environmental concentrations.

Figures 6 and 7 show the frequency distributions for AS and AES, respectively, below the 11,500 POTWs in the United States under mean-flow and low-flow conditions. AS concentrations are expected to be less than 0.010 mg/L under low-flow conditions and less than 0.002 mg/L under mean-flow conditions for 90% of the locations. AES concentrations are expected to be less than 0.063 and 0.015 mg/L under low-flow and mean-flow conditions, respectively, for 90% of the locations. As was the case with LAS, the concentrations predicted at the 90th percentile (only 10% of the treatment plants are expected to have higher concentrations) are the highest concentrations that are expected to be encountered in the environment.

PG-ROUT is a deterministic river model applicable throughout the United States (57). Predictions are based on more than 500,000 United States river miles. This model also predicts concentrations under 7Q10 and mean-flow conditions. The model is driven by several large EPA databases. Predictions are made below each of the 11,500 POTWs, at drinking water intakes and at any desired mile points in the river systems. The model output includes a frequency distribution by river mile and a detailed PC database.

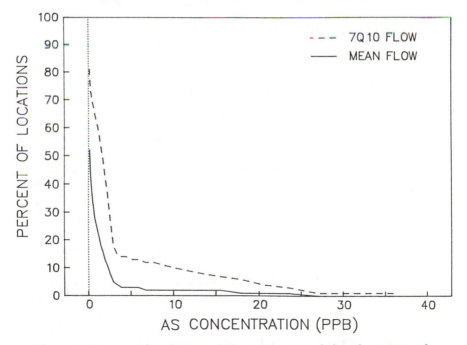

Figure 6. Frequency distribution of AS concentrations below the 11,500 publicly owned treatment works in the United States under mean-flow and low-flow conditions.

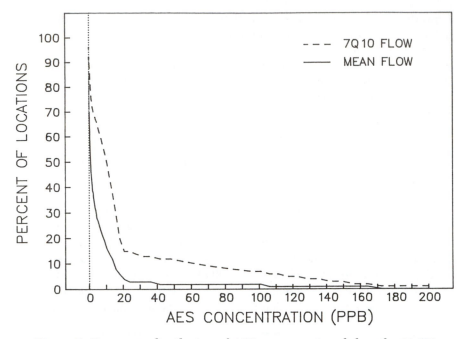

Figure 7. Frequency distribution of AES concentrations below the 11,500 publicly owned treatment works in the United States under mean-flow and low-flow conditions.

For comparison, biological oxygen demand (BOD), ammonia, and dissolved oxygen concentrations are also predicted.

As an example of how the model works, an individual river reach can be considered. The model first locates POTWs, drinking water intakes, and industries. Then the load for each particular POTW that discharges into the stream is determined. The resulting concentration is a function of the per capita usage rate for the chemical, the type of treatment that the wastewater-treatment plant employs, the effluent flow rate, and the river flow rate at the discharge point. The treatment type and flow rates are accessed from the EPA databases. The initial concentration is allowed to decay by first-order kinetics determined from biodegradation studies to the next point of interest. This process continues for selected mile points, industries, and POTWs. At each POTW a new load is added to the stream. For both industries and POTWs, ammonia and BOD loading are simulated. This process continues for all 500,000 river miles in exact hydrologic sequence.

Figure 8 shows a comparison between LAS levels predicted by PG-ROUT and actual levels obtained by monitoring for the west branch of the Trinity River (Texas). This river reach is especially difficult to model because there are a number of impoundments. However, predicted LAS concentrations ranging from 0 to near 3.0×10^{-2} mg/L were similar to concentrations

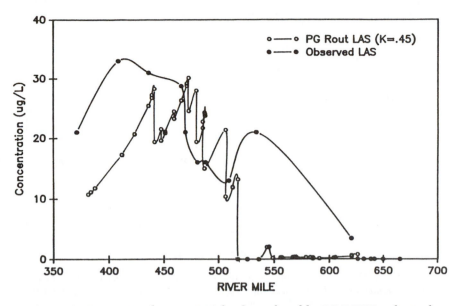

Figure 8. Comparison between LAS levels predicted by PG-ROUT and actual levels from monitoring the west branch of the Trinity River (Texas).

determined by actual monitoring. Much of the variation between the model and monitoring data can be explained by discrepancies in the discharge mile-point location data.

PG-ROUT was also used to predict LAS river-water concentrations as a function of United States river miles. Figure 9 shows the LAS concentrations predicted by PG-ROUT on a log scale versus the percentage of United States river kilometers with carbonaceous BOD (CBOD) and LAS as CBOD for comparison. LAS concentrations are predicted to be less than 0.004 mg/L for 90% and less than 0.020 mg/L for 95% of United States river kilometers. By comparison, CBOD is 9.8 mg/L for 95% of United States river kilometers. LAS expressed in terms of CBOD is 0.5% of the total CBOD at the 95th percentile. PG-ROUT predictions have not been conducted for AS or AES.

Aquatic Toxicity

LAS.   The toxicity of C_{12}LAS to aquatic organisms has been widely studied and is reported to span a wide concentration range (Table II; *see* ref. 65 for a more extensive review; the references we report in this chapter were selected for applicability to the United States and Europe and for data quality). The toxicity mechanism of LAS and other surfactants (AS and AES) is unknown but is suspected to be polar narcosis. Acute toxicity to invertebrate species (48-h LC_{50}) range from 1.7 mg/L for the oligochaete, *Dero*,

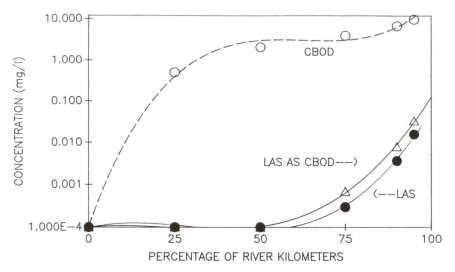

Figure 9. LAS concentrations predicted by PG-ROUT versus United States river kilometers with carbonaceous BOD (CBOD) and LAS as CBOD for comparison.

to 270 mg/L for the isopod, *Asellus* (58). For the invertebrates used most commonly in aquatic toxicity testing (the daphnids, *Daphnia* and *Ceriodaphnia*, and the amphipod, *Gammarus*) acute toxicity values are reported to range from 3.3 to 8.6 mg/L. Because daphnid sensitivity to $C_{12}LAS$ is similar to or lower than the sensitivity of other invertebrate species, acute toxicity values for the daphnids are used to assess risk to all invertebrate organisms.

The available $C_{12}LAS$ acute toxicity data for fish indicate little intra-species variability (Table II). For the four species tested, the 96-h LC_{50} concentrations of $C_{12}LAS$ range from 1.2 mg/L for the fathead minnow *Pimephales promelas* (64) to 6.2 mg/L for the minnow *Phoxinus phoxinus* (68). Although the variety of species tested is not as broad as for the invertebrates, the available data suggest that sensitivity of fish to $C_{12}LAS$ is similar to that of the most sensitive invertebrates.

No studies on the acute toxicity of $C_{12}LAS$ to algae and aquatic plants were found in the literature. Typical toxicity-testing protocols use a test duration of 4–7 days. This duration of testing represents a significant portion of the organisms' life span for the taxa tested. Thus these toxicity tests are appropriately considered chronic-toxicity tests.

The chronic toxicity of $C_{12}LAS$ to invertebrates, fish, and algae span a relatively narrow toxicity range (Table III). Toxicity to fish and one algal species is comparable and slightly greater than toxicity to invertebrates.

No-observed-effect concentrations (NOEC) for fish range from 0.3 to 1.1 mg/L for the fathead minnow (64, 66). The most appropriate chronic-toxicity value from the Fairchild et al. (64) study is the 0.7 mg/L value, as

Table II. Acute Toxicity of C_{12}LAS to Invertebrates and Fish

Organism	Test Duration (h)	LC_{50} (mg/L)	Ref.
Invertebrates			
Dero	48	1.7	58
(oligocheate)			
Daphnia magna	48	1.8	59
(water flea)		3.5	60
		5.9	61
Dugesia	48	1.8	58
(flatworm)			
Gammarus	48	3.3	58
(amphipod)			
Daphnia pulex		8.6	61
(water flea)			
Ceriodaphnia dubia	48	5.3	62
(water flea)			
Rhabditis	48	16	58
(nematode)			
Paratanytarsus	48	23	58
(midge)			
Goniobasis	24	19	63
(snail)	48	92	62
Asellus	48	270	58
(isopod)			
Fish			
Pimephales promelas	96	1.2	64
(fathead minnow)	96	3.8	65
	96	4.1	66
	96	4.7	60
Lepomis macrochirus	96	1.7	67
(bluegill sunfish)			
Oncorhynchus mykiss	96	2.5	62
(rainbow trout)			
Phoxinus phoxinus	48	6.0	68
(minnow)	48	6.2	

this was the 28-day early-life-cycle test. The other values reported by Fairchild et al. are 7-day chronic estimator tests. These tests are of shorter duration, more highly variable, and thus not considered as good a measure of toxicity as the longer-term test from the same study. Similarly, the most appropriate chronic-toxicity value from the Holman and Macek (66) study is 1.1 mg/L, as this value was based on results from a life-cycle study. The other chronic-toxicity values for the fathead minnow reported by Holman and Macek (66) are from 30-day early-life-stage tests.

Algal median effective concentrations (EC_{50}) range from a low of 0.9 mg/L for the blue-green alga *Microcystis* to 29 mg/L for *Selenastrum* (71).

Table III. Chronic Toxicity of $C_{12}LAS$ to Invertebrates, Fish, and Algae

Organism	Toxicity Text	$NOEC^a-LOEC^b$ (mg/L)	Ref.
Invertebrates			
Hyalella azteca (amphipod)	7-day survival	0.9–2.1 1.4–2.9 0.9–2.0 0.6–1.5	64
Daphnia magna (water flea)	21-day survival; reproduction	1.2 (NOEC) 1.5–2.25 4.9 (NOEC)	65 69 70
Ceriodaphnia dubia (water flea)	7-day survival; reproduction	1.4^c	62
Fish			
Pimephales promelas (fathead minnow)	life cycle (growth, survival, reproduction)	1.1 (NOEC)	66
	early lifestage (40-day survival, growth)	0.48–0.49	
	early lifestage	0.65–0.98	
	early lifestage	0.7–1.8	64
	7-day growth	0.9–2.1	
	7-day growth	0.3–0.6	
	7-day growth	0.3–0.9	
	7-day growth	0.6–1.5	
Plants			
Microcystis aeruginosa (blue-green algae)	4-day pop. growth 2-day pop. growth	0.9 10–20 0.09	71 72 73
Lemna minor (duckweed)	7-day growth	2.7	59
Nitzschia fonticula (diatom)	2-day pop. growth	20–50	72
Selenastrum capricornutum (green algae)	4-day pop. growth 2-day pop. growth	29 50–100	71 72

^{a}NOEC indicates no-observed-effect concentration.
^{b}LOEC indicates lowest-observed-effect concentration.
^{c}Mean of eight toxicity tests.

The EC_{50} value of 0.09 mg/L reported by Lewis (73) is believed to be in error. Evidence available in proprietary reports indicates the correct value to be 0.9 mg/L, as reported by Lewis and Hamm (71). This value is further corroborated by the study of Yamane et al. (72).

For invertebrates, Fairchild et al. (64) reported results of 4- and 7-day survival studies with *Hyallela*. NOEC values ranged from 0.6 to 1.4 mg/L in these studies. The $C_{12}LAS$ 21-day NOEC for *Daphnia* range from 1.2 (65) to 4.9 (70), and for *Ceriodaphnia* the 7-day NOEC was 1.4 mg/L (62).

Acute- to chronic-toxicity ratios based on one or several organisms can be used to extrapolate from acute data to chronic data for other groups of organisms. Although a number of assumptions are inherent in this approach, the method generates chronic values that can be used in initial risk assessments. By using the data from Tables II and III, and calculating geometric mean values where more than one toxicity value exists, acute-to-chronic ratios for *Daphnia magna*, *Ceriodaphnia dubia*, and *Pimephales promelas* were determined to be 2.2, 3.8, and 3.5, respectively.

AS.   The acute toxicity of $C_{12}AS$ has received a significant amount of research (Table IV), in part because of the recommendation that sodium dodecyl sulfate (SDS) be used as a reference toxicant (U.S. EPA, unpublished reference literature). Results of toxicity tests with three invertebrate species are reported, with acute 24- and 48-h EC_{50} values ranging from 1.4 mg/L for the rotifer *Brachionus rubens* to 10.3 mg/L for *Daphnia magna*. In fish, 96-h acute LC_{50} values reported range from 4.5 mg/L for the bluegill sunfish (*Lepomis macrochirus*) to 18.3 mg/L for the adult guppy (*Lebistes reticulatus*). Newsome (79) studied $C_{12}AS$ toxicity to three life stages of fish—fry (3 weeks of age), juvenile (12 weeks of age), and adult (20 weeks of age)— and observed that fry were generally the most sensitive and adults the least sensitive to the acute toxicity of AS.

Any given organism appears to have a fair amount of variability in the acute toxicity of $C_{12}AS$. Variability occurs within studies (tests conducted by a single author), as well as between studies. For example, in the study of LeBlanc (75) the eight 48-h EC_{50} values for *Daphnia magna* were reported to range from 3.3 to 9.8 mg/L. Between studies, the 48-h EC_{50} value for *Daphnia magna* ranges from 1.8 mg/L (59) to a high of 15.2 mg/L (77).

A portion of the between-study variability may indicate the direct effect of water hardness on toxicity (83, 84). This effect appears to result from an increased availability of AS in hard water (84, 85), not all of which is attributable to changes in AS environmental chemistry. The toxicity of AS appears to increase with the hardness of the acclimation water when organisms are tested at the same water hardness (84). This study suggests that acclimation conditions affect the susceptibilities of organisms to AS toxicity. Some of the variability in the toxicity estimates, both within and between studies, is expected to stem from the rapid biodegradation or other loss processes of AS. Acute toxicity testing is usually conducted by using a static exposure system without test substance renewal. Under these conditions we expect AS to degrade and thereby introduce an unquantified level of uncertainty into the acute toxicity test results.

Published reports of the chronic toxicity of $C_{12}AS$ are summarized in Table V. LeBlanc (75) exposed *Daphnia magna* to $C_{12}AS$ in a four-generation toxicity test. *Daphnia* (<24 h old) were exposed to $C_{12}AS$ for 10 days; then some of the young were isolated from each concentration and exposures

Table IV. Acute Toxicity of $C_{12}AS$ to Invertebrates and Fish

Organism	Test Duration (h)	LC_{50} (mg/L)	Ref.
Invertebrates			
Brachionus rubens (rotifer)	24	1.4[a]	74
Daphnia magna (water flea)	48	1.8	59
	48	7.0[b]	75
	48	6.3	76
	48	10.3[b]	77
Daphnia pulex (water flea)	48	8.9[c]	77
Fish			
Lepomis macrochirus (bluegill sunfish)	96	4.5	59
	96	4.8	65
	96	20.3	
Oncorhynchus mykiss (rainbow trout)	24	4.6	78
	48	6.2	79
Brachydanio rerio (zebrafish)	24	7.8	78
	96	9.9 (fry)	79
	96	12.8 (juv.)	
	96	20.1 (adult)	
Jordanella floridea (flagfish)	24	8.1	78
Pimephales promelas (fathead minnow)	96	10.2 (fry)	79
	96	17.0 (juv.)	
	96	22.5 (adult)	
Lebistes reticulatus (guppy)	96	13.5 (adult)	79
	96	16.2 (juv.)	
	96	18.3 (fry)	
Salmo trutta (brown trout)	45	18	80
Phoxinus phoxinus (minnow)	24	30.5	81
Carassius auratus (goldfish)	6	60	82
	96	28.4	79
Oryzias latipes (Japanese killifish)	24	70	83

[a]Indicates mean of six toxicity tests.
[b]Indicates mean of eight toxicity tests.
[c]Indicates mean of 10 toxicity tests.

continued with the young as test organism for another 10 days. This procedure was repeated for four generations with survival and reproduction used as end points. On the basis of survival and reproduction, 2.0 mg/L was the no-observed-effect concentration. Cowgill and Williams (86) investigated the toxicity of $C_{12}AS$ to *Ceriodaphnia dubia* in 7-day chronic toxicity tests. The mean EC_{50} value for reproduction, the most sensitive measure of toxicity, was 36 mg/L. A no-observed-effect concentration was not reported.

Table V. Chronic Toxicity of $C_{12}AS$ to Invertebrates and Algae

Organism	Toxicity Test	NOEC[a], EC_{50} (mg/L)	Ref.
Invertebrates			
Daphnia magna (water flea)	10-day survival, reproduction, 4 generations tested	2.0 (NOEC)	75
Ceriodaphnia dubia (water flea)	7-day chronic reproduction & survival	36 (EC_{50})	86
Plants			
Lemna minor (duckweed)	7-day growth	18 (EC_{50})	59
Selenastrum capricornutum (green algae)	2-day population growth	60 (EC_{50}, $C_{13}AS$)	72

[a]NOEC indicates no-observed-effect concentration.
NOTE: There are no data for fish.

Bishop and Perry (59) investigated the toxicity of six compounds, including $C_{12}AS$, with duckweed (*Lemna minor*) in a 7-day multigeneration flow-through toxicity test. Test concentrations were analytically verified, and toxicities to frond count, dry weight, root weight, and growth rate were assessed. As with most of the other compounds tested in this study, the most sensitive indicator of $C_{12}AS$ toxicity was root length, with an EC_{50} concentration of 18 mg/L. Yamane et al. (72) reported a 2-day EC_{50} value of 60 mg/L based on the specific population growth rate for the green alga, *Selenastrum capricornutum*.

There are no published reports of the chronic toxicity of $C_{12}AS$ to fish. Considering the comparable acute toxicity to fish and invertebrates of $C_{12}AS$ and the other anionic surfactants reported here, fish chronic toxicity values for $C_{12}AS$ are predicted to be similar to those reported for invertebrates.

The chronic toxicity data reported here for $C_{12}AS$ is in general agreement with the publication of Steber et al. (26). These authors reported chronic toxicity values to algae (*Scenedesmus* and *Chlorella*) and daphnids for the tallow AS (a mixture of $C_{16}AS$ and $C_{18}AS$). The 21-day no-observed-effect concentration for *Daphnia* was 16.5 mg/L, and the EC_{10} levels were 14.4 and 25.5 mg/L for *Scenedesmus* and *Chlorella*, respectively. The EC_{50} concentrations were 57.6 and 28.6 mg/L for the same species.

AES.   The acute and chronic toxicity of AES has not been investigated as thoroughly as the toxicities of other anionic surfactants. However, sufficient data are available to provide a preliminary assessment of the potential hazard to aquatic life (Table VI).

Maki (70) reported a 96-h LC_{50} value of 1.17 mg/L for *Daphnia magna* exposed to a $C_{14.7}E_{2.25}S$ in a flow-through toxicity test. The LC_{50} value was

Table VI. Acute Toxicity of $C_{12-15}E_3S$ to Invertebrates and Fish

Organism	Test Duration (h)	LC_{50} (mg/L)	Ref.
Invertebrates			
Daphnia magna	48	1.17	70
(water flea)		$(C_{14.7}E_{2.25}S)$	
		5	76
		$(C_{12}AE_3S)$	
		13	
		$(C_{12}AE_5S)$	
		17	
		$(C_{12-14}AE_{2.2}S)$	
Daphnia pulex	48	20.2^a	87
(water flea)			
Fish			
Pimephales promelas	96	1.6 (fry)	79
(fathead minnow)	96	2.5 (juv.)	
	96	2.2 (adult)	
Salmo trutta	96	1.5	88
(brown trout)			
Brachydanio rerio	96	2.2 (fry)	79
(zebrafish)	96	1.9 (juv.)	
	96	2.4 (adult)	
Lebistes reticulatus	96	2.4 (fry)	79
(guppy)	96	2.4 (juv.)	
	96	2.1 (adult)	
Oncorhynchus mykiss	96	1.0 (juv.)	79
(rainbow trout)			
Carassius auratus	96	3.0 (juv.)	79
(goldfish)			

^{a}Indicates mean of two values.

based on measured concentrations of the AES. This value is less than the toxicity to a *Daphnia* species in a static test system reported by Lundahl et al. (76). These authors studied the toxicity of three samples of AES, two $C_{12}AE_3S$ samples, and a $C_{12-14}AE_{2.2}S$, and they observed 48-h EC_{50} values to range from 5.0 to 17 mg/L for these compounds (76). In tests with *Daphnia pulex*, Moore et al. (87) reported a mean 48-h EC_{50} value of 20.2 mg/L for an AES. This variability in the toxicity of AES to members of the genus *Daphnia* may be attributable, in part, to differences in species susceptibility and the specific AES tested. However, the role of biodegradation should not be underestimated. Lundahl and Cabridenc (68) demonstrated that AES toxicity to daphnids can be completely removed by biodegradation within 30 h.

The acute toxicity of AES to fish is similar across fish species and similar to the toxicity to *Daphnia magna* reported by Maki (70). LC_{50} values range from 1.5 for the brown trout (*Salmo trutto*) (88) to 3.0 mg/L for the goldfish

(*Carassius auratus*) (79). All of the fish toxicity values are based on static exposures, but in both the Newsome (79) and Reiff et al. (88) studies, test solutions were renewed at least daily to minimize biodegradation.

Chronic toxicity data for AES exist for invertebrates, fish, and algae (Table VII). Maki (70) reported a NOEC based on the most sensitive end point (total number of young produced) of 0.27 mg/L for *Daphnia magna* exposed to $C_{14.7}E_3S$ in a flow-through, 21-day toxicity test. In a flow-through life-cycle toxicity test with fathead minnows (*Pimephales promelas*), growth was the most sensitive end point to $C_{14.7}E_3S$ toxicity (70). The NOEC was 0.1 mg/L and the lowest-observed-effect concentration (LOEC) was 0.22 mg/L. The 2-day EC_{50} value for *Selenastrum capricornutum* population growth was reported as 65 mg/L for AES (72).

Structure–Activity Relationships. The acute toxicity of a variety of surfactants increased with increased chain length of the hydrophobic portion of the surfactant (7, 61). This effect was extensively studied for LAS homologs for both fish and invertebrates (5, 61, 65, 66). These data are shown in Figure 10 and are quantitatively described by eqs 2 and 3. For toxicity to fish,

$$\log (96\text{-h } LC_{50}) = (-0.46 * CL) + 5.99 \tag{2}$$

For toxicity to invertebrates,

$$\log (48\text{-h } LC_{50}) = (-0.39 * CL) + 5.41 \tag{3}$$

where CL is the chain length of the carbon atoms on the alkyl chain. The data for fish are a combination of information on the bluegill sunfish (*Lepomis macrochirus*) (65) and the fathead minnow (*Pimephales promelas*) (66). The data for the invertebrates are from studies with *Daphnia magna* (5, 61) and *Daphnia pulex* (61).

Table VII. Chronic Toxicity of $C_{12.8}E_3S$ to Invertebrates, Fish, and Algae

Organism	Toxicity Text	NOEC–LOEC (mg/L)	Ref.
Invertebrates			
Daphnia magna (water flea)	21-day survival	0.27 (NOEC; $C_{14.67}E_3S$)	70
Fish			
Pimephales promelas (fathead minnow)	life cycle	0.10–0.22 ($C_{14.67}E_3S$)	70
Plants			
Selenastrum capricornutum (green algae)	2-day population growth	65 (EC_{50})	72

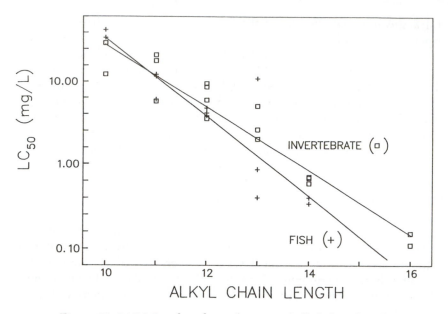

Figure 10. LAS LC$_{50}$ plotted as a function of alkyl chain length.

Addition of carbons to the alkyl chain length increases the hydrophobicity of the surfactant. The increased hydrophobicity results in greater uptake rate constants, and toxicity increases by a factor of approximately 2.7 with each additional alkyl carbon.

A limited amount of chain-length versus chronic-toxicity information exists for LAS, as well as some chain-length versus acute-toxicity information for AS and AES. These limited data also support the relationship between alkyl chain length and toxicity.

Risk Assessment

Risk assessments for anionic surfactants are obtained by comparing environmental exposure concentrations to effect levels (the quotient method). A protection factor that reflects the environmental safety of the material is calculated by dividing the exposure level by the effect concentration. If the protection factor is greater than 1, the material is deemed safe. Although this approach to assessing risk yields a numerical value that could be interpreted as the relative safety of a compound, comparisons of protection factors for different compounds should be avoided. The risk assessment for each material must be considered separately because of differences in chemical properties and differences in the database used to obtain the protection factor. In addition, the degree of uncertainty in the exposure and effect concentration determinations should be thoroughly understood so that the degree of confidence in the protection factor can be communicated.

LAS.   LAS environmental levels in United States surface water are expected to range from <0.04 to 0.14 mg/L, as predicted from modeling results and from actual monitoring data (*see* Exposure Assessment section). Because of the agreement between results from the national LAS monitoring efforts and model predictions, there is a high level of confidence in this exposure assessment.

For the effect level, application of the invertebrate acute-to-chronic ratio to the most acutely sensitive invertebrate species, *Dero*, generates a chronic toxicity value of 0.6 mg/L. This value compares favorably with the measured NOECs of 1.2 and 1.4 mg/L, respectively, for the invertebrates *Daphnia magna* and *Ceriodaphnia dubia*. The fish species most sensitive to C_{12}LAS chronic toxicity is the fathead minnow, *Pimephales promelas*. Most recently, Fairchild et al. (*64*) reported a chronic C_{12}LAS NOEC of 0.7 mg/L for the fathead minnow. This NOEC compares favorably with the NOEC of 1.1 mg/L reported by Holman and Macek (*66*) for the same species.

An exposure level of 0.02 mg/L reflects mean river flow; with a lowest no effect concentration of 0.6 mg/L it yields a 90th percentile protection factor of 30. This agrees with the risk assessment of Kimerle (*89*), who reported a LAS protection factor that exceeded 10. In addition, considering the literature and the known influence of environmental factors on the uptake and toxicity of C_{12}LAS in aquatic organisms, a risk assessment for C_{12}LAS based on laboratory toxicity studies is believed to be conservative. For example, Lewis (*73*) compared the toxicity of C_{12}LAS to the blue-green algae *Microcystis* and the green algae *Selenastrum*, which were exposed in laboratory 4-day toxicity tests, with a 10-day exposure of a natural phytoplankton assemblage (82 species obtained from a lake). These assemblages were assessed for effects of C_{12}LAS on species diversity, similarity, and number, and for changes in dissolved oxygen. The NOEC for algae exposed in this study was 27 mg/L. This NOEC is far greater than the laboratory-derived 96-h EC_{50} for *Microcystis* and *Selenastrum*; apparently laboratory algae tests provide conservative estimates of safe concentrations in the environment.

In addition, Fairchild et al. (*64*) exposed an outdoor stream community consisting of a diverse invertebrate benthic population, *Hyallela azteca*, and *Pimephales promelas*. These organisms were exposed to a mean concentration of 0.35 mg/L, a concentration not expected to cause adverse effects if the laboratory-generated NOEC data were protective of the system. End points assessed during the 45-day exposure included a variety of periphytic and benthic invertebrate community measurements. These authors found no effects on the biota contained in this study at 35 mg/L of C_{12}LAS.

Finally, Pittinger and Kimerle (62) calculated hypothetical water-quality criteria for $C_{12}LAS$. The values calculated are 0.23 mg/L for continuous concentration and 0.625 mg/L for maximum concentration. Therefore, laboratory and field data confirm that current concentrations of $C_{12}LAS$ in the aquatic environment are safe.

AS. AS concentrations in United States surface waters are expected to be less than 0.01 mg/L, based on USTEST predictions and actual measured concentrations at STP effluents. This value will serve as the AS exposure level for the aquatic risk assessment.

The majority of the AS toxicity studies have been conducted on $C_{12}AS$. Because the average AS chain length used in cleaning products is approximately $C_{14}AS$, a chronic NOEC for this material is needed for the risk assessment. The quantitative structure–activity relationship derived for LAS applied to the lowest $C_{12}AS$ chronic NOEC of 2.0 mg/L for *Daphnia magna* yields an estimated $C_{14}AS$ chronic NOEC of 0.27 mg/L.

The exposure to effect-level concentration ratio for AS yields a protection factor of >27 for $C_{14}AS$. Because of the rapid AS biodegradation observed in natural water and toxicity amelioration or attenuation resulting from sorption to solids, this safety factor is expected to be conservative. Additional $C_{14}AS$ chronic-toxicity studies with full analytical support to confirm exposure concentrations are suggested to validate the safety of $C_{14}AS$ in the aquatic environment.

AES. Exposure concentrations for AES are expected to be 0.015 mg/L, on the basis of USTEST predictions for mean-flow conditions. Because USTEST-predicted concentrations are for the mixing zone in rivers downstream from STP effluents, the model predictions do not take into account removal by sorption or biodegradation in the receiving stream. Therefore, the predicted AES exposure concentrations are the highest expected environmental concentrations.

The average AES chain length used in cleaning products is approximately $C_{13}E_3S$. The chronic-toxicity information was generated with $C_{14.7}E_3S$. However, the results can be extrapolated to determine the chronic toxicity of $C_{13}E_3S$ by using the LAS quantitative structure–activity relationship (QSAR). The extrapolated chronic toxicity NOECs for $C_{13}E_3S$ are then 0.53, 1.5, and 79 for fish, invertebrates, and algae, respectively. The toxicity value used in the risk assessment represents the NOEC from a test on the most sensitive species, the fathead minnow. The ratio of the effect-level concentration to the exposure-level concentration yields a a protection factor of 35 under mean-flow conditions. As is the case with LAS and AS, this risk assessment is considered conservative because of the amelioration of toxicity resulting from sorption to effluent and surface water suspended solids. In addition, AES are expected to undergo an initial hydrolysis like AS. This mechanism would rapidly remove AES from natural waters and would result in a much lower exposure. However, additional studies on the fate and effects of AES are planned to further support these conclusions.

Conclusions

Integration of use, biodegradation or removal during sewage treatment, environmental concentrations, and aquatic effect data into an aquatic risk assessment demonstrates the safety of LAS, AS, and AES in consumer products. Field investigations collaborate the risk assessments based on laboratory and model measurements. Additional investigations are underway to gain additional confidence in the risk assessments made for AS and AES.

References

1. *Chemical Economics Handbook;* Stanford Research Institute: Menlo Park, CA, 1988.
2. *Anionic Surfactants;* Linfield, W. M., Ed.; Marcel Dekker: New York, 1976; Parts I and II.
3. Schirber, C. A. *Soap Cosmet. Chem. Spec.* **1989**, *65*, 32–34, 36.
4. Beck, L. W.; Maki, A. W.; Artman, H. R.; Wilson, E. R. *Regul. Toxicol. Pharmacol.* **1981**, *1*, 15–58.
5. Kimerle, R. A.; Swisher, R. D. *Water Res.* **1977**, *11*(1) 31–37.
6. Gard-Terech, A.; Palla, J. C. *Ecotoxicol. Environ. Saf.* **1986**, *12*(2) 127–140.
7. Versteeg, D. J.; Shorter, S. J. *Environ. Toxicol. Chem.* **1992**, *11*, 571–580.
8. Huddleston, R. L.; Allred, R. C. *JAOCS* **1964**, *41*, 732–735.
9. Lawton, G. W. *J. Am. Water Works Assoc.* **1967**, *59*, 1327–1334.
10. Brenner, T. E. *J. Am. Oil Chem. Soc.* **1968**, *45*, 433–436.
11. Sigoillot, J. C.; Nguyen, M. H. *FEMS Microbiol. Ecol.* **1990**, *73*, 59–68.
12. Swisher, R. D. *Surfactant Biodegradation*, 2nd ed.; Marcel Dekker: New York, 1987.
13. Jimenez, L.; Breen, A.; Thomas, N.; Federle, T. W.; Sayler, G. S. *Appl. Environ. Microbiol.* **1991**, *57*, 1566–1569.
14. Larson, R. J.; Payne, A. G. *Appl. Environ. Microbiol.* **1981**, *41*, 621–627.
15. Larson, R. J. *Environ. Sci. Technol.* **1990**, *24*(8), 1241–1246.
16. Yediler, A.; Zhang, Y.; Cai, J. D.; Korte, F. *Chemosphere* **1989**, *18*, 1589–1597.
17. Federle, T. W.; Pastwa, G. M. *Ground Water* **1988**, *26*, 761–770.
18. Moreno, A.; Ferrer, J.; Berna, J. *Tenside* **1990**, *27*, 312–315.
19. Federle, T. W.; Schwab, B. S. *Water Res.* **1992**, *26*, 123–127.
20. Berth, P.; Gerike, P.; Gode, P.; Steber, J. *Cim. Oggi.* **1984**, *11*, 43–49.
21. Thomas, O. R. T.; White, G. F. *Biotech. Appl. Biochem.* **1989**, *11*, 318–327.
22. Vives-Rego, J.; Vaque, D.; Martinez, J. *Tenside* **1987**, *24*, 933–935.
23. Potipun, V. L. R. *Tr. Gos Okeanogr. Inst.* **1978**, *145*, 78–82.
24. Gerike, P.; Jasiak, W. *Tenside* **1986**, *23*, 300–304.
25. Pitter, P.; Fuka, T. *Tenside* **1979**, *16*, 198–302.
26. Steber, J.; Gode, P.; Guhl, W. *Soap Cosmet. Chem Spec.* **1988**, *64*, 44–50, 94, 95.
27. Hrsak, D.; Bosnjak, M.; Johanides, V. *Tenside* **1981**, *18*, 137–140.
28. Hammerton, C. *J. Appl. Chem.* **1955**, *5*, 517–524.
29. Hammerton, C. *Proc. Soc. Water Treat. Exam.* **1956**, *5*, 145–174.

30. Vath, C. A. *Soap Chem. Spec.* **1963,** *Feb.* 56–58, 182.
31. McGauhey, P. H.; Klein, S. A. *WPRC Conf. International Conferences on Water Pollution;* Vol. 1, pp 353–372.
32. Klein, S. A.; McGauhey, D. H. *J. Water Pollut. Control Fed.* **1965,** *37,* 857–866.
33. Vaicum, L.; Eminovic, A. *Rev. Roum. Biochim.* **1976,** *13,* 149–155.
34. Hsu, J. C. *Nature (London)* **1963,** *200,* 1091–1092.
35. Hsu, J. C. *Nature (London)* **1965,** *207,* 385–388.
36. Payne, W. J.; Williams, J. P.; Mayberry, W. R. *Appl. Microbiol.* **1965,** *13,* 698–701.
37. White, G. F.; Russell, N. J.; Day, M. *J. Environ. Pollut.* **1985,** *37,* 1–11.
38. Williams, J. P.; Payne, W. *J. Appl. Microbiol.* **1964,** *12,* 360–362.
39. Anderson, D. J.; Day, M. J.; Russell, N. J.; White, G. F. *Appl. Environ. Microbiol.* **1988,** *54,* 555–560.
40. Oba, K. Y.; Yoshida, Y.; Tomiyama, S. *Yukagaku* **1967,** *17,* 455–460.
41. Wagner, W. B.; Schink, B. *Water Res.* **1982,** *21,* 615–622.
42. Hales, S. G.; Dodgson, K. S.; White, G. F.; Jones, N. *Appl. Environ. Microbiol.* **1982,** *44,* 790–800.
43. Hales, S. G.; Watson, G. K.; Dodgson, K. S.; White, C. F. *J. Gen. Microbiol.* **1986,** *132,* 953–961.
44. "Assessment of Needed Publicly Owned Wastewater Treatment Facilities in the United States." EPA 430/9–84–011; U.S. Environmental Protection Agency: Washington, DC, 1984, 1985.
45. Holman, W. F. "Aquatic and Hazard Assessment." *ASTM Spec. Tech. Publ.* **1981,** *737,* 159–182
46. Rapaport, R. A.; Eckhoff, W. S. *Environ. Toxicol. Chem.* **1990,** *9,* 1245–1257.
47. McAvoy, D. C.; Rapaport, R. A.; Eckhoff, W. S. *Environ. Toxicol. Chem.* **1993,** *6,* 977–987.
48. Castles, M. A.; Moore, B. L.; Ward, S. R. *Anal. Chem.* **1989,** *61,* 2534–2538.
49. DeHenau, H.; Matthijs, E. *Int. J. Environ. Anal. Chem.* **1986,** *26,* 279–293.
50. Fendinger, N. J.; Begley, W. M. Society of Environmental Toxicology and Chemistry, Abstracts from 12th Annual Meeting, Seattle, WA, 1991.
51. Mann, A. H.; Reid, V. W. *JAOCS* **1971,** *48,* 798–799.
52. Gilbert, P. A.; Pettigrew, R. *Int. J. Cosmet. Sci.* **1984,** *6,* 149–158.
53. Kikuchi, M. *Bull. Jpn. Soc. Sci. Fish.* **1984,** *51,* 11–15.
54. Vashon, R. D.; Schwab, B. S. *Environ. Sci. Technol.* **1982,** *16,* 433–436.
55. Neubecker, T. A. *Environ. Sci. Technol.* **1985,** *19,* 1232–1236.
56. Rapaport, R. A. *Environ. Toxicol. Chem.* **1988,** *7,* 107–115.
57. Hennes, E. C.; Rapaport, R. A. *Tenside* **1989,** *26,* 141–147.
58. Lewis, M. A.; Suprenant, D. *Ecotoxicol. Env. Saf.* **1983,** *7*(3), 313–322.
59. Bishop, W. E.; Perry, R. L. In "Aquatic Toxicology and Hazard Assessment." *ASTM Spec. Tech. Publ.* 737; Branson, D. R.; Dickson, K. L., Eds.; American Society for Testing and Materials: Philadelphia, PA, 1981; pp 421–435.
60. Kimerle, R. A.; Swisher, R. *Water Res.* **1977,** *2,* 31–37.
61. Maki, A. W.; Bishop, W. *Arch. Environ. Contam. Toxicol.* **1979,** *8,* 599–612.
62. Pittinger, C. A.; Kimerle, R. A. Society of Environmental Toxicology and Chemistry, Abstracts from 11th Annual Meeting, Alexandria, VA, 1991; Abstract 55, p 112.
63. Hendricks, A. C.; Dolan, M.; Camp, F.; Cairns, J., Jr.; Dickson, K. L. "Comparative Toxicities of Intact and Biodegraded Surfactants to Fish, Snail, and Algae." R&D Report, Center for Environmental Studies, Department of Biology, Virginia Polytechnic Institute and State University: Blacksburg, VA, 1974.
64. Fairchild, J. F.; Dwyer, F.; La Point, T.; Burch, S.; Ingersoll, C. *Environ. Toxicol. Chem.* **1993,** in press.

65. Little, A. D. "Human Safety and Environmental Aspects of Major Surfactants." Report to the Soap and Detergent Association (NY); Soap and Detergent Association: Cambridge, MA, 1991.
66. Holman, W. F.; Macek, K. *Trans. Am. Fish. Soc.* **1980,** *109,* 122–131.
67. Lewis, M. A.; Perry, R. L. In "Aquatic Toxicology and Hazard Assessment." *ASTM Spec. Tech. Publ. 737;* Branson, D. R.; Dickson, K. L., Eds.; American Society for Testing and Materials: Philadelphia, PA, 1981; pp 402–418.
68. Lundahl, P.; Cabridenc, R. *Water Res.* **1978,** *12,* 25–30.
69. Taylor, M. J. *ASTM Spec. Tech. Publ.* **1985,** *854,* 53–72.
70. Maki, A. W. *J. Fish Res. Board Can.* **1979,** *36,* 411–421.
71. Lewis, M. A.; Hamm B. *Water Res.* **1986,** *20*(12), 1575–1582.
72. Yamane, A. N.; Okada, M.; Sudo, R. *Water Res.* **1984,** *18*(9), 1101–1105.
73. Lewis, M. A. *Environ. Toxicol. Chem.* **1986,** 5, 319–332.
74. Snell, T. W.; Persoone, G. *Aquat. Toxicol.* **1989,** *14,* 81–92.
75. LeBlanc, G. A. *Environ. Pollut. Ser. A* **1982,** 2, 309–322.
76. Lundahl, P.; Cabridenc, R.; Xuereff, R. *Qualites biologiques de quelques agents de surface anioniques;* Sixth International Congress on Surface Active Agents, 1972; pp 689–699.
77. Lewis, P. A.; Weber, C. I. In "Aquatic Toxicology and Hazard Assessment." *ASTM Spec. Tech. Publ. 854;* Cardwell, R. D.; Purdy, R.; Bahner, R. C., Eds.; American Society for Testing and Materials: Philadelphia, PA, 1985; pp 73–86.
78. Fogels, A.; Sprague, J. *Water Res.* **1977,** *11,* 811–817.
79. Newsome, C. S. "Susceptibility of Various Fish Species at Different Stages of Development to Aquatic Pollutants." *Comm. Eur. Communities Rep. EUR* **1982,** *7549,* 284–295.
80. Abel, P. D. *J. Fish Biol.* **1976,** 9, 441–446.
81. Lundahl, P.; Cabridenc, R. *Journal Francais d'Hydrobiologie* **1976,** 7, 143–150.
82. Gafa, S.; Lattanzi, B. *La Rivista Italiana delle Sostanze Grasse* **1975,** 52, 363–372.
83. Kikuchi, M.; Wakabayashi, M.; Nakamura, T.; Inoue, W.; Takahashi, K.; Kawana, T.; Kawahara, H.; Koido, Y. "A Study of Detergents. II. Acute Toxicity of Anionic Surfactants on Aquatic Organisms." *Annu. Rep. Tokyo Metrop. Res. Inst. Environ. Prot. Engl. Transl.* **1976,** 57–69.
84. Tovell, P. W. A.; Howes, D.; Newsome, C. S. *Water Res.* **1974,** 8, 291–296.
85. Eyanoer, H. F.; Upatham, E. S.; Duangsawasdi, M.; Tridech, S. *J. Sci. Soc. Thailand* **1985,** *11,* 67–77.
86. Cowgill, U. M.; Williams, L. R. *Arch. Environ. Contam. Toxicol.* **1990,** *19,* 513–517.
87. Moore, S. B.; Diehl, R. A.; Barnhardt, J. M.; Avery, G. B. *Text. Chem. Color* **1987,** *19,* 29–32.
88. Reiff, B.; Lloyd, R.; How, M.; Brown, D.; Alabaster, J. *Water Res.* **1979,** *13,* 207–210.

RECEIVED for review December 23, 1991. ACCEPTED revised manuscript October 6, 1992.

18

Fate of Hydrophobic Organic Contaminants

Processes Affecting Uptake by Phytoplankton

Robert S. Skoglund and Deborah L. Swackhamer

Environmental and Occupational Health, School of Public Health, Box 807 UMHC, University of Minnesota, Minneapolis, MN 55455

The accumulation of hydrophobic contaminants in phytoplankton plays a significant role in the transport and fate of these potentially toxic compounds. However, the limited amount of available field data indicate that partitioning models fail to adequately predict the distribution of these compounds in the water column. Several hypotheses have been proposed to explain these differences. In this chapter we propose additional explanations for these differences. We hypothesize that assumptions in the partitioning model about the rate of uptake, mechanism of uptake, and effect of phytoplankton growth also contribute to these deviations.

GROWING CONCERN OVER THE POTENTIAL TOXICOLOGICAL and ecological effects of hydrophobic organic compounds (HOCs) has made the determination of the environmental and human health impact of these compounds one of the important objectives of environmental science. A tool often used in solving ecotoxicological problems is the prediction of fluxes, distributions, and toxicology of compounds from a combination of physical–chemical generalizations and a limited amount of compound-specific data. The movement of dissolved HOCs into the aquatic food web is an important flux because this distribution can greatly increase the exposure of higher organisms, including humans, to these compounds.

Phytoplankton play an important role in the incorporation of HOCs into the aquatic food web. The lipophilicity of HOCs results in an enhanced association with lipid-rich phytoplankton in the water column. Once associated with phytoplankton, the distribution and fate of HOCs are controlled by that of the phytoplankton. The principal source of organic carbon to the aquatic food web is phytoplankton, and a significant portion of the HOCs found in the food web probably entered with this organic carbon (1). Thus water–phytoplankton partitioning is an important parameter in the flux of HOCs into food webs.

The most commonly held theory is that water–phytoplankton partitioning of HOCs is a passive thermodynamic process (2, 3). Mass transfer of these compounds appears to be driven by a net decrease in the free energy of the system associated with movement of HOCs from an aqueous to an organic phase. Thus, at equilibrium, partitioning of HOCs should be directly proportional to their octanol–water partition coefficients (K_{ow}) and inversely proportional to their aqueous solubilities. Partitioning is parameterized by a partition coefficient, referred to as the bioaccumulation factor (BAF), and defined as the concentration of a compound in phytoplankton divided by the dissolved concentration in ambient water in equivalent units. Several researchers (4–6) reported a correlation between log-transformed values of BAF and K_{ow}. Although the slopes of these relationships are consistently less than 1, which is the value suggested by the fugacity concept (2), these data support the theory that bioaccumulation of HOCs in phytoplankton results from thermodynamic partitioning.

However, Baker et al. (2) reported that field data suggest that several assumptions of HOC partitioning models are violated in surface waters and that these models fail to adequately predict HOC distribution in the water column. Furthermore, a consistent slope of <1 for the log BAF–log K_{ow} regression in laboratory studies and reports of species-specific differences in accumulation (1, 8, 9) indicate that accumulation appears to be influenced by other factors.

Published values of BAFs for HOCs, summarized by Swackhamer and Skoglund (10), range from 18 to 1,000,000. The variety of methods used in these studies prevents direct comparison, so it is unclear what factors are responsible for the large variation in BAFs. Several hypotheses have been proposed to explain these variations in accumulation and their deviations from K_{ow}-based predictions. One hypothesis (11, 12) proposed that a lack of complete reversibility in the partitioning process is responsible for the deviations. A second (13–16) theory attributed the deviations to the presence of a third phase (colloids or dissolved organic matter) and the inability to accurately separate the dissolved and sorbed states. A third (17) proposal is that partitioning is dependent on sorbent concentration. And finally, a fourth (18, 19) hypothesis holds that this deviation is a function of the effects of molecular size and shape on cellular transport.

The data indicate that all of these processes may affect accumulation and thus contribute to deviations from the predicted accumulation values.

In this chapter we present three additional factors that have been observed to affect accumulation. We hypothesize that the assumptions in the partitioning model about the rate of accumulation, the mechanism of accumulation, and the effect of growth are inappropriate and thus contribute to the deviations from the predicted values.

We conducted laboratory experiments designed to measure the influence of various sorbent and sorbate properties on the accumulation of HOCs by phytoplankton. The specifics of these experiments were published elsewhere (10, 20). In summary, 40 polychlorinated biphenyls (PCBs) (Table I) were allowed to partition to a unialgal culture of *Scenedesmus quadricauda* under batch conditions. At various time points between 0.02 and 30 days, the phases were operationally separated by centrifugation. Partitioning was determined by the congener-specific measurement of PCBs in each phase. At the beginning of each experiment, PCB congener concentrations ranged from 30 to 1100 ng/L, and phytoplankton mass was approximately 10 mg/L. Experiments were conducted under growth-limiting conditions (average growth rate = 0.03 doubling per day) and growth-unlimited conditions (average growth rate = 0.13 doubling per day). Parameters thought to influence partitioning were measured concurrently. Data from these studies were used to evaluate the rate and magnitude of water–phytoplankton partitioning and the factors that influence the process.

Rate of Accumulation

Equilibrium in an accumulation process is empirically defined as the point at which a partitioning coefficient becomes invariant (reaches steady state), and theoretically as the point at which the fugacity ratio equals 1 (21). With phytoplankton, most published reports indicated that equilibrium was reached in a matter of hours (6, 8, 22–26). As a result, predictions of HOC accumulation in phytoplankton have been expressed as equilibrium-based equations exclusively.

Our data demonstrate that accumulation of PCBs by phytoplankton is not as rapid as was initially thought; thus equilibrium-based equations are inappropriate in some instances. Work by Mackay (2) and Connolly and Pedersen (27) demonstrated that when the fugacity ratio is 1, the lipid-normalized BAF ($BAF_{(lip)}$) is equal to K_{ow}. Therefore at equilibrium, the $BAF_{(lip)}-K_{ow}$ relationship should have a slope of 1 and an intercept of 0. Figure 1 is a plot of log-transformed values of $BAF_{(lip)}$ and K_{ow} for five of the seven time points of our growth-limited experiment. Biomass increase was minimized by maintaining the incubation temperature at 10–12 °C to reduce the confounding effects of growth. A reduction in temperature from 22 to 12 °C was reported (1) to have little or no effect on accumulation of HOCs by phytoplankton.

Table I. Results from Applying the Modified Richards Model to Data
from the Limited-Growth Experiment with 40 PCBs

IUPAC Number	Substitution Pattern	Log K_{ow}[a]	m	F Value	p Value
001	2	4.46			
003	4	4.69			
004	2,2'	4.65			
008	2,4'	5.07			
018	2,2',5	5.24	−0.64	4.31	0.08
022	2,3,4'	5.58	−0.26	0.99	>0.25
031	2,4',5	5.67	−0.50	1.93	0.25
052	2,2',5,5'	5.84	−0.72	4.90	0.06
053	2,2',5,6'	5.62	−0.60	4.99	0.06
054	2,2',6,6'	5.21	−0.45	3.20	0.13
072	2,3',5,5'	6.26	−0.67	1.11	>0.25
077	3,3,4,4'	6.36	−0.03	0.02	>0.25
080	3,3',5,5'	6.48	−0.33	1.54	>0.25
081	3,4,4',5	6.36	−0.09	0.21	>0.25
097	2,2',3',4,5	6.29	−0.29	1.24	>0.25
100	2,2',4,4',6	6.23	−0.36	1.46	>0.25
101	2,2',4,5,5'	6.38	−0.36	1.66	>0.25
104	2,2',4,6,6'	5.81	−0.60	4.58	0.07
105	2,3',4,4',5	6.65	−0.08	0.12	>0.25
118	2,3',4,4',5	6.74	0.02	0.01	>0.25
126	3,3',4,4',5	6.89	0.30	5.86	0.04
138	2,2',3,4,4',5	6.83	0.12	0.49	>0.25
141	2,2',3,4,5,5'	6.82	−0.01	0.00	>0.25
151	2,2',3,5,5',6	6.64	−0.05	0.02	>0.25
154	2,2',4,4',5,6	6.76	0.08	0.10	>0.25
155	2,2',4,4',6,6'	6.41	−0.07	0.12	>0.25
156	2,3,3',4,4',5	7.18	0.39	9.62	0.02
169	3,3',4,4',5,5'	7.42	0.68	31.41	<0.005
170	2,2',3,3',4,4',5	7.27	0.49	20.08	<0.005
180	2,2',3,4,4',5,5'	7.36	0.46	19.83	<0.005
183	2,2',3,4,4',5',6	7.20	0.35	7.56	0.02
185	2,2',3,4',5,6,6'	7.11	0.37	9.02	0.02
188	2,2',3,4',5,6,6'	6.82	0.27	4.37	0.08
189	2,3,3',4,4',5,5'	7.71	0.69	21.56	<0.005
194	2,2',3,3',4,4',5,5'	7.80	0.52	14.25	0.007
195	2,2',3,3',4,4',5,6	7.56	0.44	7.08	0.03
199	2,2',3,3',4,5,6,6'	7.20	0.41	3.09	0.14
206	2,2',3,3',4,4',5,5',6	8.09	0.36	2.51	0.19
207	2,2',3,3',4,4',5,6,6'	7.74	0.35	2.40	0.20
209	2,2',3,3',4,4',5,5',6,6'	8.18	0.02	0.00	>0.25

[a]Values are taken from reference 41.

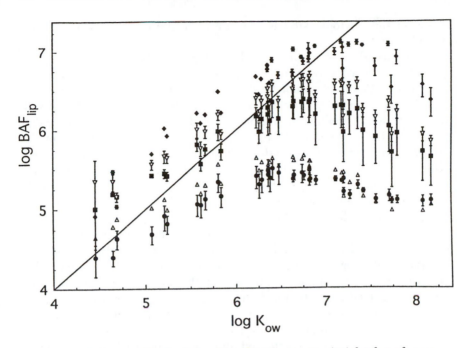

Figure 1. Log-normalized relationship of $BAF_{(lip)}$ vs. K_{ow} (41) for five of seven time points under growth-limited conditions. Key: ●, 0.02 day; △, 1 day; ■, 3 days; ▽, 6 days; and ◆, 20 days. Error bars are 1 standard deviation for duplicate samples. The data set for t = 1 was not sampled in duplicate. The reference line is the slope of 1.

Two aspects of Figure 1 indicate that equilibrium is not achieved as rapidly as was hypothesized earlier. First, many individual PCBs show a consistent increase in the measured $BAF_{(lip)}$, thus indicating a lack of steady state. The magnitude of these increases is a function both of transfer of PCBs from the aqueous to phytoplankton phases and of a decrease in lipid content of the phytoplankton (Table II). Nevertheless, a net flux of PCBs from the

Table II. Phytoplankton Lipid Content from Growth-Limited Experiment (Figure 1) and Regression Data for Groups of Congeners

Time (days)	Lipid (%)	Slopes Calculated within Groups of Congeners			
		log K_{ow}: <6	6–7	7–7.5	>7.5
0.02	16	0.64	0.00	−0.41	−0.06
0.5	17	0.45	−0.16	−0.34	−0.05
1	23	0.52	−0.02	−0.68	−0.16
3	12	0.59	0.36	−0.62	−0.47
6	10	0.62	0.43	−0.51	−0.57
12	8	0.77	0.60	0.17	−0.76
20	7	0.98	0.61	0.42	−0.77

media to the phytoplankton is seen throughout the entire experimental period.

The second aspect, which is independent of sorbate parameters, deals with the relationship between BAFs over a range of K_{ow}s. The complete data sets in Figure 1 cannot be described by a straight line, and thus a straight-line model is appropriate only for portions of these data. As a result, these data are divided into four groups according to natural gaps in the data (Table II). The slope of the $BAF_{(lip)}$–K_{ow} regression increased with time for each group except the one with the highest K_{ow} values. Changes in the slopes of these data sets confirm our original observation that steady state was not reached during the experimental period.

Although steady state was not reached, at 20 days all congeners with $\log K_{ow} < 7$ have a fugacity ratio >1. This high ratio may simply be the result of a decreasing lipid content (Table II) and a PCB depuration rate slower than that of lipids, or it could indicate an error associated with the lipid measurement. However, it brings into question the hypothesis that HOCs partition exclusively to the lipid portion of phytoplankton. These data may indicate that HOCs partition to cellular components other than lipids, and that a parameter such as the organic carbon content of the phytoplankton may be a more appropriate sorbent parameter.

These data demonstrate that equilibrium is not reached as rapidly as was previously assumed. In addition, they indicate that the time required to reach equilibrium can differ significantly between compounds. However, because the system had not reached equilibrium, they do not reveal the time required to reach equilibrium or the extent of partitioning at equilibrium. The primary ramification of slower partitioning is that measured BAFs may differ from predicted equilibrium values because the process has not reached equilibrium. In order to better predict accumulation in a system that is not at steady state, equilibrium-based equations need to be replaced with kinetic-based equations.

Mechanism of Accumulation

Traditionally, accumulation in phytoplankton is described by the model

$$\frac{dC_p}{dt} = C_w \cdot k_u - C_p \cdot k_x \tag{1}$$

where C_w is the concentration of dissolved chemical in water, k_u is the uptake rate constant, C_p is the chemical concentration in the phytoplankton, and k_x is the elimination or depuration rate constant (28–30). This model assumes that the mechanism of accumulation is diffusion into a single uniform phase. It indicates that the pattern of accumulation is an initial maximum rate, which gradually decreases until a steady-state concentration is attained (*31*).

Our data demonstrate that the accumulation of contaminants in phytoplankton deviates significantly from this classical pattern and instead shows an S-shaped or sigmoidal pattern of accumulation. This sigmoidal pattern is characterized by an initial instantaneous accumulation based on the time resolution attainable in laboratory experiments, followed by a lag period of approximately 1 day and then a slower but sustained period of accumulation. Although clear deviations from the classical model can be detected in some cases from a qualitative examination of the data, Brisbin et al. (*30*) reported a quantitative assessment in which the statistical significance of the sigmoidal nature of a data set is determined.

Brisbin et al. (*30*) described the development of the Richards sigmoidal model and presented it as an equation for contaminant uptake based on the amount of contaminant in an organism or compartment. For our data, the equation was reparameterized to use a partition coefficient rather than a concentration. It takes the form

$$P_t = \left[P_e^{(1-m)} - P_e^{(1-m)} \cdot \exp\left(\frac{-2t}{T} \cdot (m + 1) \right) \right]^{\frac{1}{1-m}} \tag{2}$$

where P_t is the lipid-normalized partition coefficient ($\text{BAF}_{(lip)}$) at time t, P_e is $\text{BAF}_{(lip)}$ at equilibrium, T is the time required for P_t to reach 90–95% of P_e, and m is the Richards shape parameter. Reparameterization was necessary because in the laboratory experiments the dissolved form of the contaminant is not an infinite source and P_0 is zero.

This model was used to test the sigmoidal nature of a data set by testing the null hypothesis $m = 0$. The data were fit (*32*) to a "complete" model in which P_t, T, and m were allowed to vary, and then to a "reduced" model in which P_t and T were allowed to vary while m was set equal to zero. An F-test evaluation of the increase in the residual sum of squares between the complete and reduced models was used to determine the acceptance of the null hypothesis (*33*). If the null hypothesis is rejected, the sigmoidal nature of the data set is statistically significant.

The results from testing the sigmoidal nature of the PCB data from our growth-limited experiments are listed in Table I. The mono- and dichlorobiphenyls are not included in the analysis because a significant portion of these congeners was lost through volatilization during the course of the experiments. The analysis indicates that with a significance level of 0.05, 10 of the 36 congeners tested show a sigmoidal accumulation pattern. An additional five congeners are included if the significance level is increased from 0.05 to 0.10. Thus, even with a small sample size ($n = 7$), the accumulation patterns of several congeners quantitatively deviated from the classical pattern. There are no easily identifiable relationships between K_{ow} and the sigmoidal shape of the uptake curve. However, this situation is consistent with the hypothesis that accumulation is not a function merely of K_{ow} but rather of several parameters.

The Richards model reduces the unexplained statistical variation in the accumulation of PCBs by phytoplankton, but it does not provide any information about the mechanisms responsible for the observed pattern. Numerous causes are possible for deviation from the classical pattern of accumulation. However, violations of assumptions associated with the classical model (i.e., constant uptake rate, instantaneous mixing within a single compartment, and a time-independent probability of depuration) are most likely the cause. With phytoplankton, several physiological mechanisms can potentially contribute to a sigmoidal accumulation curve.

1. Lack of homogeneity within the phytoplankton. The cultures used in the laboratory experiments were unialgal. However, field samples are composed of multiple species, and other studies indicated a species dependence of accumulation (1, 8, 9).

2. Multiple accumulation mechanisms. These data and other published reports (1) indicated that partitioning of PCBs to phytoplankton involves adsorption to the cell surface and incorporation into the cell matrix. However, there is no indication whether the processes operate in parallel or serial.

3. Presence of mucilaginous sheaths. These sheaths enhance the surface sorption of HOCs (9, 22, 24).

4. Time-dependent stability of sorbed HOCs. Wang et al. (34) reported that stability of a sorbed PCB, as measured by extractability, increases with time.

Although the Richards model does not provide mechanistic information, it does point out a need for further study and understanding of the uptake mechanisms. Brisbin et al. (30) reported that one consequence of sigmoidal accumulation is that there must be a period of accelerated or enhanced accumulation after the lag period in order to attain equilibrium levels similar to the classical model. In contrast to reports that cellular processes play no role in the accumulation of contaminants by phytoplankton (23, 25, 34–36), a sigmoidal accumulation curve may indicate that cellular processes such as the cycling of materials within the cell may enhance the rate of accumulation or depuration to a level above that which is attainable by diffusion alone.

Diluting Effect of Growth

The diluting effect that growth has on the chemical body burden of an organism has been addressed in accumulation models for higher organisms, but not for phytoplankton. This omission is primarily the result of the assumption that accumulation is rapid relative to other processes. However, because accumulation is not as rapid as initially thought, we hypothesize that growth can potentially have a significant effect on accumulation in phytoplankton. In a kinetic-based model, BAF at equilibrium is equal to the ratio of the net uptake and loss rate constants. Biomass increase is incorporated into this type of accumulation model by treating it as an additional loss function.

Phytoplankton community and species growth rates (GR) are generally reported in base two as doublings per day (37). Growth rate can be incorporated into the loss function by converting it to the fractional change in mass (M) per unit time or the dilution–loss rate constant (k_d), which has the units of per day. A value for k_d as a function of GR is determined by solving the growth equation for fractional change in mass per unit time.

$$M_t = M_0 \cdot 2^{t \cdot \text{GR}} \tag{3}$$

$$k_d = \frac{1}{M} \cdot \frac{dM}{dt} = 0.693 \cdot \text{GR} \tag{4}$$

where M_0 is the initial mass of phytoplankton and M_t is the mass at time t. The effects of growth on accumulation can then be seen by estimating the equilibrium $\text{BAF}_{(\text{lip})}$ for various growth conditions from the equation

$$\text{BAF}_{(\text{lip})} = \frac{k_u}{k_x + k_d} \cdot \frac{1}{f_{(\text{lip})}} \tag{5}$$

where $f_{(\text{lip})}$ is the lipid fraction of the phytoplankton. Estimates of uptake rate constants (k_u) and depuration rate constants (k_x) can be obtained from the literature. Connolly and Pedersen (27) reported that k_u and k_x for animals can be estimated from organism mass and compound K_{ow}. We chose 10^{-5} g, a factor of 100 smaller than the value used for zooplankton (27), as an estimate of the mass of a phytoplankton cell. However, the results of this exercise are not particularly sensitive to this choice. Figure 2 is a surface plot of the estimates of $\text{BAF}_{(\text{lip})}$ from eq 5, using literature values for k_u and k_x. Growth rates ranging from 0 to 1 doublings per day were used. Although Furnas (37) estimated that maximum diel-averaged growth rates of phytoplankton communities fall between 3 and 3.6 doublings per day, in situ measurements from the Great Lakes are considerably lower, reported as 0.06–0.60 doublings per day in Lake Michigan (38) and 0.15–0.65 doublings per day in Lake Superior (39).

Even low growth rates can significantly affect the measured partitioning of compounds, particularly those with high K_{ow}. When the growth rate is 0, the slope of the $\text{BAF}_{(\text{lip})}$–$K_{ow}$ relationship is 1. However, as the growth rate increases there is a rapid drop in the $\text{BAF}_{(\text{lip})}$ of high-K_{ow} compounds. The result is an apparent plateau in the $\text{BAF}_{(\text{lip})}$–$K_{ow}$ relationship. Further in-

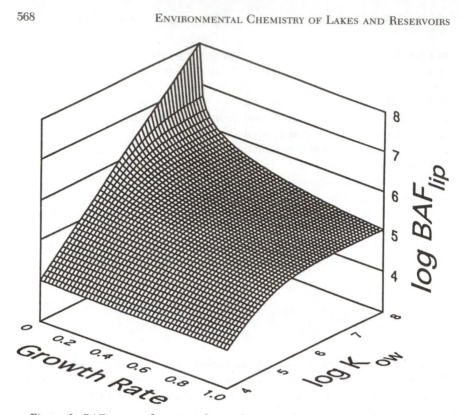

Figure 2. $BAF_{(lip)}$ as a function of growth rate and K_{ow}. Estimates of k_u and k_x are based on data from Connolly and Pedersen (27), and $f_{(lip)} = 0.13$.

creases in growth rate merely decrease the magnitude of K_{ow} at which the plateau in $BAF_{(lip)}$ begins. Within this plateau area, estimates of $BAF_{(lip)}$ are independent of compound lipophilicity.

Gobas et al. (40) reported equations for estimating k_u and k_x for aquatic macrophytes, based on K_{ow}, plant volume, and aqueous-phase and lipid-phase transport parameters. By assuming that the ratios of plant volume to transport parameters are similar for phytoplankton and macrophytes, these equations can be used to estimate a second independent set of k_u and k_x. Figure 3 is a surface plot constructed by using these estimates. The effect of growth on accumulation is similar to that in Figure 2, only the magnitude is greater.

This effect of growth dilution on accumulation was further demonstrated by our laboratory experiments in which growth averaged 0.13 doubling per day over 30 days. In these experiments the initial partitioning was similar to that seen in our growth-limited experiments. However, a decrease in the BAF for many congeners was observed between 4 and 8 days, and these lower values remained constant for the remainder of the experiment although the solids concentration increased fourfold. Figure 4 is a plot of the log-

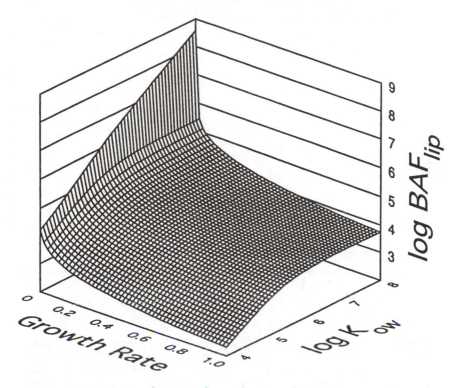

Figure 3. BAF$_{(lip)}$ as a function of growth rate and K$_{ow}$. Estimates of k$_u$ and k$_x$ are based on data from Gobas et al. (40), and f$_{(lip)}$ = 0.13.

transformed values of average BAF$_{(lip)}$ for three time points (t = 8, 16, and 30 days) and K_{ow}. Also included in Figure 4 are estimated BAF$_{(lip)}$s based on Figures 2 and 3. Figure 5 is a similar comparison between the t = 20 data set from the growth-limited experiments (Figure 1) and estimates from Figures 2 and 3 for this low-growth condition. In these comparisons between the laboratory data and predicted values, observed partitioning is best estimated by data from Figure 2. We speculate that the ratios of plant volume to transport parameters for phytoplankton and macrophytes cannot be assumed to be similar.

The similarities between the observed deviations from the predicted BAF$_{(lip)}$–K_{ow} relationship and those hypothesized to result from dilution indicate that under certain conditions growth has a significant effect on accumulation. However, the difference between the two estimates indicates that choice of rate constants is important. Analysis of eq 4 shows that it is particularly sensitive to the estimate of k_x. When k_x is small, even very low growth rates have a significant effect on accumulation. As a result, utilization of a kinetic-based model is dependent on the availability of accurate estimates of k_x.

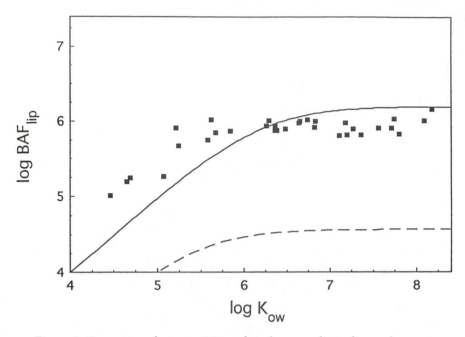

Figure 4. Comparison between BAF$_{(lip)}$ data from non-limited-growth experiments and values estimated from Figures 2 and 3. Data points represent the average BAF$_{(lip)}$ from three time points (t = 8, 16, and 30 days). The solid line is based on data from Connolly and Pedersen (27), and the dashed line is based on data from Gobas et al. (40).

Summary

Phytoplankton play an important role in the accumulation of HOCs in aquatic food webs, but equilibrium-based equations have not adequately modeled this process. In addition to presenting three additional factors that can influence accumulation, this chapter demonstrates the importance of considering the kinetics of accumulation.

Two of the factors, slow and irregular approach to steady state and dilution from growth, directly affect the relationship between observed and predicted accumulation. It is assumed that accumulation is fast enough that other parameters can be ignored. However, the data reported here demonstrate that as the rate constant for uptake decreases, the accumulation of a compound becomes increasingly susceptible to these parameters. As a result, the observed accumulation of rapidly accumulated compounds in low-growth environments will deviate very little from the equilibrium-based predictions, and that of slowly accumulated compounds in high-growth environments will deviate significantly. The third factor, pattern of accumulation deviating from that of diffusion into a single phase, indicates that

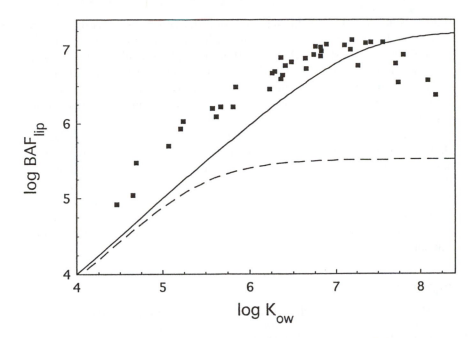

Figure 5. Comparison between t = 20 BAF*(lip) data from growth-limited experiments and values estimated from Figures 2 and 3. The solid line is based on data from Connolly and Pedersen (27), and the dashed line is based on data from Gobas et al. (40).*

multicompartment models better describe the mechanism of accumulation and that predictability will increase as the mechanisms in the model more accurately describe the process.

The accumulation of HOCs in phytoplankton plays an important role in food-web bioaccumulation. Both the increased times to steady state and the effect of dilution decrease accumulation in phytoplankton. This decrease results in a lower phytoplankton body burden and a decreased exposure in higher organisms. As a result, an equilibrium-based model will tend to overestimate concentrations in phytoplankton, and this overestimate will be evident throughout the food web.

Attempts to simulate environmental conditions in laboratory experiments consistently fall short. One particular difficulty in these experiments was the maintenance of the dissolved concentration. In the environment and large-scale experiments, dissolved concentrations are assumed to be constant. However, in these experiments the dissolved concentration decreased with accumulation. In addition, a measurable portion of many of the congeners initially partitioned to the bottle surface and then desorbed throughout the experiment. These factors presented additional variables that increased the difficulty of interpreting the data. However, it is unlikely that they significantly skewed the accumulation data because the consistent decrease in the dissolved concentration with time indicated that desorption from glassware was slower than partitioning to phytoplankton. Possibly the presence of the glassware surface helped to simulate a pseudoinfinite source by releasing PCBs over the course of the experiment. Nevertheless, a better simulation of environmental conditions should an important part of future investigations.

Further studies in the area of HOC accumulation in phytoplankton should concentrate on identifying and quantifying the mechanisms of uptake and depuration. Any descriptions of mechanisms are still speculative, although the data indicate that there are multiple uptake and loss processes. Furthermore, an understanding of the fate of HOCs within phytoplankton would be beneficial and would assist in determining the most accurate denominator for body burden (i.e., dry weight, organic carbon, or lipid fraction).

References

1. Harding, L. W., Jr.; Phillips, J. H., Jr. *Mar. Biol.* **1978**, *49*, 103–111.
2. Mackay, D. *Environ. Sci. Technol.* **1982**, *16*, 1218–1223.
3. Chiou, C. T. *Environ. Sci. Technol.* **1985**, *19*, 57–62.
4. Ellgehausen, H.; Guth, J. A.; Esser, H. O. *Ecotoxicol. Environ. Saf.* **1980**, *4*, 134–157.
5. Casserly, D. M.; Davis, E. M.; Downs, T. D.; Guthrie, R. K. *Water Res.* **1983**, *17*, 1591–1594.
6. Mailhot, H. *Environ. Sci. Technol.* **1987**, *21*, 1009–1013.
7. Baker, J. E.; Eisenreich, S. J.; Swackhamer, D. L. In *Organic Substances and Sediments in Water, Volume II;* Baker, R., Ed.; Lewis Publishers, CRC Press: Florida, 1991; pp 79–89.
8. Autenrieth, R. L. Ph.D. Thesis, Clarkson University, New York, 1986.
9. Lederman, T. C.; Rhee, G.-Y. *Can. J. Fish. Aquat. Sci.* **1982**, *39*, 380–387.
10. Swackhamer, D. L.; Skoglund, R. S. In *Organic Substances and Sediments in Water, Volume II;* Baker, R., Ed.; Lewis Publishers, CRC Press: Florida, 1991; pp 91–105.
11. Karickhoff, S. W.; Brown, D. S.; Scott, T. A. *Water Res.* **1979**, *13*, 241–248.
12. DiToro, D. M.; Horzempa, L. M. *Water Res.* **1982**, *16*, 594–602.
13. Voice, T. C.; Rice, C. P.; Weber, W. J. *Environ. Sci. Technol.* **1983**, *17*, 513–518.
14. Carter, C. W.; Suffet, I. H. *Environ. Sci. Technol.* **1982**, *16*, 735–740.
15. Hassett, J. P.; Milicic, E. *Environ. Sci. Technol.* **1985**, *19*, 638–643.
16. Gschwend, P. M.; Wu, S. *Environ. Sci. Technol.* **1985**, *19*, 90–96.
17. Connor, D. J.; Connolly, J. P. *Water Res.* **1980**, *14*, 1517–1523.
18. Shaw, G. R.; Connell, D. W. *Chemosphere* **1980**, *9*, 731–743.
19. Shaw, G. R.; Connell, D. W. *Environ. Sci. Technol.* **1984**, *18*, 18–23.
20. Swackhamer, D. L.; Skoglund, R. S. *Environ. Toxicol. Chem.* **1993**, *12*, 831–838.
21. Mackay, D. *Environ. Sci. Technol.* **1979**, *13*, 1219–1223.
22. Geyer, H.; Politzki, G.; Freitag, D. *Chemosphere* **1984**, *13*, 269–284.
23. Baughman, G. L.; Paris, D. F. *CRC Crit. Rev. Microbiol.* **1981**, *1*, 205–228.
24. Hansen, P.-D. *Arch. Environ. Contam. Toxicol.* **1979**, *8*, 721–731.
25. Sodergen, A. *Oikos* **1968**, *19*, 126–131.

26. Reinert, R. E. *J. Fish. Res. Board. Can.* **1985**, *29*, 1413–1418.
27. Connolly, J. P.; Pedersen, C. J. *Environ. Sci. Technol.* **1988**, *22*, 99–103.
28. Thomann, R. V. *Can. J. Fish. Aquat. Sci.* **1981**, *38*, 280–296.
29. Thomann, R. V.; Connolly, J. P. *Environ. Sci. Technol.* **1984**, *18*, 65–71.
30. Brisbin, I. L., Jr.; Newman, M. C.; McDowell, S. G.; Peters, E. L. *Environ. Toxicol. Chem.* **1991**, *9*, 141–149.
31. Davis, J. J; Foster R. F. *Ecology* **1958**, *39*, 530–535.
32. Wilkinson, L. *SYSTAT: The System for Statistics;* SYSTAT: Evanston, IL, 1989.
33. White, G. C.; Brisbin, I. L., Jr. *Growth* **1980**, *44*, 97–111.
34. Wang, K.; Rott, B.; Korte, F. *Chemosphere* **1982**, *11*, 525–530.
35. Rice, C. P.; Sikka, H. C. *J. Agr. Food Chem.* **1973**, *21*, 148–152.
36. Urey, J. C.; Kricher, J. C.; Boylan, J. M. *Bull. Environ. Contam. Toxicol.* **1976**, *16*, 81–85.
37. Furnas, M. J. *J. Plankton Res.* **1990**, *12*, 1117–1151.
38. Fahnenstiel, G. L.; Scavia, D. *Can. J. Fish. Aquat. Sci.* **1987**, *44*, 499–508.
39. Nalewajko C.; Voltolina, D. *Can. J. Fish. Aquat. Sci.* **1986**, *43*, 1163–1170.
40. Gobas, F. A. P. C.; McNeil, E. J.; Lovett-Doust, L.; Haffner, G. H. *Environ. Sci. Technol.* **1991**, *25*, 924–930.
41. Hawker, D. W.; Connell, D. W. *Environ. Sci. Technol.* **1988**, *22*, 382–397.

RECEIVED for review October 23, 1991. ACCEPTED revised manuscript July 13, 1992.

Differential Weathering of PCB Congeners in Lake Hartwell, South Carolina

Kevin J. Farley, Geoffrey G. Germann, and Alan W. Elzerman

Environmental Systems Engineering, Clemson University, Clemson, SC 29631

In the mid-1970s Lake Hartwell was found to be contaminated with PCBs from a capacitor-manufacturing plant located on Twelve Mile Creek, a tributary of the lake. Congener-specific field data for sediments in Twelve Mile Creek and Lake Hartwell indicated that PCB congener distributions vary with distance from the source and, for sediment cores in the lower portion of Twelve Mile Creek, with depth. Because surface sediments throughout the Twelve Mile Creek–Lake Hartwell system are expected to remain aerobic, the longitudinal variations in congener distributions provided strong evidence of physiochemical weathering. Sediment core data for Twelve Mile Creek suggested that biochemical weathering also occurs at select locations as a result of reductive dechlorination in deeper sediments. Steady-state model results for PCB congener distributions in the Twelve Mile Creek–Lake Hartwell system show that a preferential depletion of lower chlorinated congeners occurs when volatilization is the primary removal pathway and that higher chlorinated congeners are preferentially depleted when burial is the dominant process.

POLYCHLORINATED BIPHENYLS (PCBs) were originally considered to be refractory environmental contaminants because little change in the composition of Aroclors (commercially available mixtures of PCBs) was observed in environmental samples, even over prolonged time periods. Although PCBs are relatively stable compounds, it was the use of inadequate analytical technology, specifically packed-column gas chromatography (1), that prevented the detection of individual congeners and consequently of significant alterations in PCB congener composition. The advent of capillary-column gas chromatography and improved application of mass spectrometry gave researchers the ability to observe the individual behavior of the variety of congeners contained in Aroclor mixtures.

A growing body of data now indicates that PCBs are weathering (changing relative to congener distributions) in the environment and suggests that over time some congeners may be degraded and detoxified more rapidly than was previously realized (2). Thus, the focus of debate has shifted from the existence of PCB weathering to the identification of environmental processes that mediate these compositional changes. The fate and effects of PCBs are actually tied to those of the individual congeners, and analytical techniques capable of distinguishing congener composition are now available. Therefore, it no longer makes sense to talk about determining the fate of Aroclors as if they were single compounds.

Environmental weathering processes in aquatic systems are currently grouped into two distinct types: those mediated by biological activity and those caused by a physiochemical process. Brown et al. (3) and Bopp et al. (4) showed that PCBs in sediments from the upper Hudson River underwent a compositional alteration that resulted in decreased proportions of higher chlorinated congeners and increased proportions of lower chlorinated congeners relative to unweathered mixtures. Compositional changes with increasing depth in a sediment core implied the action of microbiologically mediated reductive dechlorination in the anaerobic sedimentary environment (5).

Brown et al. (2, 5) proposed a stepwise dechlorination of specific ortho-substituted dichlorobiphenyls (DCBs), trichlorobiphenyls (TrCBs), and tetrachlorobiphenyls (TCBs) to lower chlorinated congeners of the same ortho substitution patterns. The result of this stepwise dechlorination scenario is the production of 2,2'-DCB, 2,3-DCB, 2,4'-DCB, 2,2',6-TrCB, 2,3',6-TrCB, 2,4',6-TrCB, and 2,2',5',6-TCBs. Levels of these terminal dechlorination product congeners in upper Hudson River sediments were 3–8 times greater on a relative basis than in the parent Aroclor; the weight percent of the higher chlorinated congeners decreased by a factor of 2–10 (5).

Bush et al. (6), who analyzed water and sediments from the same reach of the Hudson River, also found pronounced differences in congener mixtures between the environmental PCB residues and the source Aroclors. The congener mixtures in the sediment samples were significantly skewed toward the less chlorinated congeners, but no change in PCB composition with depth in sediment was observed. Several of these sediment samples were contaminated with a PCB residue that resembled congener patterns found in the river water.

Earlier work (7) had identified similar aqueous PCB residues and attributed their presence at the study site to the preferential dissolution of the lower chlorinated congeners [primarily 2-monochlorobiphenyl (MCB),

2,2'-DCB, and 2,6-DCB] as the river flowed over contaminated upstream sediments. Thus, they hypothesized that the composition of the PCBs available for deposition in the downstream sediments had already undergone major changes. This theory led to the conclusion that the alterations in PCB residue patterns found in bottom sediments were not likely to be the result of in situ weathering processes. Instead they were seen as caused by the scavenging from the water column of previously altered mixtures of PCBs by suspended particles or water-filtering macrophytes growing in the sediments.

Theoretical predictions of physiochemical weathering of PCBs were tested by Burkhard et al. (8). Possible changes in congener composition were examined by considering successive batch equilibrations in an air–water suspended particulate matter (SPM) system with predicted physiochemical properties and the fugacity model (9). Burkhard et al. (8) found variations in Henry's law constants (K_H), and sediment–water partition coefficients (K_d) resulted in the most significant differences in congener partitioning behavior, with SPM–water partitioning as the dominant process. The tendency of individual homologous chlorination groups to behave differently in the model systems led Burkhard et al. (8) to conclude that physiochemical processes are a major factor controlling the weathering of PCBs in the environment. Dunnivant (10) came to similar conclusions on the basis of laboratory batch system investigations and computer simulations of congener fractionation in successive partitioning of congeners in three-phase water–sediment–air systems.

The environmental implications associated with physiochemical and biochemical processes make it important to determine which type dominates the weathering of PCBs in aquatic systems. Physiochemical weathering should redistribute congeners into different environmental compartments, not eliminating the contamination problem but unequally transferring it elsewhere. On the other hand, biochemical weathering should result in the destruction of congeners and possible production of others. Ideally, but not necessarily, this process would generate a PCB residue that is less toxic and more easily biodegradable.

Thus, knowledge of the dominant transformation or transport processes would lead to a more informed decision concerning remediation of PCB-contaminated systems and would improve fate predictions. For example, the best remedial action for a biologically mediated system may simply be to allow the PCBs to degrade over time into a less toxic form. Conversely, remediation of contaminated systems dominated by physiochemical modification of the PCB congeners may require an active response to prevent the problem from moving into environmental compartments over which there is little or no control.

The Twelve Mile Creek–Lake Hartwell system in northwest South Carolina provides an excellent opportunity to examine the presence and relative importance of physiochemical and biochemical weathering processes in an aquatic system. In the mid-1970s Lake Hartwell was found to be contaminated with PCBs. The source of the contamination was traced by the South Carolina Department of Health and Environmental Control (SCDHEC) (11) to a capacitor-manufacturing plant located on the Twelve Mile Creek tributary, 39 km upstream from the top of the lake.

The plant was discharging PCB-laden effluent into the creek via two serial settling basins. Sediment samples in these basins were found to have average PCB concentrations of 27,300 μg/g in the upper basin and 7970 μg/g in the lower basin (11). The average total PCB concentration of the effluent in 1976 was 30 μg/L. Subsequent studies of sediments in Twelve Mile Creek and Lake Hartwell showed maximum concentrations in excess of 150 μg/g (dry weight basis) (12). Total PCB concentrations were highest immediately downstream of the source and decreased with distance (11–14).

Differences in the hydrologic characteristics for the upper and lower portions of Twelve Mile Creek and for Lake Hartwell are believed to have a significant effect on physiochemical and biochemical weathering processes. The upper portion of Twelve Mile Creek (0–32 km) is characterized by relatively shallow waters and high water velocities. In the lower portion (32–39 km), the creek deepens and widens because of the impoundment of Lake Hartwell. Water velocities are much slower in this portion of the creek, and a significant portion of the solids load is deposited in the sediments. Flow from Twelve Mile Creek eventually enters the top of Lake Hartwell, where it mixes with uncontaminated waters from the Keowee River. The waters then move slowly down this 50-km-long reservoir toward the Hartwell Dam.

Physiochemical and biochemical weathering of PCBs in Twelve Mile Creek–Lake Hartwell was examined by using the congener-specific data set obtained by Germann (12). Results for the longitudinal variation in PCB congener distributions for surficial sediments are presented first. Because surficial sediments throughout the Twelve Mile Creek–Lake Hartwell system are expected to remain aerobic, the changing composition of PCBs in surficial sediments is used in evaluating physiochemical weathering processes. Biochemical weathering through reductive dechlorination is then examined by using the results for PCB congener distributions in sediment cores from the lower portion of Twelve Mile Creek and the upper basin of Lake Hartwell. Finally, steady-state model calculations for physiochemical weathering are presented and are used in an initial assessment of the transport and fate of PCBs in the Twelve Mile Creek–Lake Hartwell system.

Summary of Congener-Specific Field Data

Sample Collection. The samples used in this study were a subset of 85 sediment core samples collected between October 1986 and August 1987 for a larger study on the distribution of PCBs in the Twelve Mile Creek–Lake Hartwell system. Sediment core sampling was done with a Wildco sediment corer (Wildlife Supply Co., Saginaw, MI) containing 5-cm inner-diameter polycarbonate tubes. Prior to use the polycarbonate tubes were washed in a hot detergent bath and rinsed three times with distilled water.

After collection, samples were divided into 5-cm sections or into sections determined by changes in sample matrix. Only sediment from the interior of the core (not in contact with the corer) was used to avoid possible contamination from the polycarbonate and from sediment fines forced along the wall of the tube. Sectioned samples were stored in solvent-rinsed bottles at 4 °C until analyzed. Some samples used in this study were collected from reaches of the submerged riverbed (where PCBs were deposited before the reservoir was filled). Sample locations (Figure 1) were determined in relation to buoys placed and maintained by the U.S. Army Corps of Engineers. Samples used in this study were limited to those found to be contaminated with at least 1 μg/g of total PCB to minimize quantification error during weight-percent calculations.

Extraction, Analysis, and Quantification Procedures.   Extraction and analysis of the sediments for PCBs were performed by using the method developed by Dunnivant and Elzerman (*15*). It was modified for this study to minimize the volume of solvents used and the total extraction time. Sediment (3–5 g per sample) was weighed into 150-mL beakers, and 50 mL of acetone was added to each beaker. The mixture was sonicated while mixing on a magnetic stir plate in an ice-water bath for 5 min at 0.8 relative output by using a sonic dismembrator (Fisher, model 300) equipped with a standard-size probe and a 300-W generator. The sample was allowed to settle for 5 min and then passed through a vacuum filter apparatus (Millipore) containing a glass fiber prefilter (Gelman, Type A) into a 250-mL Erlenmeyer flask. The filter cake was rinsed twice with acetone to completely remove all of the PCB-containing acetone. The extract was quantitatively transferred to a 250-mL separatory funnel containing 60 mL of double-distilled water, 10 mL of saturated NaCl solution, and 15 mL of isooctane. Then it was shaken for 3 min at 300 rpm on a gyratory shaker (New Brunswick Scientific G10). After the solution settled for 1 h, the aqueous phase was collected and the isooctane was placed on a prewashed Na_2SO_4 column and collected in a 100-mL graduated cylinder. The aqueous phase was returned to the separatory funnel with 15 mL of clean isooctane and reshaken. The isooctane was placed on the Na_2SO_4 column and collected. The extracts were then composited and brought to a final volume of 60 mL. Five milliliters of extract was placed on a 1-cm-diameter column containing approximately 1.5 g of alumina (80–200 mesh, deactivated 10%), the column was rinsed with 5 mL of isooctane, and the final extract was concentrated to an appropriate volume by evaporation with nitrogen gas. Extraction efficiency was found to be 99 ± 4.9%.

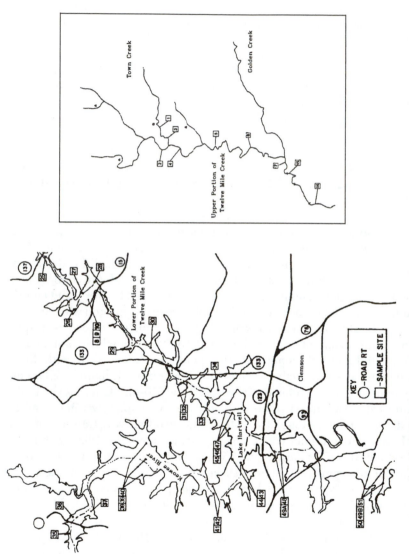

Figure 1. Map of the Twelve Mile Creek–Lake Hartwell system showing sample sites and boundaries of the different hydrologic regimes.

Analysis of extracts was performed on a gas chromatograph (GC) (5880A, Hewlett–Packard) equipped with an electron-capture detector (ECD) and a 30-m fused silica capillary column with an outer diameter of 0.25 mm and a film thickness of 0.25 μm (Durabond DB-5, J&W Scientific). The internal standard method developed by Dunnivant and Elzerman (*15*) was used, except that that only one internal standard was used (Aldrin) to minimize run time on the gas chromatograph. Daily working standards were composed of 80% Aroclor 1016 and 20% Aroclor 1254. This ratio was chosen because it matches the Aroclor distribution found in the sediments by Polansky (*13*). Quantification and collation of data were done on microcomputers with a spreadsheet program (SuperCalc 4, Computer Associates International).

Field Evidence of Physiochemical Weathering.   Figure 2 shows the weight-percent distribution (mass of congener/mass of total PCB × 100%) of the 64 peaks resolved in the GC analysis for a standard 80% Aroclor 1016 plus 20% Aroclor 1254 mixture and for surface sediment samples at varying distances from the source. Plots of weight-percent distribution were used instead of actual gas chromatograms because they eliminate any biases for or against a particular peak that may result from absolute levels of total PCB concentration in different samples. In the standard Aroclor mixture, the early-eluting peaks (peaks 1–22, representing less chlorinated congeners) account for almost 76% of the total PCB weight. The later-eluting peaks (peaks 25–63, representing the higher chlorinated congeners) accounted for the remaining 24%. All of the sample residues had significantly different weight-percent distribution patterns compared to the standard mixture.

As distance from the PCB source increased, a significant change in the weight-percent distributions was observed (as shown in Figure 2 for stations G26 and G33 in the lower portion of Twelve Mile Creek and stations G49A, G56, and G60 in Lake Hartwell). In the lower portion of Twelve Mile Creek the distribution was dominated by the lower chlorinated congeners. The distribution in Lake Hartwell sediments shifted toward the higher chlorinated congeners until an approximate 50:50 ratio was reached.

A more detailed analysis of variations of homologous chlorine groups with distance from the source is shown in Figure 3. As shown, di-CB (DCB), tri-CB (TrCB), tetra-CB (TCB), and the summation of penta-, hexa-, and hepta-CB (PHHCB) groups varied with distance from the source in different manners. The weight percent of both DCBs and TrCBs decreased with distance. The weight percent of the TCBs increased slightly with increasing distance from the source, whereas the PHHCBs initially showed a dramatic increase, followed by a slight decrease farther down the lake.

Field Evidence of Biochemical Weathering.   PCB congener composition variation with depth in sediment was analyzed for evidence of bio-

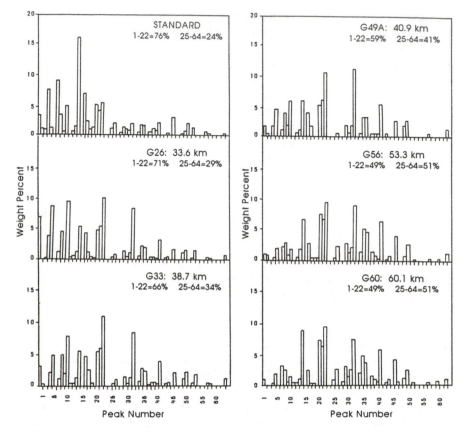

Figure 2. Weight percent (peak mass/total mass × 100%) of each peak for an 80% 1016/20% 1254 standard Aroclor mixture and for the top section of five Twelve Mile Creek–Lake Hartwell sediment cores. Sample location distances from the PCB source and weight-percent summaries for peaks 1–22 and 25–64 are given. Some peaks are not quantified because of chromatographic interferences (GC peaks 24, 27, and 64) or lack of analytical sensitivity (GC peaks 45, 48, 55, 56, 59, and 62). Peak 15 coelutes with peak 14. Peak 23 is the internal standard aldrin.

chemical alteration of the relative congener composition. Congener composition varied significantly with depth in sediments for the lower portion of Twelve Mile Creek (e.g., station G26). In these core samples, the congener composition became increasingly dominated by the lower chlorinated congeners with increasing depth in the sediment (Figure 4). This trend was absent in the less contaminated sediments farther down Lake Hartwell.

Figure 5 shows a more detailed analysis of variations of homologous chlorine groups with depth in sediment. Again two seemingly different systems are revealed for the lower portion of Twelve Mile Creek and for Lake Hartwell. The Twelve Mile Creek samples (G26 and G33) showed DCB

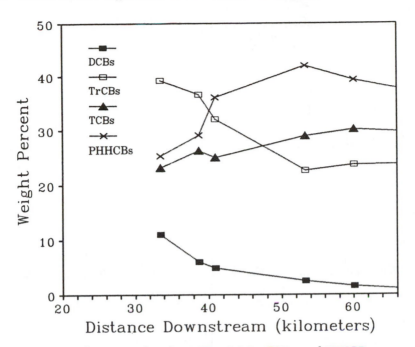

Figure 3. Weight-percent data for DCBs, TrCBs, TCBs, and PHHCBs versus distance from the PCB source for the top sections of six Twelve Mile Creek–Lake Hartwell sediment cores. Peaks containing two or more congeners were assigned to a homologous chlorination group on the basis of which congener accounted for most of the total peak mass (as reported in references 25 and 26). Peaks that could not be accurately classified (26, 28, and 31) were excluded.

levels increasing, TrCBs remaining constant, and TCBs and PHHCBs decreasing with depth. The Lake Hartwell samples (G49A and G56) did not show similar changes in congener composition with depth. Although stations G33 and G49A are less than 3 km apart, the different flow regimes (and possibly different redox conditions of sediments) in Twelve Mile Creek and Lake Hartwell appear to have a profound effect on PCB weathering processes.

To determine if higher levels of dechlorination products were present in the sediments, the sum of the weight percents of the four ortho-substituted terminal dechlorination congeners (congeners containing only ortho-substituted chlorines) was plotted versus depth in sediment at four stations (Figure 6). The presence of two separate systems was again indicated; the Twelve Mile Creek samples (G26 and G33) showed an increase in the weight percent of the terminal dechlorination congeners, whereas the weight percent of the terminal congeners in the Lake Hartwell samples (G49A and G56) remained fairly constant with depth.

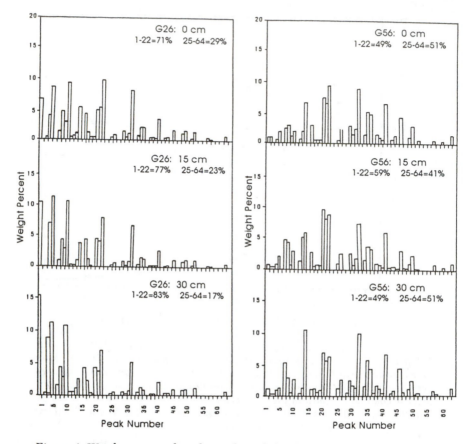

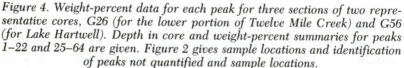

Figure 4. Weight-percent data for each peak for three sections of two representative cores, G26 (for the lower portion of Twelve Mile Creek) and G56 (for Lake Hartwell). Depth in core and weight-percent summaries for peaks 1–22 and 25–64 are given. Figure 2 gives sample locations and identification of peaks not quantified and sample locations.

Modeling Physiochemical Weathering Processes

Steady-state modeling calculations were performed to examine how congener-specific properties (such as sediment–water partition coefficients, Henry's law constants, and molecular diffusion rates) affect the transport and fate of PCBs. A basic description of the model, along with modeling results, is presented here to further explain the importance of physiochemical weathering processes in controlling the fate and distribution of PCB congeners in Twelve Mile Creek and the upper portion of Lake Hartwell.

Background Information. For model calculations Twelve Mile Creek is divided into an upper and lower reach (as shown in Figure 1). In

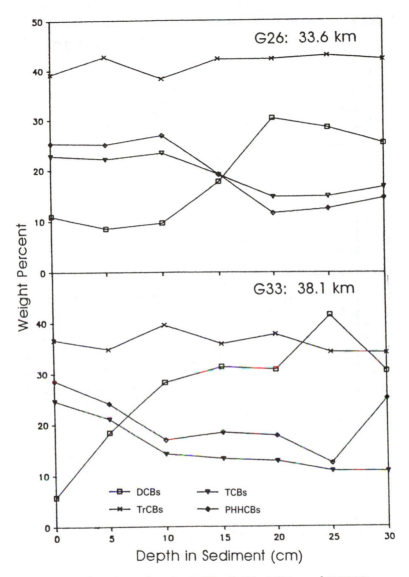

Figure 5. Weight-percent data for DCBs, TrCBs, TCBs, and PHHCBs versus depth in core for four Twelve Mile Creek–Lake Hartwell sediment cores. Distance of sample location from the PCB source is indicated. Figure 3 gives a description of homologous chlorination groups. Continued on next page.

the upper reach (0–32 km), the channel is relatively shallow (~0.4 m) with an average width of 25 m. The average annual flow in the creek is given as 6 m^3/s (*16*), and water velocities in the upper reach are relatively high (~0.6 m/s). In the lower reach (32–39 km) the creek deepens and widens because of the impoundment of Lake Hartwell. The average depth and width in

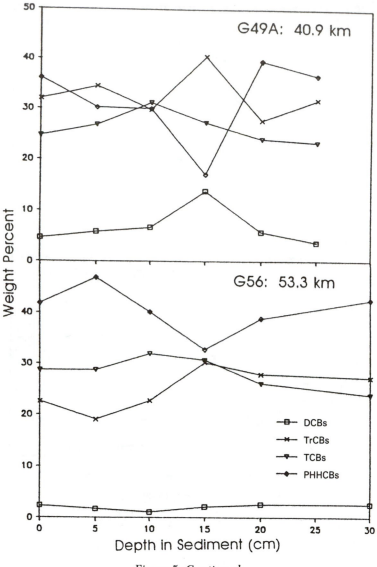

Figure 5. Continued.

this reach are 4 and 60 m, respectively, and water velocities are low
(~0.01 m/s).

The lower reach of Twelve Mile Creek empties into the upper portion
of Lake Hartwell, which is described as a series of three mixing basins (*see*
Figure 1). The uppermost basin, which receives inflows from both Twelve
Mile Creek and the Keowee River, is physically separated from the second
basin by a submerged sill under the U.S. Highway 123 bridge in Clemson,
South Carolina. The average annual flow in the Keowee River is approxi-

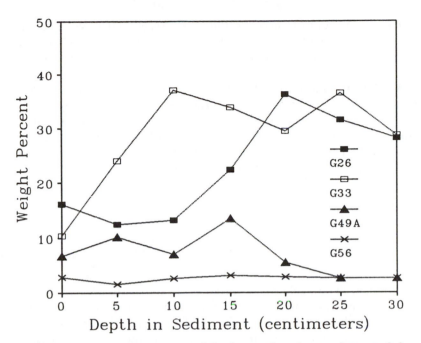

Figure 6. Sum of weight percents of the four ortho-substituted terminal dechlorination congeners (2,2'-DCB, 2,6-DCB; 2,2',6'-TrCB; and 2,2',6,6'-TCB) versus depth in core.

mately 40 m³/s (*16*). The second basin is bounded by SC Highway 123 and the land constriction at SC Highway 93. The last basin is bounded by land constrictions at SC Highways 93 and 37. Surface areas of the three basins are given as 998,000, 780,000, and 2,866,000 m², respectively. Although this portion of the lake is typically stratified during the warmer months, the water in each basin is considered well-mixed for all model calculations.

During the year suspended solids concentrations in Twelve Mile Creek vary from 20 to several hundred milligrams per liter. These solids are composed primarily of clay- and silt-sized particles with average settling velocities in the range of 0.16 to 7.8 m/day (*17*). The large variation in solids concentrations results primarily from land erosion and sediment resuspension during storm events. Organic content of suspended solids in Twelve Mile Creek and the upper portion of Lake Hartwell is approximately 2% by weight (*18*). Solids concentrations in the top sediment layers (0–10 cm) are in the range of 300 g/L (*12*). Because waters throughout the Twelve Mile Creek–Lake Hartwell system are well oxygenated and the organic carbon content of the sediments is low, surface sediments are believed to remain aerobic. Biochemical weathering through reductive dechlorination is, therefore, not expected to be significant in surface sediments. However, they may play an important role in deeper sediments. Based on bottom-profile measurements performed by the U.S. Army Corps of Engineers (*19*), little or no net sedimentation is believed to occur in the upper reach of Twelve Mile Creek. Conversely, sedimentation rates in the lower reach of the creek and in the upper portion of Lake Hartwell are estimated to be 1.5–3.0 cm/year (17).

Model Description.   The transport and fate of PCBs in Twelve Mile Creek and the upper portion of Lake Hartwell are described by a series of mass conservation equations for solids and PCBs in the water column and in the active sediment layer (given as the top 10 cm of sediment) following the approach described by O'Connor (20–22). For solids, the equations for the water column and active sediment layer are given as:

$$\frac{\partial m}{\partial t} + \text{transport fluxes} = \frac{w_s}{h} m + \frac{k_u'}{h} m_a \tag{1}$$

$$\frac{\partial m_a}{\partial t} = \frac{w_s}{\delta_a} m - \frac{k_u'}{\delta_a} m_a - \frac{k_b'}{\delta_a} m_a \tag{2}$$

where m and m_a are solids concentrations in the water column and active sediment layer, respectively; t is time; w_s is the settling velocity (m/day); h is the mean depth; δ_a is the thickness of the active bed layer; k_u' is the resuspension rate coefficient (m/day); and k_b' is the sediment burial rate (m/day).

For model calculations, the average suspended solids concentration entering the creek is assumed to be 50 g/m^3. The average suspended solids concentration in the Keowee River input is taken as 26 g/m^3 to reflect the effect of solids removal of Lake Keowee. The average settling velocity for the entire system is taken as 0.5 m/day. The settling flux in the upper reach of the creek is balanced by an equivalent resuspension flux, yielding a no-net-sedimentation condition for this portion of Twelve Mile Creek. In the lower reach of the creek and the upper portion of the lake, resuspension is not considered. The flux of solids settling to the active sediment layer is balanced by the burial term. With this set of model parameters, calculated sedimentation rates for the lower reach of Twelve Mile Creek and the upper portion of Lake Hartwell are in the range of 1.2–3.0 cm/year.

For PCBs, separate equations are written for 36 congener classes as follows:

$$\frac{\partial C}{\partial t} + \text{transport fluxes} = \frac{k_v'}{h}\left[\frac{P_g MW}{K_H} - C_{dis}\right] - \frac{w_s}{h}\Gamma m +$$

$$\frac{k_u'}{h}\Gamma_a m_a - \frac{k_f'}{h}[C_{dis} - C_{a_{dis}}] \tag{3}$$

$$\frac{\partial C_a}{\partial t} = \frac{w_s}{\delta_a}\Gamma m - \frac{k_u'}{\delta_a}\Gamma_a m_a - \frac{k_b'}{\delta_a}\Gamma_a m_a + \frac{k_f'}{\delta_a}[C_{dis} - C_{a_{dis}}] \tag{4}$$

where C and C_a represent the total (dissolved plus particulate) congener concentrations in the water column and active sediment layer, respectively; C_{dis} and $C_{a_{dis}}$ represent the dissolved concentrations; Γ and Γ_a represent the solid-phase concentrations ($\mu g/g$); k_v' is the volatilization rate coefficient (m/day); P_g is the partial pressure of the congener in the gas phase (atm); MW is the molecular weight (g/mol); K_H is the Henry's law coefficient (atm-m^3/mol); and k'_f is the dissolved exchange rate coefficient between the water column and sediment layer pore waters (m/day).

In these equations, dissolved and solid-phase congener concentrations can be expressed in terms of the total congener concentrations by using the equilibrium partitioning relationship ($K_d = \Gamma/C_{dis}$) and the total congener mass concentration equation ($C = C_{dis} + \Gamma m$). In subsequent calculations, the partial pressure of PCBs in the overlying atmosphere is assumed to be negligible (because of air transport from the source region). Finally, photochemical degradation of PCBs, which has been observed to occur in laboratory systems (23, 24), is not considered because conclusive evidence has yet to be presented for its significance in natural aquatic systems such as Lake Hartwell.

A listing of the 36 congener classes, along with corresponding values for chemical modeling parameters, is given in Table I. This choice of congener classes is consistent with analytical measurements for congener separation obtained by using the Hewlett–Packard gas chromatography system. For classes containing two or more congeners, parameter values are based on a weighted average assuming a 4:1 mixture of Aroclors 1016 and 1254. This composition closely matches the original mixture of Aroclors that was released into the Twelve Mile Creek–Lake Hartwell system (31).

Volatilization rates were determined from information given in Table I by using the two-layer model of the air–water interface where

$$\frac{1}{k_v'} = \frac{1}{k_l'} + \frac{RT}{K_H k_g'} \tag{5}$$

In eq 5, k_l' and k_g' represent the mass-transfer rate coefficients for the water and air side of the interface, respectively (m/day); K_H is the Henry's law coefficient (atm-m^3/mol); R is the universal gas constant (atm-m^3/mol-K); and T is temperature (taken as 293 K). For the upper reach of Twelve Mile Creek, k_l' was calculated by using the O'Connor–Dobbins formula for free-flowing streams (32). For the lower reach of Twelve Mile Creek and the upper portion of Lake Hartwell, k'_l was determined from the oxygen-transfer rate coefficient (taken as 0.5 m/day for light wind conditions) times the square root of the ratio of molecular diffusion rates for the PCB congener and for oxygen in water. For the entire system, k_g' was given as the water-evaporation rate coefficient (taken as 500 m/day) times the square root of the ratio of molecular diffusion rates for the PCB congener and for water vapor in

Table I. Chemical Modeling Parameters for PCB Congener Classes

GC Peak Number	IUPAC Number	Percent of Total Discharge[a,b]	Molecular Diffusion in Water[c] (cm²/s)	Molecular Diffusion in Air[d] (cm²/s)	log K_H[e] (atm-m³/mol)	log K_{oc}[f] (mL/g)	log K_{ow}[g] (mL/g)
1	4, 10	3.23	5.43×10^6	0.0543	−3.48	4.66	5.11
2	7, 9	1.06	5.43×10^6	0.0543	−3.44	5.07	5.45
3	6	0.98	5.43×10^6	0.0543	−3.49	5.06	5.45
4	8	7.30	5.43×10^6	0.0543	−3.52	5.07	5.45
5	14, 19	1.15	5.21×10^6	0.0525	−3.36	5.09	5.47
7	18	8.90	5.15×10^6	0.0520	−3.50	5.24	5.60
8	15, 17	3.33	5.21×10^6	0.0525	−3.47	5.26	5.61
9	27	0.47	5.15×10^6	0.0520	−3.39	5.44	5.76
10	16, 32	4.75	5.15×10^6	0.0520	−3.51	5.29	5.64
12	25	0.51	5.15×10^6	0.0520	−3.53	5.66	5.95
13	25	1.46	5.15×10^6	0.0520	−3.50	5.67	5.95
14	31	3.97	5.15×10^6	0.0520	−3.56	5.67	5.95
15	28	11.84	5.15×10^6	0.0520	−3.54	5.67	5.95
16	20, 33, 53	6.80	5.11×10^6	0.0516	−3.59	5.59	5.89
17	22	2.29	5.15×10^6	0.0520	−3.72	5.58	5.88
18	45	0.95	4.91×10^6	0.0493	−3.45	5.53	5.84
19	39, 46	1.19	5.09×10^6	0.0513	−3.51	5.76	6.03
20	52	4.81	4.91×10^6	0.0493	−3.50	5.84	6.10
21	43, 49	3.95	4.91×10^6	0.0493	−3.45	5.84	6.10
22	47, 48, 75	5.30	4.91×10^6	0.0493	−3.41	5.93	6.17
25	44	1.06	4.91×10^6	0.0493	−3.64	5.75	6.02

26	37, 42	1.97	5.10×10^6	0.0515	-3.76	5.82	6.08
27	41, 72	2.04	4.15×10^6	0.0520	-3.39	5.44	5.76
29	74	1.31	4.91×10^6	0.0493	-3.67	6.20	6.40
30	70, 76, 98	0.88	4.90×10^6	0.0492	-3.67	6.19	6.39
31	66, 95	0.72	4.86×10^6	0.0490	-3.64	6.18	6.38
32	55, 91, 12	1.77	4.71×10^6	0.0481	-3.37	6.41	6.57
34	84, 92	0.48	4.70×10^6	0.0481	-3.60	6.14	6.35
35	101	1.39	4.70×10^6	0.0481	-3.61	6.38	6.55
36	79, 99, 113	1.27	4.71×10^6	0.0481	-3.60	6.38	6.55
39	87	0.76	4.70×10^6	0.0481	-3.74	6.29	6.47
41	77, 110	1.72	4.70×10^6	0.0481	-3.71	6.39	6.56
44	108	0.28	4.70×10^6	0.0481	-3.76	6.71	6.82
46	106, 118, 149	2.48	4.64×10^6	0.0476	-3.79	6.72	6.83
51	153, 168	1.66	4.51×10^6	0.0465	-3.64	7.04	7.10
53	138, 158	1.02	4.51×10^6	0.0465	-3.87	6.85	6.94
SUM		95.07					

[a] Based on weighted averages assuming a 4:1 mixture of Aroclors 1016 and 1254.

[b] Based on congener composition data for Aroclor 1016 (25) and Aroclor 1254 (26).

[c] Estimated by using the method of Hayduk and Laudie, as given in Lyman et al. (27).

[d] Estimated by using the method of Fuller, Schettler, and Giddings, as given in Lyman et al. (27).

[e] Based on values given in Dunnivant (28).

[f] Based on values given in Hawker and Connell (29).

[g] Estimated from octanol–water partition coefficients by using the empirical correlation given by Karickhoff (30).

air. In the upper reach of Twelve Mile Creek, resistances on the water and air side of the interface were both important in controlling volatilization of PCBs, and calculated values for the volatilization rate coefficient ranged from 1.54 m/day (for the lower weight congeners) to 1.20 m/day (for the higher weight congeners). In the lower reach of the creek and the upper portion of Lake Hartwell volatilization rates were primarily controlled by mass-transfer rates on the water side of the interface, and coefficients ranged from 0.098 to 0.081 m/day.

For transport terms in eqs 1 and 3, the upper and lower reaches of Twelve Mile Creek were treated as one-dimensional advective (plug flow) systems. The upper basin of Lake Hartwell was modeled as a series of completely mixed reactors, with inflows into the first reactor from Twelve Mile Creek and the Keowee River. Analytical solutions for steady-state distributions of solids and PCB congeners in the water column and the active sediment layer were determined by using approaches outlined by O'Connor (20–22). Because K_d values for a specific congener class were taken to be equivalent for the water column and active sediment layer, the solutions are independent of k_f' values.

Modeling Results.   Steady-state calculations for PCB distributions in Twelve Mile Creek and the upper portion of Lake Hartwell were performed by assuming a daily loading of 6.6 kg/day from the Sangamo Electric site. This loading rate was based on the average quantity of PCBs purchased by Sangamo during 1970–1975 (17) and the assumption that 0.5% of the PCBs entered the Twelve Mile Creek–Lake Hartwell system. A 4:1 mixture of Aroclors 1016 and 1254 was considered for the input. Under these loading conditions, calculated values for sediment concentrations varied from approximately 100 μg/g near the source to 40 μg/g in the lower reach of Twelve Mile Creek (Figure 7). Sediment concentrations in the upper portion of Lake Hartwell were further reduced to approximately 5 μg/g through mixing of contaminated waters from Twelve Mile Creek with uncontaminated waters from the Keowee River. This calculated distribution of PCB contamination in sediments is consistent with field observations. The steady-state concentrations shown in Figure 7 are roughly a factor of 2 greater than the 1987 field measurements for sediment concentrations in the lower reach of Twelve Mile Creek and the upper portion of Lake Hartwell.

Calculations for the relative distribution of congeners in the system are summarized in Figure 8 in terms of weight percentages for di-CB (DCB), tri-CB (TrCB), tetra-CB (TCB), and the summation of penta-, hexa-, and hepta-CB (PHHCB) groups. Weight percentages for DCB and TrCB decreased in the upper reach of the creek, indicating preferential removal of these compounds. Because PCB removal in the upper reach resulted from volatilization (we specified that no net deposition occurs in this reach), pref-

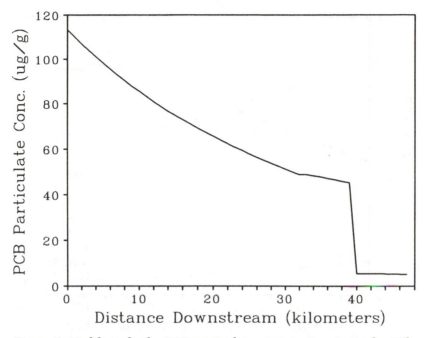

Figure 7. Model results for PCB particulate concentrations in Twelve Mile Creek–Lake Hartwell sediments versus distance from the source.

erential removal of the lower chlorinated biphenyls was primarily associated with lower K_d values. This situation translated into higher dissolved fractions for DCB and TrCB, and hence into higher volatilization rates. TCB and PHHCB were also lost from the creek through volatilization, but at much slower rates. Thus their weight percentages are shown to remain constant (for TCB) and to increase (for PHHCB).

In the lower reach of Twelve Mile Creek and the upper portion of Lake Hartwell, both volatilization and burial of PCBs in deep sediments are important removal pathways (as shown in Figure 9). Because volatilization is more effective in removing congeners with low K_d values and burial is more effective in removing congeners with high K_d values, little variation is calculated for the weight percentages in this portion of the system. In the upper portion of the lake, calculated weight percentages for the various chlorination groups are in general agreement with 1987 field observations (*see* Table II). Farther down the lake, field data (Figure 3) show a slight reduction in PHHCB, indicating that burial may be the preferred pathway for PCB removal.

Overall, the steady-state model for physiochemical weathering provided a good description of observed variations in congener distributions for surficial sediments in the Twelve Mile Creek–Lake Hartwell system. In general,

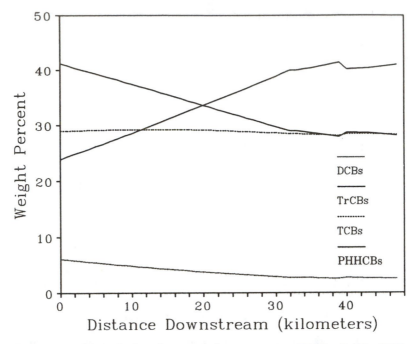

Figure 8. Model results for relative weight percentages of DCBs, TrCBs, TCBs, and PHHCBs versus distance from the source.

results showed that lower chlorinated congeners are more effectively re-moved by volatilization, whereas higher chlorinated congeners are prefer-entially removed by burial in the deeper sediments.

These trends, however, may be masked by time-variable processes. For example, "chromatographic separation" would be likely to occur in the sys-tem during periods of increased PCB loading, with lower chlorinated con-geners migrating down the creek and into the lake at faster rates because of their lower K_d values. Under these conditions, the downstream sediments would initially contain higher weight percentages for the lower chlorination groups. As steady-state conditions are approached (and higher chlorinated congeners have sufficient time to migrate downstream), the relative weight percentages of chlorination groups in downstream sediments would more accurately reflect the dominance of volatilization or burial in PCB removal.

Further studies examining time-variable behavior of PCBs in the Twelve Mile Creek–Lake Hartwell system and sensitivity of model calculations to various system parameters are presently being performed. The steady-state modeling results presented in this chapter, however, provide a reasonable base for an initial assessment of the fate of PCBs in the Twelve Mile Creek–Lake Hartwell system. The cumulative removals of PCBs from the system by volatilization and burial are shown as percents of the total PCB

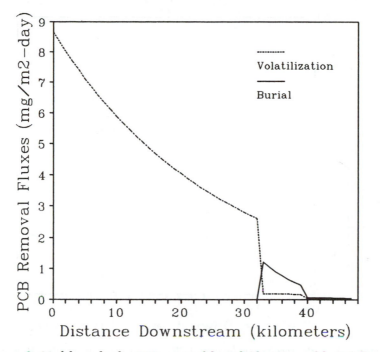

Figure 9. Model results for PCB removal by volatilization and burial fluxes in Twelve Mile Creek and the upper portion of Lake Hartwell.

Table II. Comparison of Observed and Calculated Weight Percentages for DCB, TRCB, TCB, and PHHCB in Sediments in the Upper Portion of Lake Hartwell

Compound	Source	Upper Portion of Lake Hartwell	
		Observed	Model Results
DCB	6	5	3
TrCB	41	33	29
TCB	29	25	28
PHHCB	24	37	40

input in Figure 10. Under steady loading conditions, approximately 65% of the PCB input is calculated to be removed from the system by volatilization, with most of the loss occurring in the upper reach of the creek. Removal by burial in the deeper sediments accounts for approximately 15% of the PCB input. As shown by the burial flux distribution in Figure 9, the highest values for the PCB burial flux are in the lower reach of Twelve Mile Creek.

This finding, which is consistent with our analyses of sediment cores, indicates that the lower reach of Twelve Mile Creek serves as an effective

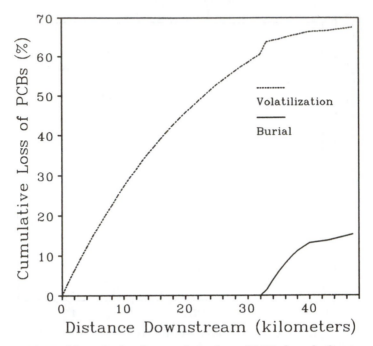

Figure 10. Model results for the cumulative loss of PCBs by volatilization and burial in Twelve Mile Creek and the upper portion of Lake Hartwell.

sediment trap for part of the contamination. However, these steady-state calculations show that approximately 15% of the PCB discharge is predicted to pass under SC Highway 37 and continue to migrate down Lake Hartwell toward the Hartwell Dam. Dilution of PCBs—particularly by inflows from Coneross Creek, Eighteen Mile Creek, Twenty-Six Mile Creek, and the Tugaloo River—is expected to be effective in reducing PCB concentrations. The remaining 5% of the PCB input is contained in congener classes that were not considered in our model calculations.

Possible Importance of Biochemical Weathering

In situ reductive dechlorination has been cited as an important pathway for biochemical weathering of PCBs in sediments (2, 5). The reductive dechlorination hypothesis is based on the premise that microbial action on PCB congeners is sterically selective for the chlorines in the meta and para positions on the biphenyl rings (i.e., 3, 3′, 4, 4′, 5, and/or 5′). Such selectivity would result in the cleaving of the chlorines in these positions (5). If this process occurs, the terminal dechlorination products of the microbially resistant ortho-substituted congeners in the anaerobic environment should be 2,2′-DCB, 2,6-DCB, 2,2′,6-TrCB, and 2,2′,6,6′-TCB. Thus, the weight percent of these congeners should increase with depth as dechlorination of the higher chlorinated congeners progresses.

This expected increase in weight percent for the terminal dechlorination congeners is clearly observed in the two Twelve Mile Creek core samples (Figure 6). For the Lake Hartwell samples (G49A and G56), however, the individual chlorine substitution groups and the terminal dechlorination congeners remain relatively constant with depth. Apparently biochemical weathering is not significant in the upper basin of Lake Hartwell.

Differences in biodegradative ability between the Twelve Mile Creek and Lake Hartwell samples may reflect differences in redox conditions of the sediments. This correlation can be explained by considering the reductive dechlorination scenario in which PCBs are used as electron acceptors by anaerobic microorganisms. In Lake Hartwell sediments, PCB concentrations are low because of mixing of PCBs with uncontaminated waters from the Keowee River. At low concentrations, the ability of PCBs to serve as electron acceptors may be diminished compared to other reducing agents (e.g., organic carbon molecules). As a result, no change in PCB congener composition would be observed with sediment depth. For higher concentrations in the Twelve Mile Creek sediments, PCBs may provide a significant fraction of the electron-receiving capacity, resulting in more cleaved chloines and hence in observable biochemical weathering of PCBs.

Similar results for biodegradative and nonbiodegradative zones were previously reported by Bopp et al. (4) for the Hudson River. Considering the occurrence of altered residues in the more highly contaminated upper Hudson River sediments, Bopp et al. (4) concluded that reductive dechlorination may not be significant at PCB concentrations less than approximately 150 μg/g. This value is much higher than the maximum concentrations seen in both the Lake Hartwell (16 and 7 μg/g) and Twelve Mile Creek (26 and 40 μg/g) cores.

Two possibilities are given here to explain why biochemical weathering may be occurring at much lower residue concentrations in the Twelve Mile Creek sediments. First, the organic carbon content in the Twelve Mile Creek sediments may be significantly lower than that of the Hudson River. This difference would cause the PCBs to make up a larger fraction of the electron-receiving capacity and result in the biodegradation of PCBs with depth. Second, any biochemical weathering occurring in the Twelve Mile Creek and the Hudson River may be mediated by different types of microorganisms and vary according to differences in temperature-dependent metabolic rates.

Other explanations for differences in Twelve Mile Creek and Lake Hartwell sediment cores are possible. For example, temporal variations in the PCB loading rate, along with chromatographic separation of congeners migrating down the creek, may also be important in explaining the congener distributions in sediments. However, it is difficult to imagine conditions that would lead to such an abrupt change in the sedimentary core records of stations G33 and G49A, which are less than 3 km apart.

In summary, no conclusive evidence exists to support the hypothesis of reductive dechlorination of PCBs in the sedimentary environment of Lake Hartwell. However, sediment core data for the lower portion of Twelve Mile Creek suggest that degradation of the higher chlorinated groups is occurring and may be a significant weathering process. Further research is needed to confirm or refute this claim and to quantify the possible significance of reductive dechlorination on the ultimate fate of PCBs in the sediments.

Conclusions

Field data for the Twelve Mile Creek–Lake Hartwell system indicated that sediments in the lower portion of Twelve Mile Creek contain significantly higher concentrations of PCBs than sediments in Lake Hartwell. PCB congener distributions in sediments were also shown to vary with distance from the PCB source and, for sediment cores in the lower portion of Twelve Mile Creek, with depth in the sediments. For surface sediments, the longitudinal variation in congener distribution was attributed to physiochemical weathering because surficial sediments are expected to remain aerobic throughout the Twelve Mile Creek–Lake Hartwell system. The sediment core data indicated that biochemical weathering also occurs at select locations in Twelve Mile Creek through reductive dechlorination.

Steady-state model calculations were performed to further examine physiochemical weathering behavior. Results were consistent with congener distributions in surficial sediments. In general, they showed that a preferential depletion of the lower chlorinated congeners occurred in the upper portion of Twelve Mile Creek, where volatilization was the only removal mechanism. In the lower portion of Twelve Mile Creek and the upper portion of Lake Hartwell, burial of PCBs in deeper sediments played a more important role. A preferential depletion of higher chlorinated congeners occurs when burial is the dominant process by physiochemical weathering.

From this initial assessment, approximately 65% of the PCB input is calculated to have been removed from the system by volatilization and 15% of the input by burial in deeper sediments, primarily in the lower portion of Twelve Mile Creek. As indicated most clearly by higher levels of specific ortho-substituted DCBs, TrCBs, and TCBs, PCBs buried in the deeper sediments of the creek have undergone further weathering by reductive dechlorination.

References

1. Erickson, M. D. *Analytical Chemistry of PCBs*; Lewis: Chelsea, MI, 1992.
2. Brown, J. F.; Bedard, D. L.; Brennan, M. J.; Carnahan, J. C.; Feng, H.; Wagner, R. E. *Science* **1987**, *236*, 709–712.

3. Brown, J. F.; Wagner, R. E.; Bedard, D. L.; Brennan, M. J.; Carnahan, J. C.; May, R. J. *Northeast. Environ. Sci.* **1984**, *3*, 167–179.
4. Bopp, R. F.; Simpson, H. J.; Deck, B. I.; Kostyk, N. *Northeast. Environ. Sci.* **1984**, *3*, 180–184.
5. Brown, J. F.; Wagner, R. E.; Feng, H.; Bedard, D. L.; Brennan, M. J.; Carnahan, J. C.; May, R. J. *Environ. Toxicol. Chem.* **1987**, *6*, 579–593.
6. Bush, B.; Shane, L. A.; Whalen, M. *Chemosphere* **1987**, *16*, 733–744.
7. Bush, B.; Simpson, K. W.; Shane, L.; Koblintz, R. R. *Bull. Environ. Contam. Toxicol.* **1985**, *34*, 96–105.
8. Burkhard, L. P.; Armstrong, D. E.; Andren, A. W. *Chemosphere* **1985**, *14*, 1703–1716.
9. Mackay, D. *Environ. Sci. Technol.* **1979**, *13*, 1218–1223.
10. Dunnivant, F. M. Ph.D. Dissertation ("Congener-Specific PCB Chemical and Physical Parameters for Evaluation of Environmental Weathering of Aroclors"), Clemson University, Clemson, SC, 1988.
11. South Carolina Department of Health and Environmental Control. "Lake Hartwell PCB Study (9/10/1976)." In Sangamo/Twelve-Mile/Hartwell PCB NPL Site Administration Record; Region IV Waste Management Division, U.S. Environmental Protection Agency; pp 1.2.54–73.
12. Germann, G. G. M.S. Thesis ("The Distribution and Mass Loading of Polychlorinated Biphenyls in Lake Hartwell Sediments"), Clemson University, Clemson, SC, 1988.
13. Polansky, A. L. M.S. Thesis ("Determination of PCB Concentrations in Lake Hartwell Sediments"), Clemson University, Clemson, SC, 1984.
14. Dunnivant, F. M. M.S. Thesis ("Determination of Polychlorinated Biphenyl Concentrations in Selected Sediments Using a Sonication Extraction Technique"), Clemson University, Clemson, SC, 1985.
15. Dunnivant, F. M.; Elzerman, A. W. *J. Assoc. Off. Anal. Chem.* **1987**, *71*, 551–555.
16. Snyder, H. S. "South Carolina State Water Assessment"; Report No. 140; South Carolina Water Resources Commission: Columbia, SC, 1983.
17. Kopf, A. "Fate and Transport Modeling of PCBs in Lake Hartwell." Department of Environmental Systems Engineering, Clemson University, Clemson, SC, unpublished manuscript.
18. "Remedial Investigation/Feasibility Study for the Sangamo Weston, Inc./ Twelve-Mile Creek/Lake Hartwell PCB Contamination Superfund Site Operable Unit Two at Pickens, Pickens County, South Carolina" (Technical Memorandum on the Results of Phase I Sampling); Prepared for Region IV Waste Management Division, U.S. Environmental Protection Agency (Work Assignment 09–4LP4); Bechtel Environmental: Oak Ridge, TN, 1992.
19. Plots of Lake Hartwell Sediment Ranges and Location Maps for 1959–1973; U.S. Army Corps of Engineers, Engineering Division, Savannah District, 1983.
20. O'Connor, D. J. *J. Environ. Eng. (New York)* **1988**, *114*, 507–531.
21. O'Connor, D. J. *J. Environ. Eng.* **1988**, *114*, 533–555.
22. O'Connor, D. J. *J. Environ. Eng.* **1988**, *114*, 553–574.
23. Hutzinger, O. S.; Safe, S.; Zitko, V. *The Chemistry of PCBs;* CRC Press: Boca Raton, FL, 1974.
24. Baxter, R. M.; Sullivan, D. A. *Environ. Sci. Technol.* **1984**, *18*, 608–610.
25. Albro, P. W.; Parker, C. E. *J. Chromatogr.* **1979**, *169*, 161–166.
26. Albro, P. W.; Corbett, B. J.; Schroeder, J. L. *J. Chromatogr.* **1981**, *205*, 103–111.
27. Lyman, W. J.; Reehl, W. F; Rosenblatt, D. H. *Handbook of Chemical Property Estimation Methods;* American Chemical Society: Washington, DC, 1990.

28. Dunnivant, F. M.; Elzerman, A. W. *Chemosphere* **1988**, *17*, 525–541.
29. Hawker, D. W.; Connell, D. W. *Environ. Sci. Technol.* **1988**, *22*, 382–387.
30. Karickhoff, S. W. *Chemosphere* **1981**, *10*, 833–846.
31. Billings, W. N. M.S. Thesis ("Polychlorinated Biphenyls Associated with Water and Sediments of a Reservoir in South Carolina"), University of South Carolina, 1976.
32. O'Connor, D. J.; Dobbins, W. E. *Trans. Am. Soc. Civ. Eng.* **1958**, *641*, 123.

RECEIVED for review April 8, 1992. ACCEPTED revised manuscript October 16, 1992.

Author Index

Affiliation Index

Subject Index

Copy editing and indexing: Colleen P. Stamm
Production: Donna Lucas
Acquisition: Cheryl Shanks
Cover design: Teddy Vincent Bell
Cover photo: Steven A. Claas

Typeset by Techna Type, Inc., York, PA
Printed and bound by United Book Press, Baltimore, MD

Bestsellers from ACS Books

The ACS Style Guide: A Manual for Authors and Editors
Edited by Janet S. Dodd
264 pp; clothbound ISBN 0–8412–0917–0; paperback ISBN 0–8412–0943–X

The Basics of Technical Communicating
By B. Edward Cain
ACS Professional Reference Book; 198 pp;
clothbound ISBN 0–8412–1451–4; paperback ISBN 0–8412–1452–2

Chemical Activities (student and teacher editions)
By Christie L. Borgford and Lee R. Summerlin
330 pp; spiralbound ISBN 0–8412–1417–4; teacher ed. ISBN 0–8412–1416–6

Chemical Demonstrations: A Sourcebook for Teachers,
Volumes 1 and 2, Second Edition
Volume 1 by Lee R. Summerlin and James L. Ealy, Jr.;
Vol. 1, 198 pp; spiralbound ISBN 0–8412–1481–6;
Volume 2 by Lee R. Summerlin, Christie L. Borgford, and Julie B. Ealy
Vol. 2, 234 pp; spiralbound ISBN 0–8412–1535–9

Chemistry and Crime: From Sherlock Holmes to Today's Courtroom
Edited by Samuel M. Gerber
135 pp; clothbound ISBN 0–8412–0784–4; paperback ISBN 0–8412–0785–2

Writing the Laboratory Notebook
By Howard M. Kanare
145 pp; clothbound ISBN 0–8412–0906–5; paperback ISBN 0–8412–0933–2

Developing a Chemical Hygiene Plan
By Jay A. Young, Warren K. Kingsley, and George H. Wahl, Jr.
paperback ISBN 0–8412–1876–5

Introduction to Microwave Sample Preparation: Theory and Practice
Edited by H. M. Kingston and Lois B. Jassie
263 pp; clothbound ISBN 0–8412–1450–6

Principles of Environmental Sampling
Edited by Lawrence H. Keith
ACS Professional Reference Book; 458 pp;
clothbound ISBN 0–8412–1173–6; paperback ISBN 0–8412–1437–9

Biotechnology and Materials Science: Chemistry for the Future
Edited by Mary L. Good (Jacqueline K. Barton, Associate Editor)
135 pp; clothbound ISBN 0–8412–1472–7; paperback ISBN 0–8412–1473–5

For further information and a free catalog of ACS books, contact:
American Chemical Society
Distribution Office, Department 225
1155 16th Street, NW, Washington, DC 20036
Telephone 800–227–5558